THE SCIENCE OF
ANIMAL AGRICULTURE

THE SCIENCE OF ANIMAL AGRICULTURE

5th Edition

Ray V. Herren

Australia • Brazil • Mexico • Singapore • United Kingdom • United States

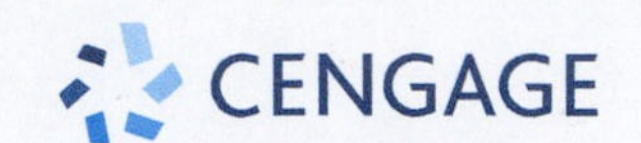

***Science of Animal Agriculture*,**
Fifth Edition
Ray V. Herren

SVP, GM Skills & Global Product Management: Jonathan Lau

Product Director: Matt Seeley

Product Manager: Nicole Robinson

Executive Director, Development: Marah Bellegarde

Senior Content Developer: Jennifer Starr

Product Assistant: Deb Handy

Vice President, Marketing Services: Jennifer Ann Baker

Marketing Manager: Abigail Hess

Senior Production Director: Wendy Troeger

Production Director: Andrew Crouth

Senior Content Project Manager: Betsy Hough

Design Director: Jack Pendleton

Cover and interior design credits: cover: Red chicken: ©spwidoff/Shutterstock.com; salmon: ©iStockPhoto.com/Hailshadow; ©cow: iStockPhoto.com/vwalakte; honey bees: ©iStockPhoto.com/shaunl; goat: ©iStockPhoto .com/michelangeloop

Bright blue background: ©Saibarakova Ilona /Shutterstock.com

interior design: white chicken: ©zhangyang13576997233/Shutterstock.com; carp: ©chinahbzyg/Shutterstock.com; sheep: ©Dmitry Pichugin/Shutterstock.com; bee on honeycomb: ©Shaiith/Shutterstock,com; cows: ©tarczas/Shutterstock.com; molecular structure: ©zhu difeng/Shutterstock.com; livestock feed: ©sharon kingston/Shutterstock.com; industrial component: ©xtrekx/Shutterstock.com

Library of Congress Control Number: 2017957240

ISBN: 978-1-337-39086-6

Cengage
20 Channel Center Street
Boston, MA 02210
USA

Cengage is a leading provider of customized learning solutions with employees residing in nearly 40 different countries and sales in more than 125 countries around the world. Find your local representative at **www.cengage.com.**

Cengage products are represented in Canada by Nelson Education, Ltd.

To learn more about Cengage platforms and services, register or access your online learning solution, or purchase materials for your course, visit **www.cengage.com.**

Printed in the United States of America
Print Number: 05 Print Year: 2021

This book is dedicated to my father, the late Banks Herren, who passed away before it was completed. He taught me in the best way a father can teach his son——by example.

CONTENTS

Preface xii

About the Author xvii

Acknowledgments xviii

CHAPTER 1 **Animal Agriculture as Science 2**

Early Legislation 4
The Scientific Method 5
Advances in Production of Food From Animals 7
Animal Immunization 9
Refrigeration 10
Artificial Insemination 11
Embryo Transfer 11
The Use of Computers 11
Summary 18
Chapter Review 19

CHAPTER 2 **The Classification of Agricultural Animals 20**

Scientific Classification 22
Classification By Breeds 25
Classification According to Use 27
Summary 30
Chapter Review 30

CHAPTER 3 **The Beef Industry 32**

Beef in the American Diet 34
The Beef Industry in the United States 36
Breeds of Beef Cattle 37
Segments of the Beef Industry 39
Summary 43
Chapter Review 43

CHAPTER 4 **The Dairy Industry 44**

Feeding 46
Gestation 48
Milk Production 49
Dairy Goats and Sheep 54
Cheese Manufacturing 56
Summary 58
Chapter Review 59

CHAPTER 5 The Swine Industry 60

History of the Industry 62
Breeds of Swine 63
Production Methods 65
Environmental Concerns 68
Summary 69
Chapter Review 69

CHAPTER 6 The Poultry Industry 70

The Breeder Industry 73
The Broiler Industry 74
Broiler Production 75
The Layer Industry 83
The Turkey Industry 84
Other Poultry 87
Summary 88
Chapter Review 89

CHAPTER 7 The Sheep Industry 90

The Wool Industry 95
Summary 100
Chapter Review 101

CHAPTER 8 The Goat Industry 102

History 104
Goat Industry in the United States 104
Production in the United States 107
Breeds of Goats 108
Anatomy and Physiology 114
Management of Goats 115
Summary 119
Chapter Review 119

CHAPTER 9 The Horse Industry 120

Classification 122
Mules 125
Anatomy of the Horse 126
Raising Horses 127
Summary 130
Chapter Review 130

CHAPTER 10 The Aquaculture Industry 132

Fish Production 134
Sport Fishing 139
Bullfrogs 140
Crayfish 140
Alligator Farming 141
Summary 142
Chapter Review 142

CHAPTER 11 The Small Animal Industry 144

The History of Pets 146
Dogs 146
Cats 147
Exotic Animals as Pets 148
Reptiles 148
Health Benefits 149
Service Animals 150
Pet Food 152
Animal Health 152
Summary 154
Chapter Review 154

CHAPTER 12 **Alternative Animal Agriculture 156**

Rabbit Production 158
Llama Production 160
Fish Bait Production 161
Large Game Animals 162
Laboratory Animal Production 162
Production of Natural and Certified Animal Products 163
Hunting Preserves 166
Summary 167
Chapter Review 168

CHAPTER 13 **The Honeybee Industry 170**

The Importance of Honeybees 172
Bees as Social Insects 173
Commercial Honey Production 177
Breeding Bees 180
Producing New Queens 182
Diseases and Parasites 183
Summary 184
Chapter Review 184

CHAPTER 14 **Animal Behavior 186**

Social Behavior 191
Sexual and Reproductive Behavior 192
Ingestive Behavior 193
Animal Communication 195
Working Safely with Animals 198
Summary 198
Chapter Review 198

CHAPTER 15 **Animal Cells: The Building Blocks 200**

The Importance of Cells 202
Cell Reproduction 208
Animal Stem Cells 210
Summary 211
Chapter Review 211

CHAPTER 16 **Animal Genetics 214**

Gene Transfer 216
The Determination of the Animal's Sex 220
Using Genetics in the Selection Process 221
Performance Data 221
Summary 226
Chapter Review 226

CHAPTER 17 **The Scientific Selection of Agricultural Animals 228**

The Selection of Swine 232
The Selection of Breeding Hogs 235
The Selection of Market Beef Animals 238
The Selection of Breeding Cattle 242
The Selection of Sheep 245
The Selection of Commercial or Western Ewes 246

The Selection of Breeding Ewes 246
The Selection of Rams 247
Judging Market Lambs 247
The Selection of Goats 249
Summary 251
Chapter Review 252

CHAPTER 18 **Animal Systems 254**

The Skeletal System 256
Bones 257
The Muscular System 262
The Digestive System 265
Ruminant Systems 267
The Respiratory System 269
The Circulatory System 270
The Nervous System 273
The Endocrine System 275
Summary 277
Chapter Review 278

CHAPTER 19 **The Reproduction Process 280**

Reproduction in Animals 282
The Male Reproductive System 284
The Female Reproductive System 285
Fertilization 288
Artificial Insemination 290
Embryo Transfer 294
Summary 298
Chapter Review 298

CHAPTER 20 **Cloning Animals 300**

Reasons for Cloning 302
Development of the Cloning Process 305
Perfecting the Process 308
Differences in Clones 309
Summary 310
Chapter Review 310

CHAPTER 21 **Animal Growth and Development 312**

Prenatal Growth 314
Postnatal Growth and Development 317
The Effects of Hormones on Growth 319
The Aging Process in Animals 320
Summary 321
Chapter Review 321

CHAPTER 22 **Animal Nutrition 322**

Water 324
Protein 325
Carbohydrates 328
Fats 329
Minerals 330
Vitamins 331
Determining Feed Rations for Agricultural Animals 333
The Digestion Process 334
Summary 338
Chapter Review 338

CHAPTER 23 **Meat Science 340**

Meat Industry 342
The Slaughter Process 342
Grading 345
The Wholesale Cuts 346
Factors Affecting Palatability 347
Poultry Processing 351
Preservation and Storage of Meat 352
Summary 357
Chapter Review 357

CHAPTER 24 **Parasites of Agricultural Animals 358**

Internal Parasites 362
External Parasites 364
Parasite Control 367
Summary 368
Chapter Review 368

CHAPTER 25 **Animal Diseases 370**

Infectious Diseases 375
The Immune System 378
Noninfectious Diseases 382
Poisoning 383
Disease Prevention 384
Summary 386
Chapter Review 386

CHAPTER 26 **The Issue of Animal Welfare 388**

Confinement Operations 390
The Use of Drugs 392
Management Practices 393
Research 396
Summary 398
Chapter Review 398

CHAPTER 27 **Consumer Concerns 400**

Meat Products 402
Poultry Products 404
Country of Origin 408
Animal Medications 408
Hormones 409
Cholesterol 409
Genetic Engineering 410
Environmental Concerns 411
The Overgrazing of Public Lands 412
Global Warming 413
Summary 414
Chapter Review 415

CHAPTER 28 **Careers in Animal Science 416**

Career Options 418
Developing Personal and Leadership Skills 422
Interview Preparation 423
Summary 426
Chapter Review 427

Appendix A 428
Appendix B 444
Glossary/Glosario 446
Index 478

PREFACE

Now in its fifth edition, *The Science of Animal Agriculture* is directed toward teaching the basic science concepts involved in the production of agricultural animals while remaining current to the latest scientific advancements and trends in the industry. All facets of modern agriculture are based on science. From the most rudimentary cultural practices to the most complicated biotechnology techniques, scientific research has produced the phenomenon known as American agriculture. The science of agriculture has brought humans from the stage of wandering and gathering food to modern civilization. Much of what we know about how living organisms reproduce and grow has come about through our quest to be more efficient in the production of food and fiber. *The Science of Animal Agriculture* contains chapters dealing with the latest concepts in animal biotechnology. Topics include animal behavior, classification, consumer concerns, animal welfare, genetics, scientific selection, reproduction, cloning, growth and development, nutrition, meat science, parasites, and disease. In addition, the scientific basis for the production of the different types of agricultural animals is presented.

NEW TO THIS EDITION

The fifth edition of *Science of Animal Agriculture* features many new exciting enhancements:

- **Alignment to National Agriculture, Food and Natural Resources (AFNR) Standards for Animal Systems**—Each chapter opens with correlations to the National AFNR Animal Systems Standards linking content to the knowledge and skills essential to preparing students for success in the industry.
- **Alignment to Precision Exams Standards for Animal Science**

  PRECISION EXAMS

  This edition of *Science of Animal Agriculture* is aligned to Precision Exams' *Animal Science* exam, part of the *Agriculture, Food and Natural Resources* Career Cluster. The *Agriculture, Food and Natural Resources* pathway connects industry with skills taught in the classroom to help students successfully transition from high school to college and/or a career. Working together, Precision Exams and National Geographic Learning (NGL), a part of Cengage, focus on preparing students for the workforce, with exams and content that is kept up to date and relevant to today's jobs. To access a corresponding correlation guide, visit the accompanying Instructor Companion website for this title. For more information about how to administer the *Animal Science* exam or any of the 170+ exams available to your students, contact your local NGL/Cengage Sales Consultant.

- **Current Information from the Industry** reports on the latest statistics from the USDA, scientific advancements in classification and breeding, the expansion of the poultry production and processing, and the causes behind hive collapse in the bee-keeping industry.
- **Expanded Information on the National FFA** includes the history and evolution of the organization and various opportunities and programs designed to further student success in the agricultural education curriculum and beyond.
- **Putting It into Practice** at the end of select chapters highlights the programs related to participation in the National FFA, including Career Development Events, Proficiency Awards, and other FFA-related activities. Appendix B at the back of the book also contains a complete list of Career Development Events and Leadership Development Events, as outlined by the National FFA Organization.
- **All-New Art and Design** allows students a practical way of connecting concepts with real-world applications through hundreds of new photos while also engaging students in the content through a visually appealing and current design.
- **New Appendix on SAE Programs**—Appendix A describes the features of a Supervised Agricultural Experience (SAE) Program and outlines various SAE project examples to spark student interest and engagement in each of the agricultural education pathways.

EXTENSIVE TEACHING/LEARNING PACKAGE

This package was developed to achieve two goals:

1. To assist students in learning the essential information needed to continue their exploration into the exciting field of agriscience.
2. To assist instructors in planning and implementing their instructional program for the most efficient use of time and other resources.

COMPANION SITE

NEW! The Companion site to accompany *The Science of Animal Agriculture* features tools to support learning and facilitate teaching:

- **Answers to Review Questions** appearing at the end of each chapter allow teachers to track and validate student learning.
- **Lesson Plans** provide an outline of the key topics in each chapter, and correlate to the accompanying PowerPoint® presentations.
- **PowerPoint** presentations align with the Lesson Plans and include photos and illustrations to visually reinforce the key points in each chapter.
- **Testing powered by Cognero,** a flexible online system, provides chapter-by chapter quizzes and enables teachers to
  - Author, edit, and manage test bank content from multiple sources
  - Create multiple test versions in an instant
  - Deliver tests from teacher/school-specific learning management system (LMS) or classrooms

- **Image Gallery** offers full-color photos and illustrations from the text to enable teachers to further enhance classroom presentations.

For these instructor-specific resources, please visit CengageBrain.com at http://login.cengage.com and follow the prompts for obtaining access to this secure site.

MINDTAP FOR THE SCIENCE OF ANIMAL AGRICULTURE 5E

NEW! The MindTap for *The Science of Animal Agriculture 5th Edition* features an integrated course offering a complete digital experience for the student and teacher. This MindTap is highly customizable and combines assignments, videos, interactivities, and quizzing along with an enhanced ebook to enable students to directly analyze and apply what they are learning and allow teachers to measure skills and outcomes with ease.

- **A Guide:** Relevant interactivities combined with prescribed readings, featured multimedia, and quizzing to evaluate progress will guide students from basic knowledge and comprehension to analysis and application.
- **Personalized Teaching:** Teachers are able to control course content—hiding, rearranging existing content, or adding and creating their own content to meet the needs of their specific program.
- **Promote Better Outcomes:** Through relevant and engaging content, assignments, and activities, students are able to build the confidence they need to ultimately lead them to success. Likewise, teachers are able to view analytics and reports that provide a snapshot of class progress, time in course, engagement, and completion rates.

FEATURES OF THIS EDITION

Objectives

Objectives are listed at the beginning of each chapter and provide both objectives in basic science and those specific to agricultural science. Those related to basic science are concepts students are expected to learn prior to graduation.

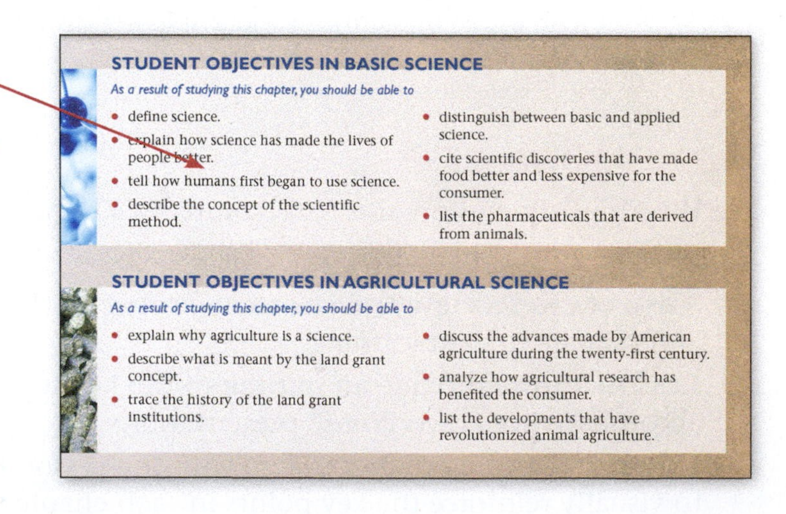

STUDENT OBJECTIVES IN BASIC SCIENCE

As a result of studying this chapter, you should be able to

- define science.
- explain how science has made the lives of people better.
- tell how humans first began to use science.
- describe the concept of the scientific method.
- distinguish between basic and applied science.
- cite scientific discoveries that have made food better and less expensive for the consumer.
- list the pharmaceuticals that are derived from animals.

STUDENT OBJECTIVES IN AGRICULTURAL SCIENCE

As a result of studying this chapter, you should be able to

- explain why agriculture is a science.
- describe what is meant by the land grant concept.
- trace the history of the land grant institutions.
- discuss the advances made by American agriculture during the twenty-first century.
- analyze how agricultural research has benefited the consumer.
- list the developments that have revolutionized animal agriculture.

Key Terms

Each chapter includes a **Key Terms** list that highlight the terms presented in the chapter and those that the student should be able to define, in context, upon completion of the lesson.

KEY TERMS

maintenance ration	herbivores	lipids	ruminant	boluses (*sing.*, bolus)
feedstuff	tankage	inorganic	cecum	rumen
anabolism	cellulose	macrominerals	esophagus	reticulum
catabolism	monosaccharides	microminerals	pepsin	omasum
amino acids	disaccharides	trace minerals	duodenum	abomasum
essential amino acids	glucose	free choice	jejunum	mucous membranes
nonessential amino acids	fructose	carotene	ileum	bloat
crude protein content	galactose	monogastric	semipermeable membrane	
carnivores	sucrose	gastrointestinal tract	diffusion	
	lactose	alimentary canal		

NATIONAL AFNR STANDARD

AS.03.01.01.a
Identify and summarize essential nutrients required for animal health and analyze each nutrient's role in growth and performance.

AS.03.01.01.b
Differentiate between nutritional needs of animals in different growth stages and production systems.

AS.03.01.01.c
Assess nutritional needs for an individual animal based on its growth stage and production system.

AS.03.01.02.a
Differentiate between nutritional needs of animal species.

AS.03.02.01.a
Compare and contrast common types of feedstuffs and the roles they play in the diets of animals.

AS.03.02.02.a
Examine the importance of a balanced ration for animals based on the animal's growth stage.

AS.03.02.03.a
Examine the purpose, impact, and mode of action of feed additives and growth promotants in animal production.

National AFNR Standard

Correlations at the beginning of each chapter link content to the **National AFNR *Animal Systems* Content Standards** for ease of reference.

From the Field

Hundreds of vibrant photos as well as informational graphics and tables encourage student engagement and keep them current with industry statistics.

FIGURE 13–3 Many crops could not survive without help from bees. © Darkness/Shutterstock.com.

FIGURE 13–4 Fruit growers hire beekeepers to bring in truckloads of bees in the spring when the trees are blooming. © GEORGID/Shutterstock.com.

Honeybees are particularly adept at pollinating. Many insects work flowers, but most go from one flower to a different kind of flower. Bees, however, work a specific kind of flower for a period of time. For example, honeybees may be working apple blossoms for several days until the flowers are gone and then go on to work a different type of flower. In doing this, they go from an apple blossom to another apple blossom and spread pollen from one apple blossom to another in that way. This process ensures that the blossoms are thoroughly pollinated.

Fruit growers hire beekeepers to bring in truckloads of bees in the spring when the trees are blooming (**Figure 13–4**). The bees live in wooden boxlike structures called **hives** with a separate colony of bees in each hive. The hives are easy to handle and can be loaded on a truck with the bees still in the hive. The owner of the bees then can move the hives from orchard to orchard for a fee from the fruit or crop producer. In addition, the producer can harvest hundreds of pounds of honey each year that can be sold at a profit.

BEES AS SOCIAL INSECTS

A characteristic of honeybees that sets them apart from many other types of insects is their social structure. They live in a highly ordered society in which each bee seems to have its job and works in concert with the rest of the bees in the hive. Within a colony of bees there are three types of bees: the **queen**, **drones**, and **workers** (**Figure 13–5**).

FIGURE 13–5 Within a colony of bees are three types of bees: the queen, drones, and workers.

PUTTING IT INTO PRACTICE

Raising Goats

Goats are becoming a very popular choice for supervised agricultural experiences (SAEs). A pair of goats can be grown in a relatively small space, as long as you have the means to feed and provide water for them. Compared with other types of livestock, goats can be raised with only a modest investment. You will need to have a small shelter for the animals and a way to keep the structure clean. As with all SAEs, you should keep accurate records on expenses such as feed, vet care, and housing costs. Be sure to keep close records on the amount of time spent caring for the animals. For further information on SAEs see Appendix A at the end of this text.

Your SAE can be showcased through the many goat shows throughout your local areas and state. This will entail making sure your animal is trained for showing and kept clean and well groomed. You will learn the proper way to show the animal and can enter the showmanship competition. Competition through Future Farmers of America can be very rewarding in terms of learning skills and having fun.

Raising goats is a supervised agricultural experience that can be conducted with a modest amount of funding. *Courtesy of the National FFA Organization*

Putting It into Practice

These engaging articles provide information on various FFA activities related to animal science, helping students connect chapter content with practical hands-on experiences.

Summary and Chapter Review

These chapter features encourage students to review and apply what they have learned in the lesson. The Chapter Review includes **Review Questions** to evaluate student knowledge, and **Student Learning Activities** to engage students in further learning.

Chapter 20 Animal Growth and Development 293

SUMMARY

Animals begin to grow at birth and in some ways continue this process until they die. All the body's systems must develop, grow to the proper size, function, and replace cells as they are lost or wear out. Growth and development are what the animal industry is all about. Ensuring that animals grow properly is the job of all those connected with the industry, from conception to the slaughter process.

CHAPTER REVIEW

Review Questions

1. Why is animal growth essential to producers?
2. By what two methods does growth occur?
3. What is the difference between the morula and the blastula?
4. What is the purpose of the placenta?
5. Name three layers of the blastula.
6. What organs arise from the ectoderm?
7. What is meant by differentiation?
8. What is oxytocin?
9. Why is it essential that newborn animals have well-developed legs?
10. At what stage in an animal's life does the most rapid growth occur?
11. What is meant by ossification?
12. List the sequence in which an animal's body deposits fat.
13. List five glands that secrete hormones that affect growth.
14. What is meant by the lean-to-fat ratio?
15. Tell the difference between chronological and physiological age.

Student Learning Activities

1. Obtain from a slaughterhouse the fetuses from animals that have been slaughtered. Try to determine the stage of maturity for each. Be sure to wear latex gloves when handling the material.
2. Visit with a producer and determine how he or she determines when animals have reached the level of maturity desired for market. Try your hand at making the determination of animals in the producer's herd.

APPENDIX A

Supervised Agricultural Experience

Supervised Agricultural Experience (SAE) is an integral part of a total program of agricultural education. It is where you plan, propose, conduct, document, and evaluate your programmatic experiential learning activities. It is where you apply and test what you have learned in your classes and FFA experiences. An SAE will help you become well versed in record keeping for your own portfolio. It will also aid in your personal leadership development and in the strategic planning process for teams, groups, organizations, and programs.

TERMS TO KNOW

- Degree program
- Portfolio
- Supervised Agricultural Experience (SAE)
- Scientific process
- Ownership/Entrepreneurship SAE
- Placement/Internship
- Research SAE
- Foundational SAE
- School-Based Enterprise SAE
- Service Learning SAE
- Improvement Project
- Proficiency awards

OBJECTIVES

After completing this summary, the student will be able to:

- define SAE
- list and explain the types of SAE
- plan an SAE program
- propose an SAE program
- conduct an SAE program
- document an SAE program
- evaluate an SAE program
- apply proper record-keeping skills
- participate in youth leadership opportunities to create a well-rounded SAE
- produce a local program of activities using a strategic planning process
- participate in a local program of activities using a strategic planning process

Supervised Agricultural Experience (SAE) is a program of experiential learning activities conducted outside of the regular agricultural education class time. The student-led, instructor-supervised, work-based learning experience is designed to help students develop and apply the knowledge and skills learned in an agricultural education classroom and/or laboratories.[1] The SAE used to be called the Supervised Occupational Experience Program (SOEP), and the focus was on the occupation. These days, the focus is on entrepreneurship and experiences rather than the occupation or the actual career.

The SAE takes place under the direction of your agricultural education teacher. Many times, individual projects (e.g., showing pigs, raising a plant in the greenhouse, etc.) are considered to be SAE, but the intent of a quality SAE is for it to be conducted as a series of career-related experiences completed during your enrollment in agricultural education.

Supervised Agricultural Experience (SAE) Programs

Information on **SAE Programs**, including how to plan, propose, conduct, document and evaluate a program, is included, as well as several examples of projects for students seeking ideas for their own SAE.

ABOUT THE AUTHOR

Ray V. Herren has been actively involved in the animal industry for most of his life. He grew up on a diversified farm, where he played a major role in the production of livestock. He obtained a Bachelor of Science degree in agricultural education from Auburn University, a Master's degree in agribusiness education from Alabama A&M University, and a Doctorate in vocational education (with an emphasis in agricultural education) from Virginia Polytechnic Institute and State University.

Dr. Herren has taught agriculture at the high school level and the university level, including at Gaylesville High School, Virginia Polytechnic Institute and State University, Oregon State University, and the University of Georgia in Athens (UGA), where he recently retired as head of the Department of Agricultural Leadership, Education and Communication. As a faculty member in agricultural education at Oregon State University, he served as the coach of the University Livestock Judging team and taught several animal practicum courses. In addition to serving as national president in the FFA Alumni organization, he has served on numerous committees from the local to international level, including a national task force to develop FFA programs for middle school and the National Committee for Career Development Events. His prolific scholarly activity includes 26 journal articles, 51 invited or refereed presentations, and 12 books and manuals. He has also earned several awards for his commitment to service, including induction into the Georgia Agricultural Teacher Hall of Fame and UGA's prestigious College of Education Outstanding Teaching Award.

ACKNOWLEDGMENTS

The author gratefully acknowledges the following for their assistance in creating the fifth edition of this text:

Dan Rollins, director of feed operations, Aviagen North America, for his technical advice on poultry; Aviagen Poultry for their help with images; Don Canerday for his help in reviewing the Entomology chapter; and James Griffith for his help in reviewing the chapters dealing with genetics and cloning.

The author and Cengage would also like to thank those individuals who reviewed the manuscript and offered suggestions, feedback, and assistance for this edition:

Lisa Hochreiter
Agricultural Educator
Warwick High School
Warwick, SD

Kori Shackelford
Agricultural Educator
Hillsboro High School
Hillsboro, OR

Keith Shane
Agriscience Teacher
FFA Advisor
Smyrna High School
Smyrna, DE

CHAPTER 1

Animal Agriculture as Science

STUDENT OBJECTIVES IN BASIC SCIENCE

As a result of studying this chapter, you should be able to

- define science.
- explain how science has made the lives of people better.
- tell how humans first began to use science.
- describe the concept of the scientific method.
- distinguish between basic and applied science.
- cite scientific discoveries that have made food better and less expensive for the consumer.
- list the pharmaceuticals that are derived from animals.

STUDENT OBJECTIVES IN AGRICULTURAL SCIENCE

As a result of studying this chapter, you should be able to

- explain why agriculture is a science.
- describe what is meant by the land grant concept.
- trace the history of the land grant institutions.
- discuss the advances made by American agriculture during the twenty-first century.
- analyze how agricultural research has benefited the consumer.
- list the developments that have revolutionized animal agriculture.

KEY TERMS

agriculture
domesticated
scientific method
hypothesis
experiment
control group
experimental group
lactation
agricultural animals
basic research
hormones
applied research
omnivorous
poultry
broiler
beef
veal
vaccinating
immunity
serum
vaccines
environment
genes
offspring
artificial insemination
sires
semen
dams
gestation
progeny
pharmaceutical

SCIENCE is the study or theoretical explanation of natural phenomena. Nature has given us a wonderfully complex world in which to live. Our environment is governed by natural laws that control everything from gravity to the weather. These laws also control the way plants and animals live and grow on our planet. Through science, these laws are investigated and new ways are found to use them to make our lives easier and better.

Americans enjoy one of the highest living standards in the world. Our per-capita income ranks near the very top among all the nations of the world. Most of our comforts and enjoyment have come about as a direct result of science.

Agriculture is the oldest and most important of all sciences. Without a sound basis of agricultural knowledge, all humans would be struggling to find enough food, shelter, and clothing to survive. Because we have discovered so much about the world around us, our food, shelter, and clothing are produced so efficiently that only a small percentage of the population is required to produce these necessities. That leaves the rest of us to develop new knowledge in different areas that will allow us to have better transportation, communications, and recreation.

From the earliest time, humans studied the environment in which they lived. Their initial thoughts dealt with how to survive, and this meant that first they had to find food. Early people ate the fruits, seeds, and animals they found in their environment. As these supplies of food became exhausted, the people were forced to leave the area and find new places where plants and animals were more abundant. They undoubtedly observed the behavior patterns of animals and from these observations devised means to locate and kill them. As they continued to study patterns of animal movement and responses, people reasoned that if they could domesticate animals, the need for moving with the herds and hunting as a group could be eliminated.

Close observation of the animals determined which ones would be best to tame. As animals were **domesticated**, people reasoned that they could use certain methods to raise the animals more efficiently. Through trial and error, they discovered the best ways of caring for the animals, and they passed these methods along to their children. Only relatively recently have people studied animals in a systematic way.

Progressive scientific research began in the United States about the middle of the 1800s. At that time, universities in the United States taught a curriculum known as the Classics, in which the main subjects studied were Latin, Greek, history, philosophy, and mathematics. People began to realize a need for institutions of higher learning where students could study areas that had practical applications. The nation was emerging as an industrial- and agricultural-based economy. To achieve progress in these areas, young people needed to be taught how to produce food and manufacture goods more efficiently.

EARLY LEGISLATION

In the late 1850s, Vermont Senator Justin Morrill introduced a bill that would provide public land and funds for establishing universities to teach practical methods of manufacturing and producing food and fiber (**Figure 1–1**). The bill passed in 1862 and became known as the Land Grant Act, or the Morrill Act. During that same year, President Abraham Lincoln signed into law a bill that established the U.S. Department of Agriculture (USDA). Soon, almost all of the states in the country established land grant colleges.

As the students enrolled and classes began, a severe problem was recognized: No one really knew anything about agriculture! Most of the knowledge about growing plants and animals had been passed from generation to generation and represented people's beliefs rather than proven knowledge. To solve this problem, Congress passed the Hatch Act in 1872. This law authorized the establishment of experiment stations in different parts of the states that had land grant colleges. The purpose was to create new knowledge through a systematic process of scientific investigation. These experiment stations put to use what has come to be known as the *scientific method of investigation*.

FIGURE 1–1 Senator Justin Morrill of Vermont introduced a bill to provide land grant colleges. *Source: Library of Congress*

In 1914, Congress passed the Smith-Lever Act, setting up the Cooperative Extension Service, which serves to disseminate information learned from new research to the population. This completed a system known as the land grant concept, which held that the purpose of a land grant college is to teach, conduct research, and carry the new information to the people in the state through the extension service.

The Smith-Hughes Act was passed in 1917, a law that established vocational agriculture as a program in the public high schools as a means of teaching new methods of agriculture.

THE SCIENTIFIC METHOD

The **scientific method** is a systematic process of gaining knowledge through experimentation. This method is used to ensure that the results of an experimental study did not occur just by chance and that something caused the change. This process involves formulating a hypothesis, designing a study, collecting data, and drawing conclusions based on analysis of the data (see **Figure 1–2**).

A scientist begins by identifying a problem that needs to be solved. The scientist may have an idea or suspicion of what causes the problem or what might solve the problem. This suspicion, called a **hypothesis**, serves as the basis for investigating a problem. The hypothesis then is subjected to a test called an **experiment**, which attempts to isolate the problem in question and determine the solution.

For instance, if a scientist wants to know whether milking cows twice a day would result in obtaining more milk than milking only once a day, he or she might milk a group of cows twice a day and record the results. Just milking twice a day and recording the results, however, would not prove much. The real question is whether milking twice a day produces more milk than milking only once a day. If two groups of cows were used—one

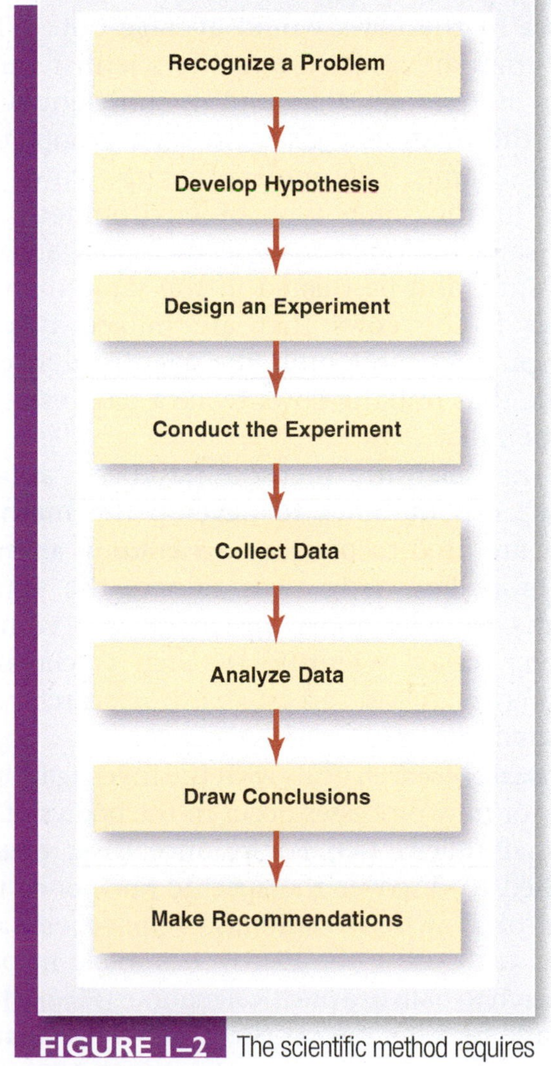

FIGURE 1–2 The scientific method requires several steps.

group that is milked twice a day and one group that is milked only once a day—the results would be more meaningful. The cows milked once a day are called the **control group**, and the cows milked twice a day are called the **experimental group**.

Other conditions also must be met to ensure that the data collected in the experiment are valid. The cows must be of the same breed, the same age, the same stage of their **lactation** cycle, and the same size. Also, each group should have enough cows so the differences in the cows would average out. When the cows are selected and the scientist thinks they are enough alike and of sufficient number, the treatment given to each group must be the same. This means that the cows will stay in the same type of environment, eat the same ration, consume the same amount of water, and be treated in the same manner. Then if the cows that are milked twice a day produce more milk, the scientist may conclude that milking cows twice a day results in more milk.

FIGURE 1–3 Basic research investigates why or how processes occur. © Alexander Raths/Shutterstock.com.

The scientific method has been applied thousands of times to develop the methods that are used to produce **agricultural animals**. As more knowledge was gained, the experiments became more involved and complicated. Today, scientific research is classified into two broad areas: basic research and applied research.

Basic research deals with the investigation of why or how processes occur in the bodies of the animals (**Figure 1–3**). For instance, basic research is used to discover the specific **hormones** that control growth in animals. **Applied research** deals with using the discoveries made in basic research to help in a practical manner (**Figure 1–4**). If basic research discovers the growth-regulating hormones, applied research uses this knowledge to develop ways of using natural or artificial hormones to increase the growth efficiency of agricultural animals.

In a very real sense, agriculture is the application of almost all the research and knowledge associated with plants and animals. Aside from the field of medicine, about the only application of basic research in the life sciences is that of agriculture.

Basic research carried through to application in animal science has benefited humans

FIGURE 1–4 Applied research investigates how basic research can be put to use. *Source: USDA, Agricultural Research Service (ARS)*

in many ways—the most obvious of which is food. Humans are **omnivorous** animals. This means that we eat both plants and animals. The production of animals provides people with a reliable, abundant source of high-quality food. Advances made through scientific research have resulted not only in an abundance of animals for food but also in relatively low prices for food. Better production methods result in higher efficiency in raising agricultural animals. This greater efficiency means a constant supply of affordable food.

FIGURE 1–5 The weight of weaned calves has more than doubled since 1925. *© iStockphoto/Mercedes Rancano Otero.*

ADVANCES IN PRODUCTION OF FOOD FROM ANIMALS

As a result of scientific research and the application of the research findings, gigantic strides have been made in the efficient production of food from animals. The Council for Agricultural Science and Technology has compiled a list of some of the advances since 1925, as highlighted in the following paragraphs.

Beef cattle liveweight marketed per breeding female increased from 220 pounds to 482 pounds. This means that for every cow raised, we now are selling more than twice the beef that we did in 1925 (**Figure 1–5**). These increases have come about as a result of the scientific selection of breeding animals, a better understanding of beef cattle nutrition, and better control of parasites and diseases.

A better understanding of all phases of the lives of the animals has led to higher quality and less-expensive beef for the consumer. The lower cost no doubt accounts to a large extent for the tremendous increase in the annual per-capita consumption of beef. Since 1925, consumption has doubled from 60 pounds of carcass weight equivalent to 120 pounds of carcass weight equivalent. Americans now have a more nutritious diet at less cost.

Sheep liveweight marketed per breeding female increased from 60 to 130 pounds. This represents an increase of 100 percent over the production in 1925 (**Figure 1–6**). Around the

FIGURE 1–6 Research has brought about a 100 percent increase in the weight of market lambs. *© Alexey Stiop/Shutterstock.com.*

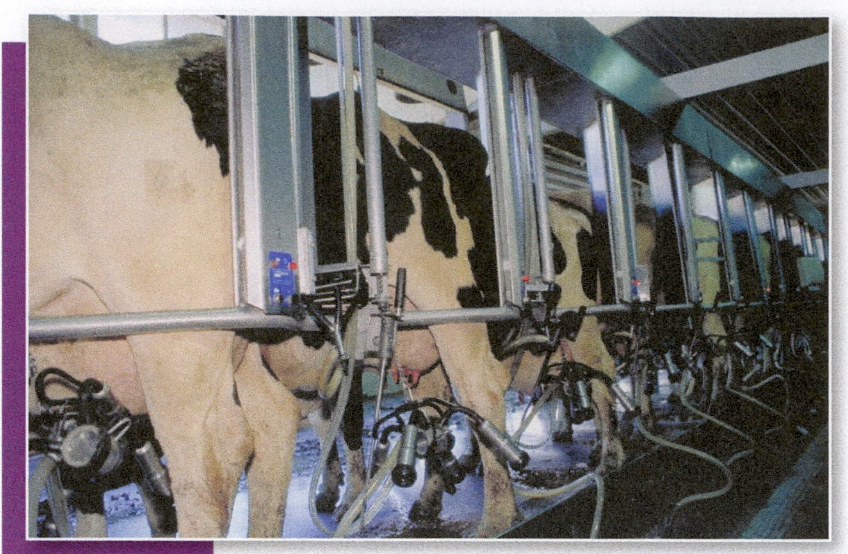

FIGURE 1–7 The same amount of milk is now produced by half the number of cows that it took in 1925. *Source: USDA*

turn of the twentieth century, sheep were raised primarily for wool. Then, through research in selection, sheep began to be raised mainly for meat. Research efforts have concentrated on raising better meat-type animals.

Since 1925, milk marketed per dairy cow increased from 4,189 to 20,625 pounds—an increase of five times the amount of milk marketed in 2008. Also in 1925, there were more than twice the number of cows as there are now (**Figure 1–7**). This means that half the number of cows now produce more than three times the amount of milk. This greater efficiency has resulted in quite a bargain for the consumer. A gallon of milk now costs quite a bit less than it would using the technology of the 1920s.

In the same timeframe, swine liveweight marketed per breeding female increased from 1,600 to 2,850 pounds. Since 1950, the amount of feed required to produce a 200-pound market hog has been reduced by about 50 pounds. Also during this time, the average time to produce a 220-pound hog has been reduced from 170 to 157 days (**Figure 1–8**). Because pigs can be raised in a shorter time on less feed, the cost of production is lower. This savings is passed along to the consumer in less-expensive pork.

No other segment of agricultural industry has made more dramatic advance than that of the **poultry** industry. In 1925, producers required about 112 days to produce a market broiler. The **broiler** weighed about 2½ pounds and required 4.7 pounds of feed to gain a pound. Currently, a 10-pound broiler can be produced in 45 days. This is due to genetics and better nutrition. Contrary to what many believe, broilers are produced without any form of supplemental hormones. The amount of feed required to produce a pound of growth has been reduced to 1.92 pounds. This means that we now can produce a heavier broiler in half the time on

FIGURE 1–8 Since 1950 the average time required to produce a 220-pound hog has been reduced from 170 to 157 days. *Source: USDA, Agricultural Research Service (ARS)*

FIGURE 1–9 Modern broilers are more than a pound heavier and are raised in half the time on half the amount of feed than in 1925.
© iStockphoto/wikoski.

less than half the feed than we could in 1925 (**Figure 1–9**). These gains in efficiency have resulted almost entirely from better genetics and better nutrition. This has come about through a tremendous amount of research. Consumers get a better broiler at much less cost and also get a bird that is completely dressed and ready to cook.

Since 1925, the annual production per laying hen tripled from 112 to around 300 eggs, and the feed required to produce a dozen eggs decreased from 8.0 to 4.0 pounds. Advances through scientific research have made possible the production of eggs in an enclosed building. This allows the hens to be much better managed because the producer can control the environment in which the eggs are produced.

The weight of marketed turkeys increased from 13.0 to 18.4 pounds. This gain was achieved on less feed (from 5.5 to 3.1 pounds) and in about half the time (from 34 to 19 weeks).

The countries in North and Central America, Europe, and Oceania (Australia, New Zealand) have only 29.9 percent of all the world's cattle, yet these countries produce 68 percent of the world's **beef** and **veal**. Similar statistics are true for the other areas of animal agriculture as well. These are countries where almost all of the scientific knowledge about raising agricultural animals has been discovered. The people in these areas are the best fed and enjoy the highest living standard of any people in the world as a result of the new knowledge acquired through basic and applied research. Before countries can develop and prosper, they must achieve a sound basis for producing food. Many of the poorer countries are attempting to imitate the agricultural systems of the more advanced countries.

Through the years, many discoveries and developments have aided the advancement of animal agriculture. Some of the progress has been the result of many small discoveries that go together to provide greater efficiency in the production of animals. Other advances have come about as the result of great strides in scientific breakthroughs. The following are some of the milestones that have served to revolutionize animal agriculture.

ANIMAL IMMUNIZATION

Until the last half of the 1800s, diseases devastated herds of all types of agricultural animals all over the world. Once disease started in an area, all the animals in the surrounding countryside contracted the disease because there was no method of preventing the spread.

During the 1870s and 1880s, a French scientist named Louis Pasteur developed a means of **vaccinating** animals to make them immune to disease. Using the scientific method,

Pasteur hypothesized that animals that had contracted a disease and survived must have built up **immunity** to the disease. Using the blood from sheep that had contracted and survived the deadly disease of anthrax, he developed a **serum**.

Pasteur's experiment used two groups of healthy sheep. One group was injected with the serum and was later injected with anthrax organisms; the other group received only an injection of the anthrax organisms. The group that had received the serum remained healthy; the group that had not received the serum died.

Following Pasteur's discovery, other scientists began to conduct research on other diseases. During the next century, numerous new **vaccines** were developed to control most of the diseases contracted by agricultural animals. Now, animals can be raised in a disease-free **environment** and at a much lower cost and with quite a bit less risk to the producer (**Figure 1–10**).

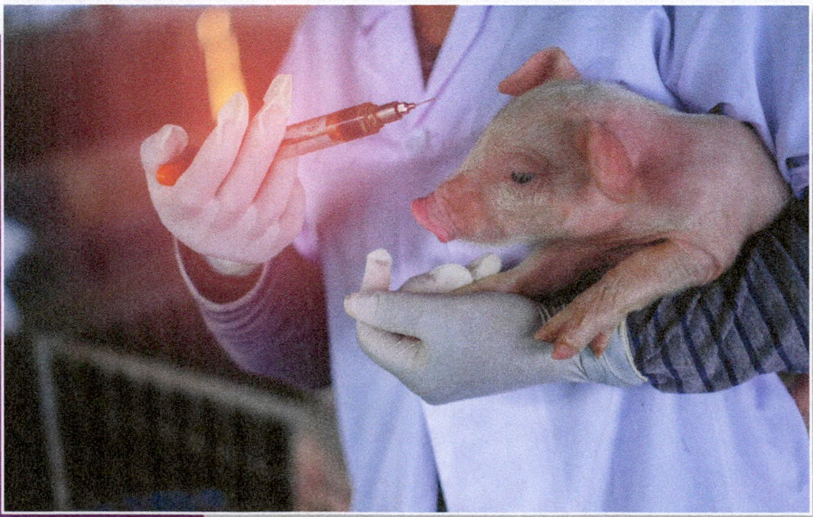

FIGURE 1–10 Through the use of vaccinations, animals can now be raised disease-free. © krumanop/Shutterstock.com.

REFRIGERATION

A problem that plagued the producers of animals since the time humans first began raising them was how to preserve the meat and other products. When a large animal was slaughtered, not all of the meat could be eaten at once. Particularly in the summer months, the meat spoiled very quickly, and if it wasn't consumed quickly, most would go to waste. In colder climates, the animals were killed in the winter, and the freezing temperatures preserved the meat until the spring thaw. About the only other way of preserving the meat was by salting or drying the meat. Both of these methods were time-consuming and did not produce a very palatable product.

Another problem that occurred later was that of getting the meat to market. Until about the turn of the twentieth century, live animals had to be delivered to the population centers. This meant driving the animals to market or to a railhead where they could be transported live to the market, slaughtered, and sold as fresh meat. If the meat didn't sell quickly, it was lost to spoilage.

The first attempt at cooling meat was to use ice that had been cut from frozen lakes during the winter and stored in icehouses. The ice blocks were suspended from the ceiling in meat storage rooms in an effort to keep the meat cool. This effort was not very successful. During the 1880s, mechanical refrigeration was developed and used in slaughterhouses to store meat.

A few years later the refrigerated boxcar was invented, which allowed meat to be transported anywhere in the country at any time during the year (**Figure 1–11**). Now, not only could animals be slaughtered at any time of the year but the meat also could be stored for long periods of time. Because meat could be distributed to everyone in the country, a larger supply of meat was needed.

FIGURE 1–11 The development of refrigeration has allowed the preservation of meat in warm weather. *© Tyler Olson/Shutterstock.com.*

ARTIFICIAL INSEMINATION

Advances in the type of animal produced are brought about through the transfer of superior **genes** from parents to their **offspring**. With the advent of **artificial insemination** in the 1930s, the transfer of genes from superior **sires** was greatly multiplied. Through modern techniques of **semen** collection, storage, and distribution, almost any producer can access the best genes in the industry. This innovation is one reason for the phenomenal advancement of the dairy industry. Most of the dairy animals born in the United States are the results of artificial insemination.

EMBRYO TRANSFER

Although artificial insemination increased access to superior sires, advances through the use of superior **dams** were slow because of the female's **gestation** period. With the development of the embryo transfer process, one superior dam can produce many offspring in one year (**Figure 1–12**). Combined with artificial insemination, this allows producers to make extremely rapid gains in the quality of their herds at a relatively low cost.

FIGURE 1–12 With the development of the embryo transplant process, one superior dam can produce many offspring in one year. *Courtesy of James Strawser Agricultural Photography Inc*

THE USE OF COMPUTERS

Computers were developed during the 1940s, but it was not until the 1980s that the use of the computer became such an integral part of our lives. The computer has had a profound effect on many aspects of the animal industry. For example, the computer has made all areas of research move more rapidly. Data that once took days and even weeks to analyze can now be computed in a matter of seconds. Computer-simulated experiments and models have helped decrease the cost and time involved in scientific research.

The selection of superior dams and sires has become more convenient and accurate through computerized production records of **progeny**. Sires for artificial insemination can be matched

FIGURE 1–13 Modern feed mills such as this use computers to balance and mix feed rations. *Courtesy of Dan Rollins*

with dams for embryo transfer through the use of computers. Many breed associations keep detailed records of all the animals in their registry on computer file.

Feed formulation now is done by computer. This includes the balancing of feed rations and also controlling the mixing and regulation of ingredients in the feed (**Figure 1–13**). This allows much more accurate blending of the nutrients needed by animals.

Perhaps the greatest impact of computers is in the area of information retrieval. Through the Internet, producers can access an incredible amount of information without leaving home. This system connects the producer to information from all over the world. Producers obtain summations of the latest scientific research; find sources for feed, supplies, and breeding animals; locate markets; and even buy and sell animals on-line. New ideas on management practices and production can be gathered from other producers throughout the world. Decisions on buying and selling have become easier because the producer can obtain immediate price quotes and predictions over the Internet. In fact, all banking and financial transactions can be done electronically.

In addition, most major breed associations place information on the Internet so producers can have up-to-the-minute information on events within the associations. Production and progeny records on the most valuable sires and dams are available to aid in decision making on breeding programs. Information on problems, such as disease outbreaks, can be instantly passed on to the producer over the Internet. Notices of sales, show schedules, and upcoming events can be placed on the Web for the use of the producers. Computers have brought the information of the world into the homes of the producers.

Efficiency in the production of agricultural animals is not the only benefit derived from scientific research (**Figure 1–14**). Also, the lives of humans have been greatly enhanced by research dealing with animals through the development of pharmaceuticals from animal by-products. A **pharmaceutical** is a substance that is used as a drug to make the life of a person better. Many of the drugs routinely prescribed by doctors are derived from animals (**Figure 1–15**). For instance, cortisone, which relieves the suffering of people with arthritis, is derived from the gallbladder of cattle. Until recently, when a synthetic form of cortisone was developed, the only source for this drug was cattle.

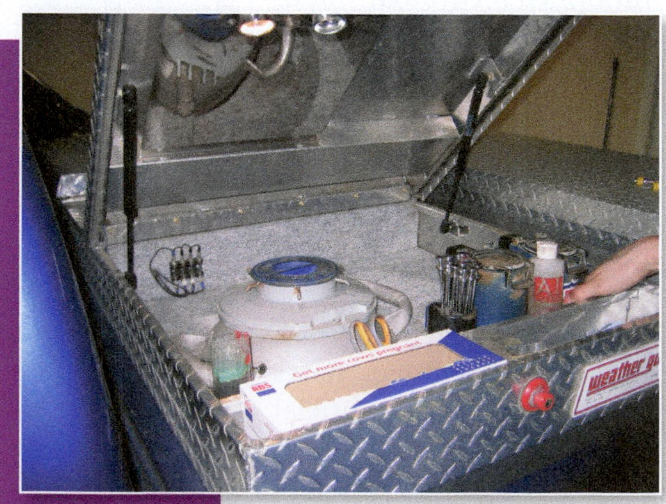

FIGURE 1–14 The development of artificial insemination techniques has allowed a superior sire to sire hundreds of offspring. This equipment is used in the process. *Courtesy of Frank B. Flanders*

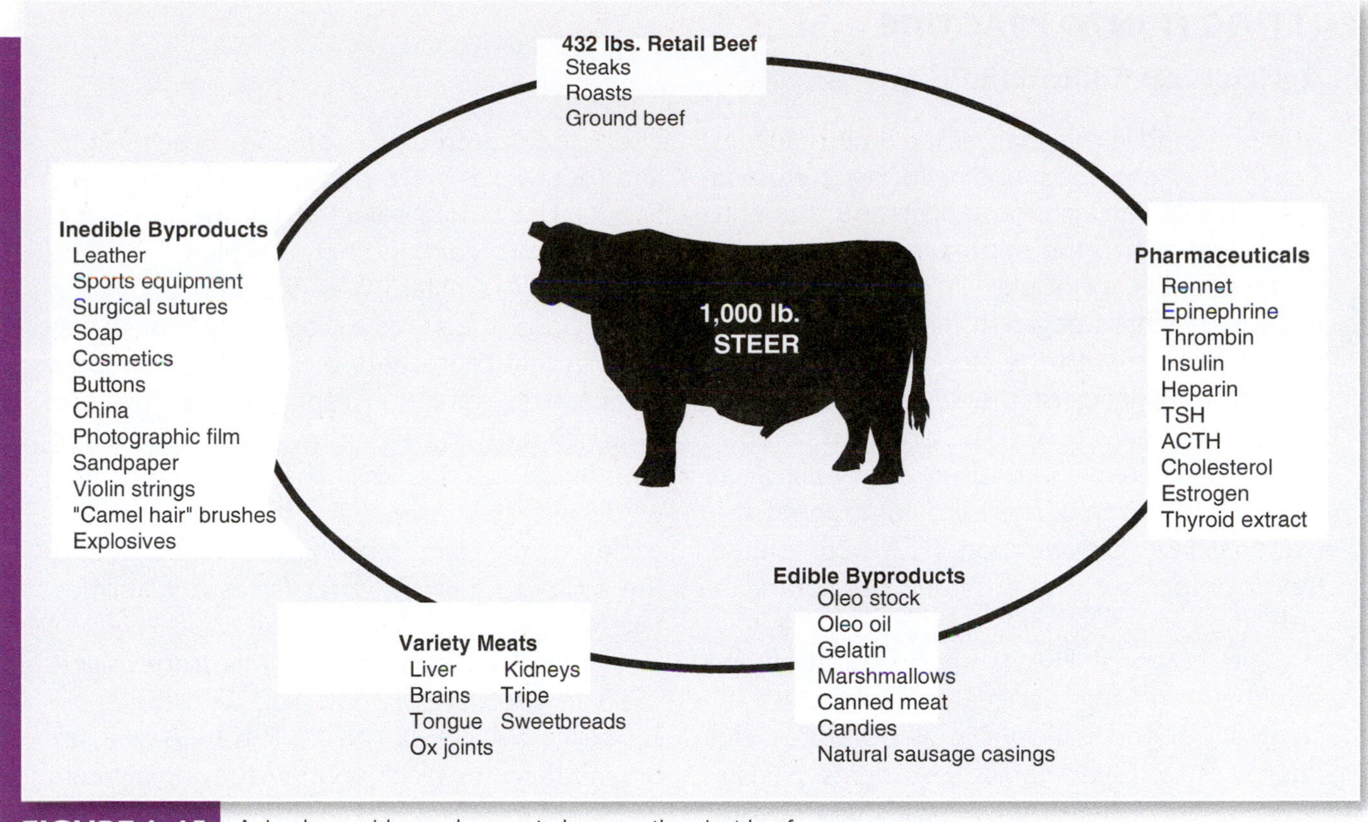

FIGURE 1–15 Animals provide much more to humans than just beef. *National Cattlemen's Beef Association*

As another example, a periodic intake of insulin is necessary for people who have diabetes. Before a synthetic form of insulin was developed, it was made from the pancreas of animals. Insulin from hogs is of particular value because this form most closely resembles human insulin. Some diabetics are allergic to synthetically produced insulin and can take only the form of insulin derived from hogs.

Many of the hormones used to treat human disorders come from animals that have been slaughtered for food. When, for some reason, the human body does not produce enough hormones to control or stimulate a body function, the hormone must be supplied from an outside source. Without the ready supply of drugs derived from animals, many people's lives would be shortened, and the quality of their lives would be lessened.

Another result of scientific research is that animal parts can be used as replacements for human parts. Pigs have been particularly important in this regard. Heart valves from pigs have been used for about 20 years as replacements for faulty human heart valves. Pig heart valves are highly superior to mechanical valves because the mechanical valves tend to accumulate residue from the flow of blood. As a result, the mechanical valve sticks and the result can be fatal. In contrast, a pig heart valve does not accumulate as much residue because it came from a living, functioning animal. Although the valve is treated to make it into an inanimate object before it is placed in a human heart, the traits that allow its use are retained. Skins from pigs are now being used to treat humans who have incurred severe burns. The pig tissues are used as skin covers until the new human skin has had a chance to grow.

PUTTING IT INTO PRACTICE

Agricultural Education Programs

As mentioned in this chapter, the Smith-Hughes Act of 1917 provided funding for agriculture in high schools. Through these programs, students in rural areas had the opportunity to learn the latest scientific principles in agricultural production. Teachers began visiting students after school to help with the student projects. This was the beginning of the supervised agricultural experiences (SAE). This program component is explained in detail in Appendix A at the back of this text. The third component, the National FFA Organization (FFA), completed the three parts of the program: classroom/lab, SAE, and FFA.

The National FFA is an organization for students studying agricultural education in high school and is active in all 50 states and Puerto Rico. Over 450,000 students participate in activities and programs sponsored by the FFA.

This organization began in 1928 when students studying agriculture met in Kansas City for the purpose of establishing a national association. Prior to this, many states had created similar organizations for rural young people. The state organizations began to compete in livestock judging and looked for a place to hold a national competition. The American Royal in Kansas City agreed to host the event. This brought the students together for the first time, and they decided there was a need for a national organization. In 1928, the Future Farmers of America came into existence.

The FFA was patterned after the Future Farmers of Virginia. This state association was developed by Henry Groseclose, Walter Newman, Edmund Magill, and Harry Sanders. They saw the need to give rural youth opportunities to develop social graces and leadership skills. Ceremonies for the FFA were designed after the ceremonies of the Freemasons, and the awards were patterned after the Boy Scouts. The colors selected for the organization are corn gold and national blue.

The FFA emblem was designed by Henry Groseclose. It features a cross section of an ear of corn that represents the nationwide scope of FFA. The rising sun represents a new era in agriculture and the plow represents labor and tillage of the soil. The owl stands for wisdom that is necessary for progress. The eagle at the top symbolizes freedom and the ability to fulfill to go as far as our abilities take us.

The FFA has passed many milestones since 1928. In 1965 the FFA merged with the New Farmers of America (NFA). The NFA was an organization for black youths studying agriculture. In 1969, girls were allowed to join the organization. In 1988, the name was changed from the Future Farmers of America to the National FFA Organization. That same year the program name was changed from Vocational Agriculture to Agricultural Education. The name change was reflected on the FFA emblem.

Today, the National FFA and its state chapters are thriving, and continue to educationally enrich the lives of academic students nationwide and uphold the mission described in the FFA Creed:

The FFA Creed

I believe *in the future of agriculture, with a faith born not of words but of deeds—achievements won by the present and past generations of agriculturists; in the promise of better days through better ways, even as the better things we now enjoy have come to us from the struggles of former years.*

I believe *that to live and work on a good farm, or to be engaged in other agricultural pursuits, is pleasant as well as challenging; for I know the joys and discomforts of agricultural life and hold an inborn fondness for those associations which, even in hours of discouragement, I cannot deny.*

I believe *in leadership from ourselves and respect from others. I believe in my own ability to work efficiently and think clearly, with such knowledge and skill as I can secure, and in the ability of progressive agriculturists to serve our own and the public interest in producing and marketing the product of our toil.*

I believe *in less dependence on begging and more power in bargaining; in the life abundant and enough honest wealth to help make it so—for others as well as myself; in less need for charity and more of it when needed; in being happy myself and playing square with those whose happiness depends upon me.*

I believe *that American agriculture can and will hold true to the best traditions of our national life and that I can exert an influence in my home and community which will stand solid for my part in that inspiring task.*

By E.M. Tiffany. Adopted at the Third National FFA Convention. Revised at the 38th and 63rd Conventions.

To that end, each state chapter includes a program of activities (POA) at the local level that defines chapter goals, outlines steps in order to meet those goals, and acts as a written guide for events in the upcoming year pertaining to committee members, including administrators, advisory committees, alumni members, and other stakeholders. A well-planned POA includes divisions, or the types of activities a chapter conducts, and quality standards, which provide guidance for the planning and execution of those activities.

Chapter officers are a vital component of ensuring the success of chapter activities. These officers include the president, who presides over meetings, appoints committees, coordinates activities, and represents the chapter; the vice president, who works closely with the president to achieve chapter goals and may assume duties of president if necessary; the secretary, who is responsible for many administrative duties; the treasurer, who oversees the finances of the chapter and maintains treasury records; the reporter, who participates in public relation activities; the sentinel, who assists the president in official meetings and other activities; and the advisor, who supervises activities year-round and acts a leader and guide for other FFA members and stakeholders. Other chapter officers may include a historian, parliamentarian, and chaplain.

To learn more about the National FFA Organization, refer to the Official FFA Manual *and* FFA Student Handbook. *Visit* www.ffa.org.

Agricultural Education programs consist of three parts: classroom/lab, SAE, and FFA. *Courtesy of the National FFA Organization.*

PUTTING IT INTO PRACTICE

The FFA Degree Program

FFA members can earn degrees as they progress through the phases of their leadership, academic and career skills development. The Discovery Degree, Greenhand Degree and Chapter FFA Degree are all earned at the chapter level. State FFA Associations recognize their top members with a State FFA Degree. The National FFA Organization awards top members from each state with the American FFA Degree.

The Discovery Degree

The Discovery Degree is the first degree available to FFA members. It is designed for middle school programs. To earn the Discovery degree, members must:

- Be enrolled in agricultural education for at least a portion of the school year while in grades 7-8.
- Have become a dues paying member of the FFA at local, state and national levels.
- Participation in at least one local FFA chapter activity outside of scheduled class time.
- Have knowledge of agriculturally related career, ownership, and entrepreneurial opportunities.
- Be familiar with the local FFA chapter program of activities.
- Submit a written application for the degree.

The Greenhand Degree

The Greenhand degree is available to first year FFA members. To earn the Greenhand degree members must:

- Be enrolled in agricultural education and have satisfactory plans for a supervised agricultural experience program.
- Learn and explain the FFA Creed, motto, salute and FFA Mission Statement.
- Describe and explain the meaning of the FFA emblem and colors.
- Demonstrate a knowledge of the FFA Code of Ethics and the proper use of the FFA jacket.
- Demonstrate a knowledge of the history of the organization, the chapter constitution and bylaws and the chapter Program of Activities.
- Personally own or have access to the *Official FFA Manual* and the *FFA Student Handbook*.
- Submit written application for the Greenhand FFA Degree.

The Greenhand degree pin is made out of bronze for its strength and durability.

The Chapter Degree

The Chapter degree is the highest degree a local FFA chapter can bestow upon its members. To earn the Chapter degree, members must:

- Have received the Greenhand FFA Degree.
- Have satisfactorily completed the equivalent of at least 180 hours of systematic school instruction in agricultural education at or above the ninth grade level, have in operation an approved supervised agricultural experience program and be enrolled in an agricultural education course.
- Have participated in the planning and conducting of at least three official functions in the chapter Program of Activities.
- After entering agricultural education, have:
  - earned and productively invested at least $150 by the member's own efforts; or worked at least 45 hours in excess of scheduled class time; or a combination thereof; and

  - developed plans for continued growth and improvement in a supervised agricultural experience program.
- Have effectively led a group discussion for 15 minutes.
- Have demonstrated five procedures of parliamentary law.
- Show progress toward individual achievement in the FFA award programs.
- Have a satisfactory scholastic record.
- Have participated in at least 10 hours of community service activities. These hours are in addition to and cannot be duplicated as paid or unpaid SAE hours.
- Submit a written application for the Chapter FFA Degree.

The Chapter degree pin is made out of silver.

State FFA Degree

The State degree is the highest degree a state FFA association can bestow upon its members. It is typically awarded at the state association's convention. The minimum requirement to earn the State FFA degree include:

- Earning the Chapter Degree.
- Completing at least 24 months of active FFA membership and at least 360 hours of agricultural education class time at or above the 9th grade.
- Earning and productively investing at least $1,000 or working at least 300 hours or a combination of the two from their SAE program.
- Demonstrating at least 10 procedures of parliamentary law.
- Giving a 6-minute speech on a topic relating to agriculture or FFA.
- Participating in the planning and completion of the chapter Program of Activities.
- Serving as an officer, committee chair or committee member at the chapter level.
- Participating in at least 5 different FFA activities above the chapter level.
- Performing at least 25 hours of community service through at least two different activities not during class time or included as part of their SAE.

States can also set additional requirements for this degree. Your teacher can provide the requirements for your state. The State degree recipients earn a charm that can be worn on the official jacket or on a gold chain.

The American Degree

The American FFA Degree is awarded to FFA members who have demonstrated the highest level of commitment to FFA and made significant accomplishments in their SAE. To be eligible to receive the American FFA Degree from the National FFA Organization, the member must meet the following minimum qualifications:

- Have received the State FFA Degree.
- Have been an active member for the past three years (36 months), and have a satisfactory participation in the activities on the chapter and state levels.
- Have satisfactorily completed the equivalent of at least three years (540 hours) of systematic secondary school instruction in an agricultural education or 360 hours of systemic secondary school instruction in agricultural education and one full year of enrollment in a postsecondary agricultural program or have completed the program of agricultural education offered in the secondary school last attended.
- Have graduated from high school at least twelve months prior to the national convention at which the degree is to be granted.

- Have in operation and have maintained records to substantiate an outstanding supervised agricultural experience program through which a member has exhibited comprehensive planning, managerial and financial expertise.
- Have earned at least $10,000 and productively invested at least $7,500 or have earned and productively invested at least $2,000 and worked at least 2,250 hours in excess of scheduled class time. Any combination of hours, times a factor of 3.56, plus actual dollars earned and productively invested must be equal to or greater than the number 10,000. Hours used for the purpose of producing earnings reported as productively invested income shall not be duplicated as hours of credit to meet the minimum requirements for the degree.
- Have a record of outstanding leadership.
- Completed at least 50 hours of community service in a minimum of three different community service.activities. All hours are cumulative. (i.e. – the 25 hours used to obtain the State degree can be used towards the American degree.) The community service hours must be in addition to hours spent on paid or unpaid SAE projects. The hours cannot serve a dual purpose in fulfilling both community service and SAE requirements. These hours also may not be duplicated to fulfill the Directed Lab requirement. All hours must take place outside of regularly scheduled class time.
- Have achieved a high school scholastic record of "C" or better as certified by the principal or superintendent.

Participation in the National FFA Organization can lead you down the path to success. *Photo Courtesy of the National FFA Organization.*

SUMMARY

No branch of science touches our lives more than that of the science of agriculture. It has revolutionized every aspect of our lives. Although other countries are fierce competitors in electronics, automotive, and manufacturing, none even come close to competing with American agriculture. Our research, combined with our free enterprise system, has made us the envy of the world in agricultural production.

CHAPTER REVIEW

Review Questions

1. Why is agriculture considered to be the oldest of the sciences?
2. How has strong knowledge about agriculture helped advances in other areas?
3. What five acts passed by Congress have helped make advances in agriculture?
4. What is the three-part purpose of the land grant institution?
5. What does the scientific method involve?
6. What advances have been made in the beef industry since 1925? In the sheep industry? In the dairy industry? In the swine industry?
7. List five developments that have had a great impact on animal agriculture.
8. What drugs are derived from animals?
9. Why are pig heart valves superior to mechanical valves in replacing faulty human heart valves?

Student Learning Activities

1. Interview your parents or an older adult (and, if possible, your grandparents). Make a list of the advances they perceive as having made their lives better since they were your age. Determine how many of these advances came about as a direct or indirect result of agriculture.
2. Interview a livestock producer. Ask him or her what improvements he or she has seen during the past five years. Also ask what problems he or she has that might be solved through scientific research.
3. Design a scientific experiment to solve a problem. Include your hypothesis, how you would conduct the experiment, and how you would use the results.
4. Go to your school computer lab and access the Internet. Locate a research study on animal science. Also locate four breed associations and print the available information.

CHAPTER 2

The Classification of Agricultural Animals

STUDENT OBJECTIVES IN BASIC SCIENCE

As a result of studying this chapter, you should be able to

- explain the importance of scientifically classifying animals.
- define and explain the use of the binomial system of classification.
- list the five kingdoms that are used to classify all living organisms.
- explain the different categories used in the scientific classification of animals.
- list characteristics of animals that place them in different classifications.
- describe methods of classifying animals by means other than scientific classification.

STUDENT OBJECTIVES IN AGRICULTURAL SCIENCE

As a result of studying this chapter, you should be able to

- explain how agricultural animals are classified scientifically.
- explain how breeds of livestock were developed.
- explain the purposes of breed associations.
- outline the classification of agricultural animals according to use.

KEY TERMS

organisms
binomial nomenclature
genus
species
kingdoms
animalia
plantae
protista
fungi
phyla
phulon
notochord
classes
orders
cud
families
polled
breed
breeding true
underline
selective breeding
purebreds
breed associations
blood typing
meat animals
work animal
draft horses
dual-purpose animals
wether

NATIONAL AFNR STANDARD

AS.06.01.01.a
Explain the importance of the binomial nomenclature system for classifying animals.

AS.06.01.01.b
Explain how animals are classified using a taxonomic classification system.

AS.06.01.01.c
Assess taxonomic characteristics and classify animals according to the taxonomic classification system.

AS.06.01.02.a
Compare and contrast major uses of different animal species.

AS.06.01.03.a
Identify and summarize common classification terms utilized in animal systems.

AS.06.01.03.c
Apply knowledge of classification terms to communicate with others about animal systems in an effective and accurate manner.

THERE ARE MILLIONS of different types of animals and other living things in the world. Most of these **organisms** have been identified, grouped, and classified in an attempt to more effectively study them and communicate about them. Plants, animals, and other organisms are classified or grouped by characteristics they have in common. They may be characterized by their physical characteristics, by the uses that people make of them, or by other categories used to put similar animals together.

Agricultural animals are classified in several ways. First, these animals are part of the overall population of animals in the world. Therefore, agricultural animals are classified scientifically just as all other animals are. A big difference is that agricultural animals have been domesticated for some type of human use. These domesticated animals have been developed into breeds having distinctive characteristics and distinctive uses. Other ways of classifying and identifying animals are by breeds and by the uses of the various types of animals.

FIGURE 2–1 Scientists all over the world recognize this animal as *Sus scrofa.* © Dmitry Kalinovsky/Shutterstock.com.

SCIENTIFIC CLASSIFICATION

Animals are given names according to a scientific classification system, known as **binomial nomenclature**. Binomial means two names; nomenclature refers to the act of giving a name. This system was developed by Swedish botanist Carolus Linnaeus, who grouped organisms according to similar characteristics. Linnaeus used two Latin names for identifying each individual organism: The first of the two names is the **genus**; the second part of the name denotes the **species**.

At the time of Linnaeus, Latin was the international language of scholars. Many of the languages of the world are based on Latin, so people from many different countries who speak many different languages can recognize the scientific names. A *Sus scrofa* will be recognized as a domesticated pig by people who speak German as well as by those who speak English (**Figure 2–1**). The two-name system also helps people from different areas of the same country accurately identify animals and other living things.

Common names are often confusing and can be misleading. For instance, in Oregon a gopher is a burrowing animal having long claws and long teeth that eats insects and roots. In Alabama, a gopher is a large rat that inhabits barns and other outbuildings. To people who live in Florida, a gopher is identified as a type of burrowing terrapin. Without a standard naming system, people from outside one of these areas would be confused about which of the three animals was being referred to. However, if the gopher in Oregon is referred to as *Thomomys species*, if the gopher in Florida is referred to as *Gopherus polyphemus*, and if the gopher in Alabama is referred to as *Rattus norevegicus*, the identity of the animal is not in doubt. A scientific system of classification allows the exact identity of an animal to be recognized anywhere in the world.

The scientific classification of all living things is an orderly and systematic approach to identification. Broad groups of animals are classified together in categories of common characteristics. Then each of these broad groups is broken down into categories of animals having similar characteristics. Each of these groups is further broken down into smaller categories.

The process is repeated until the groups cannot be categorized into smaller groups. Following is an explanation of the system and how agricultural animals fit into the classification system.

Kingdoms

All living things are first classified into seven broad categories called **kingdoms**, although some scientists recognize only six and don't include Chromista. In fact, for many years, scientists recognized only two kingdoms: the plant kingdom and the animal kingdom. As new discoveries were made and more organisms were classified, scientists realized that some animals could not be classified as either plant or animal; they didn't fit into either of the two existing kingdoms. In revising the system, new kingdoms were added. Scientists now recognize the following kingdoms of living organisms:

- **Animalia**—all multicelled animals
- **Plantae**—multicellular plants that produce chlorophyll through photosynthesis
- **Protista**—paramecia and amoebae
- **Fungi**—mushrooms and other fungi
- **Eubacteria**—complex single-celled organisms including most bacteria
- **Archaebacteria**—single-celled organisms such as bacteria that live in undersea thermal vents that create boiling water
- **Chromista**—certain types of algae

Obviously, all agricultural animals are large multicellular animals and belong to the kingdom Animalia. The kingdom Animalia includes all animals ranging from a tiny gnat to the huge whales that inhabit our oceans. Because of this great diversity, the animals in the kingdom Animalia were placed into smaller groups known as **phyla**.

Phyla

The kingdom Animalia is divided into 35 different phyla (-*singular* phylum) according to the animals' characteristics. The word -phylum comes from the Greek word **phulon**, meaning race or kind. Several of the phyla are divided into subphylas. Animals in phyla or subphylas are grouped by broad characteristics shared by the animals.

For instance, the phylum Arthropoda consists of animals that have a hard, external skeleton called an exoskeleton. This phylum includes insects, spiders, crayfish, crabs, centipedes, and so on. Another phylum, the Mollusca, includes the animals that have soft bodies protected by a hard shell—for example, starfish, snails, and clams. The segmented worms, such as the earthworm, belong to the phylum Annelida.

All agricultural animals (with the exception of certain specialty animals such as earthworms and oysters) belong to the phylum Chordata. Animals in this phylum have a stringy, rodlike structure called a **notochord**, made of tough elastic tissue that is present in the embryo. This phylum is divided into subphylas, one of which is the subphylum Vertebrata. This subphylum includes animals with backbones. The jointed backbone that supports the animal is developed from the notochord of the embryo. The animals with backbones belong to the subphylum Vertebrata and include the animals generally found on farms and ranches. However, the subphylum Vertebrata comprises animals as diverse as sharks and monkeys.

Classes

The phyla and subphyla are further divided into **classes**. Examples of the classes in the subphylum Vertebrata are:

- Amphibia—includes frogs, toads, and salamanders
- Reptilia—includes turtles, snakes, and lizards
- Aves—includes the birds
- Mammalia—includes animals that have hair, nurse their young, and give live birth

Agricultural animals such as horses, cattle, goats, sheep, pigs, and dogs belong to the class Mammalia. This class is another expansive classification group, including mice, elephants, tigers, whales, and humans.

Orders

Classes are divided into smaller groups that categorize animals within a class that possess certain characteristics. These groups are called **orders**. The class Mammalia contains 29 different orders including Primates, to which humans belong. Cattle, goats, sheep, and pigs belong to the order Artiodactyla. Animals are placed in this order because they have an even number of toes. Sometimes referred to as hooves, the feet on these animals all have an even number of divisions (usually two). Within this order are three suborders:

- Suiformes—includes pigs and hippopotami
- Tylopoda—includes camels and llamas
- Ruminantia—includes deer, cattle, sheep, and goats

Common characteristics of animals in the suborder Ruminantia are that they chew a **cud** and have several compartments to their digestive system, which allows them to eat grass, hay, and other roughages.

Horses and donkeys have only one toe (hoof) and belong to the order Perissodactyla. Also in this order are zebras, tapirs, and rhinoceroses.

Families

At this point in the classification system, the characteristics of the animals that are grouped together begin to narrow and the animals have much more in common. Still, there are considerable differences between a cow and a deer (both of which belong to the suborder Ruminantia) and between a rhino and a horse (both of which belong to the order Perissodactyla). Orders and suborders are broken down further into **families**. Each order and suborder contain many families. The suborder Ruminantia is divided into six families:

- Cervidae—includes deer, elk, and moose
- Antilopinae—includes the antelopes
- Tragulidae—includes certain types of goats
- Giraffidae—includes the giraffe
- Bovidae—includes cattle, buffalo, sheep, and domestic goats
- Moschidae—includes musk deer

However, sheep and goats are put in the subfamily Caprinus (**Figure 2–2**).

Genus and Species

The final categories of the scientific classification system are genus and species, which together comprise what is known as the scientific name. All identified animals have been given this two-part classification. Families are broken down into genera, and each genus is further divided into species. For instance, sheep are placed in the genus *Ovis* and goats are classified in the genus *Capra*. Domestic sheep are separated from the various types of wild sheep by species. The species of domestic sheep is *aries*. Within the family Bovidae, cattle are classified in the genus *Bos* and are further separated by species. Cattle of European origin

FIGURE 2–2 Goats belong to the same family (Bovidae) as cattle but are classified in a different subfamily. © NikkiHoff/Shutterstock.com.

Common Name	Pigs	Cattle	Horses	Sheep	Chickens	Turkeys	Rabbits	Honeybees	Catfish
Kingdom	Animalia	Animalia	Animalia	Animalia	Animalia	Animalia	Animalia	Animalia	Animalia
Phylum	Chordata	Chordata	Chordata	Chordata	Chordata	Chordata	Chordata	Arthropoda	Chordata
Class	Mammalia	Mammalia	Mammalia	Mammalia	Aves	Aves	Mammalia	Insecta	Osteichthyes
Order	Artiodactyla	Artiodactyla	Perissodactyla	Artiodactyla	Galliformes	Galliformes	Lagomorpha	Hymenoptera	Siluriformes
Family	Suidae	Bovidae	Equidae	Bovidae	Phasianidae	Meleagrididae	Leporidae	Apidae	Ictaluridae
Genus	Sus	Bos	Equus	Ovis	Gallus	Meleagris	Oryctolalgus	Apis	Ictalurus
Species	scrofa	taurus, or indicus	caballus	aries	domesticus	gallopavo	cuniculus	mellifera	Furcatus

TABLE 2–1 Scientific Classification of Agricultural Animals.

are classified in the sub-specie *Bos Taurus tarus*; cattle that originated in India are classified as *Bos tarus indicus*. **Table 2-1** summarizes the scientific classification of agricultural animals.

CLASSIFICATION BY BREEDS

Species of animals have many differences. For instance, a Great Dane and a Poodle are both classified as *Canis familiaris*. Just think of all the differences between these two breeds of dog! Breeds of agricultural animals can show almost as much difference within the species. Color patterns, size, horned or **polled**, and country of origin—all can be characteristics used to distinguish different breeds of cattle.

A **breed** of animal is defined as a group of animals with a common ancestry and common characteristics that breed true. **Breeding true** means that the offspring almost always will look like the parents. For instance, the Hereford breed of cattle is characterized by being brownish red with a white face and white **underline** (**Figure 2–3**). If a male and a female Hereford are mated, the offspring are expected to be red and have the characteristic white face and white underline.

FIGURE 2–3 The Hereford is unique in that it is brownish red in color with a white face and white underline. © dcwcreations/Shutterstock.com.

Selective Breeding

All breeds of hogs probably came from a common ancestor; all breeds of sheep probably came from a common ancestor; all breeds of cattle probably came from a common ancestor. After the animals were domesticated, breeds were developed by the people who took care of the animals. The characteristics of the different breeds probably were developed because the people who raised them wanted those particular characteristics. Those animals showing the desired traits were kept for breeding, and the others were slaughtered and eaten.

A group of producers may have liked the black color of some of their beef animals and therefore bred only those that were black. After a few generations of this **selective breeding**, only black calves were produced. During this process, someone may have noticed that some of the calves did not develop horns. Subsequently, if only black cattle with no horns were used for breeding, after several generations a group of cattle developed that were always black and had no horns. From this group came the modern Angus breed of cattle.

Purebreds

Animals whose ancestors are of only one breed are referred to as **purebreds**. **Breed associations** have been developed to promote certain breeds of animals. These associations usually set the standards for animals that are allowed to be registered as a purebred animal of that particular breed. If breed associations did not set standards for their animals, the breed might disappear in a few years. For example, the American Duroc Association specifies that, for an animal to be registered as a Duroc, the animal must be red in color with no white on the body. By allowing only a certain type of animal to be registered as a purebred, the breed association is assured that the characteristics that designate the animals as a certain breed will continue.

Blood Typing

Beyond physical characteristics used in breed identification, a process known as **blood typing** is also used to determine the ancestry of animals. Individual animals and humans have different types of blood known as blood groups. The blood of different types or groups has different characteristics that are passed on genetically from parents to their offspring. As the blood is analyzed, these differences show up. This process is useful in determining the parentage of a particular animal.

Because the black color of Angus cattle is dominant, the sire of a black calf from an Angus cow would be difficult to determine just by looking at the calf. However, the blood type of a calf sired by an Angus bull would be different from the blood type of a calf sired by a bull of a different breed. The blood type of two Angus bulls might even be different. Determining the parentage of animals using blood typing is usually considered to be about 90 percent accurate.

New breeds of animals are constantly being developed by combining animals of different breeds. The Brangus breed was developed by systematically breeding Brahman cattle and Angus cattle until the offspring had the desirable characteristics and bred true (**Figure 2–4**). Just as in earlier history, a breed was developed with the characteristics that certain people wanted.

Crossbreeding

Sometimes species can be successfully crossed to produce new breeds, a process called crossbreeding. For example, the breeds of cattle developed in Europe are scientifically classified as *Bos tauras taurus*. These cattle have characteristics that make them desirable as beef animals

FIGURE 2–4 The Brangus breed was developed by systematically breeding Brahman and Angus cattle.
© Mr_Jamsey/Shutterstock.com.

FIGURE 2–5 The Brahman is a species of cattle (*Bos indicus*) that developed in India. © Aumsama/Shutterstock.com.

that do well in the climates of Europe. In the tropical climate of India, another sub species of cattle (*Bos tauras indicus*) developed, with characteristics that made the animal useful in that part of the world (**Figure 2–5**). Cattle breeders in the subtropical regions of the United States (the Southwest and the Southeast) recognized the need for an animal that had characteristics of both the *Bos tauras taurus* and the *Bos taurua indicus*.

One of the first successful breeds of this type was the Santa Gertrudis, which was developed by systematically crossing the Shorthorn breed of cattle (*Bos tausus taurus*) with the Brahman breed of cattle (*Bos indicus*) (**Figure 2–6**).

FIGURE 2–6 The Santa Gertrudis was developed from the Shorthorn and Brahman breeds. Courtesy of Santa Gertrudis Breeders International

FIGURE 2–7 The mule was developed by breeding a mare with a jack. © iStockphoto/DS70.

This new breed of cattle combined the growth and carcass quality of the *Bos Taurus* with the hardiness of the *Bos Taurus indicus*. Since that time, many other breeds have been developed using these two species.

Another example is the mule, which was developed by breeding a mare (a female horse, *Equus caballus*) with a jack (a male donkey, *Equus asinus*). The resulting animal—the mule—combined the size and strength of the horse with the toughness and surefootedness of the donkey (**Figure 2–7**).

CLASSIFICATION ACCORDING TO USE

Breeds of domesticated animals are sometimes grouped together because of the uses that humans make of them. Agricultural animals are raised for several different reasons and therefore are classified by their uses.

Meat Animals

Meat animals are raised primarily for slaughter and human consumption. For instance, with the exception of those raised as laboratory animals, almost all pigs are raised for pork and have little use otherwise. Sheep, however, may be raised for various purposes. Breeds such as Rambouillet and Merinos are

FIGURE 2–8 Some sheep are raised for their wool, some for their meat, and some for their milk. © Natelle/Shutterstock.com.

FIGURE 2–9 Ayrshires were developed for milk production instead of meat production. © iStockphoto/MargoJH.

grown primarily for their wool; Suffolk and Hampshire breeds are grown primarily as meat animals; and other breeds are produced for their milk, which is used in such products as Roquefort cheese (**Figure 2–8**).

Another example can be found in cattle. Hereford cattle are raised for beef because they are produced with a lot of muscle and have only enough milk to feed their calves. Ayrshire cattle, in contrast, have considerably less muscle than Herefords but produce a tremendous amount of milk. Because of this, Ayrshires are raised for their milk and not for beef (**Figure 2–9**). Likewise, some breeds of chickens produce a lot of eggs but little meat, and these breeds are used as layers. Other breeds produce a lot of meat and are raised for slaughter.

Work Animals

Another classification according to use is the **work animal**. In the past, work animals have been an essential part of agriculture. Even today in some parts of the world, animals are the primary means of transportation and tillage of the soil. Donkeys provide power to pull carts and are ridden. Camels provide means of bearing heavy loads. Oxen, camels, water buffaloes, and donkeys are all used to pull wagons and plows (**Figure 2–10**).

FIGURE 2–10 In some areas of the world, animals such as the water buffalo are still a valuable source of power. © gionnixxx/Shutterstock.com.

Horses

In the United States, animals are still used to assist humans in work. On U.S. farms and

FIGURE 2–11 In the United States, horses are still used to work cattle. © Marquicio Pagola/Shutterstock.com.

FIGURE 2–12 Many horses work to provide people with recreation. © marikond/Shutterstock.com.

ranches, horses are still a valuable means of working cattle. In some instances, they are still used as draft animals. Other horses are used for recreational purposes. Given all of these uses for horses, it is easy to see that they are classified according to the type of work they do. Cutting horses, such as the American Quarter Horse, are used to herd and work cattle (**Figure 2–11**). Larger breeds, such as Belgians and Clydesdales, are used to pull wagons and heavy loads and are classified as **draft horses**. Some breeds, such as the Morgan, Tennessee Walker, and the American Saddlebred, are used for riding and are classified as saddle horses (**Figure 2–12**). Others, such as the Hackney and the Standardbred, are used for pulling sulkies or light carriages and are known as harness horses.

Dogs

Dogs also are used to herd cattle, hogs, and sheep. On a sheep ranch, a good sheepdog is quite a valuable asset. Not only are they used to round up and sort sheep, but they also protect sheep from predators (**Figure 2–13**).

FIGURE 2–13 Dogs are classified as work animals when they are used to herd other animals. © Feraru Nicolae/Shutterstock.com.

Dual-Purpose Animals

As animals were domesticated, many were developed to be **dual-purpose animals**. Cows, for example, could provide milk and also serve to pull plows, carts, or other implements. Also, the surplus young could be slaughtered and eaten. On most modern farms and ranches, agricultural animals are specialized; that is, they serve only one purpose. For instance, cattle are raised either for milking or for beef but seldom for both. Exceptions do exist, however. Most sheep that are raised for meat are also shorn for their wool. Although the wool is not as high quality as that of sheep raised primarily for wool, the producer does obtain some income from the sale of the wool. Likewise, calves from dairy cattle often are slaughtered for veal or beef.

FIGURE 2–14 In the Middle East, camels are used for work, milk, and meat.
© Vixit/Shutterstock.com.

In many parts of the world, dual-purpose animals still play a major part in the agricultural economy. For example, in the harsh deserts of the Middle East, camels provide a source of power for carrying or pulling loads and also are a source of milk and meat (**Figure 2–14**).

SUMMARY

Animals can be classified in many ways, and agricultural animals are no exception. Whether they are classified according to size, use, color, or breed, the scientific classification system is an organized method of helping to identify types of animals. Without such a system, there would be much confusion surrounding the names and identification of animals.

CHAPTER REVIEW

Review Questions

1. What is the binomial nomenclature system of classification?
2. Why is the scientific classification of animals essential in studying and communicating about them?
3. List the five kingdoms, and describe the type of organisms included in each.

4. What are the scientific names for the following agricultural animals: cattle, pigs, horses, sheep, dogs?
5. Explain how and why breeds of animals were developed.
6. What are the purposes of breed associations?
7. How is blood grouping used to classify animals?
8. Give examples of two different species of animals that have been bred to produce a new breed or type of animal.
9. What are three classifications of animals according to their use?
10. What are three classifications of horses according to their use?
11. Explain the term "dual-purpose animal" and give some examples.

Student Learning Activities

1. Choose a specific type of animal (a Hereford cow, a Duroc boar, a Suffolk **wether**). List all of the different ways this animal could be grouped or classified. Discuss your classification methods with others in your class.
2. Choose two types of agricultural animals (such as a sheep and a pig), and list all of their common characteristics you can think of. Also make a list of all the ways in which the animals are different. Compare your lists with others in the class.
3. Write to three breed associations and ask for information on the disqualifications of animals for those breeds. Compare the requirements of the different associations. As a class project, try to determine which breed associations are the most restrictive about their qualifications.
4. Talk to several purebred livestock producers to determine the characteristics they like best about the breeds they raise. Compare your findings with those of others in your class.

CHAPTER 3

The Beef Industry

STUDENT OBJECTIVES IN BASIC SCIENCE

As a result of studying this chapter, you should be able to

- explain the importance of beef in the human diet.
- explain how the environment helps determine where animals are produced.
- define ecological balance.
- describe how cattle make use of feedstuff that cannot be consumed by humans.

STUDENT OBJECTIVES IN AGRICULTURAL SCIENCE

As a result of studying this chapter, you should be able to

- specify the per-capita consumption of products from beef animals grown in the United States.
- explain the importance of the beef industry to the economy of the United States.
- justify the use of agricultural land to produce beef.
- describe the various segments of the beef industry.

KEY TERMS

veal
sire breeds
dam breeds
exotic breeds
purebred operations
cow-calf operations
stocker operations
feedlot operations
seed stock cattle
stocker
finish

NATIONAL AFNR STANDARD

AS.01.01
Evaluate the development and implications of animal origin, domestication, and distribution on production practices and the environment.

AS.01.02.02.a
Research and examine marketing methods for animal products and services.

BEEF IN THE AMERICAN DIET

AMERICANS ARE A NATION of beef eaters. Each year the average person in this country consumes 56.6 pounds of beef and **veal**. In fact, beef accounts for about 6 percent of all supermarket sales (**Figure 3–1**). In addition to the large amount of beef consumed in the United States, almost a million metric tons of beef are exported each year, which represents a value of nearly $2.5 billion. Despite this, over the past few years, the consumption of beef has been decreasing both in supermarket sales and as meals in commercial restaurants. This is due to the rising popularity and lower cost of poultry.

Yet few nations in the world even come close to the United States in the per-capita consumption of beef and other meats. To a large extent, this is an indication of the prosperity of the American people. In the past, livestock ownership was a sign of prosperity, and in many cultures, even today, a person's wealth is measured by the number of cattle owned.

Lean beef is very dense in nutrients. A pound of beef may equal or surpass the nutritive content of the feed consumed to produce the meat. Of all the foods that humans consume, meat is among the most nutritionally complete (**Figure 3–2**). Food from animals supplies about 88 percent of vitamin B12 in our diets, because this nutrient is difficult to obtain from plant sources. In addition, meats and animal products provide 67 percent of the riboflavin, 65 percent of the protein and phosphorus, 57 percent of vitamin B_6, 48 percent of the fat, 43 percent of the niacin, 42 percent of vitamin A, 37 percent of the iron, 36 percent of thiamin, and 35 percent of magnesium in our diets.

FIGURE 3–1 Beef accounts for about 6 percent of all supermarket sales. *© iStockphoto/97.*

FIGURE 3–2 Meat is among the most nutritionally complete foods that humans consume. *© Lisovskaya Natalia /Shutterstock.com.*

Types of Beef

Included in the consumption of beef are various types of beef that meet the needs of different consumers. With such a diverse population, there are different likes and dislikes in almost every commodity produced, and beef is no exception. The following sections outline some of the choices in beef offered to the American consumer.

Veal

According to the United States Dairy Association (USDA), veal is meat from a calf that weighs about 150 pounds. Calves that are mainly milk-fed usually are younger than 3 months old. Veal is pale pink and contains more cholesterol than beef but is also very tender. Veal is often from dairy calves that are not raised as replacements.

FIGURE 3–3 Baby beef is from young cattle weighing about 700 pounds that have been raised mainly on milk and grass. *© science photo/Shutterstock.com.*

FIGURE 3–4 Grain-fed beef comes from animals that may be around 3½ years old and weigh 1,000 pounds or more. *© Michael Zysman/Shutterstock.com.*

Baby Beef

The USDA defines "baby beef" and "calf" (two interchangeable terms) as beef from young cattle weighing about 700 pounds that have been raised mainly on milk and grass (**Figure 3–3**). The meat cuts from baby beef are smaller, and the meat is light red and contains less fat than beef. The fat may have a yellow tint as a result of vitamin A in the grass.

Grain-fed beef

Grain-fed beef is the type of beef that is sold most often in the grocery store. This beef is from animals that have been fed a high-concentrate feed (a high percentage of grain such as corn) until achieving the desired grade. These animals may be as much as 3½ years old and weigh 1,000 pounds or more (**Figure 3–4**). As explained in Chapter 22, grain-fed beef is graded according to the degree of fat and the age of the animal at slaughter. Consumers prefer a Choice grade of beef, which is the second highest grade. The top grade of beef, Prime, generally goes to the better restaurants.

Grass-fed beef

A growing demand in the beef market is for grass-fed beef—from animals that are fed grass almost exclusively. Some people think this type is healthier, but research has not yet confirmed a significant health benefit over grain-fed beef. However, grass-fed beef does contain less fat than grain-fed beef. The USDA has set the following standards for beef that is labeled as grass-fed:

> Grass and forage should make up the animal's diet for its entire lifetime, with the exception of milk consumed prior to weaning. The diet shall be derived solely from forage and animals cannot be fed grain or grain byproducts and must have continuous access to pasture during the growing season.

Because the animal must be fed grass or hay, the beef may be more expensive because the seasons may not allow year-round grazing (**Figure 3–5**). Also, grass-fed animals may take a

FIGURE 3–5 Because the animal must be fed on grass or hay, grass-fed beef may be more expensive because the seasons may not allow year-round grazing. *© Gozzoli/Shutterstock.com.*

longer time to reach maturity than animals fed grain in a feedlot.

Natural Beef

Beef is labeled "natural" if no artificial flavor, coloring, chemical preservatives, or any other artificial or synthetic ingredient are added to the meat. "Naturally raised" beef is from animals that have never been given growth promotants (such as hormones), have never been given antibiotics, and were never fed animal by-products.

Certified Organic Beef

Chapter 12 discusses the rules and regulations regarding the labeling of meat as "organic." Grain-fed, grass-fed, and naturally raised beef may be labeled organic if additional requirements are met. The biggest difference is that the feed must be certified organic feed.

THE BEEF INDUSTRY IN THE UNITED STATES

In the history of the United States, the beef industry has played a prominent role in the development of U.S. economy. Cattle have been in the New World almost as long as the European settlers. The animals were brought across the oceans to feed the settlers in their new homes. Until around the time of the Civil War, most beef was raised on family farms for the purpose of feeding the family. As the population became more urbanized, people had more difficulty raising their own meat. Also, they became more affluent and could afford to buy their food rather than raise it. The large, grassy areas of the West were being settled, and cattle were a natural product to raise on the vast plains of native grasses.

Currently, more than 92,000,000 head of beef are being raised on about 619,172 farms and ranches in the United States (**Table 3–1**). The number of operations far exceeds any other segment of animal agriculture, with the cattle industry accounting for the largest segment of all the agricultural industry in the United States. Most cattle are raised on family-owned farms and ranches. In fact, about 80 percent of all cattle businesses have been in the same family for the past 25 years. Annually, the United States produces nearly 25 percent of the world's beef supply with less than 10 percent of the world's cattle. The beef industry contributes more than $88.25 billion to the U.S. economy each year.

The United States is well suited for the production of animals that supply beef. In the West, vast areas of land are used to graze cattle. Throughout the Midwest, millions of acres of corn are grown on some of the most productive farmland in the world. In the southern portion of the country, beef producers take advantage of the mild climate to produce grass and hay to help feed the millions of head of cattle raised there.

When compared with the rest of the world, Americans spend only a small percentage of their annual income for food. This means that they can afford to buy the type of food they

Item	Beef cattle ranching and farming (112111)	Cattle feedlots (112112)	Dairy cattle and milk production (11212)	Hog and pig farming (1122)	Poultry and egg production (1123)	Sheep and goat farming (1124)	Animal aquaculture and other animal production (1125, 1129)
Farms .... number	664,431	55,472	72,537	33,655	44,219	43,891	228,152
percent	31.2	2.6	3.4	1.6	2.1	2.1	10.7
Land in farms .... acres	419,821,930	25,984,434	27,351,777	8,317,127	6,153,409	17,910,791	44,633,545
Average size of farm .... acres	632	468	377	247	139	408	196

TABLE 3–1 Of the many producers of livestock in the United States, cattle producers are by far the most numerous. *Courtesy of Cooperative Extension Service, University of Georgia.*

prefer—and they prefer meat. Critics of the beef industry contend that feeding several pounds of feed to animals in return for a pound of meat is wasteful. They say that the grains fed to animals could be better used to feed people, and 6 to 9 pounds of feed are required to produce a pound of beef. Beef producers counter the argument by saying that land used to graze agricultural animals would be of little use for other agricultural purposes. Almost half of the land in the United States is classified as land that is not practical for growing cultivated, or row, crops. Without the production of grazing animals, this land would be wasted instead of being used as a food-producing resource (**Figure 3–6**). Furthermore, beef producers point out that livestock are finished (fattened) using grains that are not considered good for human consumption. The better grades and types of grains are used for human to eat, while the lower grades of corn and grains such as grain sorghum are fed to livestock.

Beef animals also make use of by-products such as meal resulting from the cooking oil market. The harvested crops such as soybeans and cottonseed are pressed until most of the oil is removed, and the resulting cake is ground into feed for livestock. Also, by-products such as beet pulp from the sugar industry and citrus pulp from the orange and grapefruit juice industry are fed to cattle. If not fed to livestock, these valuable by-products might go to waste.

As outlined in Chapter 1, other products are obtained from animals as well. Most of the hides from the animals are used in making pharmaceuticals, leather for belts, shoes, and other articles of clothing and upholstery materials.

BREEDS OF BEEF CATTLE

More than three-quarters of the cash receipts for marketing meat animals comes from the sale of beef. These cattle are produced on almost a million farms and ranches across the United States. Contrary to popular belief, most of the beef animals do not originate from large ranches raising vast herds of thousands of cattle. The average size of the beef herds in this country is around 100 head. These producers represent a wide variety of different breeds and types of beef animals. In the United States, there are over 40 different breeds grown, besides all of the different combinations of crosses of these breeds. Livestock producers choose the breed to grow based on the type of market where the animals will be sold, the environmental conditions in which the animals will be produced, and the personal likes and dislikes of the individual producer.

It is necessary to understand the external parts of the beef animal, **Figure 3–7**. However, it is also important to understand differences between breeds. Some breeds are large and produce a large carcass; some mature at a smaller size and produce a smaller carcass. Both have a place in the

FIGURE 3–6 Cattle can make use of land that is not suited for growing cultivated or row crops. *© Arinad P Habich/Shutterstock.com.*

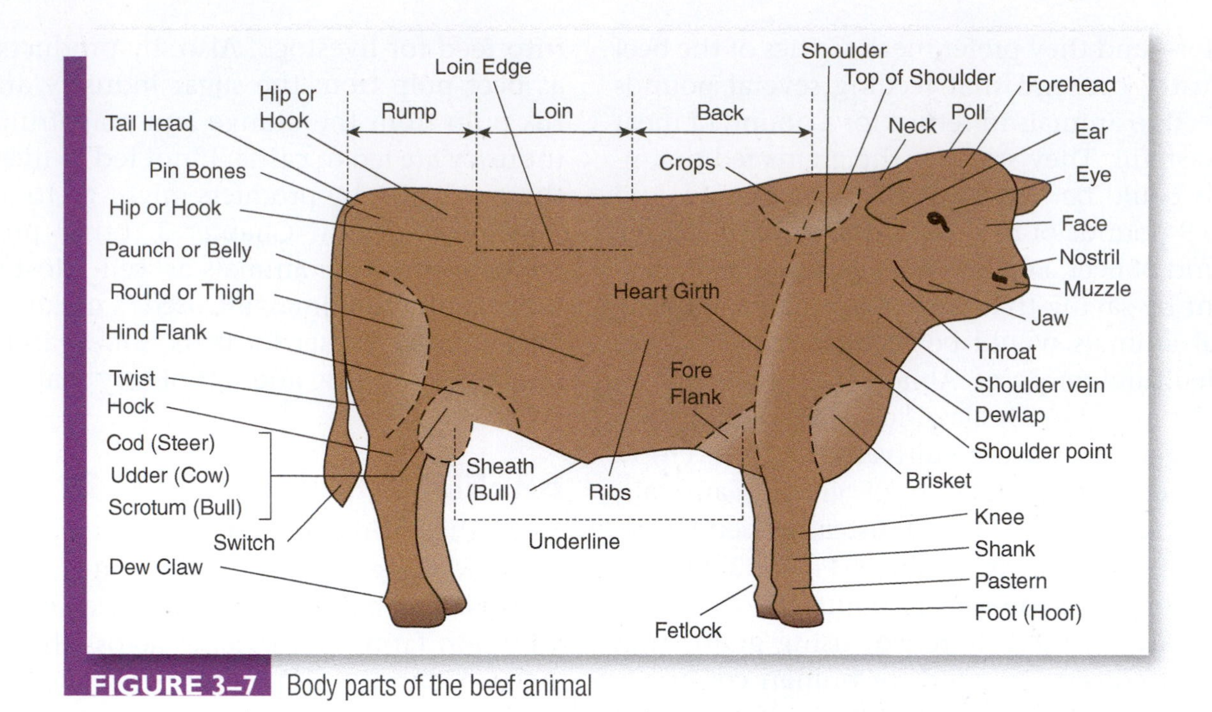

FIGURE 3–7 Body parts of the beef animal

market, and both are produced. Some breeds are better adapted to hot, humid climates, and some breeds tolerate cold and snow better than others. A producer may like the color pattern or the docile nature of a particular breed and prefer to produce that breed.

Some breeds make excellent mothers, and other breeds grow rapidly and produce high-quality, meaty carcasses. Because of this, some breeds are referred to as **sire breeds**, and some are referred to as **dam breeds**. A crossbreeding program helps producers take advantage of the good points of both types of animals. Three broad classifications of beef breeds are grown in the United States: the British breeds, the continental European breeds, and the Zebu breeds.

The *British breeds* include the Angus, Hereford, and the Shorthorn (**Figure 3–8**). These animals are generally of a docile nature and produce high-quality carcasses at a medium size. They were the first breeds brought to the United States, and there are more of this class than any other.

The *continental European breeds* include the Limousin (**Figure 3–9**), the Simmental, the Charolais, and the Chianina. These breeds, once known as the **exotic breeds**, were brought to this country because of their size and ability to grow. At maturity, most of the breeds in

FIGURE 3–8 The Shorthorn is classified as a British breed. *Courtesy of American Shorthorn Association.*

FIGURE 3–9 The Limousin is a good example of the large, meaty, continental breeds. *Courtesy of North American Limousin Foundation*

this class become quite large. The largest of the breeds, the Chianina, may reach the weight of 4,000 pounds for the bulls and 2,400 pounds for the cows (**Figure 3–10**). They are generally crossed with the British breeds.

The Zebu breeds are those that are scientifically classified as *Bos Taurus indicus*, a separate sub species from the traditional *Bos taurus* of the other breeds. The most common Zebu type of cattle in the United States is the Brahman (**Figure 3–11**), characterized by a large, fleshy hump behind the shoulder and loose folds of skin. They tolerate heat and humidity quite well and are resistant to insects. These characteristics make them well suited to the hot, humid climate of the Southeastern part of the United States and the hot, dry climate of the Southwest. Brahmans have been used as the basis for developing several breeds such as the Santa Gertrudis, Brangus (**Figure 3–12**), Simbrah, and Beefmaster. These developed breeds combine the ruggedness of the *Bos Taurus indicus* with the carcass quality and docile nature of the *Bos taurus*.

FIGURE 3–12 The Brangus is an example of a breed developed from Brahmans. *© JNix/Shutterstock.com.*

FIGURE 3–10 The Chianina are the largest of all the breeds of cattle. *© Emanuele Mazzoni Photo/Shutterstock.com.*

SEGMENTS OF THE BEEF INDUSTRY

The beef industry has four major segments: **purebred operations**, **cow-calf operations**, **stocker operations**, and **feedlot operations**. Purebred cattle are produced in the first phase of the industry (**Figure 3–13**). The purpose is

FIGURE 3–11 The Brahman is a type of Zebu cattle that is a different species from the British and continental breeds. *© Sherjaca/Shutterstock.com.*

FIGURE 3–13 The purebred breeders produce animals that will be used as dams and sires. *© iStockphoto/Mr. Jamsey.*

to produce what is known as the **seed stock cattle**. These represent the cattle that are to be used as the dams and sires of calves that will be grown out for market. As mentioned earlier, different breeds have different advantages, and the growing of purebred stock allows breeders to concentrate on improving and accentuating the advantages of a particular breed.

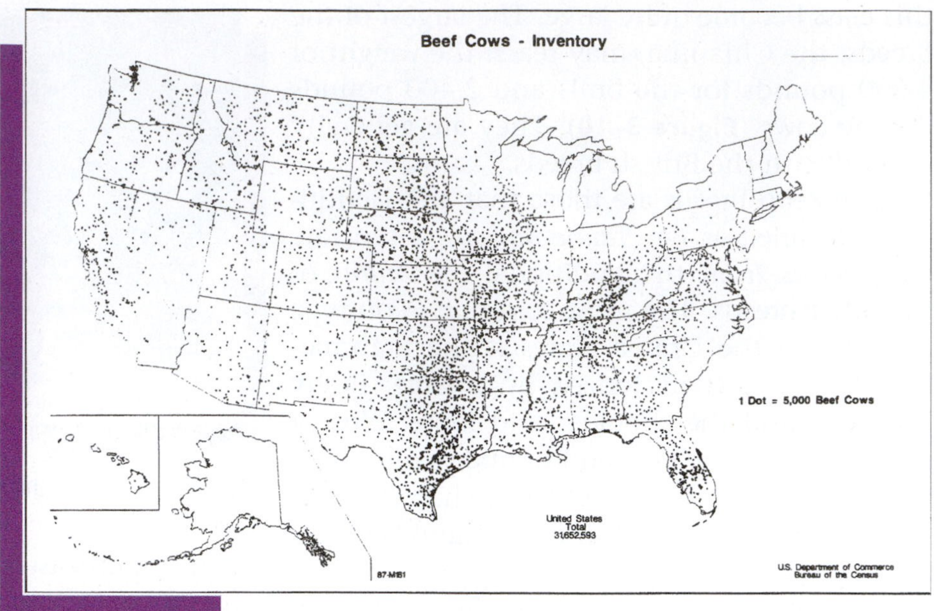

FIGURE 3–15 Beef cattle are grown all across the country. *Source: USDA, Census of Agriculture*

Each year, at numerous shows across the nation, purebred cattle breeders compete with each other by displaying their animals in the show ring. Expert judges select the animals they consider to be the best type for that breed. Shows serve both as a means of education and as a way of implementing change in the industry as economic conditions change and new research reveals new insights into the type of animals that should be selected (**Figure 3–14**).

The second phase is the cow-calf operations, where the calves are produced that eventually will be grown out and sent to market (**Figure 3–15**). Most of these calves are crossbred animals from purebred parents of different breeds. A large part of this industry is centered in the Southern and Western states. The mild winters of the South are ideal for calving in the winter. In most areas of the South, calves are born in January and February to take advantage of the weather that is too cold for flies and parasites but not too cold for the calves to thrive. In addition, the calves will be old enough to begin grazing in the spring when the grass begins to grow again.

FIGURE 3–14 Cattle shows serve to educate breeders and to make improvements in the cattle industry. *© LandFox/Shutterstock.com.*

Cows are fed primarily roughage in the form of grass or hay. The ample rains and mild temperatures of the South provide ideal conditions to produce large amounts of green forage (**Figure 3–16**). Some grasses, such as fescue and rye, grow very well in the winter months and supply a good source of feed for cows as they gestate or produce milk for their young. Much of the cropland of the hill country of the South has been converted to pasture and forest. These lands were so susceptible to erosion that it was no longer practical to produce row crops, and the growth of woodlands and pastures offered a way to make the land useful and productive.

FIGURE 3–16 Climatic conditions in the South are ideal for growing grass for cows and calves. © tviolet/Shutterstock.com.

Although the largest numbers of cow-calf operations are in the South, cow-calf operations are found all across the country. In the West, producers can take advantage of the vast amounts of government lands that are open to grazing for a small fee. Often, cows are left on free range (not fenced in) to have their calves, which then are rounded up, weaned, and sold.

Calves usually are sold upon weaning. They are weaned in the weight range of 300 to 500 pounds. Buyers prefer calves that have been castrated and vaccinated and are in good enough condition to move to a new environment.

The next phase of the industry is that of the **stocker**. **Stocker operations** provide a step between the weaning of the calves and their finishing (or fattening) prior to slaughter. For an animal to start depositing fat in the right places, the animal must be mature enough to have stopped growing. Weaned calves that weigh between 300 and 500 pounds are placed on pasture land and fed a ration designed to allow for skeletal and muscular growth. The stocker purchases the animals from the cow-calf producer and sells them to the feedlot operator. The stocker's job is to provide a transition period for the calves between the time they are weaned from their mothers and before they are put in the feedlot. During this time, the animals are fed a relatively high roughage diet and supplied with the proper balance of protein, carbohydrates, vitamins, and minerals that will ensure that they make sufficient gains to be placed in the feedlot, where they will be finished.

It is not uncommon for feedlot owners to also be the operators of stocker operations. This arrangement is economical because fewer transportation costs are incurred if the two types of operations are *close* together. The trend in the industry has been away from the stocker industry since recent research has developed production methods that allow cows to wean heavier calves. A calf that is weaned weighing 700 pounds may well go directly into the feedlot without going through a stocker operation.

The feedlot operation is the final phase before the animals are sent to slaughter (**Figure 3–17**). Here the animals are fed a high-concentrate ration designed to put on the proper amount of fat cover. The producers usually want their animals to be marketed when the cattle reach a

FIGURE 3–17 Cattle are finished for market in feedlots. © rthoma/Shutterstock.com.

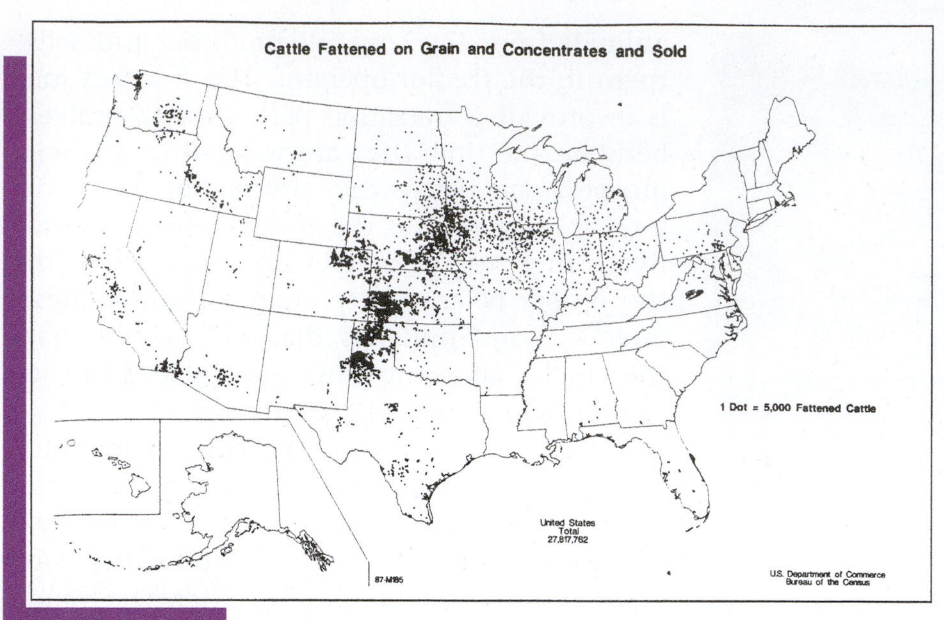

FIGURE 3–18 Feedlots are concentrated in the central part of the United States. *Source: USDA, Census of Agriculture*

sufficient fat cover to allow the animals to grade Low Choice. Many feedlots in the United States are situated in the Midwest (**Figure 3–18**). The reason is that this is the section of the country that produces the most grain, and it is usually more economical to feed the animals there rather than ship the grain across country. An exception is the state of Texas; it has more feedlots than any other state.

Some feedlots are located in other parts of the country to take advantage of byproduct feeds. For example, feedlots in Idaho take advantage of the potato industry and feed the cattle by-products from the processing of potatoes. Likewise, in Florida, cattle are fed citrus pulp that is left over from the processing of orange juice.

Feedlots range in size from a hundred or fewer head to feedlots that feed thousands of cattle every year. Long bunker feeders are automatically filled by automated systems or from trucks (**Figure 3–19**). The animals are supplied with all the high-quality feed they will ingest. They also are given medicines to prevent disease and to ward off both internal and external parasites.

When the animals have reached the proper degree of **finish**, they are quickly moved from the feedlot to the slaughterhouse. When the animals are slaughtered, they are generally around 18 to 24 months in age and can weigh from 800 to 1,500 pounds (**Figure 3–20**).

This age and size offer consumers the type of beef they prefer.

FIGURE 3–19 Feedlot cattle are fed from a long trough filled by a truck. *Source: USDA, Agricultural Research Service (ARS). Photo by Brian Prechtel.*

FIGURE 3–20 This is a properly finished steer that is ready for market. *Courtesy of North American Limousin Foundation.*

SUMMARY

The beef industry represents a large part of our diet and our economy. The food that comes from cattle provides nutrients that are difficult to obtain from other foods. Our vast continent provides an ideal environment for the production of beef cattle. Often, these cattle make use of feedstuff that otherwise would go to waste. The many phases of the industry provide jobs for millions of people all over the country. The beef industry is a dynamic, growing sector of our country and should remain so for many years to come.

CHAPTER REVIEW

Review Questions

1. Why is the United States an excellent location for raising livestock?
2. What are the reasons for using agricultural land to produce beef?
3. Discuss the nutritional value of beef in the diet.
4. List and describe the four major segments of the beef industry.
5. Why are most feedlots in the United States located in the Midwest?
6. How does the environment determine where animals are produced?
7. What is the difference between veal and baby beef?
8. Under what conditions may beef be labeled "natural"?
9. What is meant by a high-concentrate feed?
10. Why do we have a market for grass-fed beef?

Student Learning Activities

1. For the period of one week, keep a list of the amounts of all the different meats that your family eats. Which type of meat does your family eat the most? Ask your parents to tell you the reasons why they buy the type of meat they do.
2. Determine which breed of cattle would be most appropriate for your area. Give several reasons for your choice.
3. If you become a cattle producer, what type of operation (feeder calves, stockers, feedlot, purebred, etc.) will you prefer? Give the reasons for your choice.
4. Conduct an Internet search for information on a breed of cattle. Report to the class.

CHAPTER 4

The Dairy Industry

STUDENT OBJECTIVES IN BASIC SCIENCE

As a result of studying this chapter, you should be able to

- describe the process by which milk is produced.
- identify the hormones that control lactation.
- describe the composition of milk.
- explain the process of pasteurization.
- trace the biological processes used to produce cheese.

STUDENT OBJECTIVES IN AGRICULTURAL SCIENCE

As a result of studying this chapter, you should be able to

- identify the major areas of dairy production in the United States.
- explain how the producer uses the reproductive process to maintain milk production.
- trace the steps used to milk cows in the modern dairy.
- list the uses made of milk.
- tell how milk is processed and marketed.
- explain how cheese is made.

KEY TERMS

yogurt
balanced ration
silage
linear evaluation
heifers
embryo transplant
colostrum
antibodies
alveoli
prolactin
lumen
lobule
tertiary ducts
gland cistern
sphincter muscle
teat
pituitary gland
oxytocin
letdown process
epinephrine
milking parlors
stanchion
mastitis
specific gravity
homogenization
homogenized milk
pasteurization
starter culture
fermentation
enzyme
rennet
curd
whey

NATIONAL AFNR STANDARD

AS.01.01
Evaluate the development and implications of animal origin, domestication, and distribution on production practices and the environment.

AS.01.02.02.a
Research and examine marketing methods for animal products and services.

THE DAIRY INDUSTRY is a large component of American agriculture. The sales of dairy products account for about 13 percent of all receipts for farm commodities. The dairy industry is different from other segments of animal agriculture in that the product harvested is intended by nature for no other purpose than to be used as food. Cows are raised and cared for to obtain milk that is produced as food for young calves (**Figure 4–1**). As indicated in an earlier chapter, scientific research has advanced dairy cows to the point where they can produce many times more milk than is needed for calves.

Milk is often described as nature's most perfect food because of its nutritive value. Although milk is 87 percent water, the other 13 percent consists of solids that contain proteins, carbohydrates, and water-soluble vitamins and minerals. Because of the nutritive value and rich flavors, Americans consume large quantities of dairy products. Each year on the average we each consume 22.3 gallons of milk, 37 pounds of cheese, 19.6 pounds of ice cream, 5.6 pounds of butter, and 14.7 pounds of **yogurt**. This adds up to a lot of milk production. In addition, milk comes from the cow as a processed food and requires little additional processing.

FIGURE 4–1 Cows produce milk as food for their young. *© fotoslaz/Shutterstock.com.*

Milk is produced and processed in every state in the United States. The five leading milk-producing states are California, Wisconsin, New York, Minnesota, and Pennsylvania (**Figure 4–2**). These five states produce more milk each year than all of the other states combined. Unlike the meat industry, the dairy industry relies more on forage than grain to produce a product. These states produce a lot of forage. A high concentration of their state's population is in large cities.

About 85–90 percent of dairy cattle in the United States are Holstein (**Figure 4–3**). These large, docile animals with the familiar black and white markings produce a larger amount of milk with a smaller amount of milk fat than other breeds. The lower milk fat was once considered to be a disadvantage but now is considered to be an advantage because of modern consumer demand for low-fat and skim milk.

FEEDING

In the past, dairy cows usually were kept on pastures, where they could make use of grass, which is converted into milk (**Figure 4–4**). However, the modern trend is for large dairies to keep cows in lots or barns and to feed the animals a **balanced ration**. One of the main feeds of dairy cattle is **silage** (**Figure 4–5**), consisting of corn, grain sorghum, or other forage that is chopped—stalk and all—while the plant is green and growing. The chopped silage is placed in a silo or ground bunker, where it undergoes a fermenting process. This means that while the green chopped silage is stored, a chemical process takes place in which complex compounds in the forage are broken down into simpler compounds. This helps to preserve the feed and maintains the palatability, or eating quality, of the feed. The feeding of silage is timed so the milk from the cows will not have an off-flavor, which can occur if the silage is

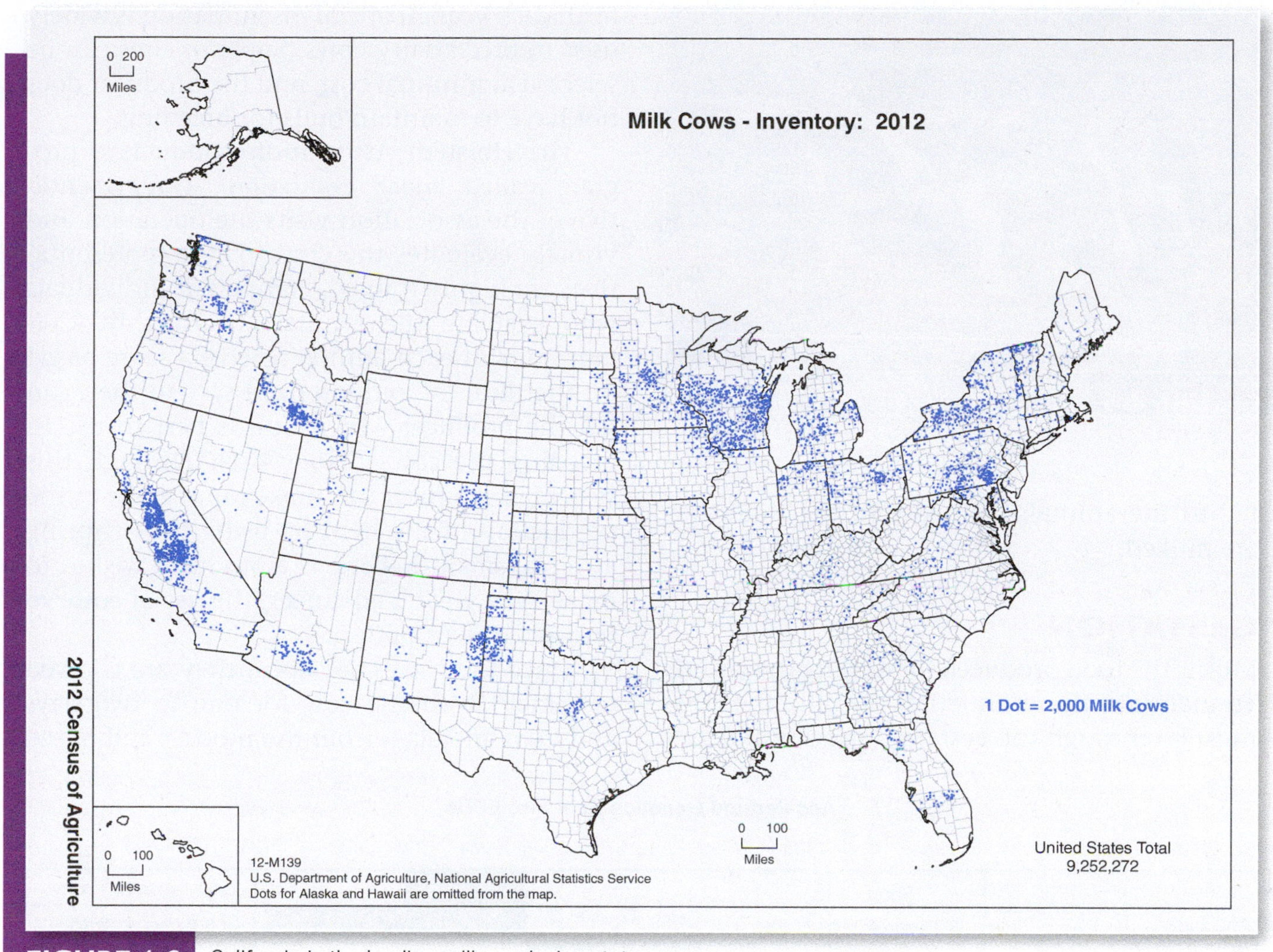

FIGURE 4–2 California is the leading milk-producing state. *Source: USDA, Census of Agriculture*

FIGURE 4–3 Most of the dairy cattle in the United States are Holstein. *© Gerard Koudenburg/Shutterstock.com.*

FIGURE 4–4 In the past, almost all dairy cattle were kept on pasture. *© Gozzoli/Shutterstock.com.*

FIGURE 4–5 The main feed of dairy cattle is silage.

Source: USDA, Agricultural Research Service (ARS). Photo by Scott Bauer.

fed to the animals too close to the time they are milked.

GESTATION

Milk is the food produced for feeding the young. To maintain the production of milk, the cows must go through the gestation process and give birth each year. Artificial insemination is widely used to breed dairy cows. Superior sires can be selected at minimal cost, and the producer does not have to maintain bulls for breeding.

The Holstein Association conducts a program called **linear evaluation**. A representative of the association visits the operation and visually evaluates the cows. These representatives are highly trained, competent individuals who evaluate the animals thoroughly. Certain traits of each animal receive a score based on the ideal. A computerized system then can tell the producer which bull is best to use in breeding the cows (**Table 4–1**). Through this system, a producer can make rapid gains in the production of the herd by using the offspring as replacement **heifers**. If a producer wishes to make even greater advances, the use of **embryo transplant** is an option.

Once the calves are born, they are allowed to remain with the cow for one to two days and then are taken from the mother and raised

Accelerated Genetics Beef Sire EPDs

Trait		3 2 1 0 1 2 3		Sta
Stature	Short		Tall	1.94
Angularity	Coarse		Sharp	3.75
Body Depth	Shallow		Deep	2.69
Strength	Frail		Strong	.89
Rump Angle	High Pins		Sloped	.41
Rear Leg Set	Posty		Sickle	.95
Foot Angle	Low		Steep	2.15
Fore Udder Attachment	Loose		Strong	.41
Teat Length	Long		Short	1.11
Teat Placement	Wide		Close	2.34
Rump Width	Narrow		Wide	1.64
Rear Udder Width	Narrow		Wide	2.15
Rear Udder Height	Low		High	1.67
Udder Depth	Deep		Shallow	.42
Udder Support	Broken		Strong	3.15
Rear Leg Rear View	Close		Wide	.42
Milkout	Slow		Fast	1.79
Disposition	Alert		Docile	.40

255 DAUS. 128 HERDS

TABLE 4–1 Accelerated Genetics Beef Sire EPDs

Courtesy of Accelerated Genetics

separately. The female calves often are raised as replacements, and the male calves are raised and sold for slaughter.

Milk from a cow that has just given birth is called **colostrum**. Colostrum is a milk containing a concentration of **antibodies** that are passed to the young from the mother. Because the young calf can absorb these antibodies only during the first 24 hours of life, it is important that the calf be allowed to suckle often during that period. Also, the milk is not generally considered fit for human consumption, so it is not allowed to enter the milk designated for market.

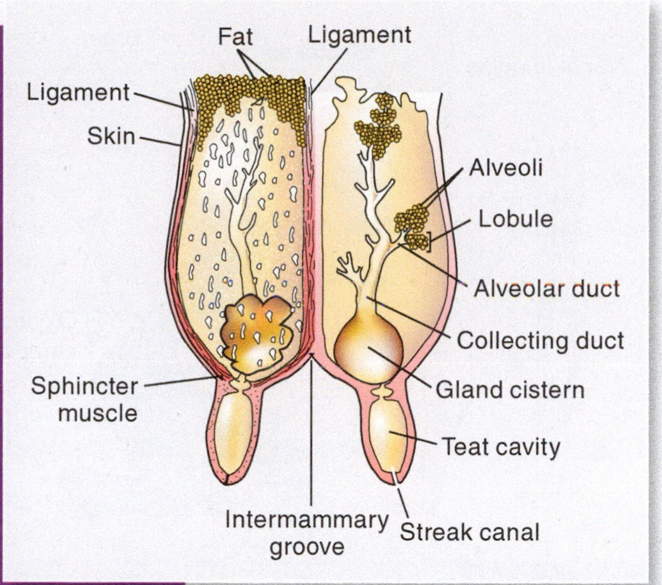

FIGURE 4–7 Cross section of an udder.

MILK PRODUCTION

Milk is produced in the udder of the cow in small clusters of grapelike structures called **alveoli**. Blood from the cow circulates through the udder. The alveoli take raw materials from the bloodstream and synthesize these materials into milk (**Figure 4–6**). For every pound of milk produced, 300 to 500 pounds of blood are circulated through the udder.

When a cow nears the time of giving birth, a hormone called **prolactin** causes the alveoli to begin to secrete milk. As long as the cow is milked or the calf nurses, prolactin stimulates the alveoli to produce milk. The longer the period from birth, the less prolactin is produced. Over a period of time, milk production decreases, so the cow is bred again to restart the process. As milk is secreted by the alveoli, it is drained into the **lumen** (a hollow cavity) in the alveoli.

The lumens (or *lumina*) are connected to the stem that connects the cluster of alveoli. This cluster is called the **lobule**. The lobule contains ducts—called the **tertiary ducts**—that drain into larger ducts, which carry the milk to an area called the **gland cistern**, where the milk is stored. A circular muscle called the **sphincter muscle** prevents the milk from leaking into the **teat** (**Figure 4–7**).

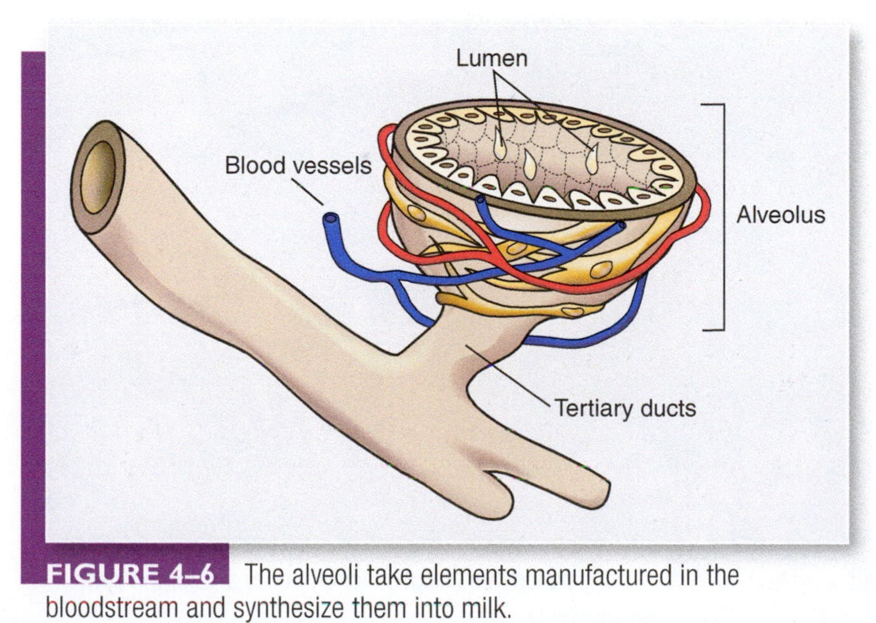

FIGURE 4–6 The alveoli take elements manufactured in the bloodstream and synthesize them into milk.

The Letdown Process

As the mother prepares to be milked or to nurse, the **pituitary gland** releases a hormone called **oxytocin** into the bloodstream. Oxytocin causes the alveoli to release milk into the ducts and

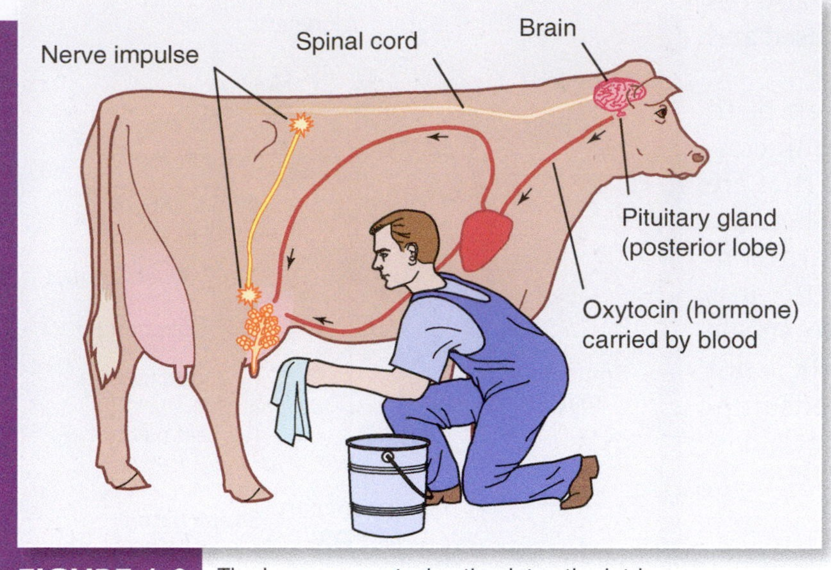

FIGURE 4–8 The hormone oxytocin stimulates the letdown process.

cisterns, and it causes the sphincter muscle in the teat to relax. The teat is relatively hollow, allowing the milk to pass out as the calf sucks or the milking machine pulsates. The release of oxytocin is caused by stimuli such as a calf rubbing the cow, the washing of the udder prior to milking, or other pleasant stimuli associated with milking. This is called the **letdown process** (**Figure 4–8**).

Milking Parlors

If the animal becomes frightened or upset, a hormone called **epinephrine** is released that inhibits milk from being let down. For this reason, it is essential that the milking area be clean and comfortable for the cows. Milkers must handle the cows as gently as possible to prevent them from becoming upset. Most milking areas, called **milking parlors**, are designed for easy handling of the cows and for the cows' comfort (**Figure 4–9**). Milking parlors are designed so the cow can enter a **stanchion**, where she stands while being milked. In some modern dairies, a computer chip in a tag around the cow's neck activates the dumping of the cow's ration into her trough as she enters a stall. The computer is programmed to recognize each individual cow by the chip around her neck, and it gives her the specific amount of ration designed for her.

A common type of arrangement in the milking parlor is the herringbone design. In this design, the cattle stanchions are arranged side-by-side at an angle resembling the pattern of the rib bones on the skeleton of a herring fish (**Figure 4–10**). The milkers work in an area below the cows so they don't have to bend to place the milkers on the cows' udders.

Modern parlors and lots where the cows are kept are designed with the cows' comfort and safety in mind. They contain items such as mats for the cows to lie on and/or rubber feed bins that prevent injury to the cow. As the cow comes into the parlor and the feed is dropped into a trough in front of her, the milker manually milks a small amount of milk into a cup called a strip cup. This procedure serves two

FIGURE 4–9 Milking parlors are designed for efficiency and for the cows' comfort. *© Syda Productions/Shutterstock.com.*

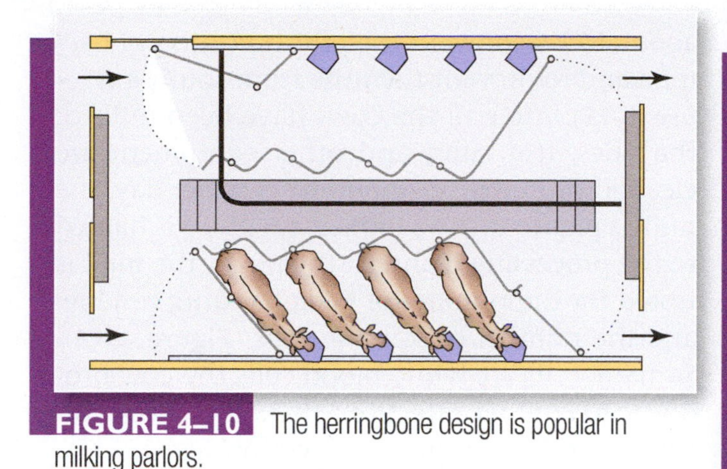

FIGURE 4–10 The herringbone design is popular in milking parlors.

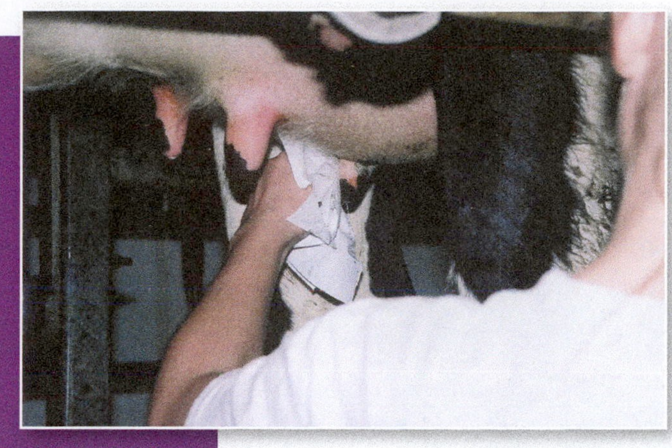

FIGURE 4–11 The udder is washed and dried prior to milking. *Courtesy of Gary Farmer*

purposes. First, the milker can check for a disease called **mastitis**, which is caused by injuries to the udder. Symptoms of mastitis are lumps or blood that come out in the milk. If evidence of mastitis is found, the cow is moved aside, where she can be treated, and her milk is not used. Second, stripping the first two or three squirts of milk removes milk that may have a high bacterial count because it is near the teat opening and more exposed to bacteria from the outside world.

The udder then is washed using a warm water solution and is dried thoroughly (**Figure 4–11**). Washing and massaging the udder helps to start the letdown process in the cow. The teat cups then are attached and the milking begins. The cups are lined with a soft material attached to a tube. The teat cups fit snugly on the cow and pulsate by means of a vacuum on the lining of the cup to gently draw the milk from the teat (**Figures 4–12** and **4–13**). The milk is removed in 3 to 6 minutes, depending on the individual cow and the amount of milk she gives. Care is taken to leave the cups on for the proper amount of time. If they are left on for too little time, the udder will not be milked out; if they are left on for too much time, injury to the udder can result. The teats then are treated with a disinfectant, and the cows are released. The teat cups are kept clean to prevent the spread of disease.

Good milkers time the operation so that as the first cow in the parlor is milked out, they will have just attached the teat cups to the last cow to enter the parlor. Then the milkers can remove the teat cups from the first cow, then the second, and so forth.

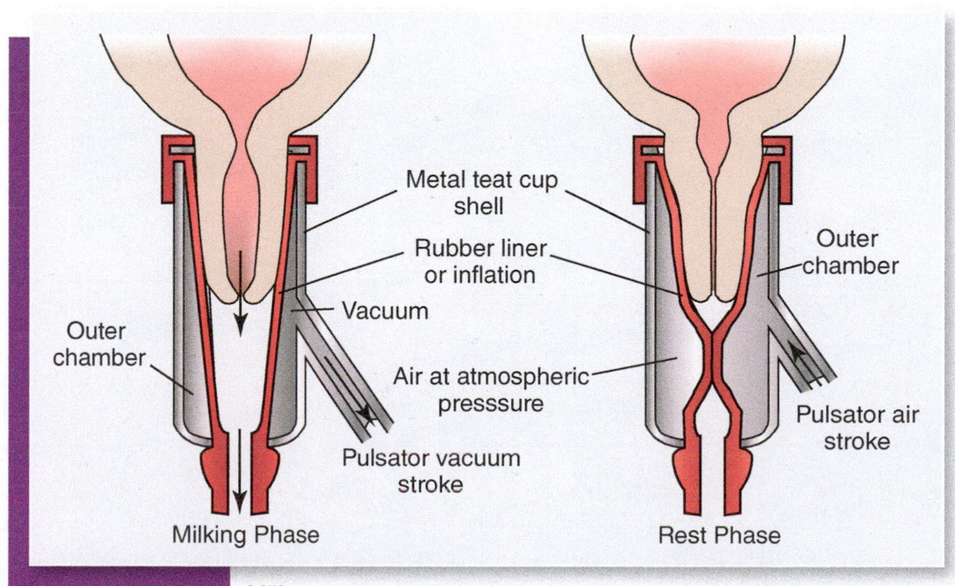

FIGURE 4–12 Milking is accomplished using a vacuum system that pulsates on the udder.

FIGURE 4–13 The teat cups are placed on the cow's teats. *Courtesy of Gary Farmer*

A newer type of milking parlor is designed in a circular pattern that allows the cows to rotate in a circle. The cows enter the stations and are given feed while the milking machine is attached. The platform slowly rotates as the cows are milked. At the proper place, the rotation stops and a cow that has been milked is let out and another comes in. This process allows the milkers to remain in one position, and the cows are rotated to them (**Figure 4–14**).

The milk is drawn through the lines and into a holding tank, where it is cooled rapidly to about 40°F to prevent the multiplication of bacteria and to prevent the milk from souring (**Figure 4–15**). After all the cows have been milked, the lines, teat cups, and other equipment are cleaned thoroughly. About every other day, the milk is picked up by a tanker truck and is hauled to the processing plant. At the plant, the milk is tested for the number of bacteria, drug residue, and the number of somatic cells (**Figure 4–16**). Somatic cells are white blood cells the cow produces to combat infection; their presence indicates that the cow has an infection.

When the milk arrives at the processing plant, it is filtered thoroughly to remove any foreign particles. The milk is allowed to sit so the cream may be removed from milk that is to be sold as low-fat milk. As consumers are becoming more conscious of the amount of fat in their diet, they prefer milk that is lower in milk fat than whole milk. In recent years, sales of low-fat and skim milk have increased sharply. In low-fat milk, the percentage of milk fat is lowered to between .5 percent and 2 percent. Skim milk or nonfat milk contains less than .5 percent milk fat. The milk fat that is removed from the milk is used to make other products, such as ice cream and other cream products.

FIGURE 4–14 A modern type of milking parlor uses a rotating platform that turns slowly as the cows are milked. *© Evgenii Sribnyi/Shutterstock.com.*

Whole milk contains about 4 percent milk fat. The globules of fat make up the cream that floats to the top of raw, unprocessed milk. These globules are larger than the other molecules in the milk, and this size difference causes the cream to separate if the milk is left undisturbed for a few hours. Cream is said to have a lower **specific gravity** than the rest of the milk. Specific gravity refers to the density of a substance compared to the density of water. Substances with a lower specific gravity than water will float on water. Because cream has a lower specific gravity than

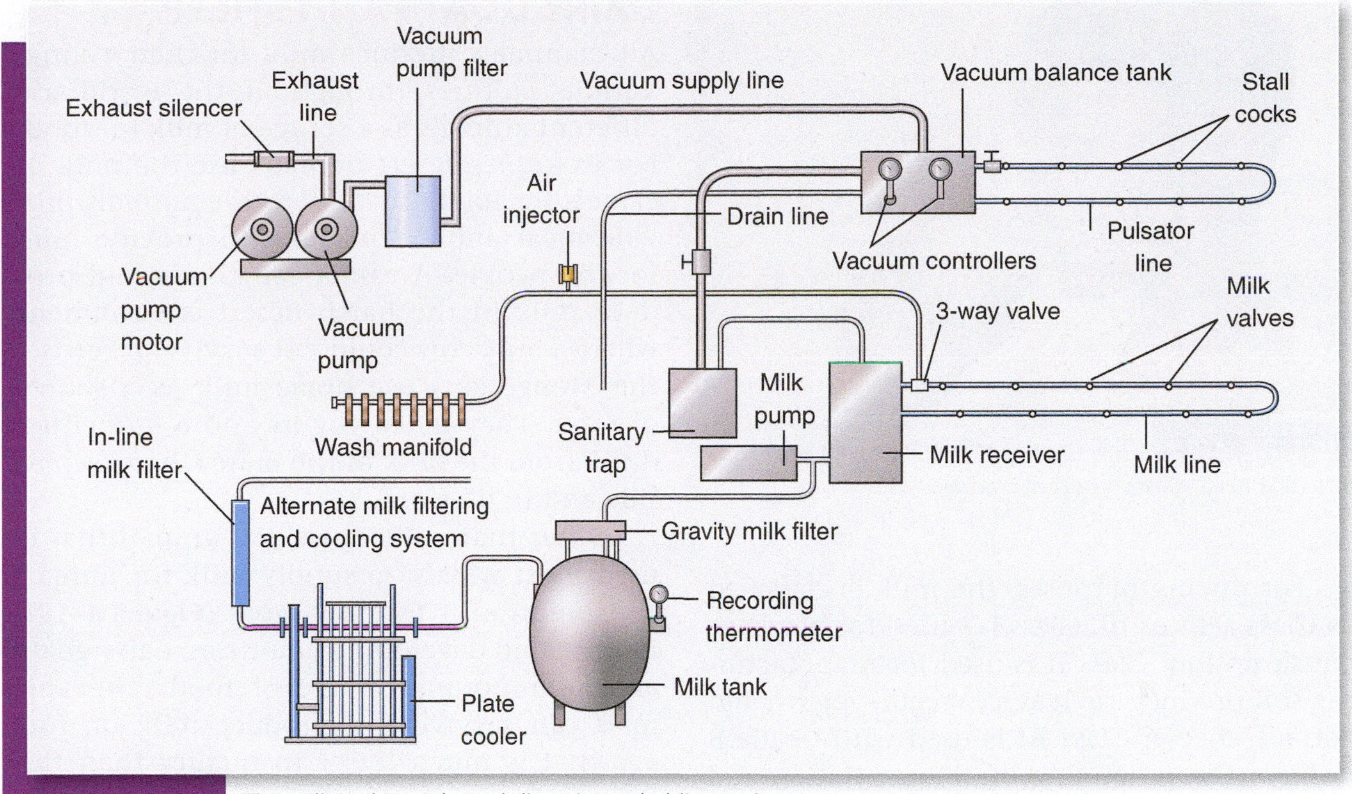

FIGURE 4–15 The milk is drawn through lines into a holding tank.

milk, the cream floats to the top. In a process called **homogenization**, the large cream globules are forced through a screen at high pressure and are reduced to the size of the milk globules. The processed milk, called **homogenized milk**, will not separate out when left sitting.

FIGURE 4–16 At the plant, the milk is tested for bacteria, drug residue, and the number of somatic cells.
© diplomedia/Shutterstock.com.

To kill any harmful organisms in the milk, the milk is heated and cooled in a process called **pasteurization**. One process of pasteurization raises the temperature of the milk to 145°F for not less than 30 minutes and then promptly cools the milk. An alternative method raises the temperature of the milk to 161°F for 15 seconds and then rapidly cools it. The time and temperature must be controlled precisely to protect the nutritive value and flavor of the milk.

Milk is graded according to the dairy from which it came. Dairies that sell Grade A milk must pass rigid standards for milk production. These involve cleanliness and other conditions under which the milk is produced. Only Grade A milk can be used for milk that is sold as fluid or beverage milk (**Figure 4–17**). Milk that is graded as Grade B milk can be used only for processing manufactured dairy products. Because the production of Grade A milk far exceeds the demand for fluid milk, Grade A milk may be used in processing as well.

FIGURE 4–17 Only Grade A milk can be sold for beverage milk. *© Africa Studio/Shutterstock.com.*

For pricing purposes, the milk is classified as Class I, II, or III. Class I is used for beverage consumption; Class II is used for manufacturing soft products such as ice cream, yogurt, and cottage cheese; Class III is used with Grade B milk in the processing of cheese, butter, and nonfat dry milk. Processing of milk into finished products such as cheese takes a lot of milk. **Table 4–2** indicates the amount of whole milk required to produce various milk products.

To Make One Pound	Requires
Butter	21.2 pounds whole milk
Whole milk cheese	10.0 pounds whole milk
Evaporated milk	2.1 pounds whole milk
Condensed milk	2.3 pounds whole milk
Whole milk powder	7.4 pounds whole milk
Powdered cream	13.5 pounds whole milk
Ice cream (1 gal.)	12.0 pounds whole milk (15 pounds when including butter and concentrated milks)
Cottage cheese	6.25 pounds skim milk
Nonfat dry milk	11.00 pounds skim milk

TABLE 4–2 Milk processing requires a lot of fluid milk.

Source: USDA

DAIRY GOATS AND SHEEP

All mammals produce milk for their young. Various cultures throughout the world use different animals as a source of milk for food. For example, desert nomads use the milk of camels for food. These versatile animals provide meat and labor and also provide milk for the people. A camel can thrive and produce milk in the harsh desert environment where a milk cow could not survive. Likewise, the Mongolians use horse milk as a source of food. They make yogurt and a fermented drink from the milk of the mares they keep to ride and to do work.

Other than milk cows, the animal that is used most widely to supply milk for human consumption is the milk goat (**Figure 4–18**). In poor and developing countries, dairy goats are an important source of food. The animals can survive and produce milk on forage that is much lower in quality than the forage necessary to sustain dairy cows. Most of the world's goat milk is produced in Africa and Asia. India is by far the world's leading producer, with about 2.7 million tons of goat milk annually.

There are more than 300,000 dairy goats in the United States, and the number appears to be increasing each year (**Figure 4–19**). The

FIGURE 4–18 In many parts of the world, goats are an important source of milk, and the Saanen is a popular breed. *© goodluz/Shutterstock.com.*

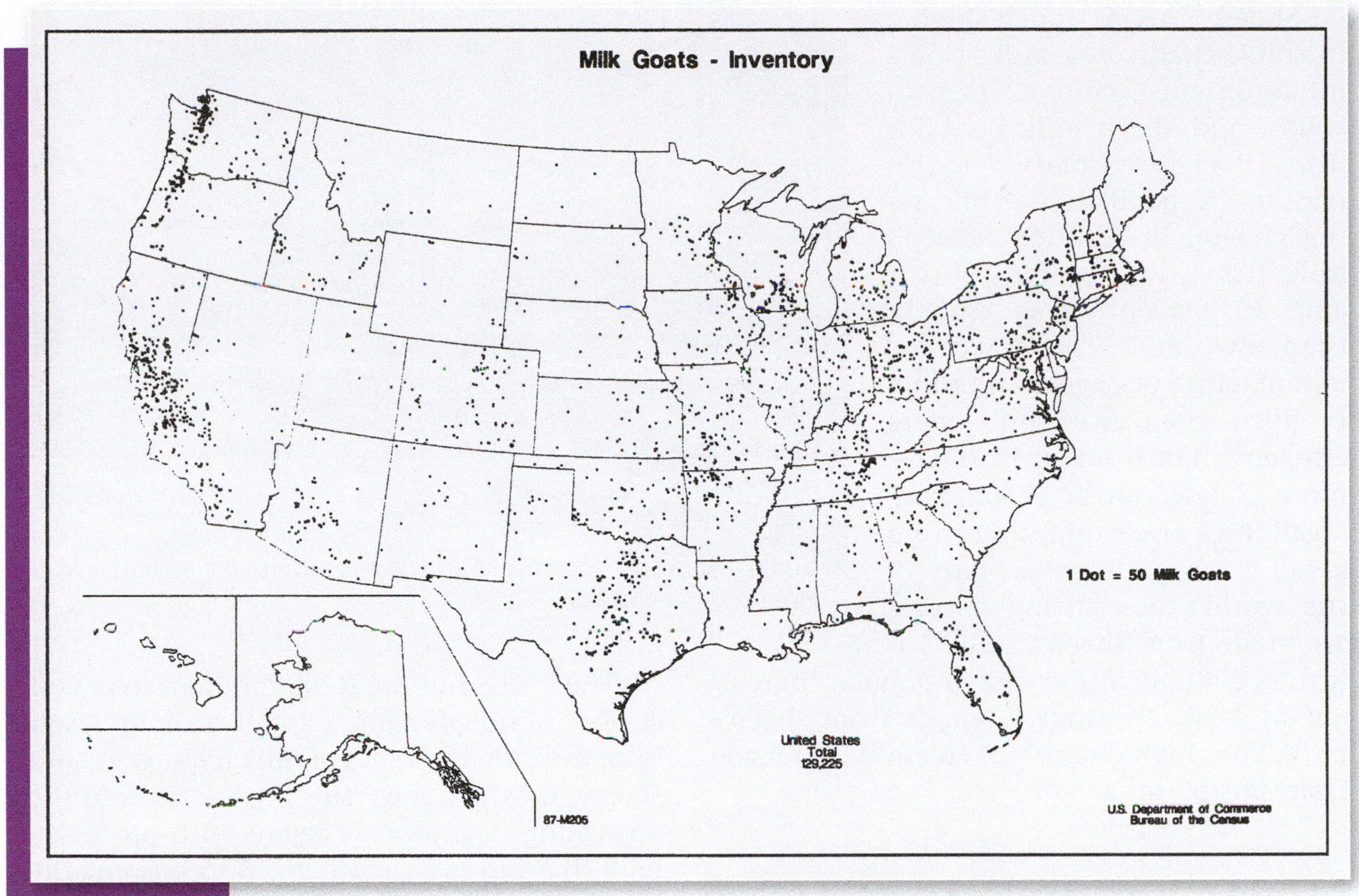

FIGURE 4–19 Milk goats are found in almost every state of the United States. *Source: USDA, Census of Agriculture*

leading states in dairy goat production are Wisconsin, California, Iowa, and Texas. Some of these goats are in large herds, but most are in small herds owned by hobbyists, and most of the milk produced is for home consumption. Goat milk is highly nutritious and is comparable to cow's milk. According to the U.S. Dairy Association (USDA), goat milk is increasing in its popularity because of its unique nutritional and biochemical properties. People with cow milk allergies and digestive disorders find goat milk superior to cow milk. Cheese, yogurt, and cottage cheese are made from dairy goat milk. In many parts of the world, cakes of goat cheese made by the producers can be bought in the local markets.

Sheep also are an important source of milk in many parts of the world; more than 100 million ewes are milked each year (**Figure 4–20**). Dairy sheep are milked in Europe, North Africa, the Middle East, and Asia. Although there are a few sheep dairies in the United States, the milking of sheep is not a large industry in this country.

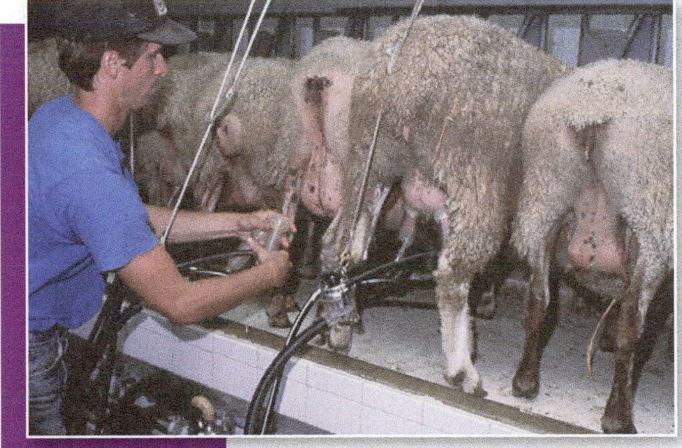

FIGURE 4–20 In some countries, sheep are milked for human use.

Sheep milk is much higher in solids than cow's milk. Cow's milk contains about 12 percent solids, and sheep milk contains over 19 percent solids. For this reason, sheep milk is used in making cheese. In addition, sheep's milk has twice the fat content and 40 percent more protein than cow's milk. This makes the manufacture of cheese and other products from ewe's milk more efficient than from cow's milk—more cheese can be made from a gallon of ewe's milk than from a gallon of cow's milk. Many of the world's best-tasting cheeses are made from sheep's milk. For instance, Roquefort cheese, a popular ingredient in salad dressings, is made from sheep's milk. Very high quality ice cream is also made from sheep's milk.

FIGURE 4–21 Milk is placed in large vats and a starter culture is added. *© Shchipkova Elena/Shutterstock.com.*

CHEESE MANUFACTURING

Of all the ways that humans process food, the processing of cheese is one of the oldest. This practice goes back thousands of years and was found in many ancient cultures. Legend has it that cheese was discovered in the deserts of the Middle East when a nomad transported milk in a bag made from a calf's stomach. The bag was thrown over a camel, and as the animal walked along, the milk sloshed and churned in the bag until the solids in the milk were separated from the liquids. This was a crude means of obtaining cheese.

Today, the world consumption of cheese continues to grow and is a large part of the diet of people in many countries. In the United States, the yearly per-capita consumption of cheese is almost 37 pounds. The manufacture of cheese accounts for almost one-third of all the milk used. Cheese can be stored easily and is a highly nutritious food that is high in protein content.

There are hundreds of different types of cheese. Although some differences may result from the different types of milk (cow, goat, and sheep), most are the result of variations in the processing. The process begins with processed milk that has been pasteurized to prevent the multiplication of harmful bacteria. The milk is placed in a large vat, where a bacteria culture is added (**Figure 4–21**). This culture is called a **starter culture** because it starts the process of **fermentation**.

Fermentation is the process that changes sugars to acids. These acids cause the proteins in the milk to coagulate (form a solid). To further the process, an **enzyme** (a substance that speeds up or stimulates a chemical process) called **rennet** (rennin) is added. Rennet is obtained from the stomachs of calves. (Remember—the discovery of cheese came about as a result of milk in a bag made from a calf's stomach.) During this step, large paddles turn the milk to ensure that the starter bacteria and the rennet are distributed evenly.

The solid resulting from this step is called **curd** (**Figure 4–22**). The liquid that is drained off is called **whey**. The curd is cut into small cubes by stainless steel wire knives that are passed through the mass of the curd to increase

FIGURE 4–22 The solid mass that results when the liquid is drained off is called the curd. © Noofoo Media Limited/Shutterstock.com.

the surface area of the curd and allow the whey to drain. After the whey is drained off, the curd sits until it forms a solid mass again. The curd then is heated, causing it to contract and further expel the whey. The amount of heat and the length of time the cheese is heated depend on the type of cheese being made. The cheese is salted and pressed into a metal form or a cloth bag.

The final step in cheese making is the curing or ripening of the cheese. The cheese is placed in an environment that is controlled for temperature and humidity; the specific conditions vary with the type of cheese being made. During this time, enzymes produced from the starter bacteria bring about changes in the flavor, texture, and appearance of the cheese. The cheese is packaged in a coating of paraffin or is wrapped in cloth or plastic (**Figure 4–23**).

FIGURE 4–23 The final step in cheese making is the wrapping and storage of the cheese. © Roman Babakin/Shutterstock.com.

PUTTING IT INTO PRACTICE

Dairy Cattle Evaluation and Management Career Development Event

The FFA offers many exciting and challenging experiences. Members who participate in team events spend many hours learning the basics in the classroom and practicing skills after school. The Dairy Cattle Judging and Management competition requires cooperative efforts of team members. If you have an interest in dairy products, plan to own dairy cattle, have a Supervised Agricultural Experience (SAE) in the dairy industry, or work on a dairy farm, you would be a good team member for these events.

The Dairy Cattle Evaluation and Management Career Development Event trains students in the selection of a quality dairy herd and well as management concepts for a modern dairy operation. Practice and experience are the only ways to develop the skills needed to select quality animals. A well-prepared team should understand what a quality dairy animal looks like and be familiar with dairy herd improvement records. Most Dairy Evaluation teams consist of four FFA members. Usually, through a competitive process, the agriculture teacher selects a team to represent the FFA chapter in competition. The FFA Dairy Evaluation event requires each team member to know the breeds of dairy cattle and traits that yield high milk production. During the event, each member is responsible for evaluating and placing groups of animals in numerical order.

Students develop leadership skills and knowledge about the dairy industry through the FFA Dairy Cattle Evaluation and Management Career Development Event. *Courtesy of the National FFA Organization*

The student must also give reasons for each animal's placement. After each team member has evaluated and placed all dairy groups, individual scores are calculated. All four scores are added to determine the total team score and placement on the national level.

Another part of the competition is the management component. The students are given a 50-question quiz on managing a dairy operation. They are also given a scenario of a dairy farm. The students study the scenario and give a presentation of their ideas for the operation. All winning teams from each state advance to the National Dairy Cattle Evaluation and Management Career Development Event.

SUMMARY

The dairy industry is almost as old as civilization. Milk and milk products have always been an important part of the human diet. Scientific research has brought about many changes in the production, processing, and storing of these products. Demand will almost certainly remain strong in the future for fluid milk, cheese, yogurt, ice cream, and all the other products made from milk.

CHAPTER REVIEW

Review Questions

1. In what way is the dairy industry different from other segments of the animal industry?
2. What are the leading states in milk production?
3. Why does a cow have to produce a calf to be able to continue producing milk?
4. Why is it important that a calf receive the first milk after birth (colostrum)?
5. List the hormones that control milk production.
6. What is meant by the letdown process?
7. What is mastitis? What causes it?
8. Regarding fat content, what are three categories of milk?
9. What is meant by homogenization? Pasteurization?
10. What is the difference between Grade A and Grade B milk?
11. Other than cows, what animals are used to produce milk for human consumption?
12. List the steps in cheese production.

Student Learning Activities

1. Obtain a cow's udder from a slaughterhouse. Using rubber gloves, dissect the udder and identify the alveoli, the lumen, the gland cistern, and the sphincter muscle.
2. Visit a large grocery store. From the dairy section make a list of all the products that are made from milk. List all the different types of cheese.
3. Prepare a list of all the processed foods in your home that contain milk. The ingredients should be listed on the food package.

CHAPTER 5

The Swine Industry

STUDENT OBJECTIVES IN BASIC SCIENCE

As a result of studying this chapter, you should be able to

- explain why pork is healthier to eat than it once was.
- tell why protein is important in the diet of a growing pig.
- list the different types of amino acids.
- define hybrid vigor or heterosis.

STUDENT OBJECTIVES IN AGRICULTURAL SCIENCE

As a result of studying this chapter, you should be able to

- explain the importance of the swine industry.
- briefly describe the history of the swine industry in the United States.
- name the predominant breeds of swine.
- distinguish between a dam and a sire breed.
- describe the production methods involved with raising swine.
- explain the environmental impacts of a large swine operation.

KEY TERMS

rendering
lard
synthetic lines
hybrid vigor or heterosis
mother breeds
sire breeds
farrowing operation
growing operation
finishing operation
feeder pigs
climate-controlled houses
castrated
docking
nursery
confinement operation
feed conversion ratio
barrows
gilts
amino acids
finished
carcass merit
lagoons

NATIONAL AFNR STANDARD

AS.01.01
Evaluate the development and implications of animal origin, domestication, and distribution on production practices and the environment.

AS.01.02.02.a
Research and examine marketing methods for animal products and services.

THE PORK INDUSTRY represents an important and dynamic component of the animal industry in the United States. In the past 30 years the number of swine operations in this country has decreased by almost 90 percent, yet the number of hogs slaughtered has actually increased. This is because producers are more efficient and because the size of the operations has increased dramatically. In 1970, there were more than 871,000 swine producers in this country; today there are just over 60,000 producers. More than half the pigs produced are from farms that raise at least 5,000 pigs per year. Each year, producers raise almost 110 million hogs, yielding more than 10 metric tons of pork, of which 2.2 million tons are exported. According to the American Pork Producer's Council, this industry supports almost 600,000 jobs and is responsible for $72 billion or more in total economic activity. The top swine-producing states are Iowa, North Carolina, Minnesota, and Illinois.

Pork is the most widely eaten meat in the world. In fact, around 42 percent of the meat consumed in the world is pork. Worldwide, the United States ranks second only to China in the number of hogs produced annually. China is a huge country and produces about 50 million tons of pork per year, as opposed to production of about 10 million tons in the United States. In per-capita consumption, we rank twelfth, with 62.8 pounds of pork consumed per person each year. Denmark leads the world in pork consumption, with more than 142 pounds consumed per person each year.

Bacon, ham, and pork chops have always been popular in the American diet; however, in recent years concern has been raised about the fat levels in pork products. The National Pork Producers Council has successfully educated consumers on the merits of pork. Although pork once was considered a fatty food, today's leaner pigs produce pork that is relatively lower in fat content and that is quite nutritious (Figure 5–1). In terms of meat, pork production and consumption rank second only to beef in this country. Pork consumption is distributed throughout the country, although certain populations, such as Moslems and Jewish people, do not eat pork for religious reasons.

FIGURE 5–1 Today's pork is much leaner than the pork produced in the past. *© iStockphoto/ZavgSG.*

HISTORY OF THE INDUSTRY

Pork production has been a part of American agriculture since the earliest Europeans settled in this country (Figure 5–2). Columbus brought pigs on his first voyage to the New World, as food for the sailors. The first pigs were introduced by the Spanish explorer Hernando de Soto when he landed on the coast of Florida in 1539. It was reported that he brought only 13 head, but in a period of only three years, this small herd had grown to more than 700 head. Native Americans developed a taste for pork and began to hunt the pigs that escaped from captivity. These escaped pigs are the ancestors of the wild pigs that are prominent in many parts of the country today.

As settlers moved west, they inevitably took pigs with them, as these animals easily adapted to differing environmental conditions. They could live off the land by eating acorns, roots, and wild plants, and they often were allowed to roam "free range" through the woods. As the settlements became denser, the free-range practice often caused problems with neighbors' crops. Roaming pigs caused such a problem in the colony of Manhattan in New York that a

FIGURE 5–2 Pigs have been a very important part of agriculture since the beginning of our country.
© iStockphoto/andipantz.

wall was built to keep the pigs out. Even after all these years, the street along this wall is still called Wall Street!

Pigs provided food for the settlers. They were proficient breeders, and each female could produce several offspring each year. When the weather turned cold, pigs were slaughtered and the meat was preserved by smoking and salting. People could eat all winter on the preserved meat. The fat from the animals was cut into chunks, placed into a large iron kettle, and **rendering**, which meant that the fat was heated until it melted and could be separated from the solid particles. The resulting fat, called **lard**, was kept for use in cooking and also as one of the central ingredients in making soap. Until about 1950, the major reason for raising pigs was to obtain fat for lard. With the advent of vegetable oils, lard became less prominent in the American diet, and hogs began to be raised primarily for meat.

At one time, most of the people who lived on farms in this country raised pigs. The animals required relatively little space and fit well into most enterprises as a sideline income. Hogs are said to have sent more farm youngsters to college than any other enterprise. Most of the feed was raised on the farm and little had to be bought. Today, many hog producers buy their feed already mixed and delivered to their farms ready to feed.

Because the gestation period is short and each litter has several pigs, the time required to build up a herd of hogs is short compared with many other agricultural animals. For this reason, an operation can be built in a relatively brief time. Also, the type of pigs produced can be changed in less time than with most agricultural animals.

For many years, most of the pork produced in the United States came from the Midwest in the states of Iowa, Illinois, Indiana, Minnesota, and Nebraska. These states produce a large amount of corn, the major grain fed to swine, and they remain leaders in the production of pork (**Figure 5–3**). In recent years, however, larger numbers of pigs are being raised in the South, where mild winters help to lower the cost of production. In fact, the state of North Carolina is now second only to Iowa in the number of pigs produced.

BREEDS OF SWINE

Although there are not as many breeds of hogs as breeds of cattle in this country, there are still several popular breeds of hogs. Modern swine are bred to be leaner and more efficient than the swine of several years ago. Being efficient means that they can grow faster, they mature at an earlier age on less feed, and they have more pigs per litter. Today, most pork producers raise one or more of nine major breeds: Yorkshire, Duroc, Hampshire, Landrace, Berkshire, Spotted, Chester White, Poland China, and Pietrain. A diagram of the external parts of the hog are shown (**Figure 5–4**).

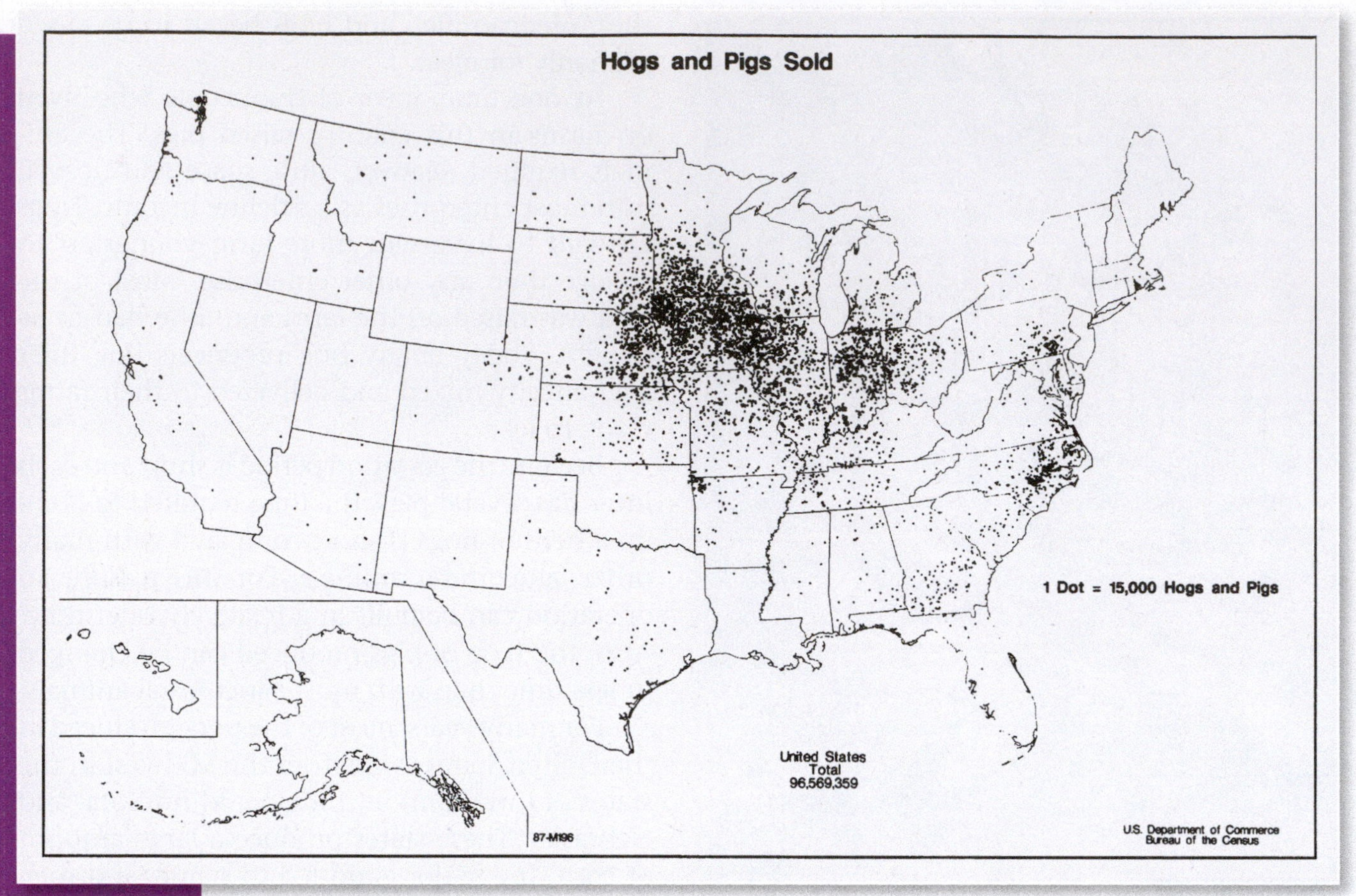

FIGURE 5–3 Most of the hogs are grown where corn is produced.

Source: USDA, Census of Agriculture

Increasingly, producers are using what are called **synthetic lines**, derived by crossing these breeds. Several commercial breeding companies develop these lines for use as breeding animals. Almost all the pigs produced for slaughter in the United States are the result of crossbreeding of purebred or synthetic lines (**Figure 5–5**). Crossbreeding makes use of a biological phenomenon known as **hybrid vigor**, or **heterosis**, which results in offspring that are superior to what might be expected of the parents.

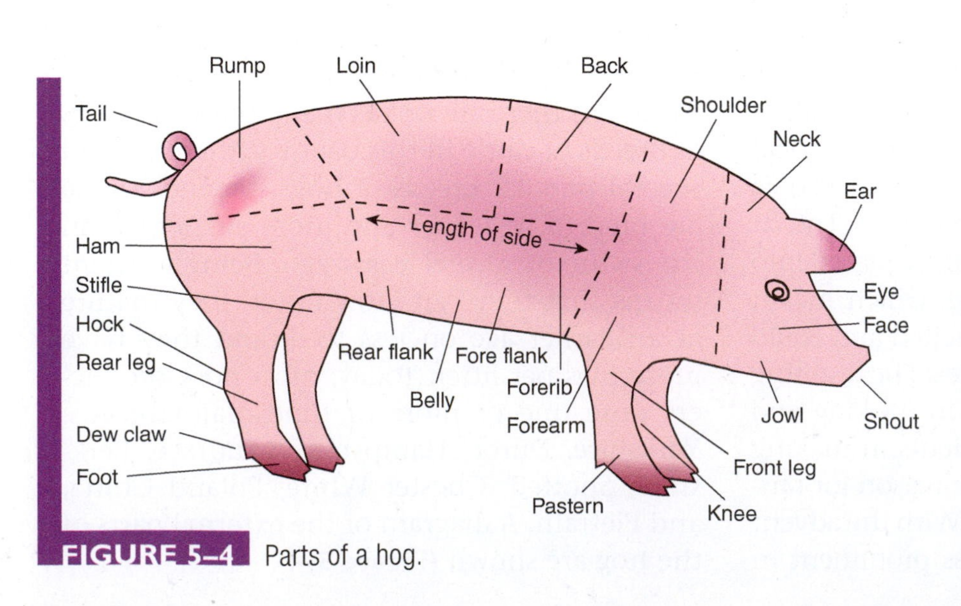

FIGURE 5–4 Parts of a hog.

FIGURE 5–5 Most pigs produced in the United States are the result of crossbreeding programs. © iStockphoto/paulobaqueta.

Breeds of swine are categorized as **mother breeds** or **sire breeds**. Mother breeds are superior in the number of pigs in a litter, the amount of milk they produce for their young, and their docile temperament. The mother breeds are white pigs that include Chester White, Landrace, and Yorkshire (**Figure 5–6**).

The sire breeds, such as the Duroc and the Hampshire, characteristically grow rapidly and produce well-muscled, meaty carcasses. They also are durable and leaner.

FIGURE 5–6 Yorkshires are considered to be a mother breed because of their large litters and their high level of milk production. © Aumsama/Shutterstock.com.

Heritage Breeds

A relatively new trend in swine breeds is the raising of the heritage breeds. These breeds are called heritage breeds because they have been around for a long time, some as long as 200 years. Red Wattle, Hereford, Mule Foot, Ossabaw Island, and American Guinea are all examples of heritage breeds. People raise them either because of the novelty or because they think the pork is higher quality. For example, the Mule Foot breed does not have toes like other pigs, but has a solid hoof. Many people think the Tamworth has much higher quality meat than most breeds. Whatever the reason, these breeds are grown all across the United States in small herds.

PRODUCTION METHODS

Pigs are unique animals. The popular perception, that they are dirty, stupid animals that constantly overeat, is erroneous and really opposite from the truth. Pigs got the reputation of being dirty animals because, if given the opportunity, they wallow in mud. The mud helps to keep them cool in hot weather and also helps to keep off parasites. If given the room, pigs will use only a certain part of their pens to drop wastes and will keep the rest of their area clean. Also, pigs are highly intelligent, ranking among the top of all agricultural animals in overall intelligence. In handling pigs, producers must understand how smart pigs are in order to move them and keep them in their pens.

Pigs are one of the few agricultural animals that will not overeat. Given the proper type of feed, they will consume only the amount of feed they need. In contrast, cattle, horses, and other animals overeat to the point where they may become ill. This is not a problem with pigs.

Scientists have studied the unique characteristics of pigs and have used the findings of these studies to design production methods to suit these animals' needs. The type of buildings, health regimens, management procedures, and diets are all designed to help pigs live healthy, comfortable, and productive lives.

Of all animals raised, pigs are the closest to humans in terms of digestive, circulatory, and other systems. Tissue from pig skin is used to replace human skin that has been badly burned. Valves from the hearts of pigs are used as replacements for human heart valves that have worn out or been damaged by disease. Also, pigs are used in many areas of research for products that eventually will be used by humans.

Three phases of the industry are the **farrowing operation**, the **growing operation**, and the **finishing operation**. The three phases can be operated separately or together. Some producers prefer to raise only **feeder pigs** (pigs that are weaned and sold shortly after weaning), and some prefer to buy feeder pigs and finish them as their only operation. Most pigs are farrowed in **climate-controlled houses** where the mother is kept in a crate to prevent her from injuring the piglets when she lies down (**Figure 5–7**). Good producers make quite an effort to provide an environment that is clean, dry, and comfortable for both the mother and the piglets.

FIGURE 5–7 Farrowing crates prevent the mother from crushing the piglets when she lies down. *© rtem/Shutterstock.com.*

At farrowing, the sow typically has 9 or 10 piglets. At birth, the piglets are dried off and their navel cords are dipped in iodine to prevent infection. Some pigs are born with long, sharp teeth called needle teeth. If left alone, these teeth may injure the sow's teats or may injure other piglets. The producer clips off the teeth to prevent these injuries. Also, the newborns are given a shot of supplemental iron to help improve the oxygen-carrying capacity of their blood. Pigs that are born and raised on the ground usually get enough iron from the soil; however, most pigs are born on slatted, concrete, or raised deck floors and need the supplemental iron.

The pigs usually are weaned from the mother at 3 to 4 weeks of age, although some producers wean the pigs at 6 weeks or as old as 8 weeks. At this time, the pigs usually weigh around 10–15 pounds and are placed in nurseries, which make use of a slotted floor that allows waste material to fall through. This helps to keep the floor cleaner and drier. The male (boar) pigs are **castrated**. In this procedure, the testicles are removed to prevent aggressiveness, avoid pregnant females, and prevent off-flavored meat when the pigs are slaughtered. All pigs have their tails removed, called **docking**. Pigs kept in confinement operations have a tendency to bite at each other's tails. Removing the tails eliminates incidents of tail biting.

In the **nursery**, the pigs are fed a scientifically balanced diet that provides the proper amount of nutrients that the animals need at this stage of their growth (**Figure 5–8**). As the animals continue to grow, their diet is changed to fit the needs of that stage of growth. By the time the pigs are moved out of the nursery at 8 to 10 weeks of age, they may have been fed five different diets, consisting of grain, protein supplements, and milk products. The protein is supplied from a mixture of plant and animal sources, and the amount and type of protein required vary as the animals grow.

After weaning, the pigs are conditioned in the nursery until they weigh 40–60 pounds and are then placed together with pigs of

FIGURE 5–8 In the nursery, the piglets are fed a scientifically balanced diet that provides exactly the proper amount of nutrients needed by the animals at a given stage of growth. © Aumsama/Shutterstock.com.

FIGURE 5–9 Pigs learn to drink from automatic waterers that ensure a constant supply of clean water. © Dmitry Kalinovsky/Shutterstock.com.

similar age, size, and sex in a **confinement operation**. This means that the hogs are kept in a pen together rather than running loose on a pasture. Sufficient space is allowed for the pigs to be comfortable and to grow at a fast rate. In this system, pigs are kept comfortable in climate-controlled houses, where they are protected from heat, cold, and rain. Although these operations are quite expensive, less labor is needed to care for the pigs. The animals drink from automatic waterers that ensure a constant supply of clean water (**Figure 5–9**). Feed is supplied from automatic feeders where the animals obtain all the feed they want. In these houses, animals are less likely to pick up parasites or contract diseases. The pens are cleaned and disinfected periodically to help protect the animals' health.

Pigs are said to be more efficient than cattle. This means they put on a pound of body weight with less feed consumed. However, pigs cannot make use of large amounts of roughage as cattle do and must be fed on a ration of grain. On the average, pigs will gain a pound for about every 5 pounds of feed consumed, compared to about 9 pounds of feed per pound of gain for a beef animal. This is known as the **feed conversion ratio**.

Scientists have developed different diets to be used in all phases of growth. As with the weaned pigs in the nursery, growing pigs receive different diets as their nutritional needs change in the maturation process. The pigs sometimes are segregated into **barrows** and **gilts**. Gilts are female pigs that have not had a litter of pigs, and barrows are males that have been castrated. By separating the pigs according to sex, the diets can be "fine-tuned" to provide even greater efficiency (**Figure 5–10**).

Pigs are fed a high-protein diet to promote their growth and muscle development. As the animals mature, the diet is switched to

FIGURE 5–10 Pigs are finished for the market in confinement operations. By separating them according to sex, the diets can be fine-tuned to provide even greater efficiency. © MARCELODLT/Shutterstock.com.

one with a lower protein content and a higher carbohydrate content. A ration that is rich in protein is needed in the early stages of growth, to build muscle and bones. Without the proper amount of protein, the muscles, bone, and internal tissues and organs will not develop properly. When the animals approach maturity, they need less protein and more carbohydrates because the carbohydrates help them develop fat as their skeletal and muscular systems mature. Some fat is required in the meat to produce the juiciness and flavor that consumers want.

At one time, protein was calculated in terms of the percentage of protein needed in the diet. Today, the building blocks of protein, called **amino acids**, are used as the basis of balancing the feed diet. Amino acids are composed of carbon, hydrogen, oxygen, and nitrogen. Swine need 10 types of essential amino acids from the feed they eat. Several other amino acids are synthesized by the animals' bodies from the essential amino acids.

The pigs should be **finished** (reach the proper market weight and condition) at about 20 weeks. Packers like to buy market hogs that weigh around 265 pounds (**Figure 5–11**). Most pigs are marketed by directly selling them to the processor, although a few are still marketed through live auctions. When sold directly to the processor, hogs are often sold on **carcass merit**. This means that premium prices are paid for pigs with low amounts of fat and high amounts of muscle.

FIGURE 5–11 Packers like to buy pigs that weigh 220–260 pounds. *© iStockphoto/curtoicurto.*

ENVIRONMENTAL CONCERNS

Strict federal, state, and local laws regulate how and where pigs are raised. Hogs in close confinement can cause problems with odor and manure disposal. The larger the operation, the greater is the problem. To dispose of the manure and odor, waste from the finishing pens is flushed into ponds called **lagoons** (**Figure 5–12**). Building and operating the lagoons are regulated to ensure that the waste material (manure) does not pose a threat to streams and water supplies. In the lagoons, bacteria help break down the waste materials into a slurry that does not have an odor as bad as untreated manure.

Periodically, the waste material is pumped from the lagoons and is spread on pastures or cropland as fertilizer. This provides a means of disposing of the manure and also supplies a high-quality, organic fertilizer for crops. This form of waste disposal is a type of recycling of nutrients that helps to protect the environment.

FIGURE 5–12 Waste from confinement operations is washed into lagoons. *Copterviews.com photo contracted by USDA/ARS*

SUMMARY

Pigs have been a part of American agriculture from the beginning and still hold a large portion of the agricultural industry. These are relatively efficient, highly intelligent animals that are often wrongly depicted. Modern pork production systems provide comfortable, clean facilities for all phases of the industry. Diets are scientifically balanced to give the animals the nutrients they need. The future of this industry is bright, and pigs will continue to play an important role in agriculture and the American diet.

CHAPTER REVIEW

Review Questions

1. What are the different characteristics of a sire breed and a mother breed? How can these characteristics be used in a crossbreeding program?
2. What are some of the popular misconceptions about hogs? Be sure to tell why these perceptions are untrue.
3. Why is pork considered to be healthier now than in the past?
4. Why are pigs often used in medical research for products that eventually will be used for humans?
5. Describe the three phases of the pork industry.
6. List the advantages and disadvantages of a confinement operation as opposed to a free-range operation.

Student Learning Activities

1. List all the pork products that your family consumes in a month. Be sure to include products such as sausage, bacon, bologna, and other processed meats.
2. Create a list of all the hog operations in your area. Define the type of operation (feeder pig, finishing, purebred), and determine which type is the most popular. Explain why this type is popular in your area.
3. Interview a purebred producer and determine why he or she grows that breed.
4. Do an Internet search and locate information on a specific breed of swine. Report to the class.
5. Choose a heritage breed of swine and research the breed. What makes this breed different? Why do producers raise them?

CHAPTER 6

The Poultry Industry

STUDENT OBJECTIVES IN BASIC SCIENCE

As a result of studying this chapter, you should be able to

- compare the process of egg development in birds and mammals.
- trace the biological processes involved in the production of eggs in birds.
- describe how the chick embryo develops in the egg.
- relate how nature protects eggs from the environment.
- describe the ideal conditions for the production of bacteria.
- tell how hatching chicks communicate.

STUDENT OBJECTIVES IN AGRICULTURAL SCIENCE

As a result of studying this chapter, you should be able to

- summarize why the poultry industry is rapidly growing.
- define *vertical integration*.
- explain how broilers are produced in modern operations.
- describe how modern hatcheries operate.
- discuss modern layer operations.
- describe modern turkey production.

KEY TERMS

broiler industry
vertical integration
cannibalism
layers
hybrid
heterosis
hybrid vigor
pigmentation
embryo
ovum
ovary
infundibulum
magnum
cells
mucin
albumen
yolk
chalazae
isthmus
uterus
shell gland
incubation
oxytocin
cloaca
sweating
fertilization
sperm
sperm nests
germinal disk
cage operations
metabolism
pullets
molting
candling
muscling

NATIONAL AFNR STANDARD

AS.01.01
Evaluate the development and implications of animal origin, domestication, and distribution on production practices and the environment.

AS.01.02.02.a
Research and examine marketing methods for animal products and services.

AS.06.03.03.a
Research and summarize the use of products and by-products derived from animals.

NOT ONLY IS THE poultry industry one of the fastest growing segments of the animal industry, but it is also one of the most rapidly changing industries. Worldwide consumption of poultry is increasing. Chickens, turkeys, ducks, geese, and other birds make up a large portion of the meat diet of people in most countries. In the United States, the per-capita consumption of broilers is over 108 pounds, which has more than tripled since 1960 (see **Table 6–1**). This increase is due to the rapid increase in the efficiency of producing

Year	Beef	Pork	Total[1] Red Meat	Chicken	Turkey	Total Poultry	Commercial Fish/ Shellfish
1980	76.6	57.3	136.8	48.0	10.3	58.3	12.5
1981	77.3	54.7	135.0	49.4	10.6	60.0	12.7
1982	77.0	49.1	129.3	49.6	10.6	60.2	12.5
1983	78.7	51.8	133.6	49.8	11.0	60.8	13.4
1984	78.4	51.5	133.2	51.6	11.0	62.6	14.2
1985	79.2	51.9	134.4	53.1	11.6	64.7	15.1
1986	78.8	49.0	131.1	54.3	12.9	67.2	15.5
1987	73.9	49.1	125.9	57.4	14.7	72.1	16.2
1988	72.8	52.4	128.0	57.5	15.7	73.2	158.2
1989	69.0	52.0	123.6	59.3	16.6	75.9	15.6
1990	67.8	49.7	120.0	61.5	17.6	79.1	15.0
1991	66.5	50.2	119.5	64.0	17.9	81.9	14.9
1992	66.5	53.1	121.9	67.8	17.9	85.7	14.8
1993	65.1	52.3	119.7	70.3	17.7	88.0	15.0
1994	67.0	53.0	122.1	71.1	17.8	88.9	15.2
1995	67.5	52.4	122.0	70.4	17.9	88.3	15.0
1996	68.2	49.1	119.6	71.3	18.5	89.8	14.8
1997	66.9	48.7	117.7	72.4	17.6	90.0	14.6
1998	68.0	52.5	122.5	72.9	18.0	90.9	14.9
1999	69.0	53.8	124.8	77.5	17.9	95.5	15.0
2000	62.7	51.2	120.7	77.9	17.4	95.3	15.6
2001	66.2	50.2	118.1	77.6	17.5	95.1	15.5
2002	67.7	51.5	120.9	81.9	17.7	99.6	15.6
2003	64.9	51.8	118.4	83.0	17.4	100.4	15.6
2004	66.1	51.3	119.0	85.4	17.0	102.4	15.6
2005[2]	68.0	51.1	120.8	87.4	17.0	104.4	15.5
2006[3]	62.5	51.5	115.5	89.1	17.3	106.4	15.4
2007[3]	62.5	51.5	115.5	90.6	17.6	108.2	15.4
2008	62.7	49.9	113.5	84.4	18.6	103.0	

[1]Includes beef/pork/veal, and mutton/lamb.
[2]Estimated by ERS/USDA.
[3]Forecasted by National Chicken Council.

TABLE 6–1 Per-capita consumption of poultry and livestock, 1980 to 2008. *Source: USDA*

poultry, which lowers the cost to consumers. Also, the industry has developed a much wider variety of poultry products that meets the needs of consumers. Just think of all the different chicken products: chicken sandwiches, chicken nuggets, Buffalo wings, and fried, stewed, and grilled breasts and thighs.

In the past several years, poultry products have expanded. The best examples are the use of parts that once were considered to be waste material. For many years, wings were considered to have so little meat that they weren't considered valuable. The trend now is that wings are valued as a snack food, particularly when grilled. Grilling may include a wide variety of sauces that give unique flavor to the meat. This has become a football tradition. In fact, we consume over 1.25 billion wings at Super Bowl parties alone. Before, the birds' feet were either ground up into pet food or fertilizer. Now new markets have opened up in places like China, where the chicken feet are considered to be a delicacy. Each year, we ship over 300,000 metric tons of chicken feet to China.

Unlike some meats, poultry is generally accepted by most cultures. For instance, the Moslem and Jewish cultures do not eat pork, and Hindu culture does not allow the eating of beef. However, almost all cultures accept poultry as a wholesome meat for human consumption. Developing countries often begin to build a sound agricultural base with poultry because birds are efficient users of feed and are easily cared for in countries where human labor is readily available. The largest producers of poultry in the world are China, the countries of the former Soviet Union, and the United States.

THE BREEDER INDUSTRY

The basis for the poultry production industry is with the breeders. These are the people who develop genetic lines of chickens or other poultry that produce chicks for certain segments of the industry. These segments include the broiler and layer industries. Within these segments are other segments. For example, consumers want chicken in different forms, and the birds are bred to conform to a particular market. Fast-food chicken that is served as takeout usually comes from a bird that will yield a 6- to 7-pound carcass. Chicken strips, which are sliced from the breasts of the bird, usually come from a larger bird that may yield a carcass as large as 10 pounds. Obviously, these different-size birds require different genetics. (**Figure 6–1**) illustrates the parts of a male chicken.

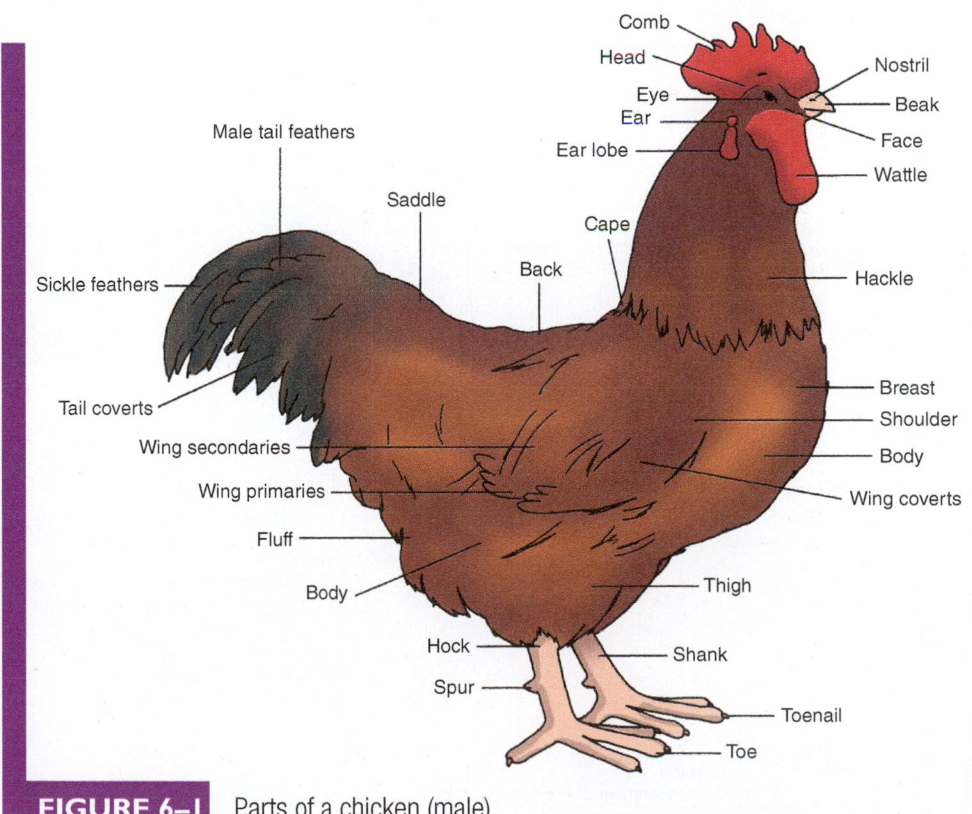

FIGURE 6–1 Parts of a chicken (male).

Most of the breeder industry is composed of large corporations that spend billions of dollars each year in research and development. These birds have to be raised in a much more stringent process than birds raised for the consumer market. When a line is developed, researchers have to be sure the advances are due to genetics and nutrition rather than any other input. This means that the birds are grown under the closest of care not to expose them to diseases and other outside factors. The rearing houses for the research and breeder birds are said to be "biocontained." In these facilities, as close to sterile conditions as can be maintained are strictly adhered to. Workers who enter the houses are required to shower and dress in sterile clothing. Even the feed that comes in has gone through a process that ensures that there are no pathogens in the feed. The buildings are sealed tightly to be rodent and wild bird proof. These strict measures are carried through complete, until the baby chicks are delivered to producers.

THE BROILER INDUSTRY

At one time in the history of our country, almost all families in rural areas had some type of poultry. Not only did chickens provide the family with fresh eggs but also with fresh meat. Today, almost all of the poultry is raised in large operations. The term **broiler industry** refers to the raising of chickens for meat. This industry is concentrated in the Southeast, where the mild winters provide an advantage to producers. The leading broiler-producing states are Arkansas, Georgia, and Alabama (**Figure 6–2**).

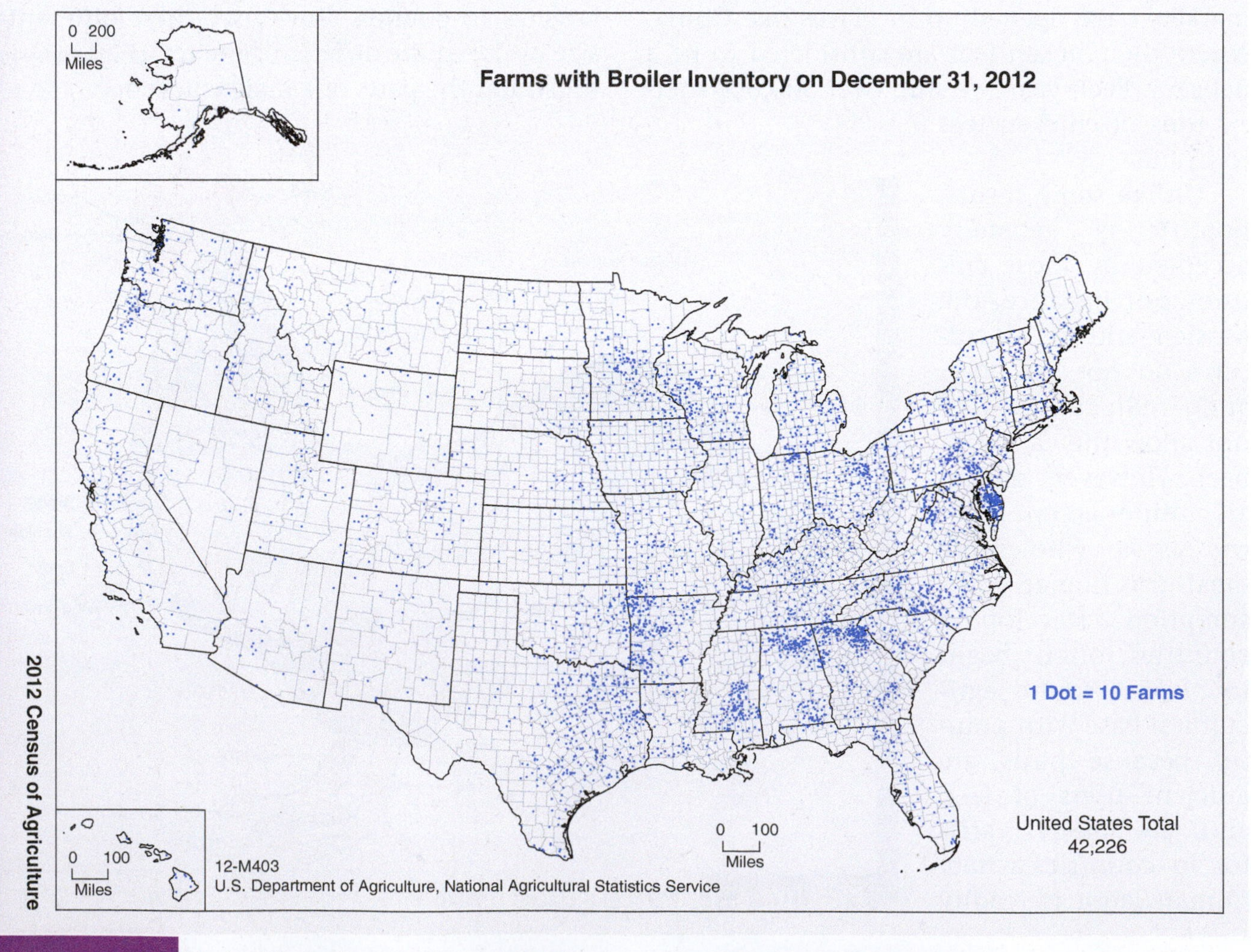

FIGURE 6–2 Much of the broiler industry is centered in the southeastern part of the United States.

A broiler is a bird that is grown out to and is dressed for market. Today it may take as little as 45 days from the time the broilers are hatched until they are marketed. The size of the birds may range from 6 to 10 pounds, depending on the market they are being produced for. The vast majority of broilers produced in this country are raised on contract. In a typical grow-out contract, the company agrees to provide the producer with chicks, feed, medications, vaccines, and other supplies. The company also agrees to pay the producer a predetermined price per pound for the broilers produced and sometimes gives an added payment to the producer as an incentive for more efficient production. The producer supplies the house, feeding and watering equipment, utilities, litter material, waste disposal, and labor. The company usually has a **vertical integration**, which means that the company owns the hatchery, feed mills, processing plants, and distribution centers.

FIGURE 6–3 Broilers are raised in large houses that provide a clean, comfortable environment for the birds. *© Kharkhan Oleg/Shutterstock.com.*

Broiler Houses

Broilers are raised in large houses, where the birds spend almost all of their lives. Broiler houses are designed to provide the birds with a clean, comfortable environment (**Figure 6–3**). The houses hold from 6,000 to 40,000 birds and are built to keep the animals warm in winter and cool in summer. The houses are insulated and have ventilators to help remove the heat. When the birds are small, heat is provided by brooders that are powered by gas or electricity. As all the birds grow larger in a well-insulated house, they usually generate ample body heat without the need for any artificial heat.

The houses usually are lighted almost around the clock. Research has shown that by leaving the lights on, incidents of **cannibalism**—the attacking of birds by other birds in the flock—are greatly reduced. This can be quite a serious problem if it is not regulated. Lights are routinely turned out for one hour each night to help prevent the birds from becoming hyperactive if a power failure occurs at night.

BROILER PRODUCTION

The process of broiler production begins with the production of eggs to be hatched for young broilers. The parents are selected from breeder lines of chickens that grow rapidly and yield a large amount of breast meat. They are quite different in appearance from hens that are used only to produce eggs. **Layers** that produce eggs for consumption are selected for their egg-laying capacity—not for the amount of muscle they produce. Most broilers are **hybrid** birds. This means that they are the result of the mating different breeds and lines of chickens to produce the type of meaty birds desired (**Figure 6–4**). Hens that lay eggs for hatching are bred either naturally or artificially. The resulting crossbred animals are generally healthier and grow faster than purebred animals. This is called **heterosis** or **hybrid vigor**.

Almost all of the broilers produced are white. Birds that are dark in color have spots of color or **pigmentation** where the feathers were removed after slaughter. These spots do not lower the

FIGURE 6–4 The bird on the left is the muscular type bred for meat production; the bird on the right is of the type that is bred for producing eggs.

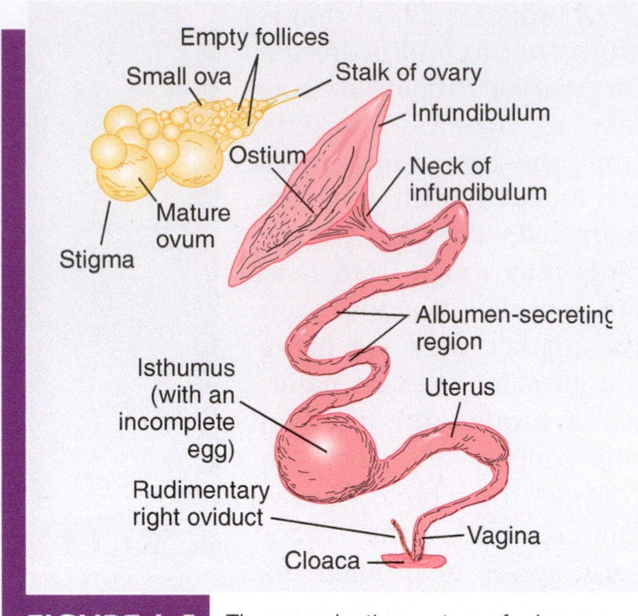

FIGURE 6–5 The reproductive system of a hen.

quality of the meat in any way, but consumers are reluctant to buy chicken with spots on the skin.

Egg Production

Eggs produced by poultry serve the same purpose as eggs produced by other agricultural animals—reproduction. Unlike the eggs of most mammals, the eggs of poultry are produced in the body of the female and then expelled from the body. Development of the **embryo** takes place outside of the mother's body. The eggs of most mammals are microscopic in size and unprotected, whereas a chicken's egg can weigh several ounces and is encased inside a hard shell.

The egg production process begins as it does in mammals, with the release of the **ovum** from the **ovary**. The follicle in the ovary ruptures and the ovum is released. The ovum falls into a funnel-shaped structure called the **infundibulum** that surrounds the ovum and holds it for about 20 minutes.

If the hen has mated or has been artificially inseminated, the ovum will be fertilized here (**Figure 6–5**). The egg then moves into a tubelike tract called the **magnum**. **Cells** in the magnum secrete a substance called **mucin** that develops into the **albumen** (the white portion) of the egg. This substance is high in protein content and serves as nourishment for the developing ovum. The albumen (the white portion of the egg) surrounds the ovum, which is the **yolk**, the yellow part of the egg. Here a ropelike substance, called **chalazae**, is formed. Later, as the egg is formed, these ropelike structures will serve to hold the yolk in position in the center of the egg.

In a process that takes about 3 hours, the egg moves through the magnum (which is more than a foot long) and enters the **isthmus**. Here, mineral salts are added and the inner and outer shell membranes are formed. These membranes lie just inside the hard shell of the completed egg. After remaining in the isthmus for about 1½ hours, the egg moves into the **uterus**, where the shell is formed around the egg (**Figure 6–6**). Because of this process, the uterus is sometimes called the **shell gland**. During this process, the shell may acquire brown or other colored pigments, depending on the breed of the hen. The shell, composed mainly of calcium and protein, serves to protect the embryo until the **incubation** process is complete and the chick hatches. Also, more

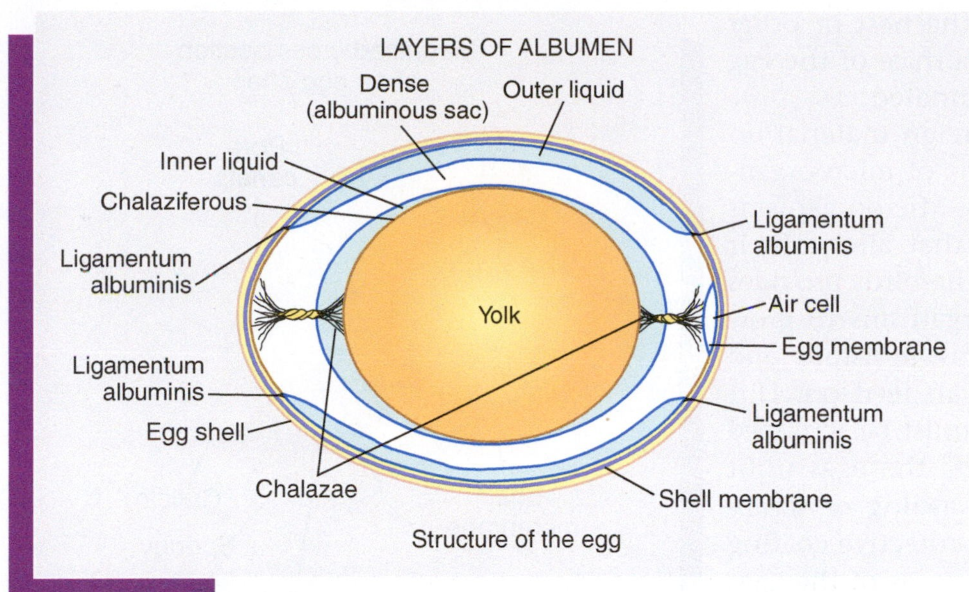

FIGURE 6–6 A cross section of an egg.

water and minerals are passed into the egg white to fill out the egg, and a waxy substance is secreted that coats and seals the pores in the shell of the egg. In 18 to 22 hours, the egg shell is completed, hardened, and moves through the vagina and out of the hen's body.

Eggs are oval-shaped, with one end smaller than the other end. In the process of development, the small end of the egg goes through the tract first. However, before the egg is laid, the ends are reversed and the large end emerges first. This turning helps the muscles of the tract to open and expel the egg. The process of laying is activated by a hormone called **oxytocin**. It causes the uterus to contract, forcing the egg through the vagina and out of the hen's body through an opening called the **cloaca**. In about a half hour, the process begins all over again with the release of a new ovum from the ovary.

Hatching Eggs

Unlike the production of eggs for consumption, eggs produced for hatching are laid by the hens in nest boxes. A lot of scientific research has gone into the design of the nest boxes to provide the type of environment the hens want. In making the hens most comfortable, the producers ensure more efficiency in the laying operation. The nests consist of boxes that have concave bottoms and are filled with bedding or artificial turf to make a comfortable nest for the hen (**Figure 6–7**).

The hens naturally prefer nests that are enclosed because this gives them a feeling of security. This behavior perhaps is passed down to the hens from their ancestors that lived in the wild and had to protect the nests from predators. Hens also prefer nests that are the gray color of galvanized metal. At one time, the nests were raised off the ground and had slatted floors that allowed manure to pass through to the ground, where it could be removed. However, in modern breeder hen operations, mechanical nests are now used. The eggs roll onto a conveyor belt, and an operator collects the eggs off the belt.

Hatching eggs must be kept as clean as possible. Any bacteria or other contamination can cause disease in the newly hatched chicks. As the egg comes from the hen, the surface of the egg is quite clean, but as the egg comes into

FIGURE 6–7 The design of a nest box is based on research that determined what the hens prefer. Note the nest boxes along the wall. *© terekhov igor/Shutterstock.com.*

contact with the surface of the nest or other areas within the house, the surface of the egg can become dirty and contaminated.

Even a small speck of foreign material on the shell can contain millions of microorganisms that can cause problems. Microorganisms must have an environment that allows their growth. Fecal material from the birds provides an excellent place for the organisms to grow because the manure contains moisture and material that the microbes can feed on. Dirt or any foreign material that must be scrubbed from the eggs usually renders the eggs unfit for hatching (**Figure 6–8**). Washing or scrubbing the eggs removes their protective coating and presses the dirt into the pores of the eggs (**Figure 6–9**). Therefore, hatching eggs are not allowed to become wet.

Before they leave the farm where they are produced, the hatching eggs are sorted to

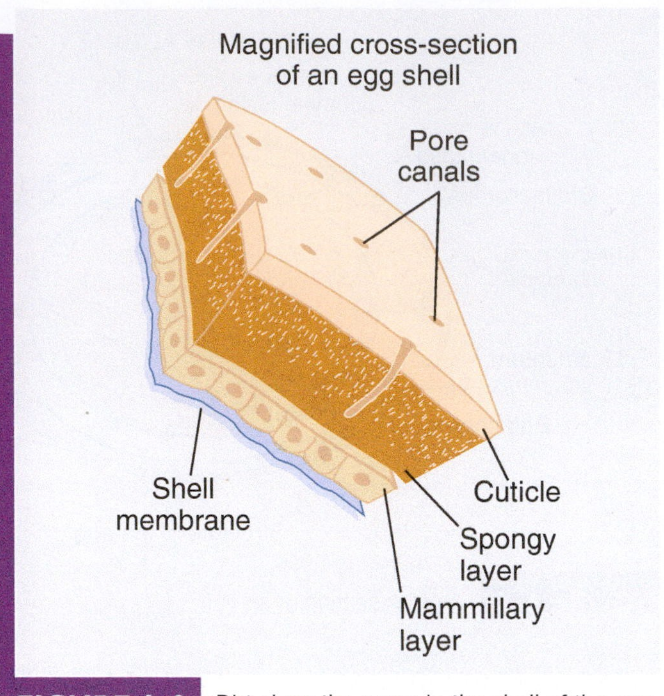

FIGURE 6–9 Dirt clogs the pores in the shell of the egg.

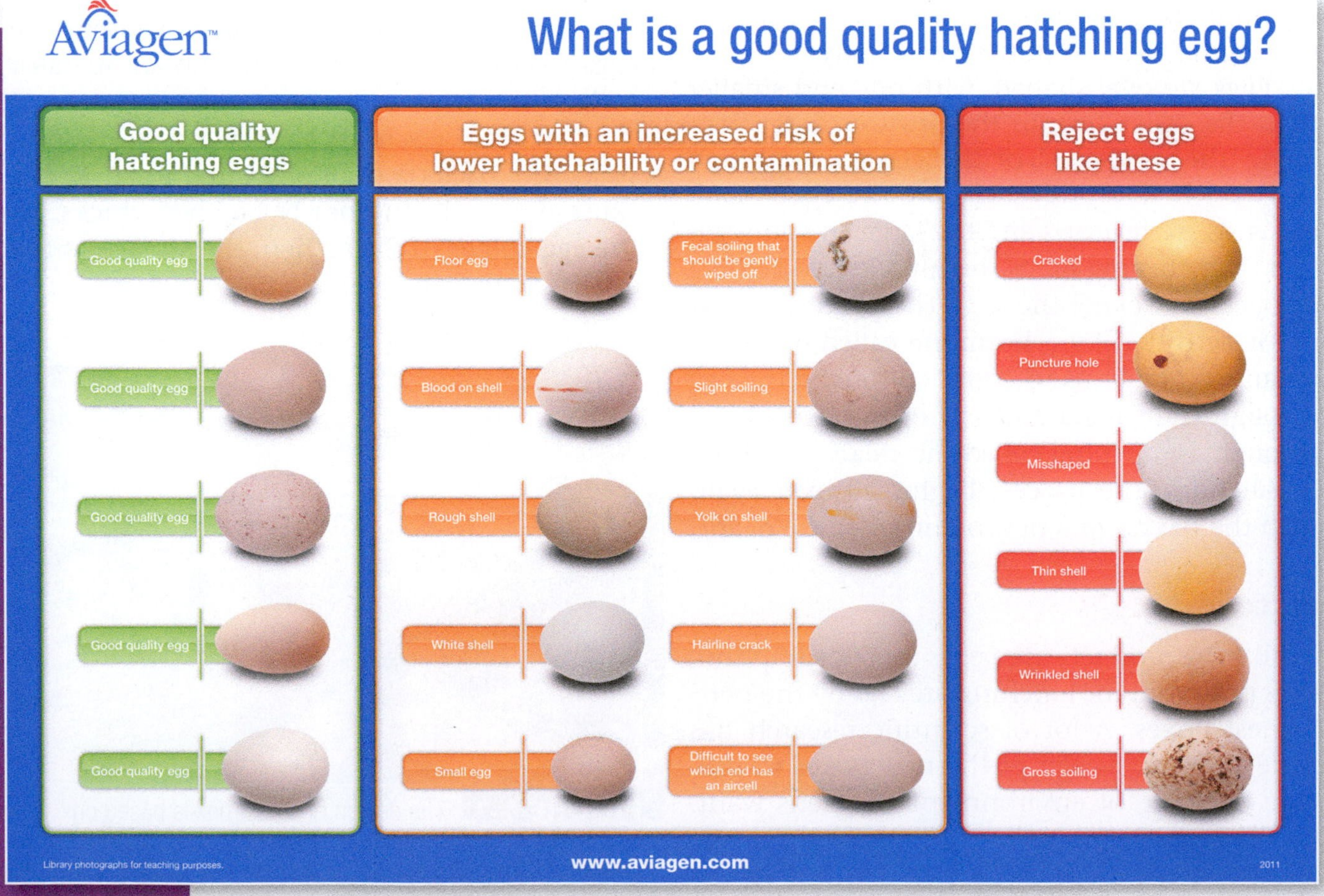

FIGURE 6–8 Hatching eggs must be free of any defects. *Courtesy of Aviagen*

remove dirty, undersized, oversized, misshapen, cracked, or defective eggs. Then they are fumigated to kill harmful organisms on the surface of the eggs by a process that is precisely regulated to prevent harm to the eggs. The eggs are never allowed to become chilled and are stored at 70°–80° F until they are placed in the hatchery.

The eggs are carefully placed on racks that fit into carts designed to prevent damage to the eggs during transportation. The carts are then loaded into trucks for transportation to the hatcher (**Figure 6–10**). The carts, trays, and trucks are all kept clean and sanitized to prevent contamination of the eggs.

At the hatchery, the eggs are removed from the carts and placed in the incubator (**Figure 6–11**). The temperature of the eggs is never allowed to be lowered and is gradually increased to prevent the eggs from sweating (**Figure 6–12**). **Sweating**, the condensing of water vapor on the surface of the eggs, occurs when the temperature of the eggs is raised too rapidly. Cold air holds more moisture than does warm air. As the temperature of the air is raised, the water vapor in the air begins to condense. If moisture is allowed to collect on the surface of the egg, it creates an environment that allows bacteria to grow and thrive. In a warm, moist environment, a single bacterium can reproduce into two bacteria every 20 minutes. If this is allowed to continue, the single bacterium can become 16 million bacteria in only 8 hours.

FIGURE 6–10 Eggs are loaded onto racks or trays for transportation to the hatchery. *© Alf Ribeiro/Shutterstock.com.*

Relative humidity is another factor that causes eggs to sweat. Relative humidity is the amount of moisture in the air relative to the amount of moisture possible at that temperature. The temperature and relative humidity are carefully controlled in the hatchery (**Figure 6–13**).

Embryo Development

The embryo begins to develop before the egg is laid. As noted earlier, **fertilization** occurs early in the formation of the completed egg. In contrast to most agricultural animals, **sperm** can remain viable in the hen's body for as long as 32 days, but fertility is highest if insemination occurs at least once a week. Sperm are stored in pockets inside the oviducts called **sperm nests**. The yolk portion of the egg contains a spot

FIGURE 6–11 At the hatchery, the eggs are removed from the carts and placed in the incubator on trays that periodically rotate from side to side. *© plumdesign/Shutterstock.com.*

FIGURE 6–12 Hatching eggs must be held at a constant temperature. They are monitored closely. *Courtesy of Aviagen.*

FIGURE 6–13 In the hatchery, relative humidity and temperature are carefully monitored. *Courtesy of Aviagen.*

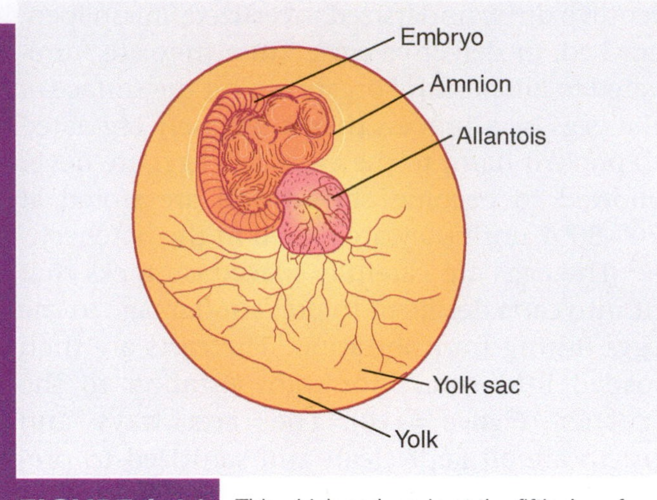

FIGURE 6–14 This chick embryo is at the fifth day of development.

called the **germinal disk**, which contains the genetic material from the female. The sperm fertilizes the egg within this germinal disk, and the embryo begins to develop. If a newly laid egg is broken open, the germinal disk is visible to the naked eye and appears as a white spot in the yolk. After the egg is laid, the embryo remains dormant until it is stimulated by heat to grow. In nature, this heat is generated by the hen's body as she sits on the nest, but in modern operations the eggs are heated by artificial means in commercial incubators.

Within 48 hours after incubation begins, the embryo has developed a circulatory system that sustains life by carrying nourishment from the yolk to the embryo. At the end of the third day of incubation, three layers of membranes have developed. The first—the allantois—serves as a place to store the waste generated by the embryo. This membrane later merges with the second membrane—the chorion—to form a type of respiratory system until these organs are developed in the embryo. The third membrane—the amnion—is filled with the fluid that surrounds the embryo and protects the developing embryo from shock (**Figure 6–14**). To prevent the embryo from sticking to the outer membranes of the egg, the eggs must be turned several times a day. In nature, the hen turns the eggs in the nest. In the incubator, the eggs are either turned automatically by a time-controlled turning device or are turned by hand.

At the end of the first week of incubation, the embryo is recognizable as a chick embryo. Most of the chick's systems, such as the lungs, nervous system, muscles, and sensory systems, are developed. By the end of the second week, the chick is covered with down. At the end of three weeks, the chick is fully developed. When the first chicks in the incubator begin to hatch, they make clicking sounds as they break the egg shell. This signals the other chicks and stimulates them to begin hatching. This behavior is a carryover from the days in the wild when all the chicks had to hatch at the same time for survival. The producer can provide slow, clicking sounds mechanically to accelerate the hatching process.

In commercial hatcheries the eggs are incubated in two separate rooms: the setting room and the hatching room. The eggs are placed in the setting room incubator and are monitored closely for temperature and relative humidity. The eggs are turned every day to ensure that a high percentage of eggs hatch. The eggs remain in the setting room incubator until one to two days before the eggs are ready to hatch. They are then placed in the hatching room, where the temperature is lowered slightly and the chicks hatch into chick holding trays (**Figure 6–15**).

FIGURE 6–15 One to two days before hatching, the eggs are placed in the hatching room. © branislavpudar/Shutterstock.com.

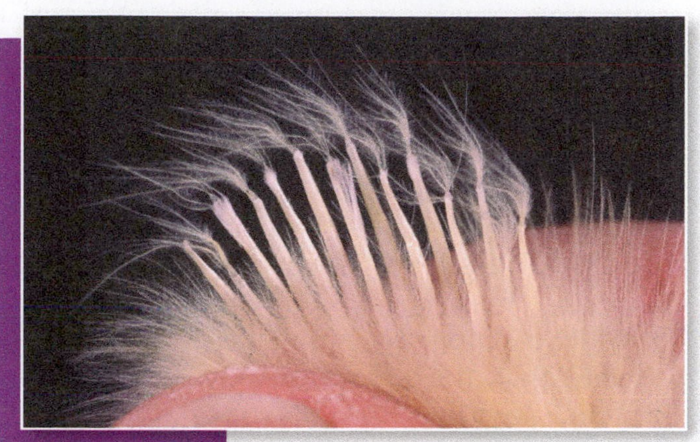

FIGURE 6–17 The sex of a chick is determined by examining the feathers; this chick is a male. Courtesy of Joe Mauldin/Poultry Science Department, University of Georgia.

When the Chicks Hatch

The chicks are removed from the incubator and are cleaned, dried, and placed in a warm, dry environment (**Figure 6–16**). The chicks are sexed by examining their feathers. If the hatchery is producing chicks that are to become laying hens, the females have to be separated from the males (**Figure 6–17**).

At one day of age, the chicks are vaccinated and their beaks are trimmed to help prevent cannibalism. The procedure, which is done with an electric knife, causes the chicks no harm because only a small portion of their beaks is removed. A recent development is the use of a chemical that is applied to the tip of the chicks' beaks. This process safely and painlessly dissolved the sharp end of the beak. The day-old chicks then are placed in ventilated cardboard or plastic boxes and are transported to the broiler houses (**Figure 6–18**).

At the Broiler House

Prior to arrival of the chicks, the producer has cleaned, disinfected, and placed clean litter in the house. Litter is the material placed on the floor to absorb moisture and to keep the birds clean and dry. This material usually consists of shavings or sawdust obtained from a sawmill.

FIGURE 6–16 Chicks are removed from the hatchery trays and placed in a warm, dry environment. © franz12/ Shutterstock.com.

FIGURE 6–18 Chicks are transported to the broiler houses in cardboard or plastic boxes. © Lano Lan/Shutterstock.com.

Brooders are used to keep the chicks warm during the first days in the house. They are usually suspended from the ceiling and can be raised or lowered, depending on the temperature and the size of the chicks (**Figure 6–19**).

FIGURE 6–19 Young chicks are kept warm by brooders that are raised higher as the chicks grow. The brooders are the round, cylinder-shaped objects suspended from the ceiling. *© terekhov igor/Shutterstock.com.*

Water is supplied by suspended waterers that the chicks quickly learn to use. They peck the nipple on the bottom of the waterer and obtain as much water as they need. Feed is given to the baby chicks by hand when they are small. But as they get older, they are fed through automatic feeders. The feed is brought to the birds by means of an auger in a tube that fills the feeders throughout the day and night to ensure that the birds always have plenty of feed, (**Figure 6–20**). Both the waterers and feeders are raised as the birds get larger. Every day the equipment and birds are continually checked to ensure that the equipment is functioning properly and that the birds are doing well.

The birds commonly are kept in the broiler house from 6 to 7 weeks. At this time, they weigh about 5 pounds and are ready for market. They usually are caught at night when they are less active and are put in cages and loaded on a truck for transportation to the processing plant. The producer then begins to get the house ready for the next batch of chicks. New litter consisting of sawdust or shavings is placed on the floor to absorb moisture from the manure (**Figure 6–21**). The litter must be removed periodically because it becomes filled with manure, and the disposal of the litter can be a large problem. A broiler house that holds 20,000 broilers produces about 180 tons of litter per year. Because the manure in the litter has a high concentration of nitrogen and other elements necessary for plant growth, the litter is a valuable source of fertilizer (**Figure 6–22**).

FIGURE 6–20 The broilers are fed by automatic feeders that are filled by augers or chains that carry the food through the tube. *© terekhov igor/Shutterstock.com.*

FIGURE 6–21 When the broilers are sent to market, fresh litter is spread on the floor. *© terekhov igor/Shutterstock.com.*

FIGURE 6–22 The manure in broiler litter is a valuable source of fertilizer. *© Valerii Iavtushenko/Shutterstock.com.*

At the Processing Plant

When the birds reach the processing plant, they are slaughtered and prepared for market. Some plants process the chickens to be sold whole, and some cut the broilers into parts, such as breasts, thighs, and drumsticks (**Figure 6–23**). Other plants process the chicken further into more complex prepared food such as chicken franks, chicken bologna, or complete frozen dinners.

THE LAYER INDUSTRY

The per-capita egg consumption in the United States has decreased sharply over the past 30 years. The consumption of whole eggs in the shell has greatly decreased while the consumption of egg products has increased. This reflects changing dietary habits and consumer preference for processed foods. Even with the decrease in demand, the layer industry in this country is still quite large.

Cage Operations

More than 90 percent of eggs are produced by layers in cages. The hens live in the cages in groups of 2 to 12 hens, depending on the operation. The most common grouping is that of four hens per cage. The birds used for **cage operations** have been developed to tolerate the confinement operation and to produce eggs efficiently. As mentioned earlier, the type of hen used to produce eggs is quite different from the hens used to produce meat. The former are smaller and are much less muscular because the smaller, trimmer birds use a large portion of their **metabolism** in producing eggs instead of developing muscles and body size. Some layers produce brown eggs, and some produce white eggs. The vast majority of the eggs sold in the United States are white eggs because consumers simply prefer to buy white eggs.

Modern cage operations are scientifically designed to provide the hens with adequate

FIGURE 6–23 Chickens may be cut up or packaged whole at the processing plant. *© Alf Ribeiro/Shutterstock.com.*

room, proper ventilation, correct temperature, plenty of food, and fresh water. In addition, lighting is carefully controlled. Hens naturally lay eggs in the spring and summer months. In the wild, the chicks would have a much greater chance of survival if the eggs were laid and hatched during the warmer months. As spring approaches, the days have more hours of light and fewer hours of dark. The longer periods of light stimulate the hen's hormonal system into producing eggs. In a commercial cage operation, the lighting is carefully controlled to allow 14 to 15 hours of light every day. As the young hens, called **pullets**, reach maturity, the light is increased gradually until they have 15 hours of light per day. This causes the hens to be in full production.

As the hens get older, they lay fewer eggs. Some producers sell the hens when production decreases; others submit the hens to a process called **molting**. In this process, feed is withheld, and no artificial light is used for a period of time. The lighting then is increased until reaching the normal 15 hours of light. The hens lose their old feathers, grow new ones, and seem to regain some of their youthful vitality.

The hens are fed by means of an automated conveyor that carries the feed directly in front of them (**Figure 6–24**). Water is supplied through a narrow, free-running trough or a nipple waterer that the hens learn to peck to obtain water. As the eggs are laid, they roll onto a conveyor that periodically moves the eggs to a collection point, where a worker gathers them and places them in flats (**Figure 6–25**). The dirty eggs are separated out, and the clean eggs are refrigerated.

FIGURE 6–24 Hens are fed by means of an automated conveyor that carries the feed directly in front of them. *© Enrico Jose/Shutterstock.com.*

FIGURE 6–25 When laid, the eggs roll onto a conveyor. *© Rony Zmiri. Image from BigStockPhoto.com.*

At the processing and packing plant, the eggs receive a thin coat of light mineral oil to prevent carbon dioxide from escaping from within the egg. The eggs are graded according to shape and size and are checked for cracks and interior spots in a process called **candling**. The eggs are passed over an intense light in a dark room, where any blood spots or cracks in the shell will show up in the light (**Figure 6–26**). The eggs are packaged and sent to the retail market or are sent to a processing plant, where they are broken and processed.

THE TURKEY INDUSTRY

The production of turkeys for meat is a rapidly expanding segment of animal agriculture. The sale of turkeys is second only to that of chicken in the overall sale of poultry meat. Each year, Americans consume about 17 pounds of turkey per-capita (**Figure 6–27**). Turkey represents a high-quality, low-cost, nutritious source of food protein. Although one-third of all turkey is sold in the weeks surrounding the Thanksgiving and Christmas holidays, the trend is toward steadier year-round sales.

The turkeys that are produced in the United States are the descendants of wild

FIGURE 6–26 The eggs pass over an intense light to reveal blood spots in the eggs or cracks in the shell. *© iStockphoto/AarStudio.*

FIGURE 6–27 Americans buy more than 4.5 billion pounds of ready-to-cook turkey each year. *© Monkey Business Images/Shutterstock.com.*

turkeys native to the Americas. The wild turkey is a bronze-colored bird that lacks the broad breast and overall **muscling** of the commercial turkey. Just as in broilers, most consumers demand that the turkeys they buy be white. As mentioned in the discussion of broilers, a colored bird will have specks of pigmentation left in the skin when the feathers are removed, so the carcasses of white birds look a lot cleaner. The modern white turkey is the result of a mutation or accident of heredity that left out the gene for the feather and skin pigmentation (**Figure 6–28**).

From the mutated white turkeys, a heavily muscled, broad-breasted bird was developed. A problem with this highly developed bird, however, is that it is not an efficient breeder. The physical act of mating is difficult because of the heavy muscling, and the birds seem more reluctant to breed. This problem has been solved through artificial insemination (**Figure 6–29**). Semen is collected from the males, and an extender is used to allow the sperm to be viable

FIGURE 6–28 The modern white turkey is the result of a mutation. *© Paul Wishart/Shutterstock.com.*

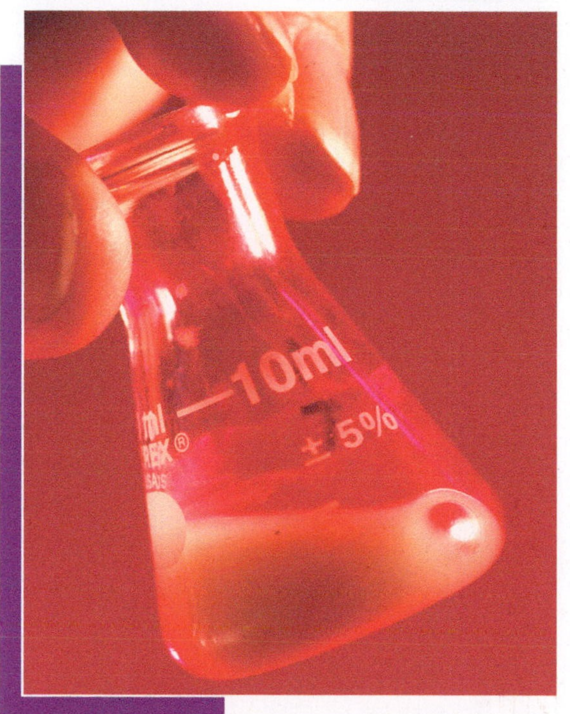

FIGURE 6–29 Turkeys are bred through artificial insemination. This is a vial of turkey semen with an extender added. *Source: USDA, Agricultural Research Service (ARS). Photo by Keith Weller.*

for about 24 hours, when it is placed in the female's reproductive tract.

Most of the turkeys in the United States are grown in the western part of the northern Central region, the south Atlantic region, and the Pacific Region (**Figure 6–30**). Turkeys seem better able to tolerate cold weather than hot weather. Most are produced by small operations of 30,000 or fewer birds.

The two ways of growing turkeys are in confinement houses and on open ranges. Confinement houses offer the advantages of environmental control of temperature and humidity, and most turkeys are raised in confinement. Although the open range offers the advantage of being less expensive, few turkeys are raised this way. Turkeys on the range can stay outdoors completely, or they are provided with housing where the birds can get shelter whenever they want (**Figure 6–31**).

FIGURE 6–31 Many turkeys are raised on the range. *Source: USDA, Agricultural Research Service (ARS). Photo by Scott Bauer.*

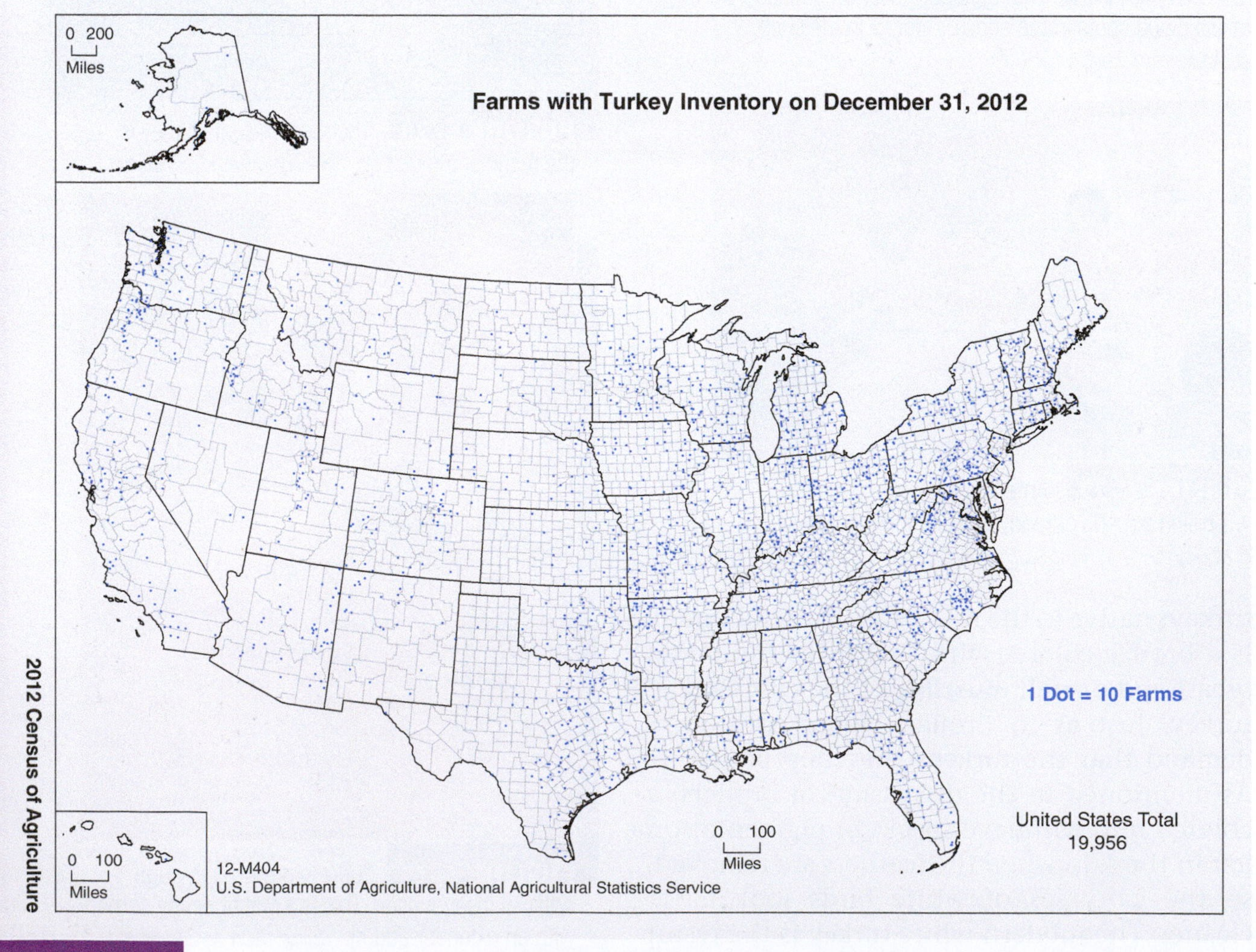

FIGURE 6–30 The turkey production industry in the United States. *Source: USDA, Census of Agriculture*

Producers usually move the range every three to four years to help reduce problems with disease and parasites.

OTHER POULTRY

In some parts of the world, poultry such as ducks and geese make up a major portion of the total poultry output. In China and Southeast Asia, ducks and geese are a main part of the overall diet. These birds are hardier than chickens because they are not as susceptible to diseases and can forage better for themselves. Also, the feathers are used to make bedding and other goods. In the United States, most ducks and geese are raised in small flocks by hobbyists or by part-time producers (**Figure 6–32**). Most of the meat sold goes to the restaurant trade or gourmet food market.

About the only other poultry production of any significance is that of growing quail and pheasant. Both of these birds are grown for the restaurant trade and the gourmet food market (**Figure 6–33**). In addition, they are raised for restocking wildlife areas. Each year, thousands of quail and pheasants are released in the wild to provide birds for hunting. This helps replenish the areas with game that is difficult to produce in the wild.

FIGURE 6–32 In the United States, most ducks and geese are raised by hobbyists.
© Frolova_Elena/Shutterstock.com.

FIGURE 6–33 Quail are grown for the restaurant trade. *© Sergey Fatin/Shutterstock.com.*

PUTTING IT INTO PRACTICE

FFA Poultry Judging Career Development Event

Students with an interest in the poultry industry can participate in the FFA Poultry Judging Event. Each of four team members judge and give reasons for classes of production hens, pullets, ready-to-cook broilers, chicken patties, interior eggs, and exterior eggs. Students also identify 10 broiler parts and place four cartons of eggs, with an explanation of how they placed them. At the end of the event, each FFA member answer a 30-question test relating to management practices used in the poultry industry. After all teams complete each section of the event, individual scores were calculated. The top three individual scores for each team are added to determine the team score and placement.

The FFA Poultry Judging Career Development Event provides students with an opportunity to strengthen their knowledge about the poultry industry. *Courtesy of the National FFA Organization*

The Poultry Production Proficiency is one of many agricultural proficiency awards presented at the FFA's annual banquet. Applicants applying for the Poultry Production Proficiency Award must be involved in activities and enterprises requiring the application of classroom lessons to the profitable management of a poultry operation. All FFA members who had a supervised agricultural experience in the area of poultry competed for this award. Maintenance of a complete set of records, including income and expenses, inventory of poultry and related equipment, and knowledge of the poultry industry helps give students the winning edge.

SUMMARY

The poultry industry is one of the most dynamic industries in agriculture. Few areas have come close to the progress made in the growing of poultry and poultry products. Most of this progress is directly due to the discoveries found through scientific research. The amount of poultry eaten by humans is on the increase. It represents a relatively inexpensive, healthy alternative to other foods. This industry has made a gigantic contribution to our food supply system.

CHAPTER REVIEW

Review Questions

1. Why is poultry production so popular around the world?
2. What is a vertically integrated company?
3. How is cannibalism reduced in the broiler industry?
4. Why are white broilers preferred over colored broilers?
5. In what ways are the eggs of poultry different from the eggs of mammals?
6. List the parts of the egg-producing tract of the hen.
7. Describe the type of nests that laying hens prefer.
8. Explain why dirty eggs are not used for hatching.
9. Why is it important that hatching eggs not get wet?
10. Describe the characteristics of a chick embryo at the end of each week of incubation.
11. Indicate two uses for chicken litter.
12. What effect does light have on layers?
13. What is meant by the molting process?
14. Why are most domesticated turkeys produced through artificial insemination?
15. What are two methods used to produce turkeys?

Student Learning Activities

1. Visit the meat counter in a large grocery store, and make a list of all the different ways in which poultry meat is offered for sale (e.g., whole fryers, cut-up fryers, breasts only, thighs only, processed, etc.). Interview the manager and determine which method of packaging and processing sells the best.
2. Break open some eggs, and locate the germinal disk. Also take note of the separation of the albumen and the yolk.
3. Using a small incubator, place eggs in the incubator for one, two, and three weeks. At the end of each day during the first week, open an egg and note the development. Write a detailed account of the differences from one day to the next.
4. During a two-week period, keep track of the type and amount of meat consumed by your family. Determine the percentage of poultry meat in the total amount. Discuss with your parents the economics of buying poultry as compared with other types of meat. Report to the class.

CHAPTER 7

The Sheep Industry

STUDENT OBJECTIVES IN BASIC SCIENCE

As a result of studying this chapter, you should be able to

- explain why sheep have been important to humans throughout history.
- list the characteristics of sheep that allow them to be good domesticated animals.
- discuss the controversy over predators of sheep.
- explain the characteristics of wool that make it useful to humans.

STUDENT OBJECTIVES IN AGRICULTURAL SCIENCE

As a result of studying this chapter, you should be able to

- discuss the importance of lamb and mutton in the diets of Americans.
- explain the importance of the sheep industry to our economy.
- list the ways in which humans use wool.
- list the uses for by-products of wool.
- discuss how wool is made into clothing.

KEY TERMS

lamb
mutton
ecological balance
cuticle
cortex
felting
crimp
suint
yolk
grease wool
scouring
scoured wool
carding
combing
pelts
mohair

NATIONAL AFNR STANDARD

AS.01.01
Evaluate the development and implications of animal origin, domestication, and distribution on production practices and the environment.

AS.01.02.02.a
Research and examine marketing methods for animal products and services.

HUMANS HAVE RAISED SHEEP for at least the past 10,000 years (**Figure 7–1**). During this time, sheep have supplied food, shelter, and clothing for human use. The meat from the carcasses and milk from the females have provided a protein-rich diet in even the poorest of societies. Because sheep can live and thrive in areas where other agricultural animals cannot, sheep have played an important role in feeding people all over the world. These animals eat plants ranging from grasses and legumes to brush. Often, they may even eat plants that are toxic to other animals.

Compared with other agricultural animals, sheep are unique in that they are very docile and easy to handle. This may be because they were one of the first animals that were domesticated and have been raised and bred for human use continuously for the past 10,000 years. In fact, they are so tame that they are largely defenseless against predators. (**Figure 7–2**) illustrates the parts of a sheep.

Compared to beef and pork, Americans eat relatively little lamb and mutton. **Lamb** refers to meat from sheep younger than a year; **mutton** is meat from sheep older than a year. In many parts of the world, lamb and mutton are a basic part of people's diet. The per-capita consumption of mutton and lamb in the United States is only about 2½ pounds. Of this consumption, about 95 percent is lamb and only about 5 percent is mutton. Americans as a whole seem to have never developed a taste for the stronger flavored mutton.

However, in certain areas of the country, lamb is a favored food. The large cities along the Eastern seaboard account for almost half of

FIGURE 7–1 Humans have raised sheep for thousands of years. *© Morphart Creation/Shutterstock.com.*

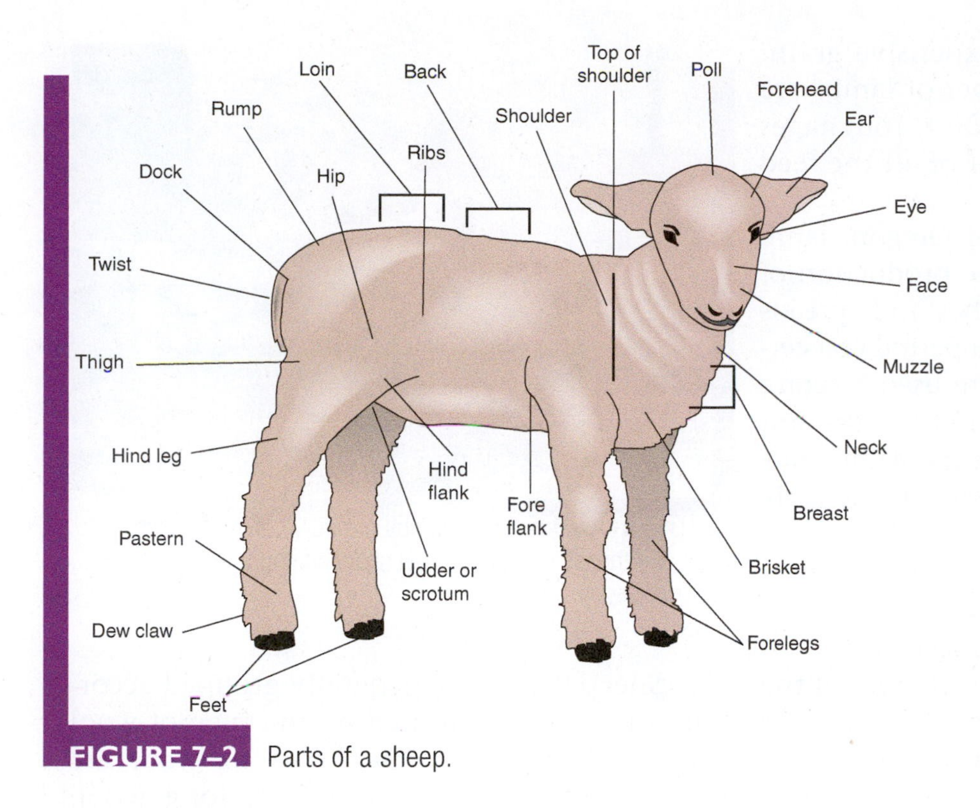

FIGURE 7–2 Parts of a sheep.

the market for lamb and mutton in this country. This presents a problem because most of the lambs are produced west of the Mississippi River (**Figure 7–3**). The leading states in sheep production are Texas, California, Wyoming, Colorado, South Dakota, Montana, New Mexico, Utah, and Oregon. Since modern refrigerated trucks and railroad cars have alleviated the spoilage problem, most of the remaining concern is the economics of transportation.

One advantage of producing lambs for market is that good-quality lambs can be produced on grass and do

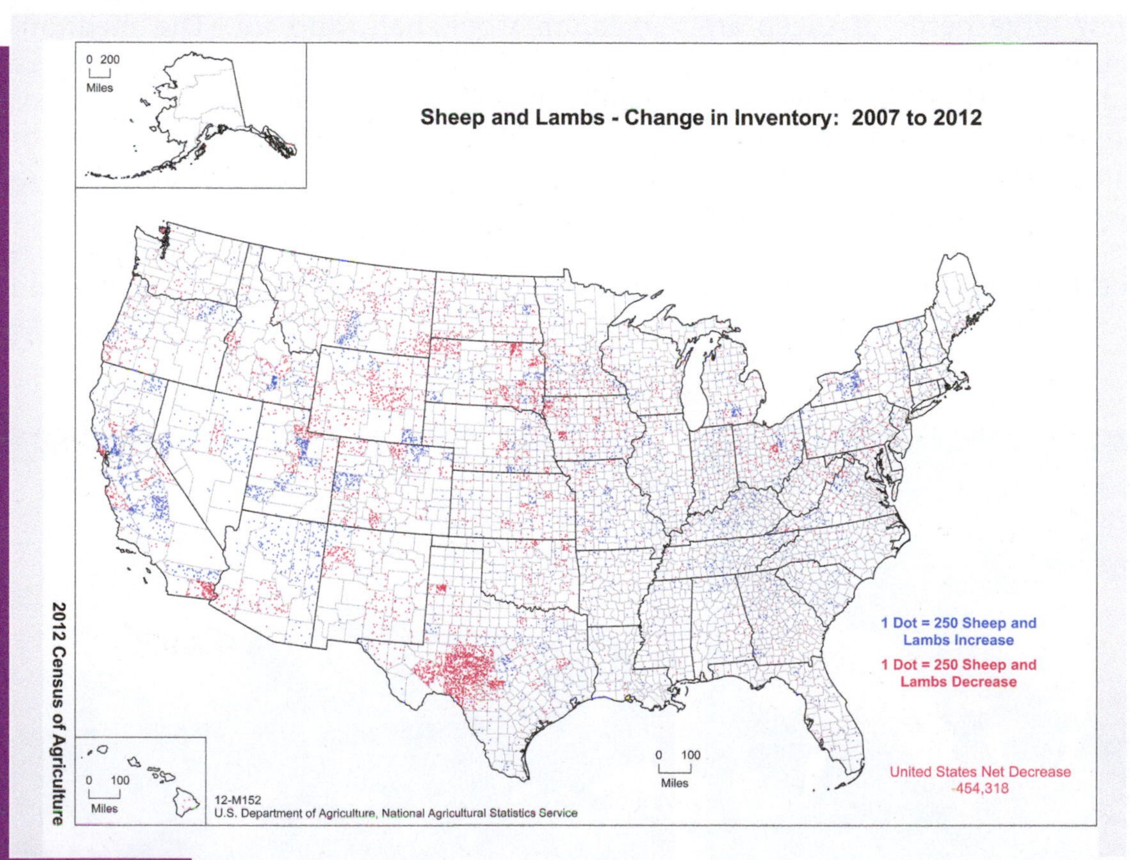

FIGURE 7–3 Most of the sheep produced in the United States are grown in the western states. *Source: USDA, Census of Agriculture*

not have to be fed a lot of expensive grain. Although an increasing number of lambs are being fed on grain in the feedlot, roughages still make up about 90 percent of all the feed consumed by sheep.

In the Willamette Valley of Oregon, lamb production fits in well with the production of rye grass seed. Rye grass bunches and spreads better if it is closely grazed for a period at a certain time of the year. Lambs are used to graze the rich green grass down, and in turn, the animals are fattened by the nutrients in the grass (**Figure 7–4**). Most lambs are born in the early spring and are called spring lambs.

Sheep also can make better use of lower quality forage than can cattle. For this reason, sheep can be grazed successfully on poorer grazing lands of the desert areas of the West. In addition, the drier climate reduces the parasites and diseases associated with sheep grown in the more humid areas. For example, very few large herds of sheep are grown in the Southeast because of problems with heat and humidity. Hot, humid weather makes a good environment for parasites and disease organisms such as foot rot; nevertheless, many lambs now are being raised in the Southeast as the National FFA Organization of America (FFA) and 4-H show animals (**Figure 7–5**).

FIGURE 7–5 Many lambs are now being raised in the Southeast as FFA and 4-H show animals. *© iStockphoto/BrandyTaylor.*

Sheep breeds are generally grouped according to use, determined by the type of wool the animals grow. The types of wool are broadly classified as fine wool, long wool, medium wool, hair, and fur. The medium-wool breeds are used most often to produce lambs for slaughter. Medium-wool breeds commonly used to produce slaughter lambs are Suffolk (**Figure 7–6**), Hampshire, Southdown, and Dorset.

A large problem facing sheep producers is that of predators. Animals such as coyotes and both wild and domesticated dogs kill large

FIGURE 7–4 Sheep are used to graze down grass raised for seed. *© 1000 Words/Shutterstock.com.*

FIGURE 7–6 The Suffolk is a medium wool breed raised primarily for meat. *© Sylvie Lebchek/Shutterstock.com.*

numbers of sheep each year. Although precise data on losses from predators are difficult to obtain, research indicates that each year, around 4 to 8 percent of the lambs and 1.5 to 2.5 percent of the ewes in the western 17 states are lost annually to predators. Some producers have reported losing as much as 29 percent of their lambs to predators in a year.

In the past, sheep producers have used measures such as trapping and poisoning to rid the area of predators. Today, such measures are closely regulated because of the potential damage to the **ecological balance** of an area. The ecological balance refers to the balance nature has regarding the number of living things in a given area. Too many or too few of a given type of animal can upset the balance of nature. If too many animals that do not prey on sheep are killed, the balance of nature can be upset.

Government programs now are in effect to help producers with losses incurred by wild predators. Guard dogs and improved fencing also have helped with the problem, but difficulties remain for sheepherders who raise animals on open government lands. The amount of control necessary to prevent predation is a subject of controversy between the environmentalists and the sheep producers. In remote areas, reintroduction of native animals such as the timber wolf and the mountain lion has prompted protest from producers who are concerned about further loss to predators.

THE WOOL INDUSTRY

Wool is made of the fibers from the hair coat of sheep and has been used as a material for making clothing for thousands of years (**Figure 7–7**). The spinning of wool is probably one of the oldest industries in which people have been engaged. In many places throughout the world, archaeologists have uncovered clothing made from wool that is well over 10,000 years old. Records of the ancient Greeks, Romans, Egyptians, and Hebrews indicate that they used wool for clothing. Several accounts of the production and use of sheep are written in the Bible and other ancient writings.

FIGURE 7–7 Wool is the fiber from the hair coat of sheep. *© Air Images/Shutterstock.com.*

Throughout history, wool is mentioned as being the standard material from which cloth was made. The wool was grown by family-owned flocks of sheep and harvested periodically by shearing the animals. The thread was made by spinning the wool on a hand-operated machine. The thread then was woven into cloth on a hand loom that was operated in the home. This cloth, called homespun, was rather coarse and plain. Almost all of the poorer people wore clothing made from this cloth.

Until the early- to mid-nineteenth century, almost all clothing was made from wool. During the first half of the 1800s, the cotton industry began to flourish in the southern portion of the United States, and it provided the world with a cheap alternative to wool clothing. In modern times, synthetic fibers from the petroleum industry have competed with both cotton and wool in the manufacture of textiles.

Even though wool is one of the oldest materials used for clothing, it is still popular today. Many of the finest, most expensive garments worn today are made from wool, as are fine carpets and tapestries. Countries in the Middle East have always been famous for their beautiful woolen carpets.

Wool has certain characteristics that make it desirable over cotton and synthetics as a fabric. Wool can be worn in the winter or the summer. Because it can absorb up to 30 percent of its weight in moisture and still feel dry, the fabric makes a good insulator from both heat and cold. Wool is also very strong; a fiber of wool is stronger than a fiber of steel of the same size. This characteristic makes wool fabric durable, a desirable trait in the manufacture of both clothing and carpets. In the manufacture of woolens, the material takes and holds dye very well, and as a result, many beautiful patterns and colors can be made from wool.

At one time, soldiers' uniforms were made from wool because of its ability to shed water and provide insulation from heat and cold, and also its resistance to burning. During the Civil War, the Confederate and Union soldiers both wore wool clothing (**Figure 7–8**). The weapons used by the troops were mainly black powder rifles and artillery, and this type of gunpowder often propelled bits of burning embers when the guns were fired. If an ember landed on the soldier's clothing, he might not be aware of it for a while. A cotton uniform would blaze up in a short time and cause a painful burn, whereas a wool uniform would only smolder and not burn. For safety reasons, then, even the soldiers of the South, who had a ready supply of the cheaper cotton, preferred wool for their uniforms.

Wool fibers are made up of two distinct layers of cells: the **cuticle** on the outside and the **cortex** on the inside. The cuticle cells of the outer layer are arranged into scales that overlap much like the scales on a pine cone (**Figure 7–9**). Because of this characteristic, the fibers of wool lock together and bond into a solid mass when they are put under pressure.

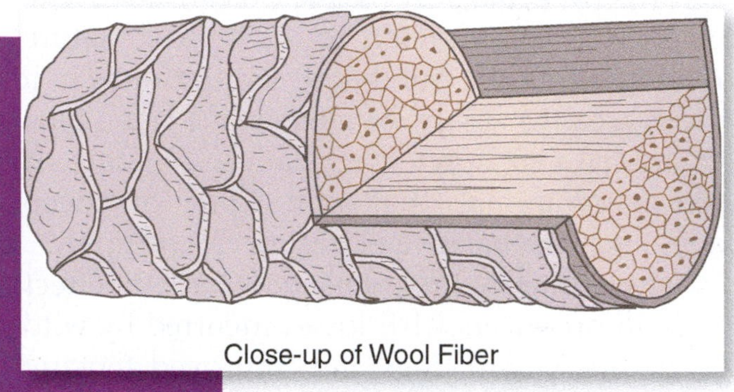

FIGURE 7–9 Close-up of a cross section of a wool fiber. *Sheep Industry Development Program*

FIGURE 7–8 During the Civil War, Union and Confederate soldiers both wore wool uniforms. *© Everett Historical/Shutterstock.com.*

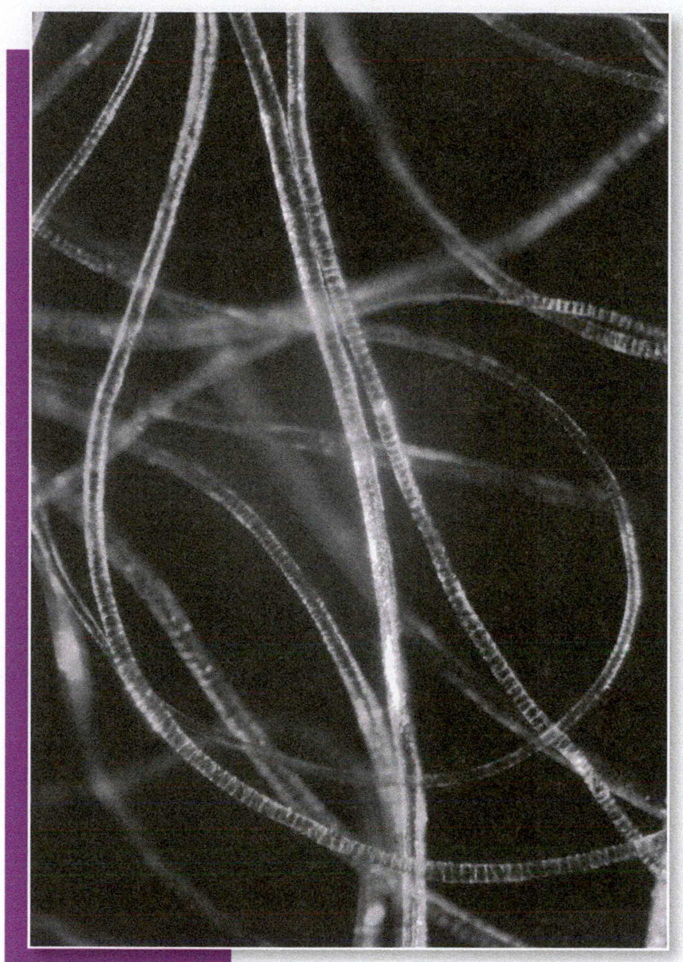

FIGURE 7–10 A microscopic view of wool fibers. Note the scales on the fibers. *© iStockphoto/Zanthra.*

The fibers form a strong, thick bond of solidly matted fibers (**Figure 7–10**). This process, called **felting**, is used to make things such as hats and other objects that require a thick layer of matting. The cortex layer that makes up the majority of the fiber gives it strength as well as elasticity. Elasticity is the ability to return to its original shape after being stretched.

The fibers are never perfectly straight. In fact, most are quite wavy and may be stretched up to 50 percent of their length and then return to their original length. This effect is called **crimp** and is used to help determine the quality of the wool. Usually, the more crimp (the wavier) and the more uniform the crimp, the higher is the quality of the wool. Wool is graded according to the diameter of the fiber, with the wool having the finer or smaller diameter receiving the highest grade. Higher quality wool is free of foreign material and is bright and white in color with no black or off-color fibers.

Wool contains oils, grease, and the salts from the perspiration of the sheep, referred to as **suint**. The wool grease, called the **yolk**, helps to hold the scales on the outer layer of the fiber together and provides the fiber with the ability to shed water. It also helps the fibers from becoming matted together in the felting process.

Wool as it comes from the sheep is called **grease wool**. The first step in wool processing is to remove the loose dirt and other particles from the fibers. This is done by a machine that opens up the fibers and dusts foreign particles from the wool (**Figure 7–11**).

The second step is called **scouring**. In this process, the wool is gently washed in detergent to remove yolk, suint, and other materials not removed in the dusting process. Almost half the weight of grease wool is removed in scouring. After the fibers have been cleaned, the wool is called **scoured wool** (**Figure 7–12**).

The water used in the scouring process is retained to remove the oils extracted from the wool. These oils are used to produce lanolin, an important ingredient in soaps and lotions.

After the wool is scoured, the wool is dried and treated to remove any remaining vegetable matter. This is done by mechanical

FIGURE 7–11 The yolk, suint, and other materials are removed from the wool. *Courtesy of American Sheep Industry Association*

FIGURE 7–12 After the wool has been cleaned, it is called scoured wool. *Courtesy of American Sheep Industry Association*

FIGURE 7–14 In the combing process, the fibers are untangled and smoothed. *Courtesy of American Sheep Industry Association*

or chemical means. Removing the matter by chemical means, called carbonizing, is the method used most often. This consists of treating the wool with acids or other chemicals to dissolve the vegetable matter.

Once the wool is clean, it is blended. This means that wool fibers of different types are mixed together mechanically to achieve a particular type of fabric or product. It is estimated that wool fibers can be blended into 2,000 combinations to produce a wide assortment of products.

At this stage, the wool can be dyed. If so, the wool is called *stock-dyed wool* (**Figure 7–13**). If the wool is dyed after it is spun into yarn, it is called *yarn-dyed wool.* Wool dyed after it is woven into cloth is called *piece-dyed wool.*

The wool fibers are untangled and laid out parallel to each other in a process called **carding.** If the wool is to be made into a type of fabric called *worsted wool,* the fibers are further untangled and smoothed by **combing** (**Figure 7–14**) and carding. If the wool is to be made into woolen fabrics, the wool goes from carding to spinning.

After the fibers are smoothed out and laid parallel to each other, they are processed into yarn by the spinning process, in which the fibers are spun around and twisted into a long, continuous thread that is used for weaving (**Figure 7–15**).

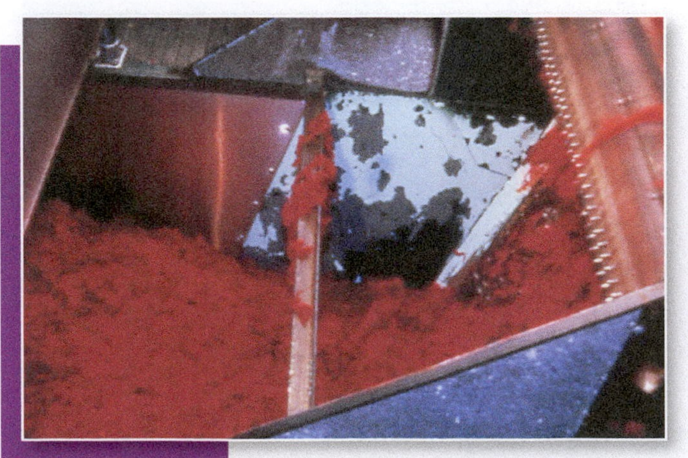

FIGURE 7–13 Wool dyed before spinning is called stock-dyed wool. *Courtesy of American Sheep Industry Association*

FIGURE 7–15 The spinning process creates long, continuous threads from the fibers. *Courtesy of American Sheep Industry Association*

FIGURE 7–16 The weaving of yarn creates fabrics in many patterns and designs. *Courtesy of American Sheep Industry Association*

The wool cloth is made by weaving the yarns together. The yarn can be woven together into a variety of patterns and designs that make up the clothing or tapestries that are so popular with consumers (**Figure 7–16**).

Each year, Americans use about 1 pound of wool per person. This means that about 100 million pounds of grease wool is used in the United States per year. Of this amount, about half is produced in this country and about half is imported. The countries exporting the most wool to this country are Australia and New Zealand. Fine wool comes from breeds of sheep such as the Merino, Debouillet, Delaine, and Rambouillet.

Karakul sheep are raised for their pelts, which have a furlike quality (**Figure 7–17**). **Pelts** are the skins of the animals with the hair left on. Most of these sheep are raised in the countries that made up the former Soviet Union, Afghanistan, and Iraq; however, some of these sheep are raised in the United States. These high-quality pelts are valuable in making coats and jackets.

FIGURE 7–17 Karakul sheep are grown for their furlike pelts. *© belizar/Shutterstock.com.*

Mohair is a fiber from the fleece of the Angora goat (**Figure 7–18**). This fiber is used to make a fabric that is resistant to wrinkles, is

FIGURE 7–18 Mohair is the fiber from the fleece of Angora goats. *© Janice Adlam/Shutterstock.com.*

FIGURE 7–19 Texas is the largest producer of mohair in the United States. *Source: USDA, Census of Agriculture*

soft and lustrous, and is unequaled in its ability to retain rich colors. This fiber differs from sheep wool in that the fibers are smooth and have less crimp. In addition, the fibers are long, often reaching a foot in length. This quality makes the fibers easier to weave and provides a wider range of usages.

The United States produces about one-third of the world's mohair, most of which is exported to England, where the fibers are processed into fabric. Most of the Angora goats raised in this country are produced in Texas (**Figure 7–19**), where the goats make good use of the sparse browse found in the western part of the state. Much of the production of mohair comes from castrated males that are raised exclusively for their hair. The goats produce about an inch of hair per month and are sheared twice a year. The freshly shorn goats must be protected from the elements until their hair can grow out enough to protect them from the cold and rain.

SUMMARY

Sheep are among the oldest animals domesticated for human use. In the United States, sheep are not raised as widely as some of the other agricultural animals. Much of our wool is imported from New Zealand and Australia; however, a significant industry does exist around the production of lamb and wool in this country. The popularity of wool continues to grow because of the unique characteristics of the fiber. Therefore, the production of wool will be with us for years to come.

CHAPTER REVIEW

Review Questions

1. Explain why growing lambs and rye grass makes a good combination.
2. What are some characteristics of wool that make it more desirable than cotton and synthetics? Why did the soldiers in the Civil War prefer wool uniforms?
3. Why is the Southeast an undesirable place to raise sheep?
4. What is yolk? What purpose does it serve?
5. Describe ecological balance. How does this affect sheep producers when dealing with predators?

Student Learning Activities

1. Determine how many different materials in your house are made from wool. List the characteristics of wool that make it a good material for each item.
2. Conduct a survey among students in your school to determine how many of them ate lamb or mutton during the past month. Also ask how many have never eaten lamb or mutton, and why.
3. Go to the Internet and locate information about a particular breed of sheep. Report to the class.

CHAPTER 8

The Goat Industry

STUDENT OBJECTIVES IN BASIC SCIENCE

As a result of studying this chapter, you should be able to

- explain why goat meat is healthier than other meats.
- explain the advantages of goat milk.
- classify goats.
- describe how the feeding habits of goats differ from those of cattle.
- explain why goats are so susceptible to parasites.
- list and define the common diseases affecting goats.

STUDENT OBJECTIVES IN AGRICULTURAL SCIENCE

As a result of studying this chapter, you should be able to

- point out the significance of goats in the United States.
- discuss the products made from goat milk.
- describe the different breeds of goats.
- explain the management practices used in raising goats.
- describe a prevention health plan for a goat herd.

KEY TERMS

chevon
cabrito
lactase
Cashmere
Angora
commercial producers
purebred producers
billy
nanny
buck
doe
wether
doeling
buckling
ruminant
browse
anthelmintics

NATIONAL AFNR STANDARD

AS.01.01
Evaluate the development and implications of animal origin, domestication, and distribution on production practices and the environment.

AS.01.02.02.a
Research and examine marketing methods for animal products and services.

HISTORY

GOATS WERE AMONG THE first domesticated animals, around 10,000 to 11,000 years ago in Asia and Africa. Humans began keeping small herds of goats for milk, meat, and fiber and used their hides for clothing and housing materials. Goats easily adapted to the nomadic lifestyle of tribal living and could survive on the sparse vegetation found in most arid areas (**Figure 8–1**). Since ancient times, goats have provided humans with food, fiber, and other products. Goat skins have been used for centuries as wine and water flasks (**Figure 8–2**).

FIGURE 8–2 Goat skins have been used for centuries as wine and water flasks. *© Sergio Foto/Shutterstock.com.*

Goats were popular meat and dairy animals aboard trade ships. Often, these ships sailed for long periods of time away from land, and the sailors had to carry food with them. The small size and good temperament of goats made them ideal for taking on long sea journeys. Because goats could provide both milk and meat, they provided a dual source of sustenance.

There were no goats in the Americas before the first settlers arrived. The first domesticated goats probably arrived when Christopher Columbus's crew left the animals they did not need behind in America. From this small herd, goats began inhabiting the New World. There is documentation of goats being on the Pilgrims' Mayflower's journey to North America, as well as the many ships that brought settlers to the New World (**Figure 8–3**). Today, there are millions of domesticated goats in the United States.

GOAT INDUSTRY IN THE UNITED STATES

More goat meat and milk are consumed than cow's milk or beef in the worldwide marketplace. The small animals are much easier to keep and are a lot less expensive to maintain than cattle. Although the exact number is difficult to calculate, it is estimated that there are

FIGURE 8–1 Goats were easy to adapt to the nomadic lifestyle of tribal living and could survive on the sparse vegetation found in most arid areas. *© posztos/Shutterstock.com.*

FIGURE 8–3 Goats came to America on the Mayflower and many other ships carrying settlers. *© iStockphoto/nicoolay.*

more than 250 million goats that are used for a variety of purposes. The goat industry is divided into three major categories, depending on the purpose of the goat: meat goats, dairy goats, and fiber goats. There are some breeds of goats that are considered dual-purpose breeds. For example, the Nubian breed is considered to be a dual-purpose breed because it can provide a family with milk and also provide a decent carcass for consumption. The leading states in goat production are Texas, Tennessee, and Oklahoma, although other states have substantial numbers of goats as well. Goats are used today in a variety of ways, from clearing out brush and undergrowth in pastures and forests to working for the government, helping to clear power lines (**Figure 8–4**).

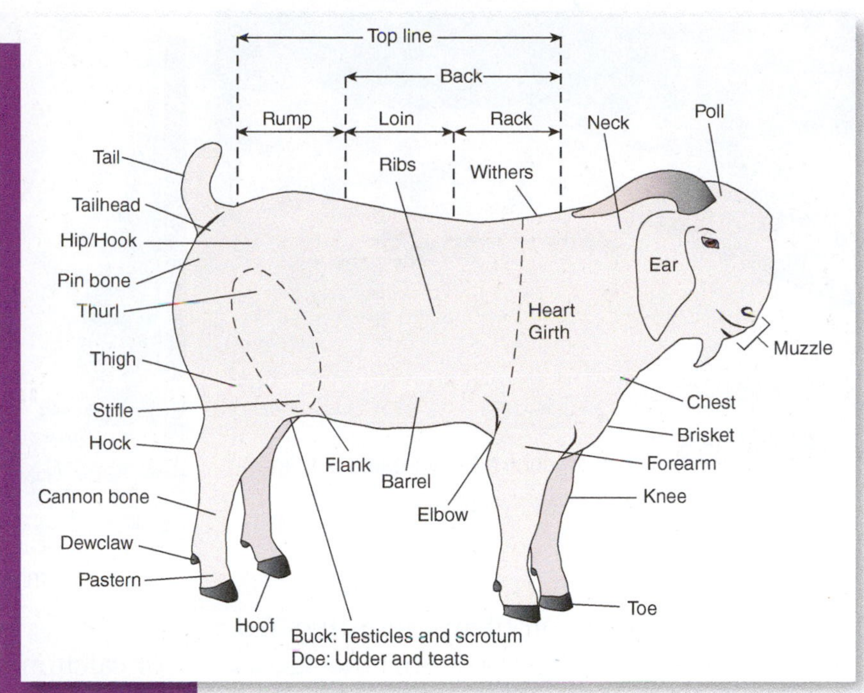

FIGURE 8–5 Parts of a goat (meat).

Meat Goats

Currently, compared with beef, swine, and poultry, the meat goat is not significant in the animal agriculture industry, although, because of the ever-changing face of America, its popularity is growing rapidly, (**Figure 8–5**). Certain religions and cultures regard cattle as sacred, and they use meat goats as an alternative source of protein. Goat meat, called **chevon** in France and **cabrito** in Spanish cultures, is a good source of protein, and the goats produce a very lean carcass. Goat carcasses have been compared with deer meat in their lean-to-fat ratio. Goats do not deposit as much subcutaneous fat as do cattle and sheep, so their meat is lower in fat and cholesterol compared with beef, pork, mutton (sheep), and poultry (**Figure 8–6**).

FIGURE 8–4 Goats are sometimes used to clear weeds and brush. *Courtesy of Shannon Lawrence*

Dairy Goats

As mentioned, goat milk is popular all over the world. The leading producer of goat milk is India, which produces almost 3 million tons of goat milk each year. In the United States, there are over 1.5 million milk goats. (**Figure 8–7**) illustrates the parts of a dairy goat.

In addition to people who enjoy the flavor of goat milk, it is a substitute for cow's milk for those with lactose intolerance and allergies. These conditions are more frequent in infants up to 3 years old, and in the elderly,

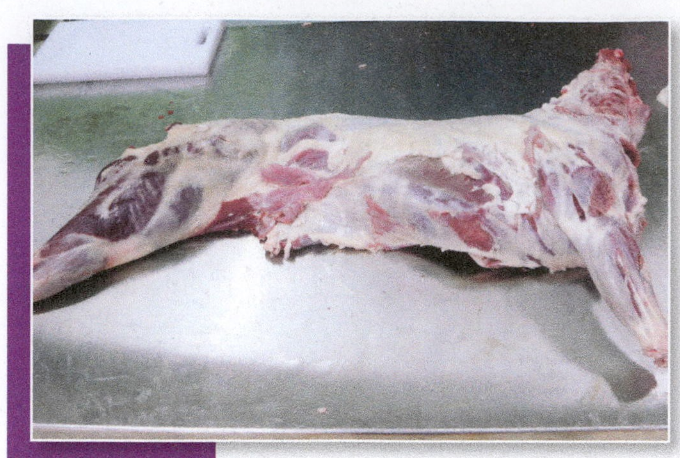

FIGURE 8–6 Goats produce a leaner carcass than sheep or beef animals. © TADAphotographer/Shutterstock.com.

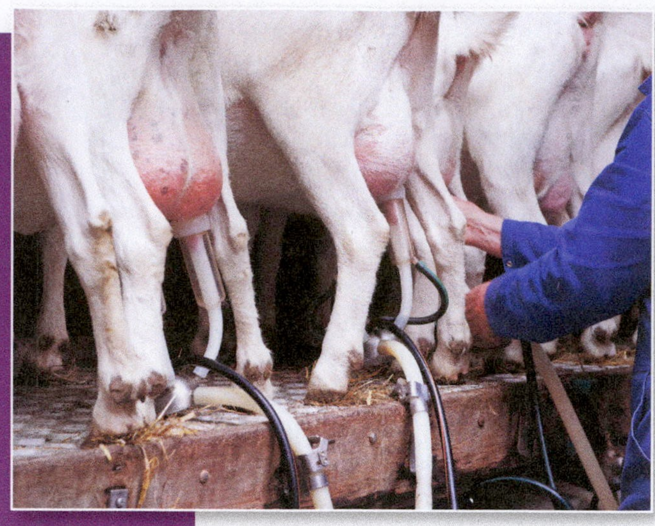

FIGURE 8–8 Goat milk has many health advantages over cow's milk. © 2xSamara.com/Shutterstock.com.

who have digestive problems resulting from lack of the lactose-digesting enzyme **lactase**. Goat milk is naturally homogenized and can be digested more easily than cow's milk. The natural homogenization in goat milk results from the smaller fat globules compared with cow's milk; smaller fat globules stay suspended in the solution.

Another advantage to goat milk is its health benefits (**Figure 8–8**). Goat milk is a good source of calcium and is a complete protein, containing all the essential amino acids and without the heavy fat content of cow milk. Goat milk has more vitamin A as well as riboflavin and other vitamin Bs; however, cow milk is higher in B6 and B12. When fed to other animal species, goat milk has been shown to enhance metabolism of both iron and copper, especially for individuals who have problems absorbing

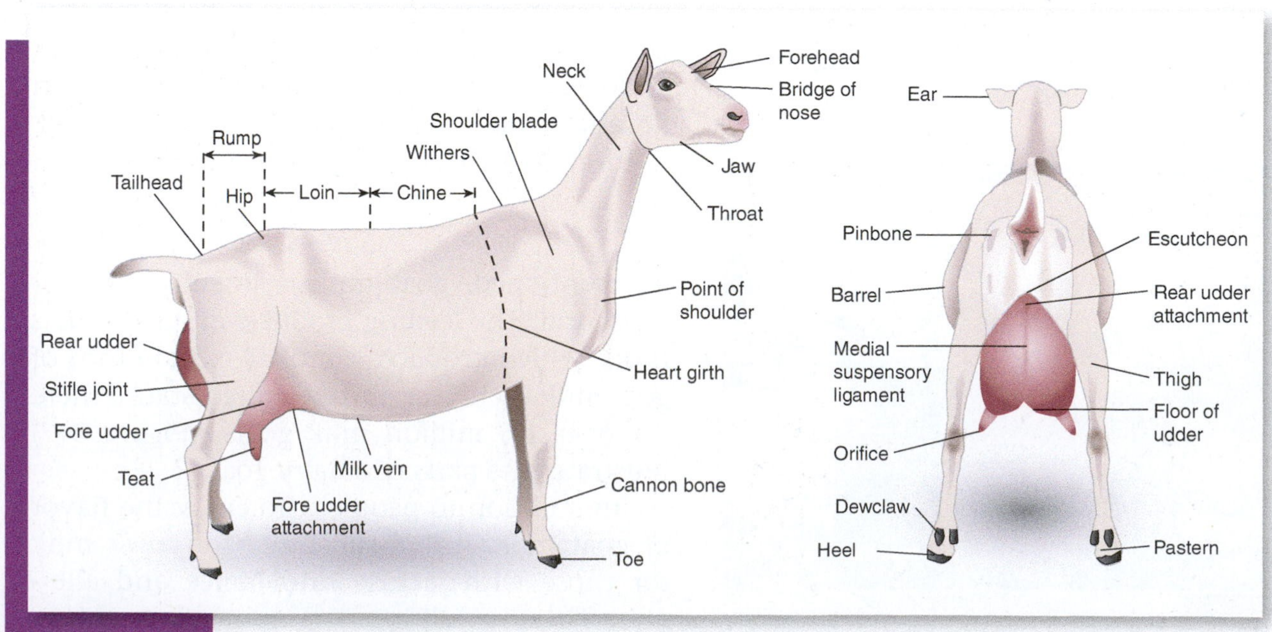

FIGURE 8–7 Parts of a goat (dairy).

minerals in the digestive tract because of compromised intestinal function.

Goat milk can be used in the same ways as cow milk in cheeses, butter, ice cream, and other dairy products. Goat milk cheese is a high-ticket item in most grocery and specialty stores. Goat milk can also be used in other forms such as soap and lotion. Ever since biblical times, the elite in Eastern societies have used goat milk in baths to holistically help maintain their beauty. Cleopatra routinely used goat milk to bathe in to maintain her beauty.

Goat milk soap is well known for its moisturizing properties. Handmade goat milk soap contains glycerin. Soap is made by combining a fat with an alkali. In a process called saponification, the fat turns into glycerin. In commercially made soaps, the glycerin is usually removed and put into other items that command a higher price, such as cosmetics and medicine. Natural goat milk soap keeps the beneficial glycerin, along with natural emollients, vitamins, and triglycerides, which moisturize the skin. Most goat milk soaps do not contain harsh chemicals that can irritate or leave the skin itchy and dry.

Fiber Goats

Some of the finest fabrics in the world come from the **Cashmere** and **Angora** fiber goats (**Figure 8–9**). The most common breeds of fiber goats are the Cashmere, the Angora, and the Pygora. Because the Angoras lacked hardiness, they were crossbred with Pygmy goats to develop the Pygora.

As their name suggests, each of the breeds produces a different kind of fiber. Some fibers are used in looming, and others are better adapted to hand-spinning and knitting. Various fibers come from the fiber goat industry, from short, coarser fibers to the long, softer fibers used to make clothing. These goats are also popular in more arid regions of the country.

Another by-product of the goat industry is goat leather. Goat hide is a popular leather material because of its strength and durability. Because it is thinner than cow hide, it is commonly used in making work and dress gloves and coats.

PRODUCTION IN THE UNITED STATES

Production of market or meat goats is divided into two major groups: commercial and purebred producers. **Commercial producers** provide high-quality carcass animals for the meat industry (**Figure 8–10**). High-quality carcasses contain a low fat-to-lean ratio. **Purebred producers** provide commercial producers with high-quality herd sires and replacement does for breeding. These producers also help to improve the breed in carcass, reproduction,

FIGURE 8–9 Angora goats such as this are a source of fine fabrics. *© iStockphoto/dmaroscar.*

FIGURE 8–10 Commercial producers raise goats for the meat industry. *© Ekaterina Belova/Shutterstock.com.*

and weight gain. Large producers of market animals are found in the southwestern United States. These producers raise goats on the vast expanse of marginal land that the goat herd can utilize but cattle will not, because goats can metabolize and utilize a lower quality of vegetation than cattle.

The industry is further divided by hobby producers and full-time producers. Hobby producers are those who do not earn all their income from the production of goats; however, both types of producers contribute to the overall market.

BREEDS OF GOATS

There are more than 200 goat breeds throughout the world. In the United States, however, only a few of these breeds are significant. These breeds are broken down by purpose: dairy, meat, dual-purpose, or fiber. Market goats are the meat goats of the industry.

Meat Goats

Several breeds of meat goats are popular in the United States: the Boer, Kiko, Tennessee Fainting, Savanna, and Spanish goat. All of the meat breeds except the Spanish goat have national associations. Meat goat people use the terminology listed in **Table 8–1** when referring to meat goats.

Boer Goats

Of the several different breeds of meat goats, the Boer is the most popular. Almost all of the market goats shown in junior shows are Boer or Boer crosses. The trademark colors of the Boer are the red head and white body, although Boer goats can be a combination of colors from solid black, to solid red, to a mixture of white plus tan, red, or black. Purebred Boer goats have a distinctive Roman nose and long, floppy ears (**Figure 8–11**). Boers are a horned breed of goat, and producers usually prefer to not remove the horns. Boers are double-muscled meat animals, which gives them their stout appearance.

Male goat, mature:	**billy**/billies
Female goat, mature:	**nanny**/nannies
Male, castrated:	wether

TABLE 8–1 Meat Goats

FIGURE 8–11 Purebred Boer goats have a distinctive Roman nose and long, floppy ears. *Courtesy of Shannon Lawrence*

Boer goats originated in South Africa, where the Dutch farmers developed the breed. (Boer means "farmer" in Dutch.) A full-blooded South African Boer buck can command $100,000 or more at auction. Herd improvements in the Boer breeding programs include artificial insemination and embryo transfer. Boer goats are known for their high fertility rates, docile nature, and rapid growth and were first imported into the United States in 1993. Since that time, Boers have been used extensively to improve domestic herds of meat goats. Boers usually have large frames to support an ample covering of meat. These goats are highly adaptable to most climates.

The breed associations recognize 100 percent Boer and 100 percent South African Boer, as well as Boer crosses. These crosses are registered as 50 percent Boer up to 100 percent Boer. Boer goats are commonly bred to Nubians as well as Spanish type goats to produce heterosis in the females or nannies. Heterosis, or hybrid vigor, comes from the crossbreeding of two or more breeds to achieve the optimal animal. These crossbred nannies also are used for implantation of embryos from full-blood Boers. Producers use the cross-bred nannies because of their excellent milking abilities and forage capabilities.

Kiko Goats

The Kiko goat is a fairly new breed of meat goat, developed in New Zealand as part of a government-funded project (**Figure 8–12**). In the 1970s, the native goat population in New Zealand was threatening crops, forests, and range land. As a result, the government paid hunters to round up and kill the overpopulation. During this hunt, some of these goats were confined for breeding purposes. While in captivity, a company called Goatex Group, LLC noticed that the goats showed enhanced characteristics for growth and meat production. These goats were saved from the kill and bred with fiber goats to produce Kiko goats.

The word Kiko means "meat" or "flesh" in Maori, a native language of New Zealand. The cross of the domestic goat and the dairy or Angora goat proved worthwhile for their hardiness and high-quality meat carcass.

Tennessee Fainting Goats

Tennessee fainting goats are also known as Tennessee meat goats, myotonic goats, stiff-legged goats, and other names. These goats gained fame (and their name) from fainting spells when they become nervous, although they are considered a meat goat. Fainting goats "faint" and fall over as a result of an epilepsy-type condition that does not originate in the nervous system but rather causes a prolonged contraction in the muscular system when the animal is startled. The extent to which these goats "faint" varies from goat to goat, from severe stiffness lasting minutes in some goats to rare stiffness in others.

The breed can be traced back to the 1880s, when a farmer moved into the Tennessee area from Nova Scotia, reportedly having four of the unusual goats with him. These goats

FIGURE 8–12 Kiko goats are a fairly new breed of meat goats developed in New Zealand as part of a government-funded project. *Courtesy of Shannon Lawrence*

became popular because they were less apt to climb out of fencing and they had a high reproductive rate and good muscular structure. The breed has enjoyed an increase in popularity recently.

Savanna Goats

The Savanna goat breed is a fairly new meat goat breed to the United States. Like the Boer goats, Savanna goats originated from South Africa. Savanna goats are the same body type as Boer goats, having a heavy frame and being heavily muscled. Savanna goats are all-white with dark pigmented skin that helps to keep the goat from being sunburned (**Figure 8–13**). These goats are rapidly gaining popularity in commercial breeding programs.

Dual-Purpose Goats

Dual-purpose goats serve as both milk and meat animals. Breeds of dual-purpose goats include the Nubian, Kinder, Spanish, and Pygmy.

Nubian Goats

Although Nubians are used extensively in the meat goat industry, their registry is still with the dairy goat breeds. The American Dairy Goat Association (ADGA) registers Nubians along with other more traditional Swiss dairy breeds. Nubians were developed from breeding English common goats with "exotic" bucks from India and Africa (**Figure 8–14**). Nubians are expected to grow fast, and they have a large frame size. In the early days of breeding in Europe, the breed was used largely as draft animals. Nubians are used in the meat goat industry to crossbreed or for embryo transfer. The milk of Nubians has a high butterfat concentration, so their milk is more nutritious than that of purebred meat goats. This makes the nannies better adapted to raise healthier babies faster, with little, if any, supplemental feed.

Kinder Goats

Kinder goats are new to the meat goat industry. They are a cross between the Nubian breed and the Pygmy breed (**Figure 8–15**). Because both the Nubian and the Pygmy are considered to be dual-purpose, their offspring are as well. Kinder goats are smaller than the large Nubians but still give a delicious rich milk, and they have a high feed conversion and superior fleshing ability.

FIGURE 8–13 The dark-pigmented skin of Savanna goats helps to keep the goat from being sun-burned. *© iStockphoto/SkyWet.*

FIGURE 8–14 Nubians were developed from breeding English common goats with "exotic" bucks from India and Africa. *© Yulia Piekhanova/Shutterstock.com.*

FIGURE 8–15 Kinder goats are a cross between the Nubian breed and the Pygmy breed. © J. McPhail/Shutterstock.com.

FIGURE 8–16 Pygmy goats are raised more for hobby and show purposes than for meat or milk in the United States; in some underdeveloped countries, however, their small size allows people with limited space to use them for meat and milk. Courtesy of Shannon Lawrence

These goats are good for smaller acreages and backyard producers who want a source of both meat and milk.

Spanish Goats

Spanish goats, also called common goats or scrub goats, are any in which the breeding is unknown. Most are a mixed-breed animal showing some traits of a different recognized breed. Spanish goats have no registration association.

Pygmy Goats

The Pygmy goat originated from West Africa and also was called the Cameroon Dwarf goat. These goats were exported into Europe for use in zoos as exotic animals. In 1959, they were exported into the United States. As their name indicates, Pygmies are a small breed or dwarf variety. Until recently, most considered this breed to be a dairy breed, but producers now are using them more for meat carcass than for milk. In the United States, Pygmies are raised more for hobby and show purposes than for meat or milk. In some underdeveloped countries, however, their small size allows people with limited space to use them for meat and milk (**Figure 8–16**).

Dairy Goat Breeds

Although they are not as popular as their meat counterparts, dairy goats are gaining in popularity in the United States. The organic movement and the self-sufficient movement have spurred the rise in dairy goat numbers. Recent data indicate that more people throughout the world are using goats as their primary source of milk and meat.

Although most meat goat producers choose to leave the horns on their animals, dairy goat breeders prefer to remove the horns or to breed for specific characteristics that produce a polled animal. Polled goats do not grow horns that have to be removed.

Of the many different breeds of dairy goats, most have a specific heritage as Swiss goats. These include the Alpine, Saanen, Sable, Oberhasli, and Toggenburg breeds. The other breeds in the dairy goat world are the Nubian, discussed earlier, the LaMancha, and the Nigerian Dwarf. The terminology listed in **Table 8–2** is used in the dairy goat industry.

Mature male goat:	**buck**
Mature female goat:	**doe**
Castrated male goat:	**wether**
Young female:	**doeling**
Young male:	**buckling**

TABLE 8–2 Dairy Goats

FIGURE 8–18 The Saanen breed tends to be calm and produces the most milk on average of any of the other breeds of dairy goat. *Courtesy of Shannon Lawrence*

Alpine Goats

The Alpine dairy goat is a Swiss origin breed. Alpines are medium- to large-size animals with upright ears (**Figure 8–17**). They are extremely adaptable to any climate conditions. They come in a variety of colors. Size and production have been stressed in the development of this breed.

Saanen and Sable Goats

Saanen dairy goats are a breed of Swiss origin. The Saanen is an all-white goat. Any coloration of this breed is registered as a Sable dairy goat. This goat is one of the largest breeds of dairy goats. The goats tend to be calm and produce the most milk on average of any of the other breeds of dairy goat (**Figure 8–18**). Saanen milk is lower in butterfat than other dairy breeds.

Oberhasli Goats

Oberhasli dairy goats tend to be smaller than other breeds. They are known for their characteristic color pattern of reddish brown with a black dorsal stripe, legs, belly, and face, although sometimes solid black Oberhasli are born (**Figure 8–19**). This breed is also of Swiss origin.

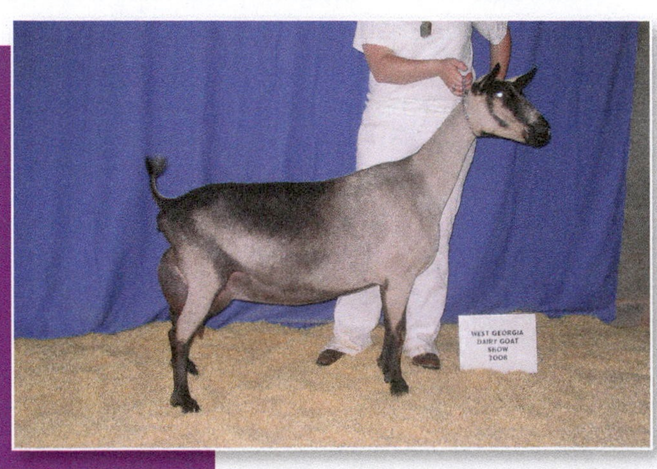

FIGURE 8–17 The Alpine dairy goat is a breed of Swiss origin. Alpines are medium- to large-size animals with upright ears. *Courtesy of Shannon Lawrence*

FIGURE 8–19 Oberhasli dairy goats tend to be smaller than other breeds. They are known for their characteristic color pattern of reddish brown with a black dorsal stripe, legs, belly, and face, although sometimes solid black Oberhasli are born. *Courtesy of Shannon Lawrence*

The breed is considered extremely hardy and thrifty, doing well on substandard pastures.

Toggenburg Goats

The Toggenburg dairy goat is a Swiss breed. These goats are medium in size, are moderate in production, and have a relatively low butterfat content (**Figure 8–20**). This breed is highly specific in coloring: The body must be fawn to dark chocolate in color, with two white stripes down the face, white stockings on the legs, and a white triangle on either side of the tail.

FIGURE 8–20 Toggenburg dairy goats are a Swiss breed, medium in size, moderate in production, and have relatively low butterfat content. *© Eric Isselee/Shutterstock.com.*

LaMancha Goats

The LaMancha dairy goat is an original American breed and is known for its apparent lack of an external ear (**Figure 8–21**). The LaMancha breed is medium in size and generally calm, quiet, and gentle. These goats are extremely sturdy and can withstand a great deal of hardship and still produce. LaMancha milk is high in butterfat. All colors are acceptable, and they have either gopher ears (little external ear) or elf ears (a small amount of external ear). This breed is becoming more popular in the meat goat industry as an alternative to crossbreeding programs with Nubians.

Nigerian Dwarf Goats

The Nigerian Dwarf dairy goat originated in Africa, but the breed did not improve until after being imported into the United States. This breed is truly a miniature dairy goat (**Figure 8–22**). The goats are not dwarfed animals, which usually show compact, big bones. This breed is optimal for backyard breeders because, despite its small size, a goat can provide enough milk daily for a family of five. Nigerian Dwarfs can be any color except the traditional Toggenburg colors.

FIGURE 8–21 The LaMancha dairy goat is an original American breed and is known for its apparent lack of an external ear. *Courtesy of Shannon Lawrence*

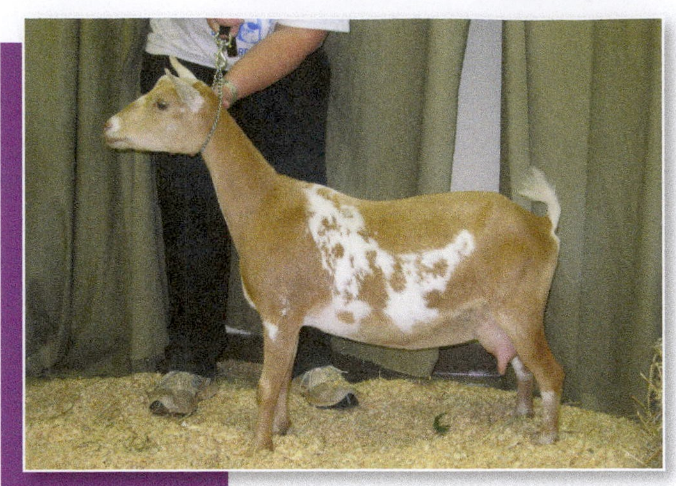

FIGURE 8–22 The Nigerian Dwarf dairy goat originated in Africa and is truly a miniature dairy goat. *Courtesy of Shannon Lawrence*

As **ruminant** animals, goats can utilize forages like other ruminants, including cows and sheep. They are agile and can maneuver well over rough terrain. This makes them a popular alternative in mountainous areas and rocky terrain where cattle won't utilize forage materials. Goats prefer woody-type plants (over grass-type plants), such as briars, bushes, and weeds, for sustenance (**Figure 8–23**). Animals that **browse** for food, termed browsers, include goats and deer. Goats have lower

FIGURE 8–23 Goats prefer a woody type of plant for sustenance, such as briars, bushes, and weeds. *© iStockphoto/Ivan_off.*

ANATOMY AND PHYSIOLOGY

Goats are well adapted to live in a variety of climates. They are members of the Bovidae family (**Table 8–3**) and are closely related to deer, antelope, and sheep. Goats and sheep are commonly grouped together, although they have some important differences.

Kingdom:	Animalia
Phylum:	Chordata
Class:	Mammalia
Order:	Artiodactyla
Family:	Bovidae
Subfamily:	Caprinae
Genus:	*Capra*
Species:	*C. aegagrus*
Subspecies:	*C. a. hircus*
Trinomial name:	*Capra aegagrus hircus*

TABLE 8–3 Scientific Classification for Goats

Source: *Linnaeus*, 1758.

incisors and both top and bottom molars, which help them to strip leaves and grass. Because they usually prefer plants that cattle and other livestock will not use, goats are good partners with cattle in mixed grazing operations.

Body Types

Meat goats and dairy goats have different body types. The dairy goat should have a body shaped like a triangle when viewed from above. This gives the body more support for the production of milk in the udder. Meat goat producers breed meat goats to have more of a rectangular shape to their body. The rectangular shape supports the development of muscle throughout the carcass. Dairy goats should have thinner, flatter bones like dairy cattle; and meat goats should have ample bone size and structure to support their double muscling as in beef cattle.

FIGURE 8–24 Goats are able to produce one to four offspring after a 5-month gestation period, although some goats have produced five or more kids in one gestational period. *Courtesy of Shannon Lawrence*

MANAGEMENT OF GOATS

Proper care and management of the goat herd is important to having healthy animals. Good record keeping helps producers track feed conversion, control parasites, conduct breeding, and control weight gain. This also helps producers with culling nonproductive and low-performing goats.

Breeding

Goats are able to produce one to four offspring after a 5-month gestation period, although some goats have produced five or more kids in one gestational period (**Figure 8–24**). Goats are seasonal breeders but can be manipulated to come into season more than once a year. Some producers' optimal goal is to have three sets of kids every 2 years, or as in other production livestock, a 100 percent kidding rate. A 100 percent kidding rate means that every doe will have at least one kid every breeding season. Females that do not produce a kid every season should be considered for culling, a practice in which the undesirable animal is removed from the herd.

Purebred meat goat producers are using the latest in technology to produce improved stock faster. To obtain maximum results, producers use a variety of methods, including artificial insemination, embryo transfer, and sonograms.

Nutrition

Contrary to popular myths, goats do not eat tin cans, clothing, or any type of garbage if they are fed a well-balanced ration. The old tale about goats eating tin cans probably came from seeing undernourished goats eating paper labels. In fact, goats are picky eaters and require nutritional standards based on performance like other forms of livestock. Commercial feeds are available for both meat production and milk production to both large and small producers, although some producers prefer to custom-blend their own feeds (**Figure 8–25**).

Housing, Fencing, and Protection

In the southwestern United States, goats usually do not have to be housed inside a building. They just need a shaded area to escape the sun and heat, especially in range-type situations. Boer goats, although they have white hair, have dark pigmented skin, so they normally do not get sunburned.

In other parts of the country, housing varies from lean-to structures to fully enclosed barn areas, depending on the climate and weather conditions. When housing is provided, it should be clean, dry, and well ventilated to prevent infection from molds and bacteria (**Figure 8–26**).

Good fencing for goats is important because goats are "escape artists" when dealing with fences. Some goats bend down fencing and may even jump over fencing. Fencing also keeps out predators such as mountain lions, coyotes, feral and neighborhood dogs, and wolves. Fencing options include electric fence, welded-wire panels, and welded-wire and knot-wire fencing (**Figure 8–27**). Meat goat producers most often leave horns on their herd animals to help them fend off predators, so getting stuck in the fence can be problematic.

FIGURE 8–26 When housing is provided, it should be clean, dry, and well ventilated to prevent infection from molds and bacteria. *© iStockphoto/LagunaticPhoto.*

FIGURE 8–25 Commercial feeds are available for both meat production and milk production to both large and small producers. *Courtesy of Shannon Lawrence*

FIGURE 8–27 Fencing options include electric fence, welded-wire panels, and welded-wire and knot-wire fencing. *Courtesy of Shannon Lawrence*

FIGURE 8–28 Goat producers often rely on livestock guardians such as dogs to protect the goats from predators.
Courtesy of Shannon Lawrence

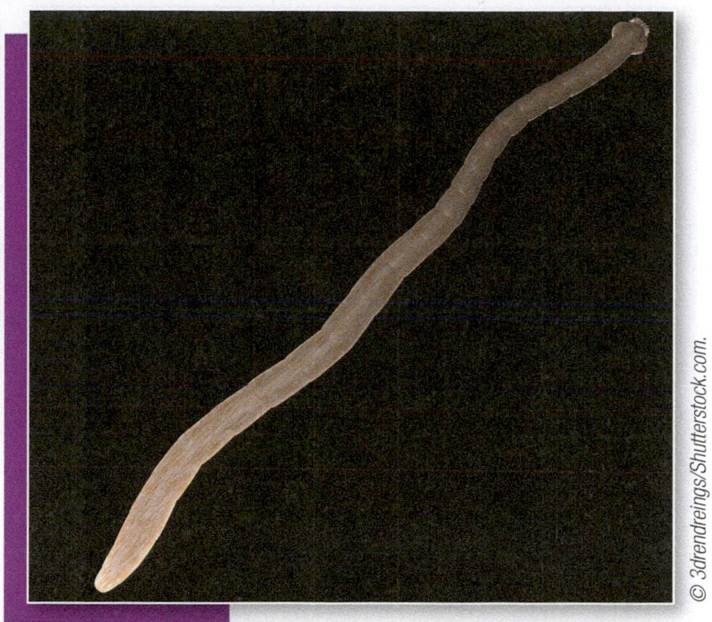

FIGURE 8–29 Goats are susceptible to internal parasites such as tape worms (top) and round worms (bottom).

© 3drendreings/Shutterstock.com.
© Carsten Medom Madsen/Shutterstock.com.

Goat producers often rely on livestock guardians for protection from predators. Some producers use donkeys, mules, llamas, and dogs to protect the goats (**Figure 8–28**). Factors affecting the selection of livestock guardians include the area, land conditions, climate, and predators in the area.

Parasites

Goats, like sheep, are susceptible to internal parasite or worm overload. Goats and sheep always have some internal parasites, but good management helps to control overpopulation of the internal parasites to a manageable level. Goats and sheep are more susceptible than other animals to internal parasites because of their grazing behavior and poor immunity. Goats that browse have fewer parasite problems, though woodland grazing may increase the risk of meningeal worm infection.

Internal parasites are controlled effectively with good management practices and **anthelmintics**, or de-wormers. Every

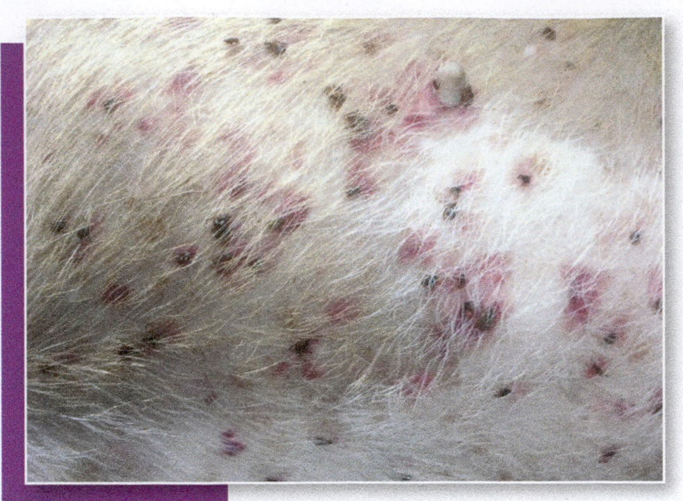

FIGURE 8–30 External parasites such as ticks can cause damage to goats. Ticks caused this damage.
© Mangpor_nk/Shutterstock.com.

de-wormer is formulated for a specific internal parasite, so routine fecal checks of the herd should be done before giving any anthelmintics. Internal parasites that are common in goats include round worms, tape worms, barber pole worms, and others (**Figure 8–29**). Barber pole worms are the number-one concern in goat management.

Other parasites that can cause problems are external parasites, which include lice, mites, flies, and ticks (**Figure 8–30**). External parasites can cause production losses and can weaken the goat herd, making it susceptible to other health issues, resulting in death.

PUTTING IT INTO PRACTICE

Raising Goats

Goats are becoming a very popular choice for supervised agricultural experiences (SAEs). A pair of goats can be grown in a relatively small space, as long as you have the means to feed and provide water for them. Compared with other types of livestock, goats can be raised with only a modest investment. You will need to have a small shelter for the animals and a way to keep the structure clean. As with all SAEs, you should keep accurate records on expenses such as feed, vet care, and housing costs. Be sure to keep close records on the amount of time spent caring for the animals. For further information on SAEs see Appendix A at the end of this text.

Your SAE can be showcased through the many goat shows throughout your local areas and state. This will entail making sure your animal is trained for showing and kept clean and well groomed. You will learn the proper way to show the animal and can enter the showmanship competition. Competition through Future Farmers of America can be very rewarding in terms of learning skills and having fun.

Raising goats is a supervised agricultural experience that can be conducted with a modest amount of funding. *Courtesy of the National FFA Organization*

SUMMARY

The goat industry is not popular in the United States today, but because of the ever-changing population structure, this industry should grow in the future. Goats typically are divided into categories based on their purpose: fiber, meat, and dairy. Some goats are considered to be dual-purpose: they can provide both milk and meat to consumers.

Goat milk can be a good alternative to cow's milk in people who have allergies or lactose intolerance issues. Goats produce a lean carcass. Depending on the region, goats can survive with a variety of housing and fencing options, and they can use forage that other ruminants may not utilize.

CHAPTER REVIEW

Review Questions

1. What are the three major categories of goats?
2. Discuss the advantages of goat milk.
3. What is meant by a browser?
4. Why do goats complement the grazing of cattle?
5. Name four meat breeds of goat and the advantage of each one.
6. Choose a meat goat or a dairy goat project, and explain in detail why you chose this project.
7. Describe the teeth of goats.
8. Explain the history of goats in the United States.
9. Why is the goat industry expected to grow?
10. How many offspring can a goat have at a time?

Student Learning Activities

1. Visit a meat goat operation and interview the owner/manager to learn of the management practices there. Report to the class.
2. Go to the Internet and research a breed of goat. Write a report including the origin of the breed, its main purpose, and advantages and disadvantages.

CHAPTER 9

The Horse Industry

STUDENT OBJECTIVES IN BASIC SCIENCE

As a result of studying this chapter, you should be able to

- explain how the anatomy of the horse makes it ideal for carrying and pulling loads.
- discuss the different ways of classifying horses.
- tell how horse behavior affects management practices.
- describe the scientific classification of the horse.
- explain the process of mating horses.

STUDENT OBJECTIVES IN AGRICULTURAL SCIENCE

As a result of studying this chapter, you should be able to

- discuss the importance of the horse industry.
- list the various uses for horses in the United States.
- describe several advantages of mules over horses.
- explain how horses are raised.

KEY TERMS

mules
light horses
draft horses
ponies
perissodactyl
cecum
pasture breeding
hand breeding
artificial insemination (AI)
colostrum
gelding
farrier

NATIONAL AFNR STANDARD

AS.01.01
Evaluate the development and implications of animal origin, domestication, and distribution on production practices and the environment.

AS.01.02.02.a
Research and examine marketing methods for animal products and services.

HUMANS HAVE USED HORSES for transportation, work, and war from the beginning of recorded history. At one time, almost all civilizations relied on horses or donkeys to provide these services. Until about 60 years ago, military history was written around the horse. From the time the ancient Assyrians used horse-drawn chariots to transport soldiers until horses and mules were used to transport supplies and to pull artillery during World War II, horses and mules have been used to wage war. From the ancient Romans to our American Civil War, generals have used cavalry to increase the efficiency of their fighting forces. Only with the advent of modern weapons and self-powered machines have horses become obsolete in warfare.

In the United States, much of our history has been built around power supplied by horses and **mules** (**Figure 9–1**). The very first explorers and settlers brought these animals to tend fields and to provide the power necessary to build an agricultural base. As settlers moved westward, horses and mules took them there and then were used to work the farms once they were settled.

FIGURE 9–1 Much of our history has been built around power supplied by horses and mules. *Source: USDA*

The number of horses and mules in the United States grew until the 1920s, when the rapid increase in cars, trucks, and tractors caused a sharp decline in their numbers. From then until 1960, their numbers steadily declined. Since the 1960s, the numbers of horses and mules have both increased dramatically. Although they no longer serve as the basis of agricultural power and transportation, horses and mules have an important role in the agricultural sector of the country. In terms of world production, the United States is the second largest producer of horses, with over 3 million head of horses (**Figure 9–2**). China leads the world with 11 million head, the majority of which are used as work animals. In the United States, there are many types and uses for horses. Many horses are used primarily for riding pleasure. Some are true work horses that are used for herding and working cattle or pulling wagons. Horses are used for entertainment in race tracks all across the country.

Often overlooked are the millions of wild horses that inhabit the western portion of the country. These herds have existed for more than 200 years and continue to thrive. They are managed by the U.S. Bureau of Land Management. Some ranchers consider them as pests, so their numbers have to be controlled. Each year, wild horses are captured and adopted by people who tame and train them.

CLASSIFICATION

The horse belongs to the genus *Equis*. Within this genus are three groups of species. The domesticated horse belongs to the species *E. caballus* and includes the animals that we generally associate with working and riding. Another group of species includes the zebras, and yet another group includes the asses or donkeys. The only true wild horse is the *E. przewalskii*, which now exists only in parts of Mongolia. Wild horses in other parts of the world are

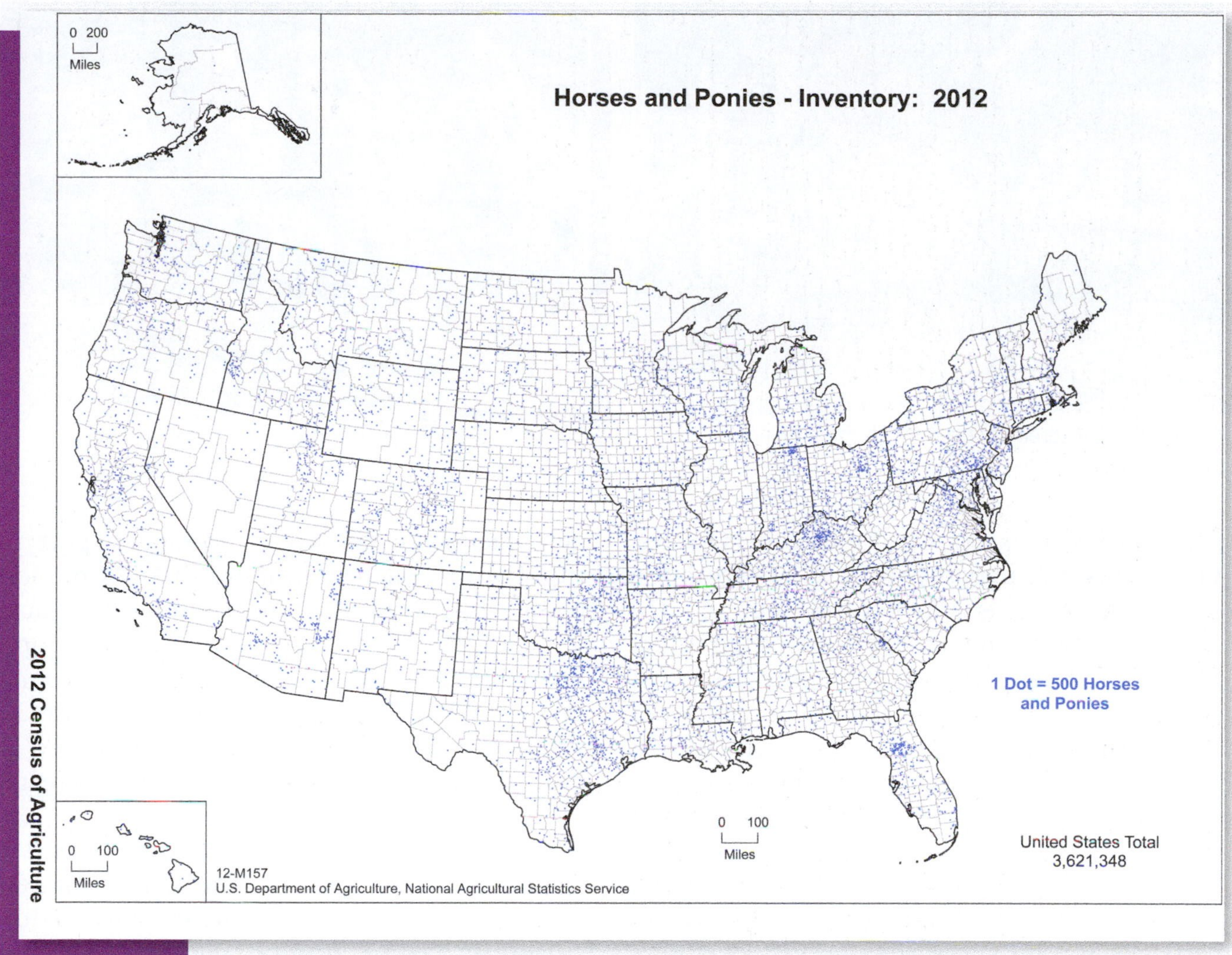

FIGURE 9–2 There are over 3 million horses in the United States. *Source: USDA, Census of Agriculture*

descendants of the domestic horse (*E. caballus*) that have escaped into the wild. Most of the species of *Equis* will interbreed to some extent, some much more successfully than others.

Although some horses and mules in the United States are used for work, most are used for recreational purposes (**Figure 9–3**). Horses today are generally categorized into one of three classes: **light horses**, **draft horses**, and **ponies**. Light horses are animals that weigh 900 to 1,400 pounds. These horses are further divided according to use. Horses can be used for many events, such as pleasure riding, trail riding, rodeos, racing, endurance riding, horse shows, fox hunting, dressage, combined training, polo, and driving. Although

FIGURE 9–3 In the United States, most horses are used for recreational purposes. *Source: USDA*

FIGURE 9–4 Gaited saddle horses such as the American Saddlebred are used for pleasure riding. © Anastasija Popova/Shutterstock.com.

FIGURE 9–6 Each year, millions of people attend thoroughbred races. © Cheryl Ann Quigley/Shutterstock.com.

some breeds are better suited than others for certain events, the majority of horse breeds in the United States are versatile (**Figure 9–4**).

Draft horses are breeds that weigh more than 1,400 pounds. At one time, these animals provided the power for pulling heavy loads such as wagons, plows, and other agricultural implements (**Figure 9–5**). Today, these animals are used in pulling competitions, shows, and parades.

Ponies are breeds of horses that weigh 500 to 900 pounds. Although some are used to pull carriages and for show, the majority of ponies are used as horses for children.

Altogether, the horse industry is a $15 billion industry in the United States. Surprisingly, horse racing ranks third behind baseball and auto racing as the largest spectator sport in this country. Each year, more than 75 million people attend thoroughbred and harness races (**Figure 9–6**). There are around 7,000 horse shows in this country ever year, where young people and adults compete with their horses in a variety of events. Many horses are owned as individual saddle horses that are used for recreation and are never raced or shown.

There are more registered quarter horses than any other breed in the United States. The American Quarter Horse Association currently has 1.8 million registered quarter horses, compared with the next most numerous breed, the Arabians, with 620,000 head. Quarter horses are still used to help herd cattle (**Figure 9–7**). No mechanized substitute

FIGURE 9–5 Draft horses are bred to pull heavy loads. © 1000 Words/Shutterstock.com.

FIGURE 9–7 Horses are still used to work cattle. © max voran/Shutterstock.com.

has been developed that is more effective in working cattle over the rough terrain of the open range. Personnel who work in remote wilderness areas rely on horses for transportation and for packing gear to areas that are inaccessible by car.

MULES

Mules have been bred and raised since ancient times, when humans recognized the special characteristics that make them so valuable. By combining the size, speed, and strength of the horse with the patience, perseverance, toughness, and agility of the donkey, a unique animal was created. The mule is a true hybrid, a cross between a male ass (jack) and a female horse (mare) (**Figure 9–8**). Because of this, the mule rarely can reproduce.

FIGURE 9–8 Mules are a hybrid cross between a male ass (jack) and a female horse (mare). *© iStockphoto/Thomas Shanahan.*

FIGURE 9–9 Mules were used in the South because they were better adapted to working in hot, humid areas. *© hutch photography/Shutterstock.com.*

The mule owns a particular place in American history. Around the time of the Revolutionary War, mules began to be bred and raised in this country to work the farms and plantations. They were particularly popular in the South, where they adapted more easily than did horses to working in the hot, humid weather (**Figure 9–9**). Mules have several additional advantages over horses: they usually have sounder feet and legs than horses. This means that mules have fewer problems with lameness, split hooves, splints, and other leg problems that are associated with horses. In rocky, hilly terrain, mules are more sure-footed and less likely to stumble than horses. For this reason, mules are used for transporting people up and down rough trails such as those found in the Grand Canyon. Tourists who journey down to the bottom of the Grand Canyon usually travel on the back of a mule.

If given the opportunity, horses will often eat so much grain that they harm themselves, but a mule will seldom overeat even if given free access to all the grain it wants. Although mules have a reputation for being stubborn, most of the stubbornness results from mules refusing to overwork themselves. When they become tired, they may balk and refuse to do any more until they are rested.

Mules are enjoying increased popularity. Each year, they are shown in various shows across the country. Mules are bred specifically for purposes such as pleasure riding, hunting, packing, and pulling wagons. Many parades in all parts of the country feature mule-drawn wagons as a part of the history of the United States.

ANATOMY OF THE HORSE

The different parts of the horse are shown in (**Figure 9–10**). It is the sum of these anatomical features that make it suitable for use by humans (**Figure 9–11**). The skeletal system is composed of strong bones that are connected with ligaments, giving the horse a fluid, gliding movement that allows a rider to sit atop the horse in comfort. Long bones in the hip and legs aid in the long stride of the horse; these bones act somewhat like a lever in propelling the horse forward.

The horse's muscular system is well advanced and is adapted to carrying heavy loads. Massive muscles down the back, over the croup (hip area), and down the legs give the horse the ability to pull loads and to sustain hard work for long periods. Horses with a relatively short back are better equipped to carry heavy loads than are horses with a long back because the muscles are concentrated in a shorter span.

The feet of horses are especially well suited for carrying loads. The horse is classified as a **perissodactyl**, an animal with only a single toe on each of its four feet. The foot is enclosed within a tough, hornlike structure that protects the tender inner structure of the hoof. Inside the hoof wall on the very bottom of the foot is the sole, which further protects the inner portions of the foot. The heel provides a flexible weight-bearing structure that also serves as a shock absorber for the foot and leg (**Figure 9–12**).

The digestive system of the horse is also highly specialized. Unlike many large animals, horses are nonruminants, meaning they lack a rumen (which produces enzymes to digest fiber). An enlargement in the digestive tract called the **cecum** provides a repository for large amounts of microbes that break down fiber into a form that is digestible to the horse. The cecum allows the animal to consume and digest grass and hay.

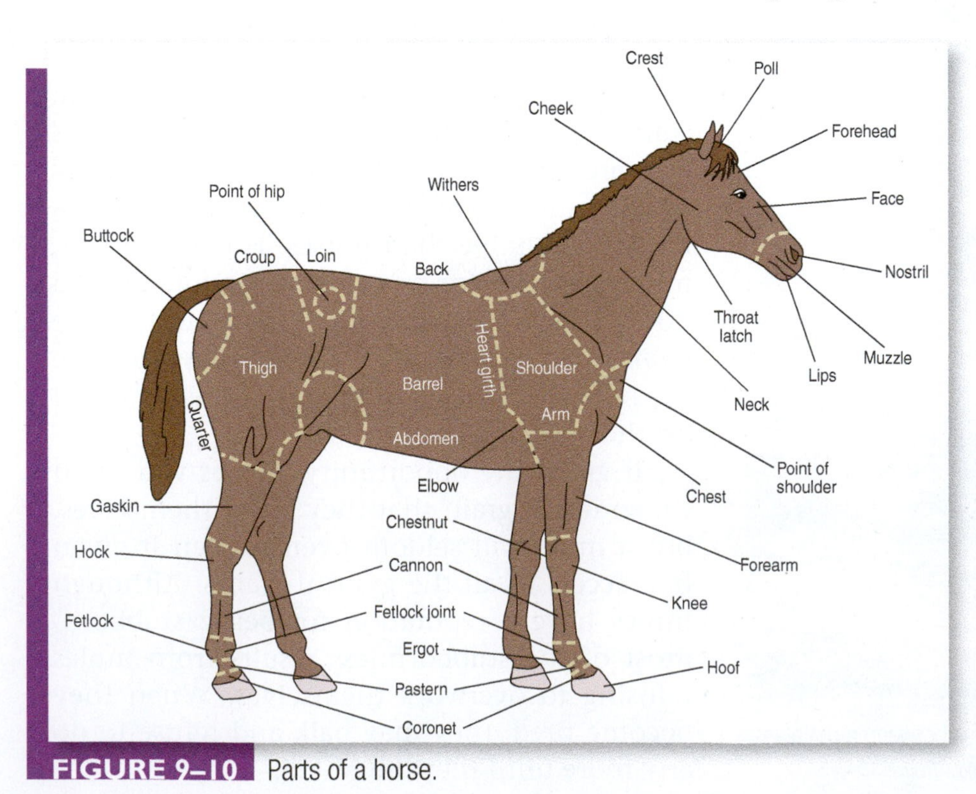

FIGURE 9–10 Parts of a horse.

Horse Conformation and Body Type

Like any other species of agricultural animal, a horse should have the proper body type to perform the required tasks. How a horse is formed—its conformation—has a direct impact on how well the horse moves, functions, or performs. Animal scientists have spent untold hours studying the best conformation for horses and have published many, many papers on the subject. Millions of dollars are spent each year buying, grooming, and showing horses that have good conformation.

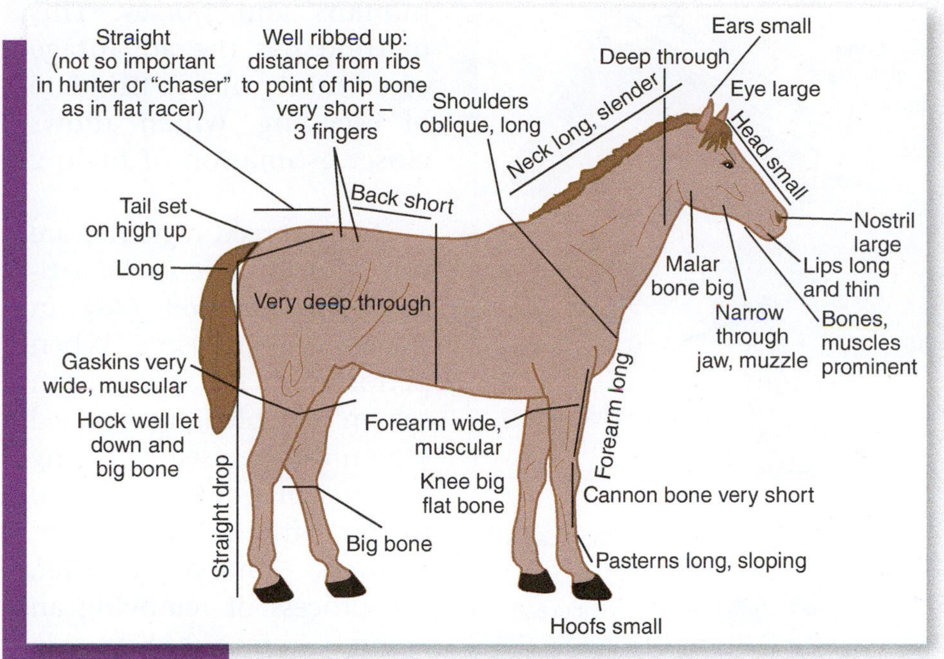

FIGURE 9–11 The horse has certain anatomical features that make it suitable for use by humans.

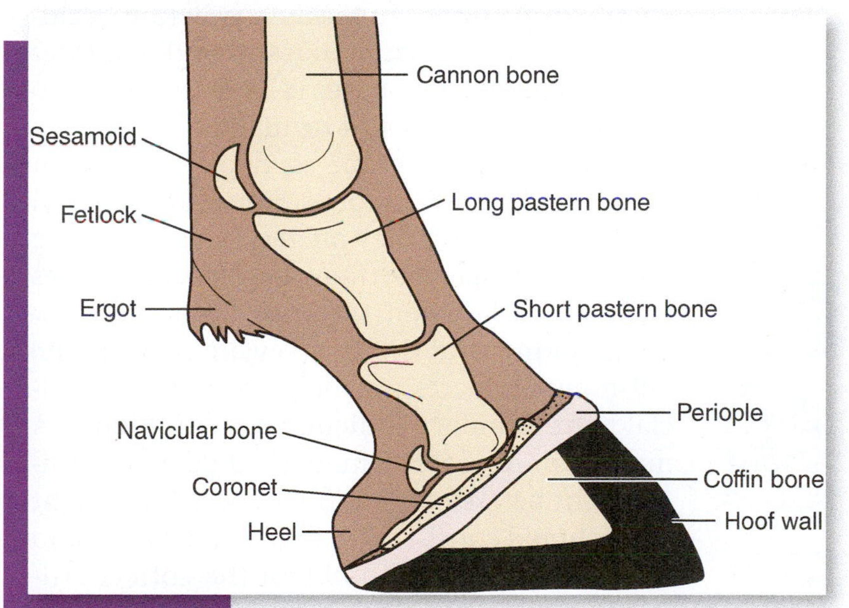

FIGURE 9–12 The feet of horses are well designed to carry heavy loads.

Although horses are used for a variety of reasons, certain characteristics are desirable in all types of horses. For example, a short back and a long, level croup are advantageous whether the animal is carrying a rider or pulling a heavy load. In addition, the neck should be long and slender to give the horse balance. Long, smooth muscles allow the animal to move freely and to work for long periods of time with less fatigue than an animal with short, "bunchy" muscles.

A horse should be able to move freely on all its legs. For a horse to function properly, its feet and legs must be structurally sound. Many horses are born with defects that make the feet and legs less than perfect, and these defects can cause problems as the horse walks or runs. If the legs are too straight, the bones will be jarred as the animal moves and the rider will not have a smooth ride. If the legs have too much curve, this will place undue strain on the muscles of the legs and the stifle (a joint in the hind leg). **Figure 9–13** shows the proper placing of the legs. Properly shaped and conformed bones and muscles allow the animal to function in the way it is expected to perform.

RAISING HORSES

Horses are generally bred using one of two methods—**pasture breeding** or **hand breeding**. Pasture breeding simply means that stallions are turned into a pasture with mares and mating takes place naturally (**Figure 9–14**). The advantage of this method is that it is less labor-intensive and usually results in a greater percentage of pregnancies. The disadvantage is that mares sometimes get rough in the mating process, often kicking and biting, which may result in blemishes in the skin of the mare or stallion.

Hand mating can be done under a variety of conditions. The stallion is usually brought

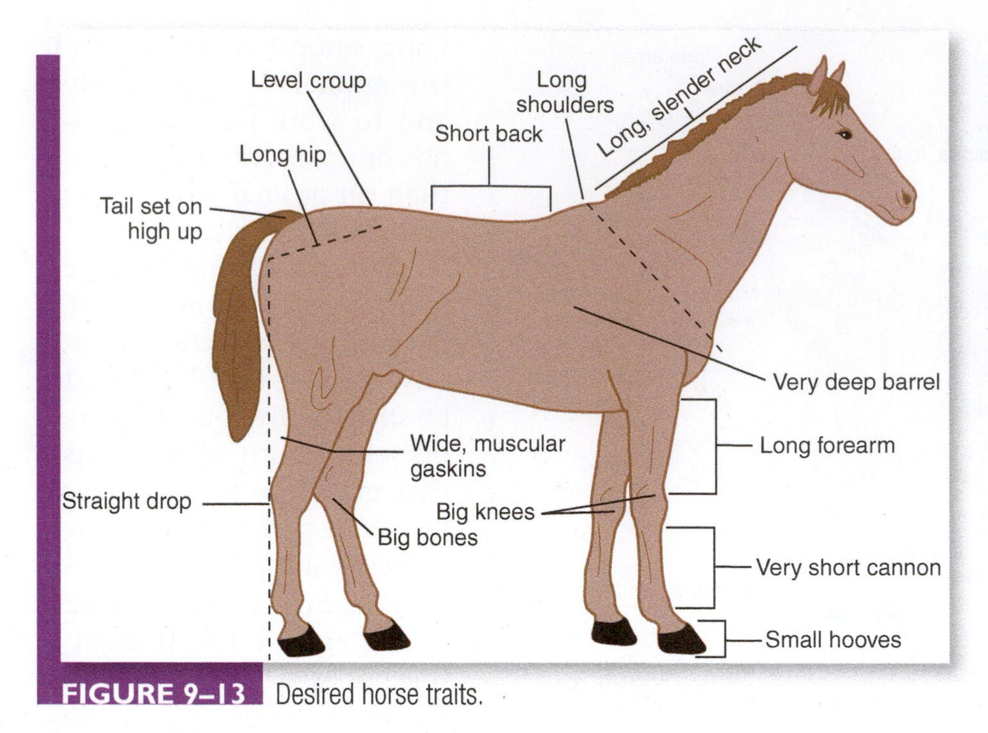

FIGURE 9–13 Desired horse traits.

FIGURE 9–14 Pasture breeding is less labor-intensive and results in a greater pregnancy rate. *© Pavel Vakhrushev/Shutterstock.com.*

to the mare, which is hobbled and restrained during mating. Another alternative is for the mare to be placed in a breeding chute, a rectangular box with short walls in the front and on both sides. Either method requires extreme caution to avoid injury to both humans and horses. This method has the advantage of certainty about the date of breeding, which allows closer estimation of foaling time.

Many breed registries are now allowing the use of **artificial insemination (AI)** in the equine industry. When performing AI, only fresh semen or cooled, transported semen can be used; shipping frozen semen, as is common in the cattle industry, is not allowed. Embryo transfer, the process of removing an embryo from one mare and transplanting it into another mare, is also becoming fairly common in the equine industry. Before deciding on either AI or embryo transfer, the breeder should contact the specific breed registry to obtain specific rules and regulations.

Equine reproduction is dependent upon photoperiod and hormones. Horses are seasonal breeders, meaning they are receptive to mating only during specific times of the year. Mares begin to cycle when the days become longer in the spring, and they stop cycling during the fall months. Mares can be placed under artificial light to induce follicular activity sooner. A mare requires approximately 60 days of artificial light before ovulation will occur. The light cycle should consist of 16 hours of daylight and 8 hours of darkness to obtain the correct artificial photoperiod. This concept is important for many breeds, as January 1 is the designated birthday for many foals. Because the gestation period is about 340 days, many breeders aim to get their mares pregnant as early as February. Having late-born foals can be a huge setback in many show and racing situations.

After a long gestation period, the foal is born (**Figure 9–15**). Following parturition, the navel cord is treated in a 10 percent iodine solution, and the foal usually is given an enema of warm, soapy water to help the foal

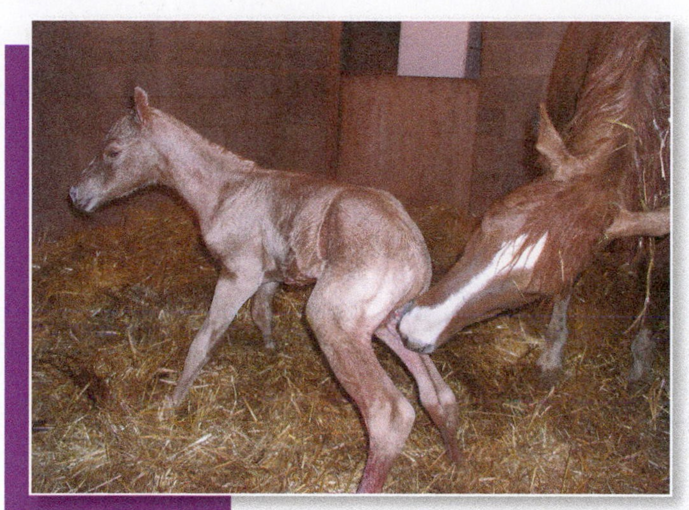

FIGURE 9–15 After a gestation period of about 340 days, the foal is born. *© Melissa E Dockstader/Shutterstock.com.*

FIGURE 9–16 Foals are generally weaned between 4 and 6 months of age. *© iStockphoto/Bob Balestri.*

pass the fetal meconium. The foal should try to stand and nurse within a few minutes. If the foal does not try within 1 to 2 hours, it should be helped to its feet and guided toward the mare's udder. As with all other mammals, the baby should receive the first milk, called **colostrum**, from the mother. This milk is rich in antibiotics and nutrients needed by the newly born animal.

Foals are generally weaned between 4 and 6 months of age (**Figure 9–16**). Males may be castrated any time from birth to 2 years of age to prevent the aggressive behavior of a mature intact male. A **gelding** (a castrated horse) is easier to handle and has a better disposition than a stallion.

Training usually begins before the foals are weaned because very young horses are easier to handle and train. Training begins by teaching the foal to lead with a halter and accustoming it to humans. The age at which a horse is trained to accept the saddle and be ridden depends on the breed of horse and its intended use. Most horses are taught this at age 2, but some breeds are taught at a later age. The amount of training and the overall time it takes to train a horse depend on what the horse is going to be used for. Care and patience are required to train the animals to respond properly to humans.

As with all other animals, many management practices must be observed when owning a horse. For example, horses have to be de-wormed and vaccinated regularly. The services of both a veterinarian and a **farrier** are required periodically. A properly trained and well-managed horse will give many years of faithful service (**Figure 9–17**).

FIGURE 9–17 Horses require a lot of management and service, including the feet. *Source: USDA*

SUMMARY

Horses are a vital part of the history of the United States. Over the years, they have served well as workers and companion animals. These animals are uniquely designed to provide service to human beings. Currently, the numbers of horses are on the increase for use as pleasure animals. This trend is likely to continue as more people become involved in the horse industry. Mules, a cross between a horse and a donkey, play an important role as draft animals. Like horses, they are making a comeback as recreational animals.

CHAPTER REVIEW

Review Questions

1. Discuss some of the uses that humans have made of horses in the past.
2. What are the three classes of horses?
3. Explain why mules make such good work animals.
4. Discuss the anatomical features of the horse that make it suitable for use by humans.
5. List at least five ways that horses are used presently in the United States.
6. Explain why horses are able to digest roughage.
7. Explain the advantages of hand breeding over pasture breeding.
8. How are foals cared for at birth?
9. Why are male horses castrated?
10. When does the training of a horse usually begin?

Student Learning Activities

1. Log onto the Internet and research a particular breed of horse. Find out about its origin, characteristics, uses, and popularity. Report your findings to the class.

2. Take a field trip to a horse show or training facility. Make notes on how the behavior and nature of the horses are used in the training. Report to the class.
3. Invite a veterinarian to visit the class to discuss good management procedures and basic first aid in caring for horses.
4. Go to the Agricultural Census site on the Internet and find out the number of horses in your state. How does your state compare with other states? Think of some reasons why there are relatively many or few horses in your state. Share your reasons with the class.

CHAPTER 10

The Aquaculture Industry

STUDENT OBJECTIVES IN BASIC SCIENCE

As a result of studying this chapter, you should be able to

- explain why fish gain more on less feed than other animals.
- discuss how fish obtain oxygen.
- explain how oxygen is dissolved in water.
- explain how oxygen is depleted from the water.
- distinguish between cool-water fish and warm-water fish.
- discuss the behavioral characteristics of bullfrogs that make them difficult to raise.

STUDENT OBJECTIVES IN AGRICULTURAL SCIENCE

As a result of studying this chapter, you should be able to

- list the reasons why aquaculture is a rapidly growing industry.
- give the advantages of fish over other agricultural animals in a production operation.
- discuss the problems associated with fish production.
- explain why catfish are the most widely produced aquatic animal in the United States.
- describe the methods used in the production of various aquatic animals.

KEY TERMS

aquaculture
crustaceans
steer
ectothermic
cold-blooded
photosynthesis
warm-water fish
cold-water fish
fry
fingerlings
seines
larvae

NATIONAL AFNR STANDARD

AS.01.01
Evaluate the development and implications of animal origin, domestication, and distribution on production practices and the environment.

AS.01.02.02.a
Research and examine marketing methods for animal products and services.

AQUACULTURE IS THE GROWING of animals that normally live in water. This production includes freshwater and saltwater fin fish; **crustaceans** (shrimp, prawns, and crayfish); mollusks (clams and oysters); amphibians (bullfrogs), and reptiles (alligators). Throughout recorded history, humans have eaten fish and other animals that live in streams, lakes, ponds, and the ocean. A ready supply of high-quality, protein-rich food could be obtained by harvesting organisms from the waters. As with other agricultural animals, humans soon discovered that by producing their own aquatic animals, the supply would be more dependable and easier to harvest.

Although it is difficult to determine just when aquaculture began, archaeologists know that people have raised fish for at least 3,000 years. The ancient Chinese and Egyptians kept fish in captivity for use as food. This is evidenced by the paintings and drawings on the walls of the tombs of the ancient Egyptians and by the writings of ancient Chinese scholars. Later, the Romans grew aquatic animals such as fish and eels.

The commercial growing of fish has increased in recent years. Each year, more than 66.6 million tons of fish are produced by fish farmers throughout the world. Asian countries grow more tonnage of aquatic animals than any other region, producing 88% of the world's production. The top-producing countries in Asia are China, India, Vietnam, Indonesia, and Bangladesh. Europe follows Asia in fish production, and North America ranks third. In the United States, fish culture is one of the fastest-growing agricultural enterprises. The demand by consumers for seafood (a term for all the aquatic animals) has increased to almost 15 pounds per-capita. Although this is far behind countries such as Japan, where the per-capita consumption is 148 pounds, seafood still accounts for more than 12 percent of the meat consumed by Americans.

Until a few years ago, the demand for fish and seafood had been easily met by the commercial fishing industry, which harvested wild fish from the sea and from freshwater sources. As the world demand increased along with the population increases, the commercial fishing industry had trouble meeting the demand because of overfishing in certain areas of the world. As a result of scientific research, aquatic animals are understood much better now than they were in the past. This allows producers to provide the type of environment that allows the animals to be produced efficiently and economically.

FISH PRODUCTION

Fish have several advantages over other agricultural animals. Although it takes about 9 pounds of concentrated feed for a **steer** to gain a pound, a fish will gain a pound on about 2 pounds of feed. This is because fish are **ectothermic**, once called **cold-blooded**, animals that do not require a large portion of their nutrient intake to go into maintaining body temperature. The bodies of ectothermic animals adjust to the temperature of their environment, so their body temperature is regulated by the surroundings. An endothermic animal—such as a mammal—regulates its own body temperature, and the internal temperature remains relatively constant. In addition, the natural buoyancy of their bodies in water helps fish move and supports their bodies; therefore, they use less energy to move about.

Fish have a higher percentage of edible meat than other animals. Typically, only about 35–40 percent of a steer's body weight is in edible meat. In contrast, a catfish is about 55 percent edible meat, and a trout may be as high as 85 percent edible meat. Thus, much more meat can be produced on an acre devoted to fish production than with any other agricultural animal. A well-managed pond can produce as much as 6,000 pounds of catfish per acre (**Figure 10–1**). With an ever-increasing world population, this could prove to be substantial in the future.

Fish producers face problems not encountered with the production of other agricultural animals. The grower must make sure that the dissolved oxygen level in the fish ponds is adequate for the fish. Like all other animals, fish must

FIGURE 10–1 A well-managed pond can produce as much as 6,000 pounds of catfish per acre. *Source: USDA, Agricultural Research Service (ARS).*

have oxygen to live. Whereas land animals obtain oxygen from the air they breathe, fish get oxygen from the water. Fish have gills that serve much the same purpose as lungs in land animals. The lungs in air-breathing animals separate oxygen from the other gases in the air and pass the oxygen into the bloodstream. In fish, the gills take oxygen from the water and pass the oxygen to the bloodstream (**Figure 10–2**). This oxygen is in the form of oxygen dissolved in water by green aquatic plants that release oxygen in the process of **photosynthesis**. Photosynthesis occurs only when the sun is shining, so oxygen is released from these plants only during the daylight hours.

Oxygen also passes into the water directly from the atmosphere through waves blown by the wind, ripples produced by moving streams, and waterfalls that allow water to drop through the air. Most human-made ponds are static and do not have the movement we see in flowing streams or waterfalls. When the oxygen level of the water is low because of atmospheric conditions, fish producers must rely on power-driven devices to fling the water into the air to absorb oxygen. Calm, cloudy days combined with a high temperature may cause the oxygen level in the ponds to fall below the level that the fish need.

Under these conditions, nights are particularly bad because aquatic plants are not undergoing the photosynthesis process that releases oxygen into the water. Without aeration, the entire population of a pond may die on a hot night. To prevent this from happening, the water is monitored periodically throughout the day and night using an oxygen meter that lets the operator know how much dissolved oxygen is in the water. When the oxygen falls below an acceptable level, large aerators are turned on, throwing the

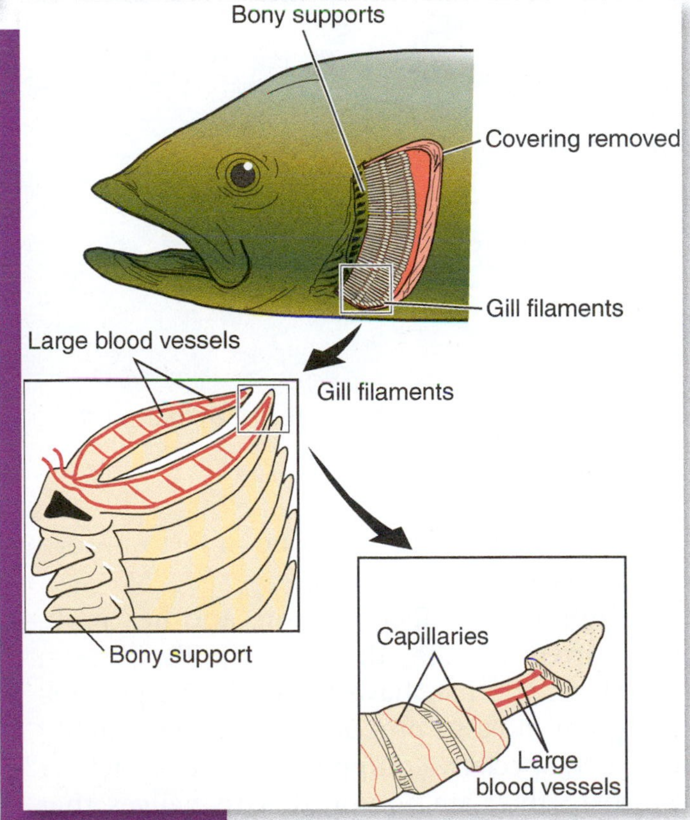

FIGURE 10–2 The gills of a fish take oxygen from the water.

FIGURE 10–3 Large, mechanically powered aerators are used to replace depleted oxygen. *Source: USDA, Agricultural Research Service (ARS).*

FIGURE 10–4 To deliver fish live to the processor, they are transported in special trucks that monitor the oxygen. *Source: USDA.*

water high into the air to be able to absorb more oxygen (**Figure 10–3**).

Shipping the fish presents another problem. After fish die, the meat spoils quickly; therefore, the fish have to reach the processing plant alive to produce the highest-quality product. Specially equipped tank trucks deliver the fish; they are equipped with gauges that monitor the water closely for temperature and oxygen levels (**Figure 10–4**). The fish are loaded and unloaded through large tubes that pump water containing fish to and from the truck tank (**Figure 10–5**).

Because fish have to be monitored so closely in all phases of their production, the operations are said to be labor-intensive. This means that people have to spend a lot of time on the job to produce the fish. The operations also are considered to be relatively high risk because the fish can be lost so rapidly.

Commercially grown fish are grouped into two broad categories: **warm-water fish** and **cold-water fish**. Warm-water fish will not thrive in water temperatures below 60°F, and cold-water fish will not thrive in water warmer than 70°F. In the United States, the most popular commercially raised fish are catfish and tilapia (warm-water fish) and the trout and salmon (cold-water fish).

Catfish Production

The most widely grown fish in this country is the catfish. Each year, producers in the United States harvest and market over 154,000 metric tons of catfish, amounting to a farm value of catfish that exceeds $300 million. These fish are gaining an increasingly wide market in all areas of the nation. Consumers are beginning to recognize farm-raised catfish as a tasty alternative to other forms of fish and seafood.

Catfish are different from most freshwater fish in that they have a smooth skin with no scales (**Figure 10–6**). They are hardy fish that

FIGURE 10–5 The fish are loaded and unloaded through large tubes.
Source: USDA

FIGURE 10–6 Catfish have a smooth skin with no scales. *Source: USDA, Agricultural Research Service (ARS).*

produce well in small ponds and survive with lower levels of oxygen than most other fish. Although there are many different varieties of catfish, the only variety of economic importance is the channel catfish. If left to grow for many years, these fish can grow to weigh more than a hundred pounds. Huge channel catfish are caught in large lake reservoirs every year in the deep South.

Because these fish do best when the water temperature is around 85°F, most are produced in the South. Mississippi is the leading producer of catfish, with about 80 percent of production, followed by Arkansas and Alabama. Most catfish are grown in open ponds in not more than 6 feet of water. In Mississippi, the acreage of ponds can be quite large. Farms with 250 acres under water are commonly found, and a number of farms in that state have as many as 1,000 acres in ponds.

Eggs are collected from female catfish by allowing them to lay in nests provided by the producers. The eggs then are collected and placed in tanks or jars in the hatchery. The eggs are moved back and forth gently by means of paddles in the tanks that slowly move the water in a wavelike action or the bottles containing the eggs are moved in a slow rocking action.

Just as with bird eggs, catfish eggs have to be turned for the embryo to grow properly. The rocking motion involved in the turning provides the necessary turning for the embryos to develop.

When the small fish, called **fry**, hatch, they are placed in a tank until they are about 2 inches long. These young fish, called **fingerlings**, then are placed in ponds or put in cages, where they will remain until they reach a weight of 1 to 2 pounds. The fish are fed a commercially processed food compressed into small pellets. The fish are fed twice a day by spreading the pellets on the water. In larger operations, labor is reduced by using a feed truck that drives to the edge of the pond and blows feed into the water.

The ponds are constructed so the producers can move through the ponds with large nets called **seines** to harvest the fish. Several passes with the seines are usually necessary to get most of the fish out of the pond (**Figure 10–7**). The fish then are placed in holding tanks until they

FIGURE 10–7 Catfish ponds are designed so the fish can be caught using seines. *Source: USDA, Agricultural Research Service (ARS). Photo by Less Torrans.*

FIGURE 10–9 Fish raised in cages are much easier to harvest. *© Dimitrina Lavchieva/Shutterstock.com.*

are pumped into trucks for transport to market. Salt is added to the water of the transport truck to keep the fish alive and well during the trip to the market. The salt water has a calming effect on the fish, lessening the stress.

In another method, the fish are grown in cages that are submerged in water (**Figure 10–8**). This method has several advantages over growing fish in open ponds: the fish are kept in a confined area where they may be inspected more closely. Less feed is wasted because the feed is spread out only within the cage. Cage raising helps solve problems of predators—such as turtles, cranes, and herons—that feed on fish in an open pond. Also, fish raised in a cage are much easier to harvest (**Figure 10–9**).

FIGURE 10–8 Fish sometimes are raised in cages that are submerged in the water. *Source: USDA, Agricultural Research Service (ARS).*

Tilapia Production

Tilapia (**Figure 10–10**) are fish native to Africa that are grown commercially all over the world. These fish resemble our native sunfish, and they reproduce prolifically, grow rapidly, and are considered to be a good-quality food fish. They are very hardy fish that survive high temperatures, low oxygen levels, and overcrowded conditions. Tilapia are a widely cultured fish and are second only to carp worldwide. Tilapia have been raised in the United States only within the past few years, but they are gaining in importance. Fish biologists consider them to have high potential as a commercially raised food fish in the United States. Because it is a warm-water fish, it grows best in the southern region of the country.

FIGURE 10–10 Tilapia are native to Africa and have potential as a food fish in the United States. *© Yossi James/Shutterstock.com.*

FIGURE 10–11 Trout are raised in concrete raceways.
© Kara Grubis/Shutterstock.com.

If the water temperature falls much below 50°F, tilapia cannot survive.

Trout Production

Trout is considered to be among the best-tasting fish and is served in restaurants all across the country. It is highly desirable as a food fish, not only for its eating quality but also because such a high percentage of the body is edible meat. These fish are cold-water fish that cannot survive in climates where the water temperature rises above 75°F. Trout are raised in smaller quantities than are catfish and are produced in the northern part of the country.

Most trout are raised in concrete raceways where the water is kept clean and moving (**Figure 10–11**). The moving water helps to keep the temperature low and oxygen in the water at an acceptable level. Diseases can be controlled more easily in concrete raceways, whereas disease organisms can be harbored in the soil of a regular pond. Harvesting in concrete spillways is much easier than in an open pond with an irregular bottom surface.

Salmon Production

Another cool-water fish that is gaining popularity as a cultured fish is the Atlantic salmon. In the coastal states of Washington and Maine, salmon are stocked in floating net cage enclosures anchored in coves and bays. The salmon are fed and cared for during an 18- to 24-month period. The fish then are harvested at around 9–11 pounds. These fish are meaty and have a flavorful taste. Although most of the salmon consumed in the United States come from ocean fishing or from Norway, many authorities think that the culture of salmon has a bright future in this country.

SPORT FISHING

Fish also are grown for sport fishing. Hatcheries all over the country raise small fish to stock ponds, lakes, and streams for the benefit of sport fishers. Almost every state in the country has large human-made reservoirs that are kept stocked with game fish for recreational purposes. Many of the fish caught were hatched in commercially operated fish hatcheries or in government-run hatcheries. Each year, these hatcheries stock lakes and streams with bass, bream, crappie, muskie, trout, and several other species of fish. This use of taxpayers' money can easily be justified because recreational fishing is a big industry in the United States. This agricultural enterprise spurs offshoot industries such as fishing tackle, boats, and guide services. People who run restaurants, hotels, and other stores all benefit from people who come to the lake to enjoy a weekend of fishing.

Some people grow fish in their privately owned ponds and make money by charging people to fish. Charges are made by the day or by the pounds of fish caught. As in most of the other types of fish-production operations, the fish are hatched, cared for, and harvested for a profit.

BULLFROGS

Frog legs are considered a gourmet food that is served in many restaurants. Although most of the frog supply comes from the wild, some cultivation of frogs occurs in Taiwan and Japan. In the United States, frogs have not been grown successfully in large numbers. Because the demand far exceeds the supply, however, small profits have been made as a sideline crop from fish ponds (**Figure 10–12**). The frogs are harvested much as they would be in the wild.

FIGURE 10–12 Bullfrogs are raised commercially in some parts of the world. *Courtesy of George Lewis, Cooperative Extension Service, Univeristy of Georgia*

In the United States, attempts have been made to produce frogs, but because of production problems, few have been successful. One of the major problems is that these animals are so territorial, which means that the animals claim a certain area and do not allow other frogs into their space. A second problem is that frogs will eat only food that is alive, so they cannot be fed processed feed. Their diet includes their own young, which can cause a real problem when frogs are raised in captivity. A third problem is predators. Almost all the predators that inhabit areas near water feed on frogs; these include raccoons, fish, herons, snakes, ducks, and cranes. As with the development of all the other agricultural animal industries, researchers will someday devise a way to raise bullfrogs profitably to meet market demand.

CRAYFISH

Crayfish (also known as crawfish and crawdads) are raised commercially in several states. The largest producing state is Louisiana, with more than 100 million pounds produced each year. Other states that produce crayfish include Oregon, California, Washington, Texas, and Mississippi. With proper management, more than 1,000 pounds per acre can be produced.

Crayfish are grown in constructed earthen ponds that are no more than 2 feet in depth. Often, they are grown along with crops such as rice. Crayfish are omnivorous—they eat both plants and animals—but most of their diet is made up of decomposing plant material. Crops such as rice leave large amounts of stubble behind when they are harvested. As the stubble decays, it creates a lot of food for the crayfish. In addition to decomposing plant material, crayfish eat worms and insect **larvae**.

The crayfish are put in the ponds in the spring. The water is slowly drained off in the late summer. As the water is lost, the crayfish burrow into the bottom, where they reproduce. The adult females lay large numbers of eggs, which hatch during the summer and early fall. These young are the crayfish that provide the harvest.

In the summer, a cover crop such as rice is planted and the pond is flooded (**Figure 10–13**). The crayfish are harvested in the late fall, winter, and early spring. Harvesting usually is done with a trap made from chicken wire that is closed on the bottom and has an inverted funnel at the top, (**Figure 10–14**). Canned dogfood, cottonseed cake, or chunks of fish are placed in the traps for bait. Trapping is more effective at night because the crayfish are searching more actively for food then. Once harvested, the crayfish are packed into porous bags and shipped. As long as the bags are kept cool and wet, the crayfish will survive and reach the processing plant in good condition.

FIGURE 10–13 Crayfish are produced in flooded rice fields. *© Yarddo/Shutterstock.com.*

ALLIGATOR FARMING

At one time, alligators were hunted to the point of extinction because of the high value of their hides. Today, because of extensive conservation efforts, the numbers in the wild have greatly multiplied until they are no longer endangered. As a result of the conservationists' efforts, techniques for growing alligators were perfected. This technology now is used for commercial production of the animals on farms located primarily in Louisiana, Florida, and Georgia.

Alligators can be harvested at about 26 months of age, when they have reached 5 to 6 feet in length (**Figure 10–15**). The hides are sold to make exotic leather goods such as bags, boots, shoes, and belts. The meat is tasty and is sold to the restaurant trade. Specialty items such as skulls and teeth also are sold.

Brood alligators are obtained from other producers or from the wild. To use wild alligators, producers have to secure a permit from the state game commission and agree to release back into the wild a predetermined number of alligators that are a year or older.

FIGURE 10–14 Crayfish are harvested by catching them in traps. *© iStockphoto/knape.*

FIGURE 10–15 Alligators can be harvested at about 26 months of age, when they reach a length of 5 to 6 feet. *© iStockphoto/Lorraine Boogich.*

The females build nests from vegetation and mud, lay around 40 eggs in the nest, and cover them. Producers remove the eggs from the nest as soon as they are laid because the eggs are a favorite food of several predators. As the eggs are removed, they are marked so the proper end will be placed in an upright position throughout incubation to ensure that the eggs hatch. The eggs then are wrapped in hay and are kept moist. The wet hay harbors bacteria that help to decompose the shell of the egg. A partially decomposed shell enables the young alligator to break through without much difficulty.

The temperature during incubation is critical in determining the sex of the young animals. Temperatures lower than 86°F produce all-female broods; temperatures above 93°F produce all-male broods. If the temperature is held at about 88°F, the brood is mixed in gender.

When trawlers harvest the sea, fish are brought up in the nets that are not desirable for human consumption. These are the fish that producers use to feed alligators (**Figure 10–16**).

FIGURE 10–16 Alligators are fed undesirable fish that are left over from the trawling industry. *© AlenaLitvin/Shutterstock.com.*

Some animals are fed by-products from poultry-processing plants, even though research has shown that this diet is somewhat too high in fat content. Other sources of food include carcasses of animals that are raised and slaughtered for their fur.

SUMMARY

Aquaculture is one of the newest components of animal agriculture. Because of the high efficiency of aquatic animals and the healthy, nutritious contribution they make to our diet, this area of animal agriculture is likely to grow in the future. The operations are highly labor-intensive and expensive to operate, but the demand for fish and other products from aquatic animals continues to grow as the oceans of the world cannot keep up with demand. Research will show us ultimately how to produce the animals more efficiently and at a lower cost.

CHAPTER REVIEW

Review Questions

1. For how long have people been engaged in aquaculture? How do we know this?
2. Why are fish more efficient users of feed than are traditional agricultural animals?

3. Name at least two significant problems associated with the production of fish.
4. Why are catfish so widely produced?
5. What are the characteristics of tilapia that give them such a high potential for production?
6. What are the advantages of raising fish in cages? Why are trout and salmon raised in concrete raceways?
7. What are three problems that make bullfrogs difficult to raise?
8. Why are crayfish grown with crops such as rice?
9. What are the commercial uses for alligators?
10. What governmental regulations apply to the raising of alligators?

Student Learning Activities

1. Visit a large grocery store that markets seafood. List all the types of food from aquatic animals that the store sells. Determine which of the aquatic animals were caught in the wild and which were produced in aquaculture operations. Record the differences in price for each category.
2. Locate and visit an aquaculture operation in your area. Determine the problems and solutions in running the operation. How are the animals marketed?
3. Talk with a local conservation officer about the types of fish that are released in lakes and streams in your area. Find out where the fish come from, the stocking rate, and plans for introducing different types of fish in the future.

CHAPTER 11

The Small Animal Industry

STUDENT OBJECTIVES IN BASIC SCIENCE

As a result of studying this chapter, you should be able to

- describe how humans first adopted pets.
- explain behavior characteristics that make some animals good companions.
- describe how different breeds were developed.
- analyze the health benefits of owning pets.
- list some diseases that may be transmitted from pets to humans.
- describe ways of preventing the transmission of diseases from pets to humans.

STUDENT OBJECTIVES IN AGRICULTURAL SCIENCE

As a result of studying this chapter, you should be able to

- describe the importance of the pet industry to the economy of the United States.
- tell how companion animals were and still are used in agriculture.
- explain how dogs are classified according to their use.
- describe how by-products are used in the processing of pet food.
- explain the regulations governing the raising and importing of companion animals.

KEY TERMS

companion animals
exotic animals
exothermic
service animals
assistance dogs
hippotherapy
zoonoses

NATIONAL AFNR STANDARD

AS.01.01
Evaluate the development and implications of animal origin, domestication, and distribution on production practices and the environment.

AS.01.02.02.a
Research and examine marketing methods for animal products and services.

THE RAISING OF AND CARING for pets is a large and rapidly growing animal industry. In the United States, almost two-thirds of households own pets. Americans spend over $69 billion a year on companion animals. Dog and cat food purchases alone account for about $8 billion of the total spent on pets. By comparison, people spend approximately $1 billion on baby food. Dogs and cats are the most common pets, and in recent years, cats have outnumbered dogs as the favorite pet in the United States. There are about 94 million cats and about 90 million dogs in American households. Many cat owners have more than one cat, so although cats are more numerous, dogs are found in more households (**Figure 11–1**). Other pets that people own include horses, rabbits, hamsters, gerbils, guinea pigs, and fish. There is a trend to own exotic animals such as potbellied pigs, reptiles, ferrets, fancy birds, and even tarantulas. This wide range of pets provides companionship for people of all ages.

The popularity of dogs and cats as companions is a result of a combination of factors, including the ability of dogs and cats to form lasting social bonds with humans, establishing a close relationship with their owners. These animals also can be house-trained fairly easily, whereas house training is impossible with many companion animals, such as horses. Dogs and cats have another advantage in being large enough for humans to interact with and play with, yet small enough to be kept in the house.

FIGURE 11–2 Americans spend more than $30 billion each year for pet food. *© Iakov Filimonov/Shutterstock.com.*

FIGURE 11–1 Dogs are in more homes than any other pet. *© foaloce/Shutterstock.com.*

Many of these pets are bred and raised in commercial operations that supply animals to be used for pets. A gigantic industry has grown up around the care of pets and companion animals. Americans spend almost $30 billion each year for pet food (**Figure 11–2**) and an additional $16.7 billion for veterinary services. As the affluence of the average person increases, the expenditures for pets and pet care likewise will increase.

THE HISTORY OF PETS

Evidence of people keeping pets dates back several thousand years B.C. The first companion animals probably were domesticated from wild animals and served purposes other than strictly as pets. Dogs were used for hunting and herding animals, and they also served as watch animals, warning humans of wild animals or strange humans approaching. By definition, **companion animals** are animals that are domesticated and can be considered a pet.

DOGS

Dogs have been associated with humans as far back as the Stone Age. All modern dogs developed from wild dogs that resembled wolves.

FIGURE 11–3 People discovered that dogs could be useful in herding animals. *© Mikkel Bigandt/Shutterstock.com.*

FIGURE 11–4 The Brittany Spaniel belongs to the Sporting Group of dogs. *© Capture Light/Shutterstock.com.*

Dogs are scavengers, and archaeologists think that dogs may have adopted humans rather than humans adopting dogs. The theory is that dogs began hanging around villages to scavenge leftover food of the humans. People probably discovered that the dogs could help in the hunts by tracking and herding animals (**Figure 11–3**), so they began to raise the animals for hunting.

A characteristic of dogs is that there can be quite a mixture of traits in the animals born in the same litter due to genetic variation. As humans began to see traits they liked in certain dogs, they began to select animals with those characteristics and breed the males and females to produce the type animal they wanted. This is how different breeds of dogs developed. Within the dog family, some breeds weigh scarcely 2 pounds and other breeds weigh more than 200 pounds. Dogs come in all sizes, shapes, colors, and temperaments. In fact, there are more than 400 recognized breeds of dogs in the world.

Breeds of dogs are divided into seven major groups. The *Sporting Group* includes hunting dogs such as the Labrador Retriever, the Irish Setter, and the Brittany Spaniel (**Figure 11–4**). The *Hound Group* is used for tracking and treeing game and includes breeds such as Beagles, the Bloodhound, and the Black and Tan Coonhound. The *Terrier Group* consists of smaller dogs including the Fox Terrier, the Welch Terrier, and the Bull Terrier. The *Working Dog Group* was developed to provide service as sled dogs, guard dogs, and messenger dogs; examples of this group are the Alaskan Malamute, the Boxer, and the Doberman Pinscher. The *Herding Dogs* were bred to help in the raising of livestock by herding the animals and protecting them from predators; these include the Border Collie, the Old English Sheepdog, and the Australian Cattle Dog. The *Toy Dog Group*, the smallest of the dogs, include breeds such as the Chihuahua, the Pekinese, and the Pug. The last group is the *Nonsporting Dogs*, composed of a wide variety of dogs that are used primarily as companion animals; in this group are the Bulldog, the Poodle, and the Dalmatian.

CATS

Archaeologists were able to distinguish the remains of the first domestic cats from the wild species in their excavations of ancient Egypt; they say the cat was bred and worshipped in Egypt (1570–1085 B.C.) to some extent. The popularity of the cat, however, may have stemmed—more than any other reason—from the protection they gave to granaries by killing mice and other rodents. The ancient Romans also valued cats for their service of ridding homes and grain storage areas from destructive vermin, as well as their use as companion animals. The Romans probably were the first people to take cats into

FIGURE 11–5 Cats are clean, quiet, intelligent animals that make good companions. © iStockphoto/sdominick.

Europe and other parts of the world. Today, cats inhabit almost every country in the world, both as wild and domestic varieties.

In the United States, cats were used to rid homes of mice and other vermin; they could be found on farms and in homes all across the nation. Most times, these animals earned their keep by catching and consuming their own food, and they often lived in barns or grain storage buildings. Today, cats are popular even though they are not used as much to catch mice. They are excellent companion animals because they are clean, quiet, and intelligent. Cats provide enjoyment for millions of people (**Figure 11–5**).

Unlike dogs, cats only relatively recently have been bred selectively, and the breeds have much less variation. Consequently, there are fewer breeds of cats than dogs. Most cat breeds can be traced back to the late nineteenth century, continuing to the present. Most of the breeds that are named after a region, such as the Persian and the Abyssinian, probably did not originate in these areas. Breeds of cats are generally divided into two groups: shorthair and longhair.

EXOTIC ANIMALS AS PETS

Many people prefer pets other than the traditional dogs, cats, and fish. A large industry has developed around raising and/or importing **exotic animals**. At one time, there was a thriving market for animals captured from the wild and imported for use as pets. Now, strict laws regulate the importation and sale of certain animals; it is illegal to import and sell most animals captured from the wild. Many endangered species are protected from sale to private individuals. These animals tend to make poor pets, and removing them from their natural habitat increases the risk of their extinction.

The importation of animals also carries a risk of bringing disease into this country. A classic example is New Castle disease, a highly contagious viral disease that decimated much of the poultry industry in the United States. It is thought to have been brought in by a parrot smuggled into New England from Mexico. All animals brought into the United States must go through a quarantine period to check for disease and parasites.

REPTILES

The fastest growing category of pets in the United States today consists of reptiles. More than 7 million pet reptiles are living in homes in the United States (**Figure 11–6**). Because reptiles are cold-blooded, they traditionally have not been considered good pets. Through better education about the characteristics of these animals, their use as pets has increased. Snakes such as boas and pythons are common as pets.

FIGURE 11–6 The fastest growing category of the pet industry consists of reptiles.

Iguanas are the most popular pet among reptiles. These lizards are the size of a domestic cat when fully grown. They are clean, odorless, and can be house-trained. Most are grown in Central and South America under unsanitary conditions, and salmonella poisoning is a danger when handling these pets. Owners should thoroughly wash their hands with antibacterial soap after handling these animals.

Reptiles require special care, as they are **exothermic** animals, which means that their internal body temperature comes from the environment. They need lights and some source of heat, such as a heat rock, to keep them comfortable.

HEALTH BENEFITS

Evidence is increasing that pets are more than just companions; they also are good for people's health (**Figure 11–7**). Scientists are discovering that living with a pet contributes to both the physical and the mental well-being of humans. There is a need in humans to have a relationship with something living. Since earliest human history, there has been the need to be around and to have dealings with other human beings. Now, scientists say that the need for companionship can in part be supplied by companion animals. According to this theory, relationships with animal companions seem to be beneficial to humans because these relationships are uncomplicated. Animals are accepting, attentive, and responsive to affection. They are not judgmental; they don't talk back; they don't criticize or give orders. Pets give people something to be responsible for and make them feel special and needed.

FIGURE 11–7 Pets are more than companion animals; they are good for people's health. *© vladgphoto/Shutterstock.com.*

FIGURE 11–8 Under the right circumstances, companion animals can have a good influence on children. *© Aleph Studio/Shutterstock.com.*

In the right circumstances, companion animals are a good influence on children of all ages because they help children develop a sense of security. Animals have been used to encourage shy or withdrawn children to open up (**Figure 11–8**). Children who are normally hyperactive often become much calmer around a companion animal. Caring for animals also helps them develop a sense of responsibility, which can be useful in all areas of life.

Evidence of the beneficial effects of companion animals on human health is continually being discovered. We now understand that pets make people feel good, and a sick person who feels good mentally is likely to get better faster. People who have pets report fewer minor health problems such as colds and flu. Studies also have shown that petting a dog or a cat or watching fish in an aquarium can help lower a person's blood pressure and heart rate.

Companion animals can play an important role in the lives of older individuals. Because of the increasing mobility in today's society, many elderly people no longer have family members living close by. Many of them would feel isolated and alone without the pets that provide them with companionship and a sense of being valued and needed. Older people may actually live longer, healthier lives because of their relationships with companion animals.

Today, about half of nursing homes use animals in some capacity to aid in the care of the elderly and invalids. Bird aviaries and aquariums, as well as cats, rabbits, and guinea pigs, are popular with nursing home residents. Some authorities do not recommend that dogs live in nursing homes full-time because they tend to be overfed by the residents, who can't seem to resist feeding them cookies and treats. The dogs can become obese and health problems result. Today, many volunteers take their dogs to visit the residents of nursing homes and hospitals in their community.

SERVICE ANIMALS

Many people in the United States and other parts of the world are challenged by some form of disability. Companion animals that are known as **service animals** have a tremendously valuable role in assisting with everyday aspects of life. Dogs serve as the eyes, ears, or legs for thousands of people who need assistance in moving about and tending to daily routines (**Figure 11–9**). A number of agencies in the United States train **assistance dogs** to give people with disabilities more independence and mobility. Training usually takes between 4 and 8 months, depending on the difficulty of the tasks that must be learned and the dog's aptitude. Although training assistance dogs can cost thousands of dollars, many agencies provide them at little or no cost to people who need them.

The best-known example of an assistance dog is the guide dog for the blind. The most commonly used breeds are German Shepherds and Labrador Retrievers. To work as a guide dog, the dog has to be exceptional. It has to be able to walk through crowds, climb stairs, take elevators, and ride on buses and in cars. Most important, the dog has to be able to think for itself (**Figure 11–10**). It must learn to disobey a command if it could bring harm to its master.

Many of the organizations that train guide dogs have volunteers who raise the puppies during the first year. Some of the volunteer puppy raisers belong to the 4-H organization. The volunteers are not expected to train the puppies as guides but are required to follow some basic rules. The puppies must be exposed to many of the activities that they will have to handle with ease as a guide dog. The puppy raisers are encouraged to take the puppies to the mall, to the park, and to nursing homes, schools, and any other setting that involves the public. The puppies must be kept on a leash in public, and they must sleep next to the volunteer's bed, just as they will when serving as guides.

The volunteers are warned not to play ball, tug-of-war, or other games with the puppies

FIGURE 11–9 Dogs can serve as helpers to people who are physically challenged. *© Rock and Wasp/Shutterstock.com.*

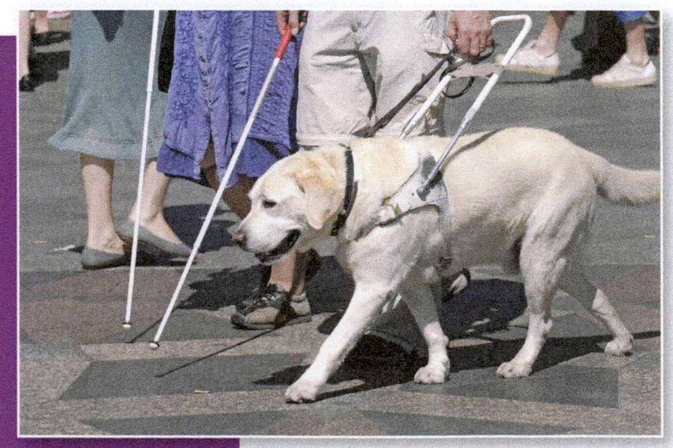

FIGURE 11–10 Guide dogs must be able to recognize dangers. *© Lars Christensen/Shutterstock.com.*

because these games could turn into bad habits when the dogs become guide dogs. The puppy raisers must not feed their pets human food, as guide dogs must not be tempted by the sight or odor of human food when they accompany their owners to restaurants, and they can't be jumping up on tables begging for food.

Only about half of the puppies raised to be guide dogs complete the training successfully. During training, the dogs learn to work in a harness, and they learn commands such as "Forward" and "Find the door." The dogs are trained through repetition and praise. They learn to ignore crowds, noises, squirrels, and cats. The dog must imagine itself to be as wide and as high as a human so it can guide a person through doorways and buildings successfully. It also must be aware that obstructions it can walk around or under easily, such as awnings or branches, would impede its owner. A guide dog wearing a harness is on duty and should never by petted by other people; however, when the dog is out of the harness, it is like any other family pet.

Hearing-ear dogs are trained to listen for those who cannot hear. Also called *signal dogs*, these animals can respond to more than 30 common household sounds, including doorbells, telephones, alarm clocks, and fire alarms. They even can be trained to respond to a crying baby. The dogs alert their deaf or hard-of-hearing owners by walking back and forth from the source of the noise to the owner. Signal dogs are taught not to bark when alerting their master, as the person would be unable to hear the barking. The dogs can also be trained to respond to sign language commands. Because the size of the dog is not important, hearing-ear or signal dogs usually are mixed breeds and often are rescued from local animal shelters.

Service dogs are trained to help people who use a wheelchair or have a spinal injury. These dogs are able to respond to more than 40 different commands. Service dogs can open doors, work light switches, pull emergency cords, and pull wheelchairs. Each dog may be trained a little differently to address the needs of the individual who will own the dog. Service dogs have to be large, and they often are retriever breeds.

Another companion animal that has been beneficial to humans is the horse. Horses are used in programs of physical therapy for people with disabilities. This type of therapy, called **hippotherapy**, can be used with people of all ages but is especially helpful for physically challenged children (**Figure 11–11**). The gait of a horse simulates the motion of humans as we walk. When we walk, our bodies move from side to side and up and down. Riding a horse re-creates that sensation in people who are unable to walk unassisted.

Through hippotherapy, physically challenged individuals improve their balance, posture, strength, coordination, and muscle flexibility.

FIGURE 11–11 Horses can aid in the recovery of physically challenged people. *© iStockphoto/tirc83.*

In addition to the physical benefits, the riders gain confidence. Hippotherapy can provide them with a whole new perspective and a sense of freedom. A horse can take its rider where no wheelchair could go. Volunteers are used in the care and maintenance of the horses, as well as for lessons; therefore, hippotherapy programs offer many opportunities to local agriculture programs and youth organizations.

PET FOOD

The pet food industry utilizes many of the by-products and surpluses of the human food industry. The main ingredient in dry dog foods is grain—corn, soybean meal, or wheat millings. The main ingredients in canned pet foods are meat by-products, which may include the waste products of meat that was processed for human consumption. Also, carcasses declared unfit for human consumption may qualify for use as pet food. The composition of pet food is formulated carefully to meet the animal's nutritional needs. Canned, semi-moist, and dry foods are equally nutritious, but the canned varieties generally contain a higher percentage of protein and fat.

ANIMAL HEALTH

Americans spend more than $10 billion every year on the health care of their companion animals. There are approximately 40,000 veterinarians in the United States, and one-third of them treat small animals exclusively (**Figure 11–12**). Like all other veterinarians, those who specialize in the care of companion animals perform a wide variety of tasks. They treat animal injuries, set broken bones, immunize healthy animals against disease, and perform surgery.

Veterinarians today can perform hip replacements and kidney transplants on companion animals. They also have performed balloon angioplasty to open clogged arteries, open-heart surgery, and dental surgery. In recent years, medical care for animals has become as highly technical as medical care for humans. New medical procedures have been perfected on animals to the extent that, in some instances, animals may get even more advanced care than humans. The costs of these health care advances, however, may be more than some pet owners are willing or able to bear.

FIGURE 11–12 Over one-third of veterinarians treat small animals exclusively. *© Nestor Rizhniak/Shutterstock.com.*

Veterinarians stress preventive measures when they counsel pet owners. They encourage vaccination programs and regular dental exams in the pet's health care plan. According to veterinarians, one of the most common problems they see in dogs and cats today is obesity. Half of American dogs and almost one-fourth of cats are obese. The animals suffer from overfeeding and lack of exercise. Some animals simply do not get enough attention. When animals get bored, they have a tendency to eat too much, just as people do (**Figure 11–13**). A sound diet and daily exercise routine should be part of the pet's overall health care plan.

Diseases and Afflictions

Unfortunately, pets can give us more than companionship. Each year, pets pass along infectious diseases to thousands of Americans. Approximately 30 varieties of pet-borne diseases and infections, called **zoonoses**, can be transmitted from animals to humans. These conditions can be contracted through direct contact with the animals or acquired indirectly through contact with animal feces or other contaminants. Most diseases passed from animals

FIGURE 11–13 Many animals have a tendency to overeat. © iStockphoto/humonia.

to humans are easy to avoid and are treatable. Good hygiene and safe handling procedures should always be practiced when working with animals. Cats and dogs are responsible for the majority of zoonoses, but birds, fish, and turtles also are culprits.

Rabies is the best known and the most feared example of a zoonosis. It is contracted through the saliva of rabid animals. Although this disease is rare in pets, rabies is increasing in the wild animal population. When pets could come into contact with wild animals, they should be vaccinated against the disease.

The common round worm in dogs is a parasite that can infect humans. The parasite is transmitted through contact with the animal's feces or with contaminated soil. Children playing in areas frequented by dogs are especially at risk. Dog owners should make sure their pets are dewormed regularly (**Figure 11–14**).

Toxoplasmosis is a parasitic disease that can be caused by contact with cat feces. Toxoplasmosis is especially dangerous to pregnant women, so most veterinarians recommend that pregnant women do not clean cat litter boxes.

Psittacosis, or parrot fever, is a disease transmitted by parrots, budgerigars, and other related caged birds. Humans can be infected by contact with the feces of contaminated birds. Those who handle birds and cages should use dust masks or protective face shields.

Ringworm is not a worm; it is a fungus that results in skin aggravation in humans. It is passed to humans primarily by kittens and puppies. The animal appears to be unaffected because the fungus infects only the animal's fur. It is passed to people when handling their pets. Ringworm is more common in children because adults seem to become more resistant with age.

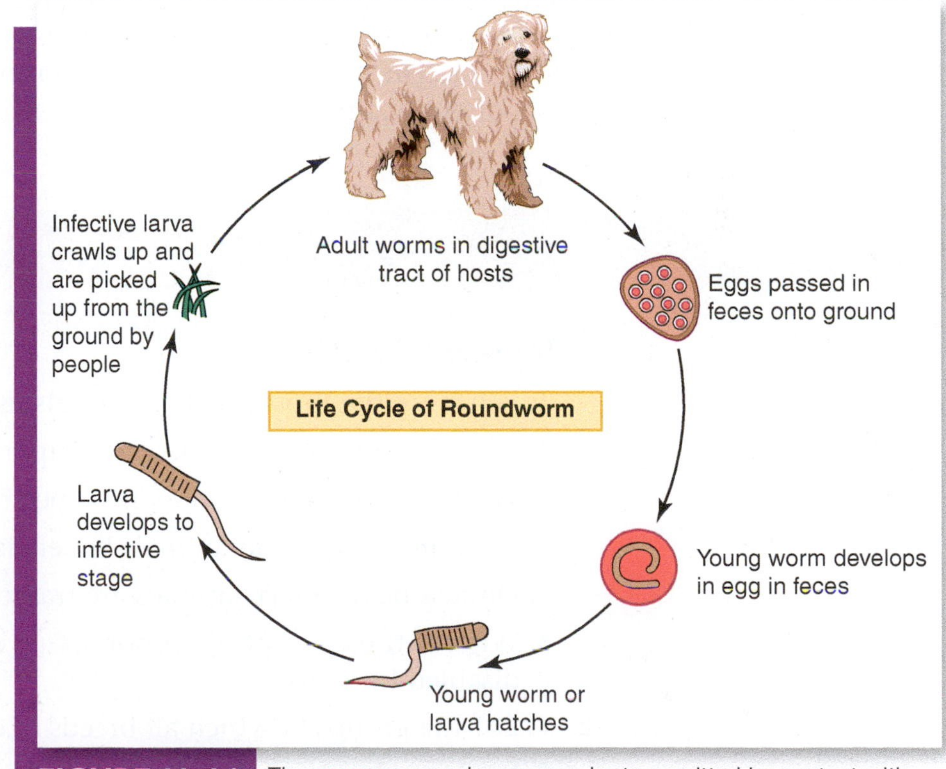

FIGURE 11–14 The common round worm can be transmitted by contact with contaminated soil.

Rocky Mountain spotted fever and Lyme disease are tick-transmitted diseases that can affect humans and animals alike. Ticks are found in grassy, wooded areas and can be brought into the house by dogs and cats that have been outside. Both diseases are treatable with antibiotics. Normal grooming of animals after they have been outside will help locate and eliminate ticks.

Infections from animal bites and scratches are another concern. The potential for infection varies. Less than 5 percent of dog bites become infected, but up to 50 percent of cat bites result in infections. Cat scratch fever is associated with cat scratches or bites. This disease is not serious and can be treated with antibiotics. Safe handling techniques are an important measure to prevent bites and scratches, as well as to prevent injury to the animal. Prompt and thorough washing of pet bites and scratches with soap and water is always recommended.

Allergies are probably the most common afflictions that result from human contact with animals. Many people develop allergies to animal hair and their dander, or flaking skin. These allergies cause hay fever–like symptoms in children and adults.

SUMMARY

The pet industry in the United State is huge and growing. Americans like pets and are willing to spend a lot of money to own and care for companion animals. These pets serve many purposes, ranging from service animals to helping humans become better adjusted and content. Types of animals range widely, and, as the result of a demand for exotic pets, the importation of animals is regulated closely by the government. The business of pets will continue to be a large and growing part of our economy.

CHAPTER REVIEW

Review Questions

1. What are the most popular animals used for pets?
2. Why did dogs and cats become important to early humans?
3. What are the benefits of owning pets?
4. Why are companion animals beneficial to elderly people?
5. Discuss how service animals are trained.
6. What is the concept of hippotherapy as it relates to horses and disabled humans?
7. List the groups to which all breeds of dogs belong.
8. Why are there more breeds of dogs than cats?
9. What by-products are used in pet food?

10. What precautions should be observed in handling pets to prevent disease?
11. List some of the common zoonoses.

Student Learning Activities

1. Choose a pet and list the characteristics of that particular animal that make it desirable as a pet.
2. Pick a breed of cat or dog and research the origin of the breed. Report to the class.
3. Conduct a survey of the members of your class to determine the number and types of pets owned. Have each person explain why they like that particular type of animal.
4. Take one or two well-behaved pets to a local nursing home and visit with elderly people. Be sure to check with the nursing home administrators before you go. Record the reactions of the people to your pets. Report to the class.

CHAPTER 12

Alternative Animal Agriculture

STUDENT OBJECTIVES IN BASIC SCIENCE

As a result of studying this chapter, you should be able to

- explain the meaning of the pH scale.
- list the animals that are most often used in scientific research.
- define *certified laboratory animal.*

STUDENT OBJECTIVES IN AGRICULTURAL SCIENCE

As a result of studying this chapter, you should be able to

- define *alternative animal agriculture.*
- list the advantages of raising rabbits as agricultural animals.
- discuss the uses of llamas.
- explain how fish bait is raised.

KEY TERMS

alternative animal agriculture
hutches
USDA
pH
castings
laboratory animals
genetic defects

NATIONAL AFNR STANDARD

AS.01.01
Evaluate the development and implications of animal origin, domestication, and distribution on production practices and the environment.

AS.01.02.02.a
Research and examine marketing methods for animal products and services.

ALTERNATIVE ANIMAL AGRICULTURE refers to the production of animals other than the traditional agricultural animals such as cattle, sheep, horses, and poultry. In modern times, producers have looked for animals beyond traditional livestock to raise for a profit. Alternative animal production is usually small in scale and provides a product for a specialty market. Producers may have alternative operations to supplement their traditional operations. People who work full time in areas outside agriculture may produce alternative animals as a hobby or as a way to make a profit in a part-time operation.

FIGURE 12–1 Domestic rabbits are raised in commercial operations. *© francesco de marco/Shutterstock.com.*

RABBIT PRODUCTION

People have raised rabbits for food for hundreds of years. The Romans produced rabbits as far back as 250 B.C., and rabbit meat was a substantial part of their diet. The Phoenicians were great sailors and trading people who are accredited with introducing domesticated rabbits throughout much of the known world as far back as 1100 B.C. Domesticated rabbits were brought into the United States around 1900 and were produced in large rabbitries in Southern California. Since that time, the rabbit industry has grown all across the country.

Although the majority of the rabbits are produced by part-time growers, there are several large commercial operations in this country (**Figure 12–1**). So many rabbits are produced in small, private rabbitries that it is difficult to determine the actual number. However, there are over 200,000 large-scale rabbit operations in the United States. China leads the world with 597,000 metric tons of rabbit meat produced each year. Venezuela comes in second with 277,000 tons, and Italy is third with 230,000 tons produced. The United States lags far behind with 35,000 tons.

The American Rabbit Breeders Association (ARBA) registers and promotes all the breeds of purebred rabbits grown in this country. The association currently has more than 23,000 members.

FIGURE 12–2 Rabbits are raised in wire cages called hutches. *Courtesy of Dr. James McNitt, Center for Small Farm Research*

Raising rabbits has several advantages over raising other agricultural animals. First, rabbits can be raised easily by anyone under almost any climactic condition. Most are raised indoors in cages called **hutches** (**Figure 12–2**), which consist of woven wire with boxes for the rabbits to sleep in and to bear their young. The facilities take up little space compared with other agricultural animals such as hogs or cattle. Rabbit houses usually are heated in the winter and cooled in the summer to provide comfort for the animals. In areas where the climate is mild, however, an adequately insulated house may provide the animals with a comfortable environment without artificial heating or cooling. This means that the producer can work with the animals in relative comfort as well.

Rabbits gain weight on a relatively small amount of feed. The feed efficiency ratio for properly fed and managed rabbits is about 2.5 to 1. This means that for every 2.5 pounds of feed fed to the rabbit, the animal gains 1 pound in body weight. Also, rabbits can be fed a lower-quality feed than some other animals, as rabbits' digestive system allows them to make use of roughage such as alfalfa and other fibrous plant material. This is quite an advantage over other agricultural animals because rabbits potentially can be raised at less expense per pound.

The demand for rabbits is greater than the supply. As mentioned, rabbit meat is imported into the United States, so the potential exists for expansion of the rabbit industry in this country. Many restaurants now offer several dishes that are prepared using rabbit meat. The **USDA** points out that rabbit meat is one of the most nutritious meats available. Not only is it high in protein and low in fat and cholesterol, but it also is easily digested and flavorful (see **Table 12–1** for a comparison with other animals).

In addition to the meat, rabbits are used for several other purposes. The fur is used in making coats, hats, liners for boots, and for toys. Scientists use the animals in experiments dealing with medical research. Manufacturers use rabbits for testing products. In addition, many rabbits are sold as pets because of their docile nature, clean habits, and cuddly fur.

Rabbits are prolific breeders. They produce their young 30 days after breeding and produce four or five litters per year, consisting of up to eight young per litter. Over a one-year period, a pair of rabbits can produce a lot of meat, considering that some breeds reach sexual maturity (the ability to have young) at about 5 months of age.

An example of how rapidly rabbits reproduce is that of the wild rabbit in Australia. Rabbits were first introduced there in 1859, when sailors released a pair of wild European rabbits. In only 30 years, more than 20 million rabbits inhabited the country. Since then, the animals have become a serious pest in both Australia and New Zealand. Harsh measures have been taken to control the wild rabbit population.

In domesticated production, the young are born in small nesting boxes that give the mother security and comfort and offer the young protection from outside stresses. The female will pull fur from her own body to make soft, warm bedding for the newborn rabbits (**Figure 12–3**). After weaning, the young rabbits are put into cages, where they are grown out to about eight weeks of age. This is considered the proper age for the animals to be slaughtered for meat.

	Calories	Protein (grams)	Fat (grams)	Water (grams)
Rabbit	136	20.05	5.55	72.82
Lamb	267	16.88	21.59	60.70
Veal	144	19.35	6.77	72.84
Beef	291	17.32	24.05	57.26
Pork	398	13.35	37.83	47.86
Chicken	215	18.60	15.06	65.99
Turkey	160	20.42	8.02	70.04

TABLE 12–1 Nutritional Value of Common Meats (per 100 gram)

Source: USDA, Human Nutrition Service

Note that rabbit meat is higher in protein and lower in fat than many other meats.

FIGURE 12–3 The female rabbit pulls fur from her body to make a nest for her baby rabbits. *© IrinaK/Shutterstock.com.*

FIGURE 12–4 A major drawback of raising rabbits for food production is that they are so cute and cuddly. *© sirtravelalot/Shutterstock.com.*

FIGURE 12–5 Llamas are well adapted to thin, cool mountain air. *© studiolaska/Shutterstock.com.*

Rabbits offer potential as a food enterprise, especially in developing countries. Because rabbits can digest roughage such as alfalfa and other plants, the feed is low cost compared with feed for other animals, and the initial outlay in beginning production is less. Research has shown that rabbits can do well on feeds containing very little grain. Because many developing countries have an abundance of roughage that rabbits can eat, the animals can provide a relatively inexpensive food supply of much needed protein.

Even though rabbits are a relatively inexpensive source of meat, Americans do not consume as much rabbit meat as do other peoples in the world. This reluctance on the part of Americans is a major drawback for the industry. While we accept the production and slaughter of cattle, hogs, and chickens, the thought of slaughtering cute, cuddly animals such as rabbits is repulsive to many (**Figure 12–4**).

LLAMA PRODUCTION

Llamas are native to South America and belong to the same family as camels. Llamas have been raised in several countries in South America for hundreds of years. In Chile, Peru, and Bolivia, the ancient Incas and other peoples raised these animals for work animals. They are well adapted to the cool, thin mountain air of the Andes mountains but can adapt to most climactic conditions (**Figure 12–5**). During the past 15 years, llamas have developed into an animal industry in the United States, with an estimated 120,000 llamas and increasing in number.

Llamas stand 3 to 4 feet high at the shoulder, weigh from 250 to 400 pounds when mature, and can carry heavy packs for long distances. Being related to the camel, llamas can last longer than many other animals between drinks of water and can subsist on low-quality forage. Their coats have two types of fiber: long guard hairs and short, fine fiber that helps to keep the animals warm. The fiber length may range from 3 to 10 inches.

Because of these characteristics, raising llamas has gained popularity in the United States. Most are produced in the western part of the country, where people use them to carry gear on hunting or camping trips into the mountains (**Figure 12–6**). They also are used to pull

FIGURE 12–6 In the United States, most llamas are used as pack animals. *© Jerry Voss/Shutterstock.com.*

carts and are raised as pets. Their hair is used for a variety of crafts, such as making rope. A close relative, the alpaca, is raised for its high-quality wool, which is made into fine rugs and blankets.

FISH BAIT PRODUCTION

One of the great outdoor hobbies of Americans is fishing. Thousands of streams and lakes in our country are well stocked with game fish. People take advantage of these waters to catch fish and enjoy a leisurely outing. A popular way to catch fish is using natural, live bait. Fish bait is grown all across the country, largely by part-time producers.

Earthworms

Earthworms are grown in beds of loose, porous materials (**Figure 12–7**), which might include shredded newspaper, shredded cardboard, garden compost, grass clippings, straw, or well-decayed manure. Usually, peat moss is added to the mixture to keep the material loose and to help hold moisture.

The **pH** of the bedding is monitored and is kept slightly acidic (pH 6.8) pH is the measure used to indicate how acidic or how alkaline a material is. On the pH scale, 7 is neutral (neither acid nor alkaline). A number less than 7 indicates an acid; the lower the number, the more acidic the material. A number greater than 7 indicates an alkaline; the higher the number, the more alkaline the material. Because the material used for bedding is usually acidic, limestone is added to help neutralize the acid.

The beds are kept moist, and lights are used to prevent the worms from crawling out of their beds. Worms are sensitive to light and normally come out only at night. As long as the worms see light when they come to the top of the bedding, they will stay near the bottom.

The worms are fed vegetable scraps and cornmeal. They mature at 1 to 2 months of age. They are packaged and marketed in small, round containers of approximately 100 worms per container for the smaller red wigglers and about 25 to 50 for the larger night crawlers. In mature worms, a broad, raised band encircles the body behind the head (**Figure 12–8**).

An alternative marketing source is that of selling the worms to gardeners. Earthworms improve the quality of the soil by creating pores as they move through the soil, which allow better movement of air and water. Manure from the worms, called **castings**, enriches the soil.

FIGURE 12–7 Earthworms are raised in beds of loose, porous materials. *© 88Studio. Image from BigStockPhoto.com.*

FIGURE 12–8 Mature earthworms have a broad, raised band encircling their bodies just behind their heads. *© iStockphoto/Ben185.*

Crickets

Crickets are raised in wooden boxes covered with screens. The floors of the boxes are covered with sand in which the adults lay their eggs. The sand is covered with fine wood shavings or other shredded material. Heat lamps are used to keep the crickets warm and to keep the sand warm for hatching the eggs. When the young crickets hatch, they are fed grain mixtures in small trays. Small trays filled with water-saturated cotton provide the crickets with a ready drinking fountain. The crickets are placed in cages and shipped to bait outlets. There, they are sold to fishers, who put them in their own cricket cages.

FIGURE 12–9 Americans consume approximately 100 metric tons of deer and elk meat each year.
Courtesy of North American Elk Breeders Association

LARGE GAME ANIMALS

A growing area of alternative animal agriculture is the commercial production of large game animals. By far the largest component of this industry is the production of elk. The growing of domesticated deer and elk goes back at least 5,000 years in China and other parts of the world. In the United States, elk have been grown commercially for the past 100 years. Currently, there are about 1,900 elk farms in North America, with around 68,000 animals in domestication.

These large animals offer several advantages to traditional grazing animals. They convert feed more efficiently than cattle on the range; this means that they gain more weight on less feed than beef animals when raised under the same conditions. Also, elk can make use of lower-quality feedstuff such as browse plants.

Americans consume approximately 100 metric tons of deer and elk meat per year, most of which is imported from New Zealand (**Figure 12–9**). The meat is relatively low in fat content, and many people prefer the flavor. In addition to the sale of the meat, producers market the antlers, which are used as ornaments, in the making of jewelry, and as an ingredient in herbal medicines.

LABORATORY ANIMAL PRODUCTION

Scientists need millions of animals each year in conducting research. Almost all materials that come in contact with humans—food, medicines, and cosmetics—have to be tested on animals to prove the effectiveness and safety of the products. Most of the advances in medicines have come about through the use of **laboratory animals**. There is considerable controversy over the use of animals in experimentation, but no one can deny the benefits to humans brought about through the use of animals in research.

These animals are produced by commercial and part-time producers. The animals most in demand for research are mice, rats, hamsters, guinea pigs, and rabbits (**Figure 12–10**). Other animals, such as primates, are used for highly specialized research.

Animals that are raised for use in laboratories have to be raised under strict conditions. Measures have to be taken to ensure that the animals have no **genetic defects** and are not harboring disease organisms. Animals of this nature could very well cause an otherwise well-designed research study to turn out wrong. The product being tested or the experiment being conducted could be tainted by a disease organism or a genetic defect that would cause the

FIGURE 12–10 Rats and other small animals are used in research studies. *© panyawat bootanom/Shutterstock.com.*

FIGURE 12–11 At one time, grass-fed beef was considered to be inferior to beef fattened in a feedlot. *© critterbiz/Shutterstock.com.*

animal to react differently than a healthy animal. Most producers raise animals that are certified for laboratory use.

PRODUCTION OF NATURAL AND CERTIFIED ANIMAL PRODUCTS

A rapidly growing niche market for agricultural animals is the production of natural or organic animal products. Some people are worried about the use of antibiotics and hormones in the production of meat or dairy products. Even though there is no credible research indicating that conventionally produced meat is any less healthy than meat produced by "organic" means, some consumers still want meat they consider to be "pure."

There is a difference between "natural" and "organic" animal products. A good example is natural beef. "Natural" refers to beef that is produced without using feed additives such as medications or hormone implants. Usually the beef cattle are raised entirely on grass and not fed grain concentrates as are most of the cattle raised for slaughter. This type of beef is sometimes called grass-fed beef, and until a short time ago was considered to be inferior to that of grain-fed beef (**Figure 12–11**).

No standards have been set for what can be labeled as natural beef, and the labeling often is left to the producers or people who market the beef. Natural beef is usually a lower-quality beef than that coming from a feedlot. The U.S. Department of Agriculture (USDA) grades beef that is sold to the public. The quality grade is determined by the age and degree of fat content in the beef. USDA graders grade beef that is marketed through conventional markets such as grocery stores. However, the USDA does not have any set standards for natural beef, and it often is sold at the farm. Consumers buy a live beef animal directly from the producer and have it custom-processed. Under these conditions, the beef does not have to be USDA-inspected or graded.

Organic animal products are those that are produced under strict requirements set by the USDA. For any meat or dairy product to be labeled organic, the producer must adhere to limitations and regulations specified by the USDA. These rules center on four main areas: origin of the livestock, livestock feed, livestock health care, and livestock living conditions. Following are some of the regulations published by the USDA:

USDA REGULATIONS FOR ORGANIC ANIMAL PRODUCTS

ORIGIN OF LIVESTOCK

(a) Livestock products that are to be sold, labeled, or represented as organic must be from livestock under continuous organic management from the last third of gestation or hatching. Poultry or edible poultry products must be from poultry that has been under continuous organic management beginning no later than the second day of life (**Figure 12–12**).

Dairy Animals

Milk or milk products must be from animals that have been under continuous organic management beginning no later than one year prior to the production of the milk or milk products that are to be sold, labeled, or represented as organic. When an entire, distinct herd is converted to organic production, the producer may

1. for the first nine months of the year, provide a minimum of 80 percent feed that is either organic or raised from land included in the organic system plan and managed in compliance with organic crop requirements.
2. Provide feed in compliance for the final three months.
3. Once an entire, distinct herd has been converted to organic production, all dairy animals shall be under organic management from the last third of gestation.

Breeder Stock

Livestock used as breeder stock may be brought from a nonorganic operation onto an organic operation at any time if such livestock are gestating and the offspring are to be raised as organic livestock. The breeder stock must be brought onto the facility no later than the last third of gestation.

(b) The following are prohibited:

1. Livestock or edible livestock products that are removed from an organic operation and subsequently managed on a nonorganic operation may not be sold, labeled, or represented as organically produced.
2. Breeder or dairy stock that has not been under continuous organic management since the last third of gestation may not be sold, labeled, or represented as organic slaughter stock.

(c) The producer of an organic livestock operation must maintain records sufficient to preserve the identity of all organically managed animals and edible and nonedible animal products produced on the operation (**Figure 12–13**).

Livestock Feed

(a) The producer of an organic livestock operation must provide livestock with a total feed ration composed of agricultural products, including pasture and forage, that are organically produced and, if applicable, organically handled. There are, however, certain nonsynthetic substances and synthetic substances allowed to be used as feed additives and supplements.

(b) The producer of an organic operation must not

1. use animal drugs, including hormones, to promote growth.

FIGURE 12–12 Poultry or edible poultry products must be from poultry that has been under continuous organic management beginning no later than the second day of life.
© FiledIMAGE/Shutterstock.com.

FIGURE 12–13 The producer of an organic livestock operation must maintain records sufficient to preserve the identity of all organically managed animals and edible and nonedible animal products produced on the operation.
© Goodluz. Image from BigStockPhoto.com.

2. provide feed supplements or additives in amounts above those needed for adequate nutrition and health maintenance for the species at its specific stage of life.
3. feed plastic pellets for roughage.
4. feed formulas containing urea or manure.
5. feed mammalian or poultry slaughter by-products to mammals or poultry.
6. use feed, feed additives, and feed supplements in violation of the Federal Food, Drug, and Cosmetic Act.

Livestock Health Care Practice Standard

(a) The producer must establish and maintain preventive livestock health care practices, including

1. selection of species and types of livestock with regard to suitability for site-specific conditions and resistance to prevalent diseases and parasites.
2. provision of a feed ration sufficient to meet nutritional requirements, including vitamins, minerals, protein, and/or amino acids, fatty acids, energy sources, and fiber (ruminants).
3. establishment of appropriate housing, pasture conditions, and sanitation practices to minimize the occurrence and spread of diseases and parasites.
4. provision of conditions which allow for exercise, freedom of movement, and reduction of stress appropriate to the species (**Figure 12–14**).
5. performance of physical alterations as needed to promote the animal's welfare and in a manner that minimizes pain and stress.
6. administration of vaccines and other veterinary biologics.

FIGURE 12–14 Organic producers must provide conditions that allow for exercise, freedom of movement, and reduction of stress appropriate to the species.
© iStockphoto/DesignSensation.

(b) When preventive practices and veterinary biologics are inadequate to prevent sickness, a producer may administer synthetic medications provided such medications are allowed under the USDA listings. Parasiticides allowed under the USDA listings may be used on

1. breeder stock, when used prior to the last third of gestation, but not during lactation for progeny that are to be sold, labeled, or represented as organically produced.
2. dairy stock, when used a minimum of 90 days prior to the production of milk or milk products that are to be sold, labeled, or represented as organic.

(c) The producer of an organic livestock operation must not

1. sell, label, or represent as organic any animal or edible product derived from any animal treated with antibiotics, any substance that contains a synthetic substance not allowed under the USDA listings, or any substance that contains a nonsynthetic substance prohibited under the USDA listings.
2. administer any animal drug, other than vaccinations, in the absence of illness.
3. administer hormones for growth promotion.
4. administer synthetic parasiticides on a routine basis.
5. administer synthetic parasiticides to slaughter stock.
6. administer animal drugs in violation of the Federal Food, Drug, and Cosmetic Act.
7. withhold medical treatment from a sick animal in an effort to preserve its organic status. All appropriate medications must be used to restore an animal to health when methods acceptable to organic production fail. Livestock treated with a prohibited substance must be clearly identified and shall not be sold, labeled, or represented as organically produced.

Livestock Living Conditions

(a) The producer of an organic livestock operation must establish and maintain livestock living conditions which accommodate the health and natural behavior of animals, including (**Figure 12–15**)

FIGURE 12–15 The producer of an organic livestock operation must establish and maintain livestock living conditions that accommodate the health and natural behavior of animals. *© iStockphoto/GenoEJSajko.*

1. access to the outdoors, shade, shelter, exercise areas, fresh air, and direct sunlight suitable to the species, its stage of production, the climate, and the environment.
2. access to pasture for ruminants.
3. appropriate clean, dry bedding. If the bedding is typically consumed by the animal species, it must comply with the feed requirements as specified by the USDA.
4. shelter designed to allow for
   (i) natural maintenance, comfort behaviors, and opportunity to exercise.
   (ii) temperature level, ventilation, and air circulation suitable to the species.
   (iii) reduction of potential for livestock injury.

(b) The producer of an organic livestock operation may provide temporary confinement for an animal because of

1. inclement weather.
2. the animal's stage of production.
3. conditions under which the health, safety, or well-being of the animal could be jeopardized.
4. risk to soil or water quality.

(c) The producer of an organic livestock operation must manage manure in a manner that does not contribute to contamination of crops, soil, or water by plant nutrients, heavy metals, or pathogenic organisms and optimizes recycling of nutrients.

Record Keeping by Certified Operations

For producers to be certified as organic producers, strict records must be kept and monitored. The USDA requirements for record keeping are outlined below.

a. A certified operation must maintain records concerning the production, harvesting, and handling of agricultural products that are or that are intended to be sold, labeled, or represented as "100 percent organic," "organic," or "made with organic (specified ingredients or food group(s))."

b. Such records must
   1. be adapted to the particular business that the certified operation is conducting.
   2. fully disclose all activities and transactions of the certified operation in sufficient detail as to be readily understood and audited.
   3. be maintained for not less than five years beyond their creation.
   4. be sufficient to demonstrate compliance with the act and the regulations in this part.

c. The certified operation must make such records available for inspection and copying during normal business hours by authorized representatives of the secretary, the applicable state program's governing state official, and the certifying agent.

HUNTING PRESERVES

During the latter half of the twentieth century, wildlife numbers in the United States increased dramatically. The numbers of deer, turkey, and other game animals reached record populations. Although these animals can be considered pests, they also can be a valuable source of income. Many hunters live in urban or suburban areas with no access to hunting land (**Figure 12–16**).

To be able to enjoy the sport, they must find landowners who will allow them to hunt on their property. Many producers have begun hunting management areas on their farms.

FIGURE 12–16 Many hunters live in urban or suburban areas with no access to hunting land. © Canon boy/Shutterstock.com.

FIGURE 12–17 The producer or manager may also plant food plots of crops such as rye, wheat, clover, or other legumes. © bikeriderlondon/Shutterstock.com.

They then sell the rights to hunt on the land to individuals or to hunt clubs.

To have adequate wildlife for hunting, producers manage the populations of game on their lands. This means that they have to create conditions conducive to game animals living and producing on the land. Producers have to make sure that the habitat for the animals is maintained. The first consideration is that the animals have proper food. This means that the producer must create areas that will provide the type of food a particular game animal may require. For example, if the producer wants to maintain a huntable population of eastern whitetail deer, enough browse must be available. This may include plants such as blackberry vines, honeysuckle, or other foliar plants that are in reach of the deer. Also, deer like acorns, and a thicket of white oaks can provide good feed for deer. The producer or manager may also plant food plots of crops such as rye, wheat, clover, or other legumes (**Figure 12–17**).

Bird hunting for turkey, quail, and pheasants is also popular. The producer not only must provide food plots for the birds but also must make sure that other management techniques do not interfere with the life cycle of the birds. For example, if row crops are harvested and all land is plowed under, the birds may have no cover or food for the winter. A better management practice might be to leave some of the crops around the edge of the fields to provide food and shelter for the birds. Hedgerows along drainage ditches, and fence rows can be left or be seeded with a crop that birds like to eat.

SUMMARY

As the industry of animal agriculture grows, many new types of animals are added, each having a unique characteristic or quality that makes it valuable to humans. These animals now may be considered as animals seen normally in the wild; however, we must realize that at one time, all the animals we now grow were wild animals. As with other aspects of agriculture, research will find new uses for these animals and new and better ways of producing them.

CHAPTER REVIEW

Review Questions

1. What is meant by an alternative agricultural animal?
2. What advantages do rabbits have over other agricultural animals?
3. What is one major drawback in the production of rabbits?
4. What uses do people have for llamas?
5. What is meant by the pH of a material?
6. Why are lights used in the production of earthworms?
7. Why are elk raised as agricultural animals?
8. Why must laboratory animals be free of disease and genetic defects?
9. Why do some people want to buy natural or organic animal products?
10. Distinguish between natural and organic beef.
11. List four main areas of organic production that are regulated by the USDA.
12. List three things that organic producers cannot put in the feed of their animals.
13. What are three regulations regarding the design of shelters for organically produced animals?
14. Explain why managing hunting areas can be a source of income for producers.
15. What type of habitat do managers provide for deer?

Student Learning Activities

1. Create a list of the alternative animal operations in your area. Choose one of these operations, and report to the class. How did the producer get started? What are the markets? What is the potential for expansion?
2. Think of an alternative animal enterprise that was not covered in this chapter. Give a report to the class on your ideas. How would these animals be raised? Where would the market be? Why are not more of these animals being raised?

CHAPTER 13

The Honeybee Industry

STUDENT OBJECTIVES IN BASIC SCIENCE

As a result of studying this chapter, you should be able to

- describe the characteristics of a honeybee.
- define *social insect*.
- trace the life cycle of a honeybee.
- explain the role of bees in pollination.
- identify the types of bees in a colony and the roles they play.
- explain how bees communicate.
- discuss how parasites affect bees.

STUDENT OBJECTIVES IN AGRICULTURAL SCIENCE

As a result of studying this chapter, you should be able to

- explain why honeybees are agricultural animals.
- discuss the importance of honeybees to the agricultural economy.
- describe how the modern hive was developed.
- explain how bees produce honey.
- explain how beekeepers collect and extract honey.
- describe how producers grow and package bees.
- discuss how queen bees are produced and installed.
- list some of the problems encountered in raising bees.

KEY TERMS

apiary
hives
queen
drone
worker
brood
brood cells
worker bees
queen cells
royal jelly
swarm
honey comb
nursery bees
scout bees
guard bees
pheromone
bee space
foundation comb
brood chamber
queen excluder
propolis

NATIONAL AFNR STANDARD

AS.01.01
Evaluate the development and implications of animal origin, domestication, and distribution on production practices and the environment.

AS.01.02.02.a
Research and examine marketing methods for animal products and services.

THE IMPORTANCE OF HONEYBEES

HONEYBEES ARE CLASSIFIED AS insects because they fit the characteristics of the insect class. They have three distinct body segments—a head, a thorax, and an abdomen—and also three pairs of legs. Even though the honeybee is an insect, it is still an animal, and it is an important *agricultural* animal. Each year Americans consume more than a pound of honey per person, consisting of honey produced in the United States as well as honey imported from countries such as Argentina, Canada, Mexico, and China (**Figure 13–1**). Although this might not seem like a lot of honey, consider that there are almost 300 million people in the United States. This adds up to a lot of honey.

The United States ranks third in the world in honey production behind China and Turkey. In this country, honey comes from bees that are raised by more than 211,600 producers all across the country who tend to approximately 3 million colonies each year. The vast majority of these producers are hobbyists with fewer than 25 hives. A group of hives is known as an **apiary** (**Figure 13–2**), a term derived from *Apis mellifera*, the scientific name of the honeybee. In order of their production, North Dakota, South Dakota, California, Florida, and Montana are the top honey-producing states in the nation. Considering the difference in climate between North Dakota and Florida, it is safe to say that the honeybee is adaptable to a wide range of weather conditions.

FIGURE 13–1 Each year Americans consume more than a pound of honey per person. © Makistock/Shutterstock.com.

FIGURE 13–2 A group of hives is called an apiary. © Darryl Brooks/Shutterstock.com.

The total value of annual honey production in the United States is around $326 million. In addition to honey production, beekeepers also produce a lot of beeswax. This wax comes from the combs produced by the bees and is used in a wide variety of products such as cosmetics, candles, and high-grade polishes. The greatest value of the honey industry, however, may be in another area. Bees, it could be argued, are the most important of all the agricultural animals because of their role in pollinating agricultural plants and crops. Research indicates that the total value of honeybees is more than $14 billion each year. Although the value of the services provided by bees each year is difficult to estimate, it is safe to say that many crops could not survive without help from bees (**Figure 13–3**).

Most other agricultural animals eat plant-derived feed and rely on bees to pollinate the plants they eat. Bees assist in the pollination of flowering plants by scattering pollen from one flower to the next as they gather nectar and pollen. This is nature's way of ensuring that the flowers are pollinated so they will produce seed. Often, the seed and/or the fruit surrounding the seed of a plant is a crop that producers raise.

FIGURE 13–3 Many crops could not survive without help from bees. © Darkness/Shutterstock.com.

FIGURE 13–4 Fruit growers hire beekeepers to bring in truckloads of bees in the spring when the trees are blooming. © GEORGID/Shutterstock.com.

Honeybees are particularly adept at pollinating. Many insects work flowers, but most go from one flower to a different kind of flower. Bees, however, work a specific kind of flower for a period of time. For example, honeybees may be working apple blossoms for several days until the flowers are gone and then go on to work a different type of flower. In doing this, they go from an apple blossom to another apple blossom and spread pollen from one apple blossom to another in that way. This process ensures that the blossoms are thoroughly pollinated.

Fruit growers hire beekeepers to bring in truckloads of bees in the spring when the trees are blooming (**Figure 13–4**). The bees live in wooden boxlike structures called **hives** with a separate colony of bees in each hive. The hives are easy to handle and can be loaded on a truck with the bees still in the hive. The owner of the bees then can move the hives from orchard to orchard for a fee from the fruit or crop producer. In addition, the producer can harvest hundreds of pounds of honey each year that can be sold at a profit.

BEES AS SOCIAL INSECTS

A characteristic of honeybees that sets them apart from many other types of insects is their social structure. They live in a highly ordered society in which each bee seems to have its job and works in concert with the rest of the bees in the hive. Within a colony of bees there are three types of bees: the **queen**, **drones**, and **workers** (**Figure 13–5**).

FIGURE 13–5 Within a colony of bees are three types of bees: the queen, drones, and workers.

The only reason the queen exists is to lay eggs for the hive. Even though this is a singular role, in her lifetime she lays thousands of eggs that hatch into workers that carry out the other tasks associated with perpetuating the hive and producing honey. A fertile queen may lay as many as 1,500 eggs per day and as many as 200,000 in a single year. Worker bees labor hard and live a short life of only about 6 weeks. Exceptions are bees that are hatched in the late fall that live all winter in a semi-dormant stage. For this reason, the continual producing and growing of **brood (young)** by the queen and bees in the colony are essential. Brood is the stage in the bees' life that lasts from the time they hatch to the time they emerge from the cell.

A hive whose queen has stopped laying or has slowed because of her age is not productive, and the entire colony may even die because there are not enough new bees to replace those that die. Throughout her life, she is truly treated as a queen. The other bees feed and care for all of the queen's needs. The queen is recognizable from the rest of the bees, even in a colony of many thousands of bees, because she is much larger and more slender (**Figure 13–6**).

The queen lays eggs in cells called **brood cells**, which are slightly larger than the cells used to store honey. A single egg is deposited inside and at the bottom of each brood cell. The eggs hatch into larvae that are fed and cared for by **worker bees** in the hive. The larvae develop into pupae and then into the adult stage, all while remaining in the cell. When the bee has reached the adult stage, it emerges from the cell and assumes its role in the social structure. The entire metamorphosis takes about 3 weeks (**Figure 13–7**).

FIGURE 13–6 Even within a colony of many thousands of bees, the queen is recognizable from the rest of the bees because she is much larger and is more slender. *© iStockphoto/proxyminder.*

FIGURE 13–7 In about 3 weeks, the metamorphoses is complete and the bees emerge in the adult stage. *© Vova Shevchuk/Shutterstock.com.*

If the queen dies or the hive becomes too crowded, the bees will produce a new queen by drawing special large cells called **queen cells**. Larvae in these cells are fed a special substance called **royal jelly** that is secreted from the bees. This food causes the larvae to develop into queen bees. When the new queen emerges, the old queen usually leaves with a portion of the bees, called a **swarm**, to form a new colony. A suitable site for the new colony is located, and the swarm moves to the new place. In the meantime, the bees cluster around the queen on a limb, on a house, or on another object until locating a new home. Beekeepers often place these swarms into a new hive if the swarms are detected in time.

Beekeepers do not want the colonies to swarm because this lowers the number of workers in the hive and lowers the amount of honey produced (**Figure 13–8**). Producers prevent swarming by making sure that the bees have

FIGURE 13–8 Beekeepers do not want the colonies to swarm because this lowers the number of workers in the hive and lowers the amount of honey produced. *© iStockphoto/fpwing.*

FIGURE 13–9 Most colonies contain around 50,000 worker bees. *© elleon/Shutterstock.com.*

plenty of room in the hive. As the bees fill the hive with honey, a new box or upper is added to the hive to give more room.

The queen is not the only type of bee that has only a single job. Drones are the male bees whose sole purpose is to mate with the queen. Although they are larger than the worker bees, they do no work in the hive and do not even have stingers. When a young queen emerges as the new queen for the hive, she goes on what is called her maiden flight, where she attracts drones with a unique scent. As many as a dozen drones may mate with a single queen, and the drones die as soon as they mate with the queen. The queen mates only once during her lifetime and retains enough sperm to fertilize her eggs for as long as 4 years. Drones that do not mate with queen are removed from the hive by the worker bees and are not allowed to overwinter in the hive. The next spring new drones will hatch.

The worker bees are all sterile females and comprise by far the largest number of bees in the colony. Although the number of worker bees present in the hive at one time varies widely, most colonies contain around 50,000 worker bees (**Figure 13–9**). These bees are socially divided into groups with specific jobs. Most of them bring nectar and pollen to the hive from the field to be produced and stored as honey to eat during the winter months.

Honey is made from the nectar that the bees gather from the flowers. Different flowers make honey that varies in color and strength of flavor depending on the source of the nectar. Bees go from flower to flower gathering nectar. A bee can visit as many as 50 to 100 flowers until it has enough nectar to return to the hive. This movement from flower to flower is the means by which bees pollinate plants. As a bee lands on the flower and extracts the nectar, pollen is attached to hairs on the bee's body and is carried to the next flower (**Figure 13–10**). Both nectar and pollen are carried back to the hive and stored for use as food.

A bee can fly with a load that is as heavy as its own body weight and may make as many as 10 trips per day. Nectar from about 5 million flowers is required to make a pint of honey. No wonder the bees wear out so soon!

The bees store the nectar in six-sided cells that are joined together to create a **honey comb**. The cells are made from wax secreted from glands on the bees' bodies. The bees work the wax particles loose from their bodies and use them to shape the cells into perfect hexagonal storage containers. The hexagonal (six-sided) shape of the cells creates the ideal storage space for the honey and pollen (**Figure 13–11**).

FIGURE 13–10 Pollen is attached to hairs on the bee's body and is carried to the next flower. Note the yellow pollen on the bee's rear leg. *© Sergey Lavrentev/Shutterstock.com.*

FIGURE 13–11 The bees work the wax particles loose from their bodies and use them to shape the cells into the perfect hexagonal storage containers. *© Tischenko Irina /Shutterstock.com.*

Although the walls of the cells are very thin, they are strong enough to bear the weight of the honey or the developing brood.

Once the bees have deposited the nectar in the cells, the bees begin to concentrate it into honey by using their wings to create air movement that evaporates the excess water in the nectar. Once the cells are full of honey, the bees mold wax across the top of the cells to cap and seal them. Some cells are filled with pollen that serves as a source of protein for the bees.

Some workers, known as **nursery bees**, care for the queen and brood. These bees feed the young brood and the queen. In addition, they keep the hive at a constant temperature by flapping their wings to create a current of air that cools the hive to an even temperature of 95°F. Bees can move their wings more than 11,000 strokes per minute. This is what causes the distinctive buzzing or humming sound in a beehive.

The wing movement also causes moisture to evaporate from the stored nectar, which helps to concentrate the nectar into the thick liquid we call honey. The concentration process may take several days to reach the proper consistency. Then the cells are capped with wax until the honey is needed. Pollen is stored in separate cells. Nursery bees also keep the hive clean by removing droppings, wax particles, dead bees, and other debris that may accumulate in the hive (**Figure 13–12**).

Other bees scout the area for nectar. These **scout bees** seek the closest source and may travel more than a mile to obtain nectar. They

FIGURE 13–12 Nursery bees keep the hive clean by removing droppings, wax particles, dead bees, and other debris that may accumulate in the hive. Here they are removing a dead larva. *Source: USDA, Agricultural Research Service (ARS).*

fan out in all directions and locate plants that are flowering. Some flowers contain more nectar than others, and these make the best food source for the bees. After locating the flower sources, the scout bees must communicate with the nectar-gathering bees.

Perhaps the most dramatic example of animal communication is that used by honeybees to tell other bees in the hive about nectar and pollen sources they have located. Through a series of elaborate moves and dances, the scout bees tell the worker bees the direction, distance, and amount of the nectar source (**Figure 13–13**). The scout bees bring back samples of the type of nectar they located. By smelling and tasting the nectar, the worker bees become aware of the type and quality of the nectar. Recent research has shown that the bees also use sound to communicate the distance to the food source. Scientists believe that the bees use the sun to orient their flight and that they communicate this orientation to the worker bees in the hive.

Others worker bees serve as **guard bees** at the hive entrance. They regulate all the insects that enter the hive. Even bees that are not a part of the colony are attacked when they enter the hives. As each worker bee enters the hive, it touches antennas with a guard bee. Bees smell with their antennas and can distinguish between bees that

FIGURE 13–13 Through a series of elaborate moves and dances, the scout bees tell the workers the direction, distance, and amount of the nectar source. Note the bee in the center. *© rtbilder/Shutterstock.com.*

FIGURE 13–14 If the guard bees sense danger, they give off an alarm that sets the other bees into a defensive mode. *© Jaroslav Moravcik/Shutterstock.com.*

belong to the hive and those that do not. This is accomplished by a pheromone secreted by each bee in the colony. **Pheromones** are chemicals that send messages by organisms to communicate. Guard bees generally are young worker bees that spend only about 1 to 2 days serving as a guard bee before going on to other duties. Not all worker bees serve as guards, and scientists have yet to determine why some workers become guards and others do not.

If the guard bees sense danger, they give off an alarm that sets the other bees into a defensive mode (**Figure 13–14**). Usually, only a small percentage of the bees in a colony will sting an intruder. A sting is almost always lethal to the stinging bee because the stinger is barb-shaped. When it penetrates the skin of an animal, the barb catches and the end of the bee's abdomen is pulled out when the bee tries to extract the stinger.

COMMERCIAL HONEY PRODUCTION

Humans have used honey for food for thousands of years. As far back as recorded history goes, accounts have been written of people gathering honey and enjoying the sweet, high-energy food. Honey was first gathered from the wild. People probably watched for bees at

FIGURE 13–15 At one time, bees were kept in crude hives such as this hollow log. © TimePRO.TV/Shutterstock.com.

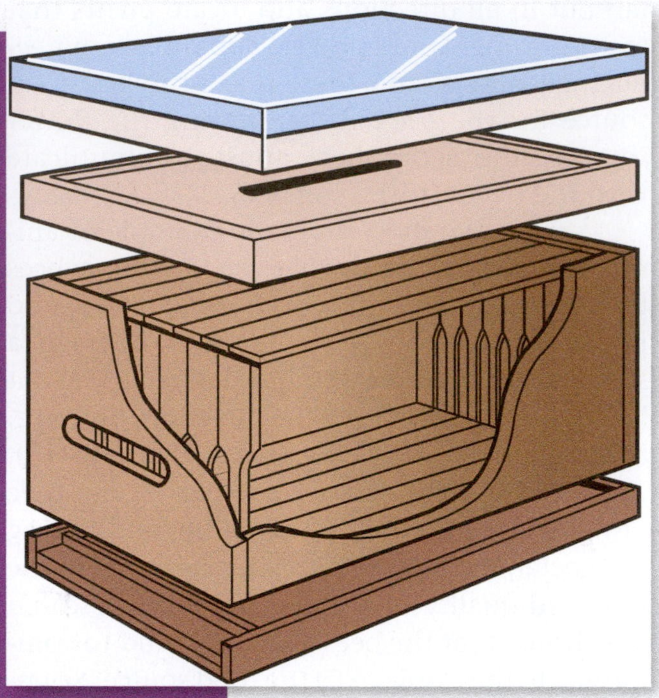

FIGURE 13–16 The Langstroth hive uses removable frames inside a box for the bees to build comb. The frames are hung into boxes called supers. Several supers stacked together compose the hive.

FIGURE 13–17 Bees are about 3/16 of an inch tall. The space between the frames is 3/8 of an inch apart, allowing the bees to work back-to-back. © aaltair/Shutterstock.com.

flowers or water sources and followed them back to their hives. Later, humans began to keep bees and tend to them. The first hives were very crude by today's standards. Often, a hive consisted of nothing more than a tree or a hollow log (**Figure 13–15**). Later, people began to build houselike structures for the bees. Early beehives ranged from simple boxes made of wood to domes made from straw and sticks. Some were even made of clay.

One of the major problems with early hives was that the hive could not be taken apart to harvest the honey without destroying most of the interior. In the 1850s, this problem was solved by a Pennsylvania beekeeper, Lorenzo Langstroth. His hive design uses removable frames inside a box for the bees to build comb. The frames are hung into boxes called supers (**Figure 13–16**). Several supers stacked together compose the hive, where as many as 80,000 bees may live during the peak of the honey season.

The space between the frames in the super is critical. If there is too much space, the bees will build across the frames and the frames will be difficult to remove. If there is too little space, the bees will fill only one side of the frame. This critical space, called **bee space**, is about 3/8 of an inch (**Figure 13–17**). This allows just enough room for two bees to work back-to-back on opposite frames. The beauty of this system is that the beekeeper can take the hive apart to inspect the bees or can take frames out for harvesting honey without disturbing the rest of the hive.

The beekeeper places sheets of comb called **foundation comb** into frames in which the bees build comb to fill with honey (**Figure 13–18**).

FIGURE 13–18 Foundation comb is installed in these frames. *© Vasileios Karafillidis/Shutterstock.com.*

Foundation comb is made by rolling beeswax into thin sheets that have the imprint of the hexagonal cells. This causes the bees to begin building cells along the frames on the foundation. The queen is kept in the lower part of the hive, the **brood chamber**, by means of a **queen excluder**, which consists of a screen with openings large enough for the workers but small enough to prevent the queen from passing. The queen excluder prevents the queen from laying eggs in the comb that the beekeeper will remove to extract honey.

When all of the honey in a super is capped, the beekeeper removes the super and replaces it with an empty one. The keeper is always careful to leave enough honey for the bees to live on through the winter. The beekeeper uses smoke from a smoker to make the bees more docile. The bees think the hive is on fire and begin to gorge on honey in preparation for their departure from the hive. This seems to make the bees calmer and more docile. The smoke also serves to disorient the guard bees at the hive entrance before they can give an alarm to the rest of the hive.

The beekeeper must pry the supers apart to remove them. The bees use a sticky gum called **propolis** to seal and glue the parts of the hive together. Propolis is made from the sap and gum of trees. It is harvested by the bees and brought back to the hive, and it is used to patch holes and cracks in the hive. It also strengthens the hive by sticking all the parts together with a strong bond (**Figure 13–19**).

FIGURE 13–19 Bees use propolis to stick hive components together. The brown material on the side of this frame is propolis. *© Dainis Derics/Shutterstock.com.*

After the supers have been removed, the frames are taken out and the caps cut from the top of the comb using a heated uncapping knife. The honey is removed from the comb using an extractor, which consists of a metal drum with racks for the frames. The racks containing the frames are revolved in the drum, and centrifugal force slings the honey from the comb without damaging the cells in the comb (**Figure 13–20**). The frames containing the empty cells are put back into the super and the super is put back on the hive for the bees to fill

FIGURE 13–20 A honey extractor revolves frames in the drum, and centrifugal force slings the honey from the comb without damaging the cells in the comb. *© shoot4pleasure/Shutterstock.com.*

FIGURE 13–21 The most popular means of sale is that of packaging the bees in containers weighing 2 or 3 pounds. *Courtesy of Dr. Keith Delaplane, University of Georgia*

again. The honey then is processed by heating it gently to prevent the honey from granulating (turning to sugar). The honey now is ready to be packaged and sold.

Beekeepers obtain their bees from producers who raise bees for sale. Many of these operations are located in the southern part of the country because of the mild winters and early spring there. This allows bees to be ready for shipment when the honey flow begins in the spring. The most popular way to sell bees is to package them in containers weighing 2 or 3 pounds (**Figure 13–21**). The bees are shaken from the frames into a large funnel, where they slide into an opening in the top of a box with screened sides. A queen is added, and the opening is closed with a can containing sugar water, which provides feed for the bees during shipping.

A newer form of purchasing bees is the bee nucleus, often referred to as a "nuc." This consists of selling four or five frames of brood along with the bees and queen that are clinging to the brood. The advantage of this method is that the queen is already accepted by the colony and there are young bees about to be hatched.

BREEDING BEES

Like most other agricultural animals, bees have been bred selectively for a long time. To be highly productive, the bees must have certain characteristics that make them easy to keep and that produce a lot of honey. One of the most important characteristics is that they must be docile. This means that they are not very prone to attacking and stinging. Although any breed or type of honeybee will sting, some are more aggressive than others. Bees also must not be prone to swarming, and some breeds swarm and split the colony more often than others. The beekeeper wants the hive to have a large quantity of bees to be able to produce large amounts of honey. If the bees have a tendency to swarm too often, large colonies will be difficult to achieve. Other desirable characteristics include resistance to disease and parasites, tolerance to cold weather, and the ability to raise large numbers of young bees. The most popular breeds of bees are described next.

Italian Bees

The Italian bee is by far the most common breed in the United States. These are gentle bees that produce large amounts of honey. Italian bees have a somewhat low tendency to swarm and are considered to be one of the best types of bees to raise (**Figure 13–22**). They are produced commercially all over the world.

FIGURE 13–22 Italian bees are one of the most widely produced breeds of bees in the world. *© Meister Photos /Shutterstock.com.*

Carniolan Bees

The Carniolan breed of bees originated in the country of Slovenia in the Alps. This dark-colored bee is one of the gentlest of all of the bee breeds and has a reputation for building up the colony quickly in the spring. This breed maintains small numbers over winter, generally needs less feeding than other breeds, and does well in cold climates.

Caucasian Bees

Caucasian bees are somewhat larger than the Italians and Carniolans. They are grayish in color and very gentle. This breed has a reputation for building up large amounts of propolis.

German Bees

German bees are smaller and very dark in color, so they sometimes are called German Black bees. This breed produces large amounts of honey but is too aggressive for most beekeepers. These bees are highly protective of their hive, and they attack when threatened. They have been crossed with other bee breeds to produce gentler bees.

Russian Bees

The Russian bee is relatively new to the United States. Because of its resistance to parasites, bee researchers hope that the Russian bee will help solve some of the serious problems that producers have with parasitic mites.

African Bees

In recent years, a real concern for beekeepers has been the threat of invasion by the Africanized honeybee. These African bees have been referred to as "killer bees" because of their extremely aggressive nature (**Figure 13–23**).

Honeybees are not native to the United States. Most of the bees in this country are descendants of bees that originated in Europe, introduced by European settlers to the New World. The European or yellow bees have a relatively docile nature and are easy to keep. In 1956, research scientists in Brazil imported bees from Africa to cross with the European bees. Their idea was to develop a new strain of hybrid bees that would be more productive than the pure European bees. Despite efforts to contain the African bees, they managed to escape into the wild. In 1990, these bees were reported as far north as southern Texas.

FIGURE 13–23 African bees have been referred to as "killer bees" because of their extremely aggressive nature. The bee on the left is an African bee. *Source: USDA, Agricultural Research Service (ARS).*

The coming of "killer bees" created much attention in the media because of their tendency to attack humans and animals. It has been suggested that the African bees eventually could destroy the bee industry in the United States by replacing the more docile European bees. The African bees take over a colony by entering a colony of European bees, killing the queen and replacing her with an African queen.

However, scientists tell us that the African bees are adapted to living in tropical areas and do not thrive in temperate climates. Problems with these bees have not been as severe as some people predicted. Even if bees with African traits do cause problems with the bee industry, the problems likely will be confined to the southernmost regions of the United States. Most scientists agree that the Africanized honeybee cannot live in areas where the weather turns cold and that their migration has progressed as far as it can.

FIGURE 13–24 The bees draw out an elongated cell called a queen cell, where the new queen is raised. Usually several queen cells are produced in a colony. *© kosolovskyy /Shutterstock.com.*

PRODUCING NEW QUEENS

Beekeepers periodically place new queens into the hives to ensure that each hive has a vigorous queen that will lay plenty of eggs. Queens are produced commercially in small hives known as "nukes." Here, a small colony of bees is formed with no queen. As discussed earlier, when a colony does not have a queen, the worker bees create a new queen by feeding larvae on a substance known as royal jelly. The bees draw out an elongated cell called a queen cell, where the new queen is raised (**Figure 13–24**). Usually, several queen cells are produced in a colony. The first new queen to emerge from the cell may kill the other queens before they can emerge. The producer must be present when the new queens emerge to separate them from the nuke before all of the other new queens are killed.

The new queens are artificially inseminated using semen that has been extracted from drones. This ensures that the queen will be fertilized by bees with the proper genetics (**Figure 13–25**). If the queen is allowed to take her maiden flight, she may mate with wild bees and the genetics may not produce the desired characteristics.

New queens are shipped to the beekeepers in small cages. Two or three worker bees are placed in the cage to feed the queen from a sugar cube stuck in the end of the queen cage. When the bees arrive, the producer takes the cork out of the opening that holds the sugar in place. Within a few days the worker bees will eat through the sugar and the new queen can

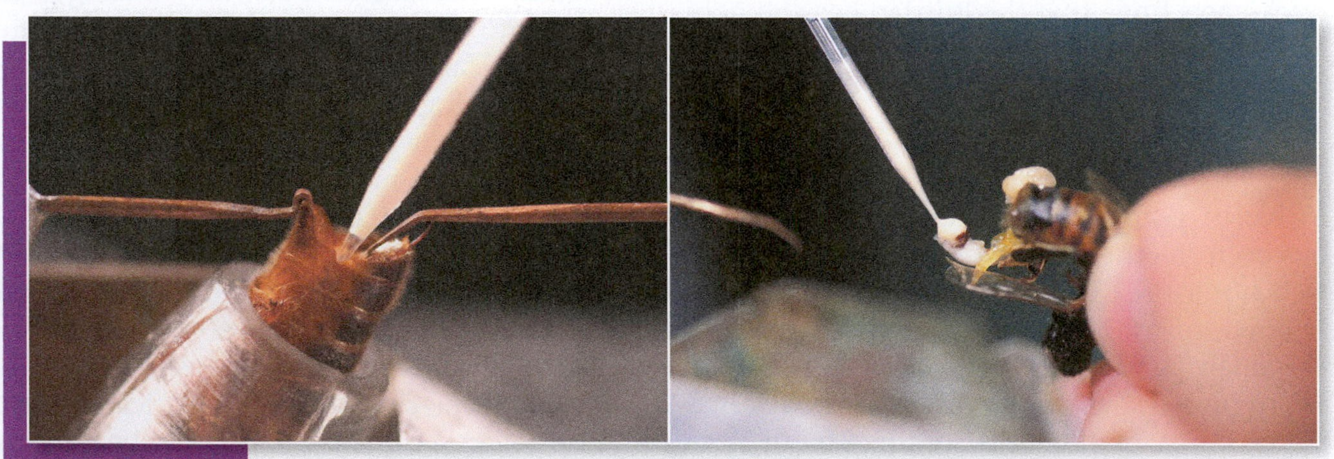

FIGURE 13–25 Semen is collected from the drone (left) and placed in the queen (right). *Source: USDA, Agricultural Research Service (ARS).*

emerge and begin her work in the new colony. The delay allows the bees in the hive to accept her. Some beekeepers replace their queens every year or every 2 years.

DISEASES AND PARASITES

Beekeepers have always been plagued with pests of honeybees. Like any other animal, bees are susceptible to a number of parasites and diseases that cause beekeepers to spend a lot of money and time trying to control the health of their hives. Further, within the past few years, new parasites have emerged that are having a dramatic effect on the bee industry. Scientists fear that if too many colonies of bees die, it will have devastating results on our crops. Without bees and other insects to pollinate flowers, fruit cannot be set on the plants.

Parasites of Bees

Two types of parasitic mites have threatened to wipe out the population of bees in this country. In 1984, tracheal mites were first discovered in U.S. beehives. These microscopic internal parasites lodge in the air passages of adult bees, restricting their ability to breathe. In addition to clogging the air passageways, the mites affect the bees' ability to fly efficiently. These symptoms decrease the bees' lifespan. The mites spend almost all of their life cycles in the trachea of bees, where they reproduce (**Figure 13–26**). They come out only when they move to a new host bee. The mites attach to the interior of the trachea and feed on the bees' body fluid. This is a serious problem for beekeepers because the treatment is difficult. Medications cannot be used in times during the honey flow because the medications can show up in the honey. The most effective means appears to be the breeding of mite resistant strains of bees.

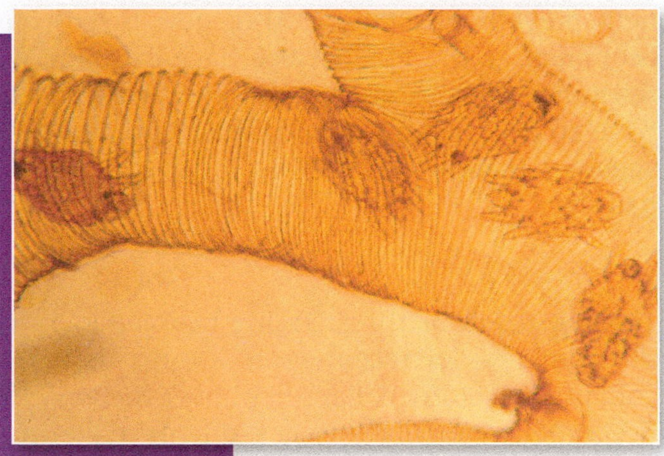

FIGURE 13–26 Microscopic trachea mites spend almost all of their life cycle in the trachea of bees. *Source: USDA, Agricultural Research Service (ARS). Photo by Lilla De Guzman.*

In 1987, the presence of varroa mites was found (**Figure 13–27**). These mites feed on the blood of pupae and adult bees. Not only is the loss of blood a problem for the bees but the damage caused by the mites also makes the bees much more susceptible to diseases. At one time, these two serious pests accounted for as much as a 60 percent loss of commercial bees, and up to 90 percent of the wild bee population was wiped out.

FIGURE 13–27 The arrow indicates a varroa mite on the thorax of this worker bee. *Source: USDA.*

The combined loss of bee colonies from these two mites has had a far-reaching effect on the bee industry as well as the pollination of many of our crops. Scientists have developed treatments for the parasites that include the use of medications in the fall and the spring to help prevent infestation. Also, new breeds of bees such as the Russian bee have introduced resistance to parasites into colonies. Beekeepers have to be vigilant and maintain strict management practices to prevent losses caused by parasites.

Hive Collapse

The most serious threat facing beekeepers is a phenomenon of recent years known as hive collapse or colony collapse. All over the world, beekeepers find hives that suddenly have no bees in them. The bees go in search of nectar and never return. Since 1947, the number of bees in the United States declined by over 40 percent. The problem seems to be getting worse instead of better. Since the bees don't die in the hive, it is difficult to find dead bees to study. Scientists think the problem is related to the parasites such as the varroa mite, mentioned previously. Insecticides used during crop blossom period lowers the bees' immunity system, making them more susceptible to parasites and diseases.

To combat the problem, producers are breeding more bees than ever before. They divide the hives, causing the bees to produce more hives. Also, new production methods, such as treating the hive for parasites and limiting the use of pesticides, are being used. Currently, many scientists worldwide are researching ways of eliminating hive collapse.

SUMMARY

Even though honeybees are the smallest of our agricultural animals, they may be the most important. Each year, bees are responsible for pollinating our crops. Without them, our entire plant industry would not be able to produce. These complex social insects are intriguing to study and fun to keep as a hobby. The honey and wax they produce add a lot to our economy and food supply.

CHAPTER REVIEW

Review Questions

1. Why is the honeybee considered to be one of our most important agricultural animals?
2. Explain how bees pollinate flowers.
3. Name the types of worker bees, and explain their jobs.
4. Why do bees swarm?
5. What is the role of the drone in the colony?
6. How do scout bees communicate?
7. What causes the characteristic humming heard in a beehive?

8. What were some of the materials used in making early beehives?
9. What improvement did Langstroth make in hives?
10. Explain the concept of bee space.
11. What function does a queen excluder play?
12. How do the bees use propolis?
13. Explain two ways in which producers buy bees.
14. How are queens raised for sale?
15. Why are varroa mites such a problem?
16. Explain the life cycle of the trachea mite and how it affects bees.

Student Learning Activities

1. Conduct an Internet search, and determine what products use honey in their production. Hint: Look at sweet baked goods. What products are made from beeswax?
2. Conduct research on the various breeds of bees. Choose one you think would be good for your area. Explain to the class why you chose that breed.
3. Observe a number of different flowers for 15 minutes. Count the number of bees that visit the flower during that time. Which flowers are most popular with the bees? Why do you think these flowers are so popular?
4. Read the labels on honey jars in the grocery store. Where was the honey produced? From what flowers was the honey produced?

CHAPTER 14

Animal Behavior

STUDENT OBJECTIVES IN BASIC SCIENCE

As a result of studying this chapter, you should be able to

- define *ethology*.
- discuss the difference between instinctive and learned behavior in animals.
- explain the concept of animal intelligence.
- define *conditioning*.
- discuss how animal behaviors are developed.
- describe the types of social behavior in animals.
- discuss the types of sexual and reproductive behaviors in animals.
- describe the types of ingestive behaviors in animals.
- explain how certain animals communicate.

STUDENT OBJECTIVES IN AGRICULTURAL SCIENCE

As a result of studying this chapter, you should be able to

- explain how animal behavior is used to protect sheep.
- discuss the social behaviors of agricultural animals.
- describe the sexual and reproductive behaviors of agricultural animals.
- list the ingestive behaviors of agricultural animals.
- describe the methods that agricultural animals use to communicate.
- discuss how the natural behaviors of agricultural animals can be used to provide the animals with a safer, more comfortable environment.

KEY TERMS

ethology
instinct
imprinting
intelligence
conditioning
docile
indiscriminate breeders
social behavior
stallion
estrus
ingestive behavior
grubs
ruminants
dental pad
cecum

NATIONAL AFNR STANDARD

AS.02.01.02a
Research and summarize the challenges involved in working with animals and resources available to overcome them.

AS. 05.01.01.a
Differentiate between the types of facilities needed to house and produce animal species safely and efficiently.

AS.05.01.01.b
Critique design for an animal facility and prescribe alternative layouts and adjustments for the safe, sustainable, and efficient use of the facility.

AS.05.01.02.a
Identify and summarize equipment, technology, and handling facility procedures used in modern animal production.

AS.05.01.02.b
Analyze the use of modern equipment, technology, and handling facility procedures and determine if they enhance the safe, economic, and sustainable production of animals.

AS.05.01.02.c
Select, use, and evaluate equipment, technology, and handling procedures to enhance sustainability and production efficiency.

FOR CENTURIES, scientists have debated the intelligence levels of animals. Although animals doubtless can be trained to do tasks, there was (and still is) disagreement about the level of animals' intelligence. Can animals reason? Solve problems? Think? Feel emotion? For many years, the answer to these questions was "no."

In recent years, scientists have begun to rethink the answer. For example, chimpanzees have been observed using tools such as sticks to collect grubs from decayed wood (**Figure 14–1**). Crows have been seen dropping nuts on the highway so cars will run over the shells and expose the edible parts of the nuts. A lot of research is ongoing in this area, and disagreements will continue. Still, we do know that all animals, whether wild or domesticated, act in certain ways. Scientists have long recognized that animals have behavior patterns. Many of these behaviors are predictable; animals usually will act in a certain way under specific conditions.

The study of how animals behave in their natural habitat is called **ethology**. The habitat may be the natural area of wild animals or the pastures, pens, or facilities of domesticated animals. The behavior of animals in the wild has been an accepted branch of biological science for many years. Only recently has the science of ethology been used to research methods of producing agricultural animals. Ethology plays a significant role in the production scheme of the modern livestock industry because the way animals behave can be better understood and, in turn, be used to create a better growing environment for the animals.

FIGURE 14–1 Chimpanzees have been observed using sticks to collect grubs. *© Norma Cornes/Shutterstock.com.*

Different types of animals behave in different ways. Most animal behavior can be divided into two categories: instinctive and learned behavior. The most basic is **instinct**, the behavior that is set in an animal at birth and causes the animal to respond automatically to an environmental stimulus. This behavior is a result of genetics, and is not something the animal learns. As a good example, newly born and young animals are able to nurse, and most agricultural animals are able to stand and nurse only minutes after they are born (**Figure 14–2**). This behavior is not taught but, rather, is with the animals at birth. Other instinctual behaviors include breeding, eating, and drinking.

Animals also can learn behaviors. One of the most basic types of learning is **imprinting**. Imprinting means that an animal will attach itself to or adopt another animal or object as its companion or parent. This usually occurs shortly after the animal is born or hatched. For instance, if a hen sits on duck eggs, the

FIGURE 14–2 Most agricultural animals are able to stand and nurse shortly after birth. *© Judy Marie Stephanian /Shutterstock.com.*

FIGURE 14–3 Dogs raised with sheep often act as guardians. *© Nadiia Korol/Shutterstock.com.*

FIGURE 14–4 Some scientists rank pigs just under chimpanzees and dolphins in intelligence. *© Dmitry Kalinovsky /Shutterstock.com.*

resulting ducklings will accept the hen as their mother. Goslings (baby geese) have been known to adopt dogs or humans in this manner if the human or dog becomes a companion within the first 36 hours after hatching. Dogs also have been known to adopt other animals as "their own." Some sheep producers make use of this behavior by placing very young pups with a flock of sheep. The dog is raised among the sheep and accepts the sheep (**Figure 14–3**). The dog then acts as guardian to its adoptive family and keeps predators away from the flock.

Several species of fish, especially salmon, are imprinted as to that location when they hatch. After several years, the mature salmon return to spawn in the place where they were hatched. This drive in the fish is extremely strong, and they will go over and through severe obstacles on their journey to the place of their hatching.

Different species of animals have differing abilities to learn. This is called **intelligence**. Obviously, the most intelligent animals are humans. Primates, such as chimpanzees, are next in order of intelligence, followed by ocean mammals such as dolphins and whales. Among agricultural animals, the pig is considered to be the most capable of learning and, therefore, the most intelligent. In fact, some scientists rank the pig just under chimpanzees and dolphins in intelligence (**Figure 14–4**).

Learning in animals comes about through several means. One is **conditioning**, which means that an animal learns by associating a certain response with a certain stimulus. A Russian scientist named Ivan Pavlov was famous for his theories of conditioned reflex. His experiments involved feeding meat to a group of dogs and ringing a bell as the animals were fed. The only time the bell was rung was when the animals were fed, so they associated the ringing of the bell with eating. After a time, the animals would begin to salivate when the bell was rung even if no food was in sight. Agricultural animals are conditioned to perform certain reflexes. For example, as cows enter the milking parlor, they let down their milk because they associate going into the milking parlor with being milked.

Animal producers can apply this principle to teach animals to respond in a certain way. Animals also can learn on their own by trial and error. For example, a horse may learn to open a gate by tinkering with the latch mechanism until it learns the proper sequence to open the latch and get where it wants to go. A pig can learn that by lifting a lid, it can gain access to feed in a self-feeder, or it may be able to get water from a self-waterer by applying pressure in the correct location. Horses are trained by receiving positive rewards when the animals respond in a manner the trainer desires. Animal trainers use animals' natural abilities and instincts to teach them to do tricks, perform work, or be more productive.

Dogs with a natural instinct to herd animals are trained to herd sheep, hogs, and cattle (**Figure 14–5**). Scientists tell us that this instinct comes from the wolves that were ancestors of the herding dogs (**Figure 14–6**). Wolves in the wild work together in a pack, circling a group of animals until a vulnerable one is singled out for the kill. Breeds such as the Border Collie retain much of this instinct and can be trained to care for and move animals instead of herding them as prey.

Other breeds, such as the Blue Tick Coonhound, are not trained as easily to herd animals, but they are easily trained to track animals and to tree them. Certain breeds of horses, such as the American Saddlebred, are trained for pleasure riding and for show. Other breeds, such as the Belgian, are trained as draft animals and are used less for pleasure riding.

When humans first started domesticating animals, they discovered that not just any animal was suitable to be tamed for raising. Animals possessing certain natural characteristics made them more desirable for domestication. Animals were chosen that had some use, as food, as workers, or as companions. The animals also had to have certain behaviors that were conducive to domestication. For humans to raise them, the animals had to be **docile**, or gentle, and easy to handle (**Figure 14–7**). The animals that were more docile were selected for breeding and eventually developed into animals that could be handled by humans. In addition, the animals had to be **indiscriminate breeders**. This meant that the animals could not be paired for life, but that any male would breed with any female. This allowed for selective breeding and

FIGURE 14–5 Some breeds of dogs have an instinct for herding animals. *© anetapics/Shutterstock.com.*

FIGURE 14–6 Scientists say that the instinct to herd comes from wolf ancestors. *© iStockphoto/KenCanning.*

FIGURE 14–7 For humans to raise them, animals have to be docile and easy to handle. *© Dita_Moya/Shutterstock.com.*

development of the characteristics that producers deemed desirable.

Livestock producers look for several types of behavior that make raising livestock easier, more efficient, safer, and more comfortable for the animals. These behaviors are based on the animals' natural instincts, but they may have been developed or enhanced through years of selective breeding. At any rate, some patterns of behavior seem to run in all species of agricultural animals.

FIGURE 14–9 Wild horses usually live in small groups. *© Viola90/Shutterstock.com.*

SOCIAL BEHAVIOR

Social behavior refers to the manner in which animals interact with each other. Most farm animals are gregarious. This means that they tend to want to herd or flock together (**Figure 14–8**). Even in the wild, cattle, sheep, and horses tend to want to group together. This is the result of a natural instinct, for defense purposes. Young, old, and weak members of the herd or flock can be better protected if the animals remain together in a group.

FIGURE 14–8 Most farm animals are gregarious; they tend to herd together. *© Dudarev Mikhail/Shutterstock.com.*

Animals seem to prefer a certain size of herd or group. In the 1830s, Charles Darwin wrote about large groups of cattle (10,000 to 15,000 head) breaking up into smaller groups. Even after being mixed together and stampeded, the cattle would come back into groups of 40 to 100 head. Many cattle producers today divide their herds into this approximate size. Sheep will band together in much larger groups, some reaching several hundred in number. Pigs in the wild tend to band together in groups of about 10. Wild horses usually live in small groups consisting of a **stallion** and his harem of mares (**Figure 14–9**). Other males may live in a small group together with one of the stallions as the leader or dominant male. Producers have made use of this gregarious behavior by moving, feeding, and caring for animals as a group. As mentioned earlier, the animals can be moved and controlled more efficiently through the use of herding dogs. Without animals' gregarious behavior, working dogs would not be nearly as efficient.

Within each group of animals is a hierarchy, or order of social dominance. Some animals within the group are recognized by the other animals as having dominance, the ability to exert social influence or pressure over others in the group. In poultry, this is known as the pecking order (**Figure 14–10**). Certain chickens in the flock are allowed to have priority for space, food, water, and the like.

FIGURE 14–10 Chickens establish a social order known as the pecking order. *Courtesy of Joe Maudlin, Cooperative Extension Service, University of Georgia.*

FIGURE 14–12 Male animals often fight to establish dominance. *© Johnwoodkim/Shutterstock.com.*

FIGURE 14–11 Each piglet has a favored teat to which it will go. *© D McKenzie/Shutterstock.com.*

Research has shown that there is a social dominance order in other agricultural animals, too. For instance, according to some research studies, baby pigs have been competitive about which teat they will nurse. The teats closer to the front seem to be preferred. Each piglet has a certain teat it will go to for nursing (**Figure 14–11**). Research has indicated that the social dominance established during nursing is carried over when the pigs are weaned. Dominant pigs are allowed to be first to the feed trough, and they establish where they want to sleep in the pen.

Social dominance patterns also have been noted in most other agricultural animals. For example, if two or more males are in the same flock or herd, one will be the dominant male that will mate with most of the females. This dominance usually is established by fighting among the males, with the strongest and most vigorous emerging as the dominant male (**Figure 14–12**). This is nature's way of ensuring that the heartiest animals are the ones that will breed and raise the next generation of the species.

Producers must consider social dominance as they plan to grow their livestock. For example, males must be kept separate to prevent injury. If animals are on a limited ration, they must be separated or the dominant animals will get too much feed and the subordinate animals will get too little.

SEXUAL AND REPRODUCTIVE BEHAVIOR

All animals have certain behaviors associated with mating and reproducing. (The process of

FIGURE 14–13 Males actively seek out females that are in estrus. *© RichSouthWales. Image from BigStockPhoto.com.*

reproduction is discussed in Chapter 19.) Most female agricultural animals come into **estrus** (heat) in preparation for mating. As this happens, the females engage in behavior that indicates their condition. Cows may bellow, mill around restlessly, allow other cows to mount, or mount other cows. Sows will mount other sows, appear restless, urinate frequently, and grunt loudly. Sheep and other female agricultural animals show signs of estrus in a similar manner.

Males actively seek out females that are in estrus to complete the mating process (**Figure 14–13**). Often during mating, the males become more aggressive or belligerent toward other animals and humans. Some breeds of animals display a behavior that differs from other breeds of the species. For example, Brahman cattle usually prefer to breed at night rather than in daylight.

As the end of the gestation period approaches, females display behavior that indicates the approach of birth. Sows, if in open pasture, usually try to build a nest from grass, soil, or other materials they may find. Cows that are about to give birth generally appear nervous and isolate themselves from the herd. Sometimes they may even hide if there is enough cover from trees or undergrowth in the pasture. A mare usually bites at her flanks, switches her tail, and lies down and gets back up repeatedly.

After the offspring are born, the mothers' behavior changes. They almost always become more aggressive and protective of their young. Even females that normally are docile can become belligerent after the birth of offspring. This is nature's way of protecting the young from predators.

Most mothers of agricultural animals recognize their own offspring and will allow only that individual to nurse. An exception is pigs. A sow usually accepts orphan pigs if she has enough teats for all of the pigs to nurse. With cattle and sheep, it is more difficult to get them to accept young that are not their own. The cows or ewes recognize the scent of their newborns and will accept only that smell. Sometimes producers can fool the mothers into accepting an orphan by changing the way it smells. A cattle producer may rub both the mother's calf and the orphan in a strong-smelling solution. Because the cow cannot distinguish which calf is hers, she may accept both calves. A sheep producer may take the skin from a dead lamb and cover an orphan with it to trick the ewe into thinking the orphan is indeed her lamb. After nursing begins, the ewe soon will accept the orphan.

INGESTIVE BEHAVIOR

Ingestive behavior refers to the manner in which animals eat and drink. Different animals have different habits or ways in which they take food. Obviously, most of these differences reflect the way the animals are made—their digestive system and type of food they prefer.

Pigs that run outside in a pasture or in a lot tend to root or dig in the ground for food (**Figure 14–14**). This is a carryover from the

FIGURE 14–14 Pigs tend to root or dig in the ground for food. *© iStockphoto/LordCastle.*

time before they were domesticated, when their diet consisted of roots, **grubs**, insects, seeds, and nuts. Even the most modern breeds of pigs will revert to the habit of digging in the ground with their snout. Their digestive system contains a simple stomach, and they are not capable of digesting large amounts of fiber as the **ruminants** do. Therefore, the type of food found by their rooting action is suited for their digestive system.

People who are not familiar with agricultural animals think of pigs as animals that overeat. Expressions such as "eat like a pig," "pig out," and the like, suggest that pigs eat so much that they make themselves sick. Contrary to this belief, pigs will eat only what they need and do not make themselves sick by overeating (**Figure 14–15**). Actually, given the opportunity, pigs will eat the right amounts of given feeds and balance their own diet. As mentioned earlier in this chapter, pigs are among the most intelligent of agricultural animals.

Ruminant animals, such as sheep, goats, and cattle, have digestive systems designed to handle large amounts of roughage such as grass or other plants. Even though ruminant agricultural animals eat basically the same type of food, they gather and ingest food in different ways. Cattle, goats, and sheep have no upper front teeth and must rely on a thick **dental pad** in the top of the mouth to tear off plants as they graze (**Figure 14–16**). Cattle wrap their tongue around the plants and tear them off between their lower teeth and upper dental pad. For this reason, cattle prefer to graze in forage at least 6 inches high. Sheep, in contrast, cut off the forage by nipping it with their teeth and dental pad and gathering it into their mouth with their lips. This is why sheep can graze much closer to the ground than cattle can (**Figure 14–17**).

FIGURE 14–16 Sheep and cattle have a dental pad instead of upper teeth. *© Eric Isselee/Shutterstock.com.*

FIGURE 14–15 Pigs eat only what they need and don't make themselves sick by overeating. *© Igor Stramky/Shutterstock.com.*

FIGURE 14–17 Sheep graze by nipping off grass between the dental pad and the teeth, cutting off the grass shorter than cattle do. *© iStockphoto/real444.*

This behavior of sheep was the basic cause of the range wars of the late 1800s in the U.S. Western frontier. Sheep that were moved into cattle country grazed the land so close that in some areas the grass would not grow back. This angered the cattle ranchers because they thought that cattle grazing protected the land better. Later research revealed that a system of grazing sheep and cattle on the same ground could be beneficial if managed properly. Cattle prefer to eat grasses, and sheep prefer to eat plants that are leafy and coarser. If the two species are control-grazed on land that contains both types of plants, the cattle and the sheep both benefit.

Sheep tend to graze for a longer period of time per day than cattle do. Cattle usually graze from 4 to 9 hours a day, whereas sheep graze from 9 to 11 hours per day (**Figure 14–18**). As ruminants graze, periods of eating are followed by periods of rest. This allows time for rumination, or digestion of the plants. During these periods, the animals regurgitate and chew the plant material they have swallowed.

Although horses eat large amounts of plant material, they do not ruminate. Instead, their digestive system has a large section, called a **cecum**, which processes the roughage. Because horses have both upper and lower front teeth, they bite off the plants as they graze. They usually prefer pasture forage but will eat brushy plants if no other forage is available.

FIGURE 14–18 Cattle typically graze 4 to 9 hours per day.

ANIMAL COMMUNICATION

The ability to communicate means that animals are able to pass information from one to another. We as humans tend to think of communicating as being able to speak and express ourselves within the context of a broad and diverse vocabulary. Obviously, animals do not talk as we do, but, nonetheless, they do pass information between them. Their form of communication may be through body motions, through sounds they emit, or through smell.

As pointed out in Chapter 13, perhaps the most dramatic example of animal communication is that used by honeybees to tell other bees in the hive about nectar and pollen sources they have located. Within the bee **colony**, certain bees serve the purpose of looking for food sources and for this reason are called **scout bees**. Through a series of elaborate moves and dances, the scout bees tell the workers the direction, distance, and amount of the nectar source. The scout bees also bring back samples of the type of nectar that they located. By smelling and tasting the nectar, the worker bees become aware of the smell and taste of that specific nectar. Recent research has shown that the bees also use sound in communicating the distance to the food source. Scientists believe that scout bees also use the sun to orient their flight and communicate this orientation to the worker bees in the hive.

Other agricultural animals communicate as well. Through sounds, chickens call other chickens to feed. A mother hen uses a certain cluck to call her chicks. Different stances depict social standing within the flock. On the one hand, a chicken that stands in a crouch with unruffled feathers and tail feathers held close together indicates a submissive animal. On the other hand, a chicken that stands tall, head held high, tail feathers spread wide, and body feathers ruffled is a dominant chicken. The other chickens in the flock recognize this form of communication and act accordingly.

When a sow lies on her side for the piglets to nurse, she grunts in a certain manner. This distinctive grunt calls the piglets to come and nurse. In addition, pigs emit specific sounds that warn the others of danger. In the wild or in the pasture, pigs may rub trees, stones, or other objects as a way of marking their territory. The rubbing leaves an odor that other pigs can detected.

Horses communicate through several different means as well. For example, the direction the ears are pointed can transmit a horse's mood. Anger is expressed by laying the ears straight back toward the neck. Ears that are pulled forward show that the horse is interested in something (**Figure 14–19**). Horses also communicate through the sounds they make. A neigh or whinny may indicate that the horse is frightened or concerned. A snort may be used to warn other horses of approaching danger. A squeal may indicate that a horse is angry. Of course, people are able to communicate their commands to horses. Horses are trained to respond to verbal commands of riders, or they may respond to a tug of the halter or a nudge with the knee. Good riders or horse handlers are able to determine the mood or attitude of their horses.

Cattle communicate by using their voice. A cow that is in estrus will bellow to find a mate. A cow communicates vocally with a calf to call it to her or to warn it of impending danger. Cows

FIGURE 14–19 A horse will point its ears forward when it is interested in something. *© Callipso/Shutterstock.com.*

FIGURE 14–20 A bull pawing the ground is a sign of aggression. *© Taina Sohlman/Shutterstock.com.*

may even hide their calves in a wooded area. Cattle also use body stance to relay messages. A lowered head with the horns or the top of the head thrust forward indicates that the animal is ready to fight. A bull pawing the ground is a sign of aggression (**Figure 14–20**). Twisting or slinging the head issues a warning. A head held high with the back swayed and the tail head raised indicates that the animal is about to take flight.

Much of the modern ways of handling livestock are based on the research and findings of Dr. Temple Grandin of Colorado State University. Dr. Grandin started observing animal behavior when she was in Graduate School. She observed that some cattle would loudly "moo" while others were quiet. This began a study of what frightened animals. Her findings led to ways of designing handling facilities that kept the animals quieter. The best producers study the behavioral activities and requirements of agricultural animals and use their studies to provide safer and more comfortable environments for the animals. Producers adjust space requirements to provide the animals with the proper amount of room. The desirable space for all animals has been well researched. For example, research indicates that pigs like to be in contact with each other and that touching other pigs is important to them (**Figure 14–21**). Pigs raised in isolation do not do as well as those raised with other animals. And chickens

FIGURE 14–21 Pigs like to be in contact with each other. © yevgeniy11/Shutterstock.com.

are given enough space in cage operations to make them comfortable.

Producers also can consider the instincts of animals when designing facilities. For instance, cattle and sheep urinate and defecate indiscriminately. These animals deposit waste materials any place within their living space. Horses and pigs, though, eliminate waste only in a certain part of their space. As facilities for these different animals are designed and constructed, the animals' instincts have to be taken into consideration.

Among other traits discovered by Dr. Grandin is that cattle have a natural tendency to follow a curved passageway and usually want to circle to the right. Therefore, handling facilities should be designed to accommodate this behavior. Cattle can be handled more effectively and safely in corrals and chutes that are completely opaque (the animals can't see through them) and that circle to the right (**Figure 14–22**). Horses should not be kept in

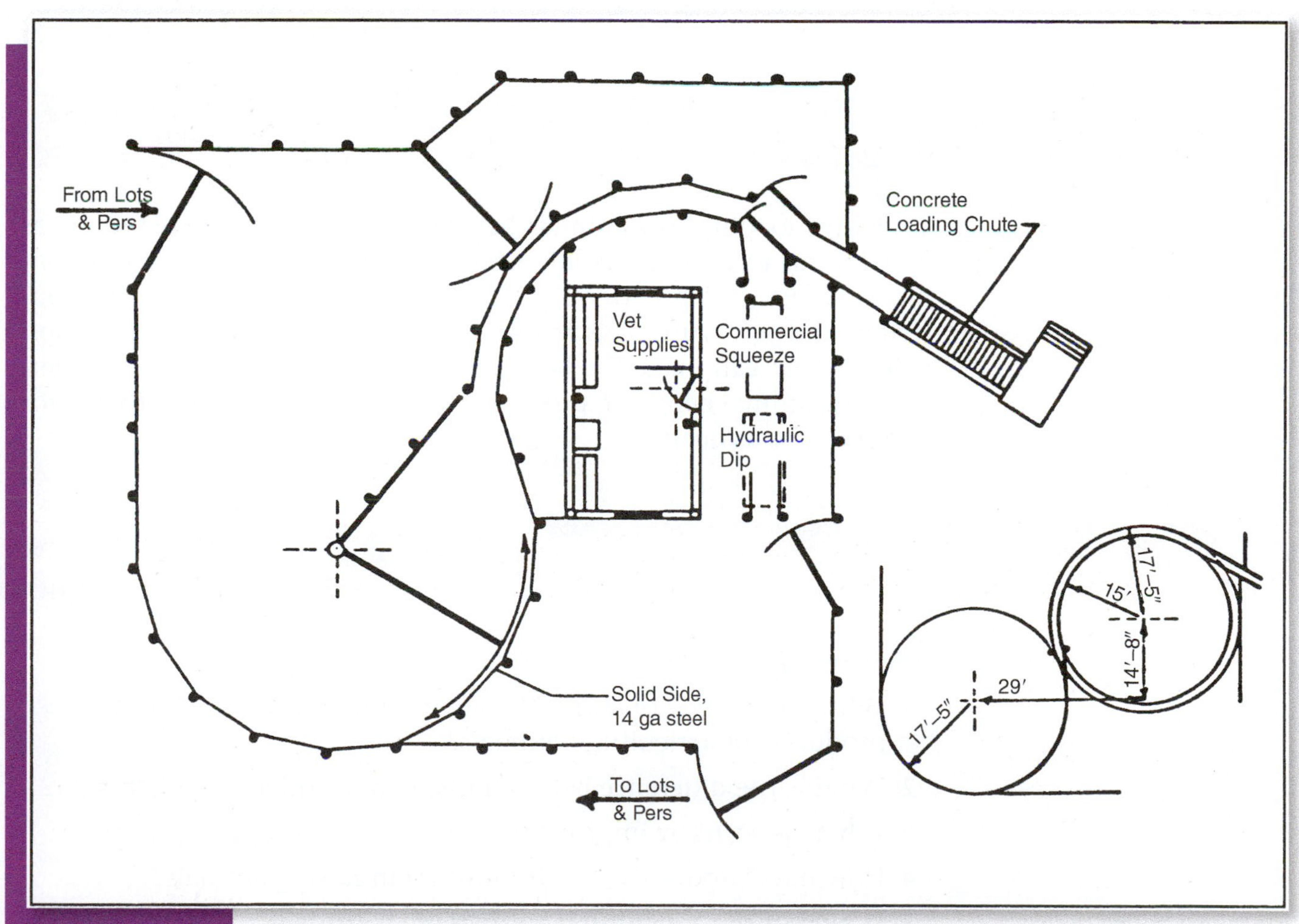

FIGURE 14–22 Handling facilities are designed to take advantage of cattle's tendency to follow a curved passageway. *Cooperative Extension Service, University of Georgia.*

barbed wire fences because they have a tendency to cut themselves on the wire if they become spooked or excited. Cattle are kept well in barbed wire fences and seldom stampede into the fence.

WORKING SAFELY WITH ANIMALS

Animals can be dangerous and should anyone working with animals should be cautious. Even though the animal can seem like a pet, they can cause injury. Any male animal is by nature aggressive and wants to maintain his territory. A large animal such as a bull may seem playful, but his playfulness lead to broken bones or worse. Anytime you approach a bull, make sure you have an escape route where you can get away from him. Never try to pet a bull and remember that he is several times your size. The same can be said for rams, boars, and stallions. Always be alert and keep a watchful eye out for these animals.

Female animals are extremely protective of their young and may see you as a threat to her offspring. When caring for young animals keep a barrier between you and the mother. A calf may call out or a piglet may squeal. This is a signal to the mother that her baby is in danger.

Management practices such as vaccinations must always be done when the animal is secure in a catch pen or other means of restraint. Remember that the animal does not know that you are helping. It only has a sense of discomfort. By keeping the nature of the animals in mind, modern producers can raise healthier, more productive animals at a more profitable rate.

SUMMARY

To a large extent, how animals behave determines their usefulness as agricultural animals. By studying how animals act in their environment, scientists can better understand how to keep the animals contented and safe. Also, this knowledge helps in designing production systems that can make the best use of the animal's nature. By understanding animal behavior, producers are better equipped to provide for and produce animals.

CHAPTER REVIEW

Review Questions

1. Why is the study of animal behavior (ethology) important to producers of agricultural animals?
2. What is the difference between instinctive and learned behavior?
3. What is meant by imprinting?
4. How have producers used imprinting in raising animals?
5. Which of the agricultural animals are considered to be the most intelligent?

6. Give two examples of how conditioning is used in animal agriculture.
7. What were the characteristics of animals that early humans looked for in selecting animals to tame and raise?
8. What are some ways in which animals determine social dominance?
9. Briefly discuss at least two sexual and reproductive behaviors of agricultural animals that are important to producers.
10. What aspect of animal ingestive behavior caused range wars during the 1800s?
11. Explain how the differing ingestive behaviors of sheep and cattle make them beneficial to each other.
12. List some ways in which agricultural animals communicate.
13. Explain why cattle chutes should be built in a curving pattern.

Student Learning Activities

1. Choose a herd or flock of agricultural animals to observe. During a period of several hours, list all of their behaviors. Decide which of the behaviors are learned and which are instinctive. Compare your list with others in the class.
2. Visit a livestock producer and obtain permission to tour the operation. Make a list of all of the aspects of the operation (facilities, work schedules, feeding times) that deal with animal behavior.
3. Choose one animal and observe it for several hours. Record all attempts by the animal to communicate with other animals. This communication might include the animal's attempts to communicate with you!
4. Visit the local animal shelter and interview the workers. Determine what they have learned and observed about animal behavior. Report how they use the animals' behavior in their work.
5. Research the life of Dr. Temple Grandin. Share your findings with the class. You might also want to see the movie, *Temple Grandin*, about her life and work.

CHAPTER 15

Animal Cells: The Building Blocks

STUDENT OBJECTVES IN BASIC SCIENCE

As a result of studying this chapter, you should be able to

- explain the concept of cells as building blocks.
- describe the different types of cells.
- name all the components of an animal cell.
- explain the process of meiosis.
- explain the process of mitosis.
- discuss the process of diffusion.
- discuss the process of osmosis.

STUDENT OBJECTIVES IN AGRICULTURAL SCIENCE

As a result of studying this chapter, you should be able to

- describe how livestock producers work with animal cells.
- discuss why an understanding of animal cells helps producers raise animals.

KEY TERMS

cell
prokaryotic cells
eukaryotic cells
nucleus
chromosomes
genes
cytoplasm
organelles
plasma membrane
diffusion
osmosis
homeostasis
mitochondria
vacuoles
enzyme
microfilaments
Golgi apparatus
endoplasmic reticulum
lysosomes
meiosis
mitosis
centrioles
cytokinesis
cleavage
blastula
placenta

NATIONAL AFNR STANDARD

AS.06.02.01.a
Research and summarize characteristics of a typical animal cell and identify the organelles.

AS.06.02.01.b
Analyze the functions of each animal cell structure.

AS.06.02.02.a
Examine the basic functions of animal cells in animal growth and reproduction.

AS.06.02.02.b
Analyze the processes of meiosis and mitosis in animal growth, development, health, and reproduction.

AS.06.02.03.a
Identify and summarize the properties, locations, functions, and types of animal cells, tissues, organs, and body systems.

AS.06.02.03.b
Compare and contrast animal cells, tissues, organs, body systems, types, and functions among animal species.

THE IMPORTANCE OF CELLS

LIVESTOCK PRODUCERS, whether they realize it or not, work with animal cells. This is because most of the life processes that take place in an animal's body, such as reproduction, growth, disease immunity, and nutrient utilization, take place at the cellular level (**Figure 15–1**). **Cells** are the building blocks of life and the basis for all animal and plant systems. When producers balance a feed ration, give a vaccination injection, or plan a breeding program, they actually are working at the cellular level because that is where the life processes begin and/or function.

Literally hundreds of different types of cells perform different functions within the animal's body. Cells that make up the skin are different from cells that make up bones and serve different functions. In animals, the cells start to differentiate as the fetus develops in the uterus. This means that cells grow and divide into different tissues such as muscle, bone, and nerves. Growth in all organisms comes about as a result of cell division. This chapter will discuss how cells operate and how they are similar and different.

Types of Cells

Cells come in a wide variety of types and sizes (**Figure 15–2**). The smallest of cells are less than a micrometer (1 millionth of a meter) in diameter and require the use of the most powerful microscopes to be able to observe them. In contrast, some cells can weigh a couple of pounds. Just think of an ostrich egg! It's a single cell (**Figure 15–3**). Cells may be round like a ball, square like a box, some are long and thin like a string, and some are shaped like a plate.

FIGURE 15–1 Life processes such as reproduction, growth, digestion, and disease immunity take place at the cellular level. *© iStockphoto/percds.*

Within an animal, each type of cell has a specific role to play, and the shape of the cell is related to that role. Cells are broadly grouped into two types: **prokaryotic cells** and **eukaryotic cells**. The two types are similar in that they both contain genetic material and are filled with a watery substance called cytoplasm.

Prokaryotic Cells

One basic difference in the two types of cells is that the genetic material in a eukaryotic cell is contained within the confines of a membrane-enclosed nucleus. Prokaryotic cells contain genetic material, too, but this material is not confined to a nucleus. The genetic material (deoxyribonucleic acid, DNA) is contained within a single molecule and is in contact with the cytoplasm. Prokaryotic cells are the smallest of all cells and are generally considered to be neither plant nor animal. They include one-celled organisms such as bacteria and blue-green algae (**Figure 15–4**).

Viruses are not generally considered cells because they lack the ability to reproduce on their own. They must become parasites to other cells in order to translate their genetic code. Viral cells also have simple internal structures surrounded by a stiff cell wall that shapes and protects the cell. Although prokaryotic cells are not part of animal systems, they play an important role in the life of animals. Other chapters discuss the role of bacteria in the digestive process, as well as in disease outbreaks.

Eukaryotic Cells

All plants and animals are made up of eukaryotic cells. Even though plant and animal cells

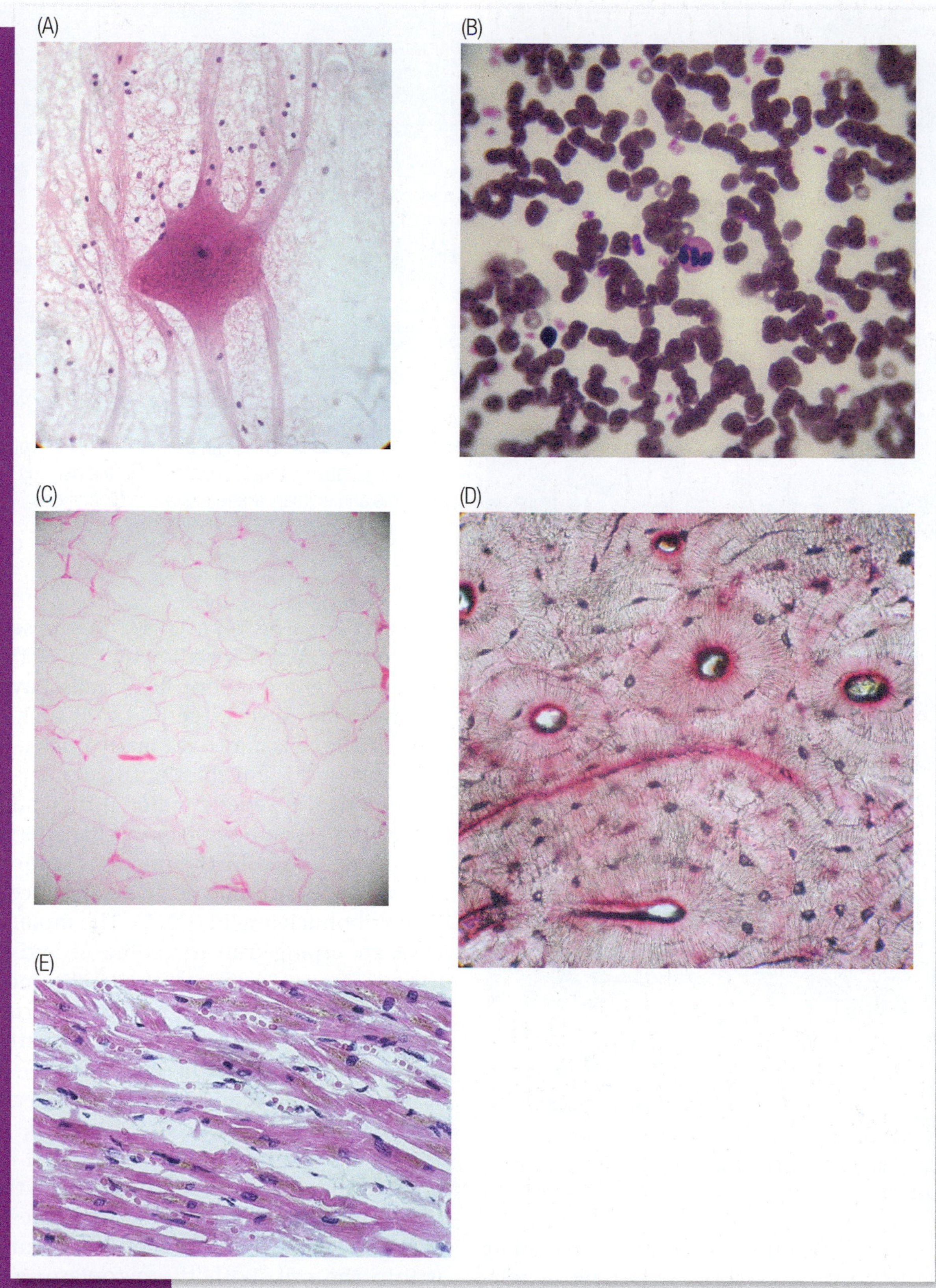

FIGURE 15–2 Cells come in a wide variety of shapes and sizes: (A) nerve cell; (B) blood cell; (C) fat cell; (D) bone cell; (E) muscle cell.

FIGURE 15–3 An ostrich egg is a large single cell. © iStockphoto/hidesy.

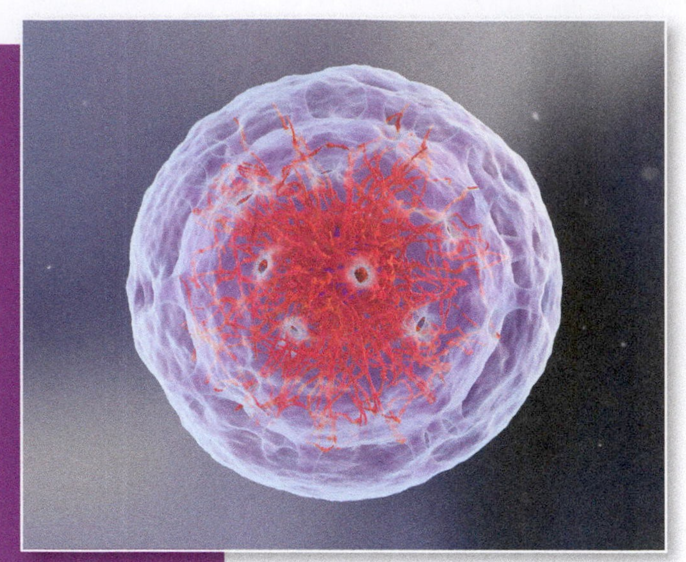

FIGURE 15–5 This is an example of a eukaryotic cell. The genetic material is contained in the nucleus. All plants and animals are made of this type of cell. © YuriiHrb/Shutterstock.com.

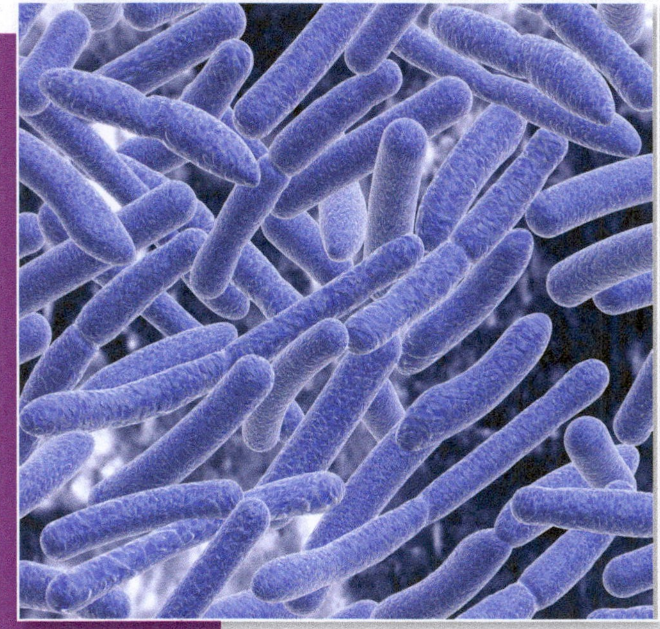

FIGURE 15–4 Bacteria are prokaryotic cells. The genetic material is not confined to a nucleus. © Fedorov Oleksiy/Shutterstock.com.

have many differences, they also have some similarities. All eukaryotic cells have a relatively large structure called a **nucleus**, composed primarily of nucleic acids, proteins, and enzymes (**Figure 15–5**). This structure serves as the control center for all activities of the cell, including reproduction. Most cells have only one nucleus; however, certain cells such as animal muscle cells may have many nuclei. One of the most important roles of the nucleus is that it contains the genetic material that translates a code to give an organism its characteristics.

As mentioned in the previous section, this genetic code is contained in a substance called deoxyribonucleic acid (DNA). The molecules of DNA are arranged in threadlike strands called **chromosomes**. Segments of the chromosomes called **genes** are responsible for transferring the genetic code. Gene transfer is dealt with more completely in a subsequent chapter.

Animal cells contain a thick, clear fluid that surrounds the nucleus. This fluid, known as **cytoplasm**, contains all the material the cell needs to conduct life processes. Cytoplasm contains membrane-enclosed structures called **organelles** that perform specialized functions within the cell. Both the cytoplasm and the nucleus are contained within a cell membrane composed of proteins and lipids (fats).

Eukaryotic Cell Components

Eukaryotic cells have many components, all of which serve specific functions and must work in concert with all the other parts of the cell. The interactions of the cell parts are not only complex but also essential to the well-being of the entire organism. If one part of the cell does not function properly, the other parts cannot function properly. Obviously, when this happens the organism cannot properly function.

Cell Membranes

Every eukaryotic cell contains a cell membrane, also known as the **plasma membrane**, which serves three purposes (**Figure 15–6**). First, it encloses and protects the cell's contents from the external environment. Second, it regulates the movement of materials into and out of the cell, such as the taking in of nutrients and the expelling of waste. Third, the cell membrane allows interaction with other cells. In animal cells, the membrane is known as a plasma membrane. Plasma is the liquid part of the cell.

All material that passes into and out of the cell must pass through the cell membrane. The membrane is said to be selectively or semipermeable, which means that it allows only certain materials to go through; not all substances are allowed to pass through the membrane. The substances that are allowed through are usually small molecules and ions (charged molecules). This membrane serves the purpose of allowing material, such as water and other nutrients needed for the life processes, to pass through into the cell. It also gets rid of the waste materials left over from these processes that otherwise would accumulate and harm the cell. The materials pass through the membrane in a process called **diffusion**. In diffusion, molecules in solution pass through the membrane from a region of a higher concentration of molecules to a region of lower concentration of molecules (**Figure 15–7**).

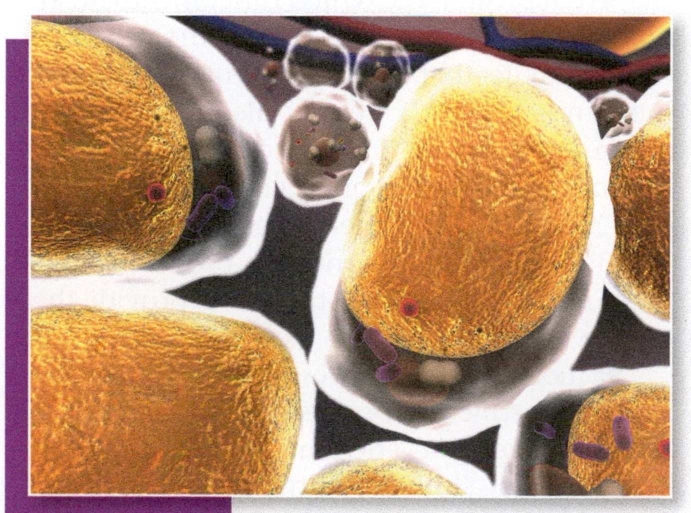

FIGURE 15–6 These are animal fat cells. The enclosure around each cell is the cell membrane. *© UGREEN 3S/Shutterstock.com.*

For example, in an animal's cell there are fewer molecules of oxygen inside the cell than there are outside the cell. Also cells usually have more carbon dioxide molecules inside than there are outside. As the cell uses up oxygen molecules, more oxygen is allowed through the membrane because the molecules try to equalize the number without and within the cell. Likewise, the carbon dioxide cells move out of the cell to an area that is less concentrated with carbon dioxide molecules. Through diffusion, the cell constantly takes in needed molecules such as oxygen and expels unwanted molecules such as carbon dioxide.

Water also is passed through the semipermeable cell membrane in a process called **osmosis** (**Figure 15–8**). As in diffusion, the water moves from a region of high concentration of water to a region of low concentration, so the more material that is dissolved in water, the less it is concentrated. If the cell has relatively little water inside, the solution tends to "draw" water from outside into the cell through the cell membrane.

The processes of diffusion and osmosis regulate the materials moving from one part of the cell to another and in and out of the cell. In all organisms this is essential because these processes allow the cell to remain constant even

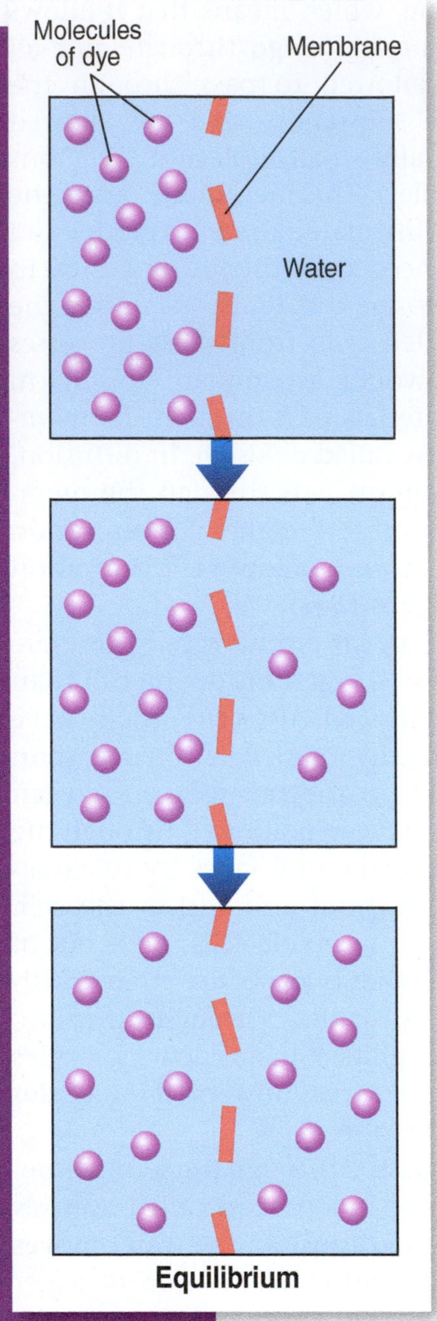

FIGURE 15–7 In the process known as diffusion, molecules in a solution pass through the cell membrane from a region of higher concentration to a region of lower concentration of molecules.

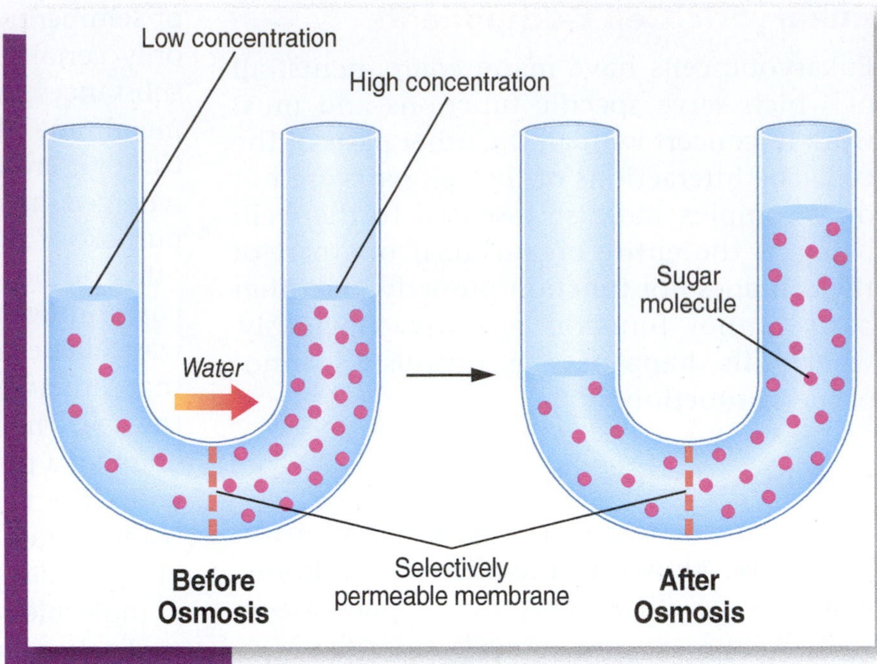

FIGURE 15–8 Water is passed through the cell membrane in a process called osmosis.

though conditions in the environment may change. The ability of an organism to remain stable when conditions around it are changing is called **homeostasis**.

Organelles

Within the cytoplasm of cells are small structures that serve different roles. In much the same way as the organs of a body support an animal, these structures, called organelles, support the cell (**Figure 15–9**). One of the most important of the organelles is the peanut-shaped **mitochondria,** which functions to break down food nutrients and supply the cell with energy. Cells that use more energy contain more mitochondria than cells that are less active. For example, muscle cells contain more mitochondria than bone cells because bone cells require far less energy than muscle cells. This is simply because muscles provide movement for the animal, and the bones provide the framework for the body.

Vacuoles are organelles that serve as storage compartments for the cell. Consisting of a membrane that encloses water and other material, they store the nutrients and enzymes that animals need. An **enzyme** is a type of protein found in all living organisms that causes, speeds up, or slows down a chemical reaction.

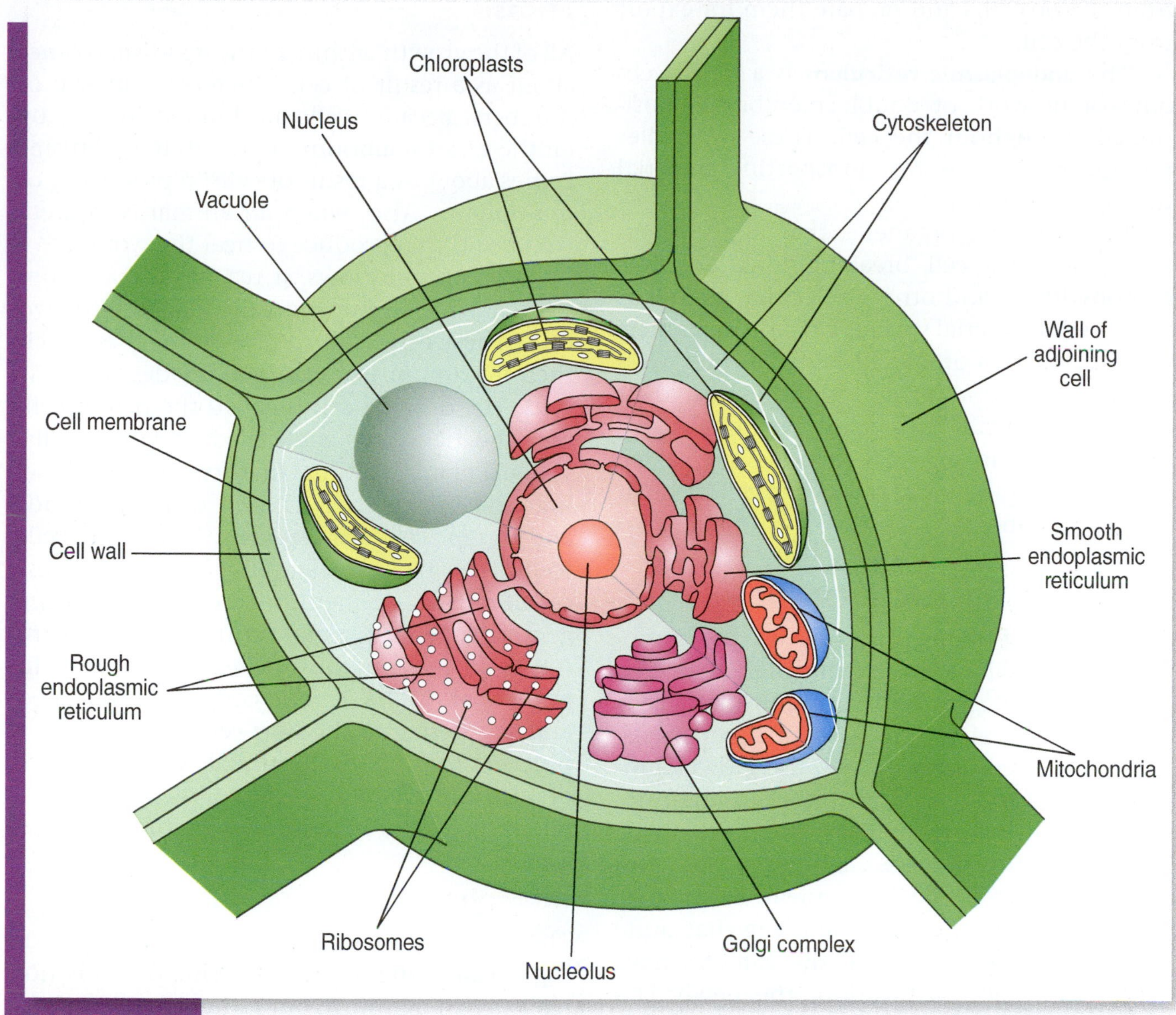

FIGURE 15–9 Animal cells have small structures called organelles, each of which serves a specific purpose.

Vacuoles also provide a storage space for the waste materials that the cell gives off.

Some cells contain organelles called microtubules. They are shaped like small, thin, hollow tubes, composed of protein, that act as the "bones" of the cell. These structures, found in animal cells, provide support to cells and give the cells their shape and also assist in moving chromosomes during cell division.

Microfilaments are fine, fiberlike structures composed of protein. These organelles help the cell to move by waving back and forth. Cells contain thousands of tiny structures called ribosomes. These organelles are the sites where protein molecules are assembled in the cell. All cells need proteins for growth and other important functions, and enzymes that regulate the chemical process in the cell are composed of protein molecules.

The **Golgi apparatus** in the cell is an organelle that is shaped like a group of flat sacs bundled together. Its function is to remove water from the proteins and prepare them for export from the cell.

The **endoplasmic reticulum** is a large webbing or network of double membranes positioned throughout the cell. These organelles provide the means for transporting material throughout the cell.

Lysosomes are organelles that are the digestive units of the cell, breaking down proteins, carbohydrates, and other molecules, as well as any foreign material such as bacteria that enters the cell. Also, as other cell parts become worn out and nonfunctional, lysosomes use their digestive enzymes to break down these parts. Products of the digestive actions are passed into the cytoplasm and out of the cell through the cell membrane.

CELL REPRODUCTION

Continuation of life depends on the reproduction of cells. Even in the higher-ordered animals, life begins with the uniting of cells known as gametes from each of two parents. Once these cells have united, the growth process begins, and cells multiply until an entire new animal is formed. The cells must divide throughout this process. Gametes, or sex cells, are formed in a process known as **meiosis**, which produces the sperm and egg that unite to form an embryo. When the gametes have merged through fertilization, the newly created cell begins to divide in a process known as **mitosis**, in which growth and cell replacement take place.

Meiosis

In meiosis, cells are divided into cells that contain only half of the chromosomes needed to form the young animal. Sperm cells are formed in the testicles of the male, and the eggs are formed in the ovaries of the female. When the gametes unite, the full number of chromosomes is accomplished by each parent contributing half. This process is discussed in detail in Chapter 19.

Mitosis

All of the growth within living organisms comes about as a result of cells' increasing in size or numbers. Because cells are limited in size, by far the greatest amount of growth in organisms comes about as a result of cells' reproducing or multiplying. Also, when an animal is injured, cells begin to reproduce to heal the wound.

When a cell grows, it reaches a maximum size, at which time the cell divides into two cells. These cells in turn grow until they reach their maximum size and each divides into new cells. The original cell is called the parent cell, and the new cells are called daughter cells. When a plant or animal matures, it stops growing, and cell division is used to heal wounds and to replace worn-out cells. Eukaryotic cells (cells that have nuclei) divide by a process called mitosis. As mentioned earlier, the entire genetic code for passing on traits of an organism is located in the nucleus of the cell. In mitosis, all of the genetic coding is duplicated and transferred to the new cells. Although the process of mitosis is continual, scientists have divided the events into the following four different phases for better understanding.

Interphase

The period when the cell is not actively dividing is called the interphase. This phase is not really a part of mitosis but rather is a time when the cell is carrying on processes such as synthesizing materials and moving them in and out of the cell (**Figure 15–10**). This is the time when the cell grows. As the cell reaches its maximum size, the DNA replicates and forms two complete sets of chromosomes. The threadlike molecules of DNA that make up the chromosomes, called chromatin, are spread throughout the nucleus. Animal cells have strands of genetic material outside the nucleus called **centrioles**. (Most plant cells do not contain centrioles.) At the end of the interphase, the cell is the correct size, the chromosomes are duplicated, and the cell is ready to divide.

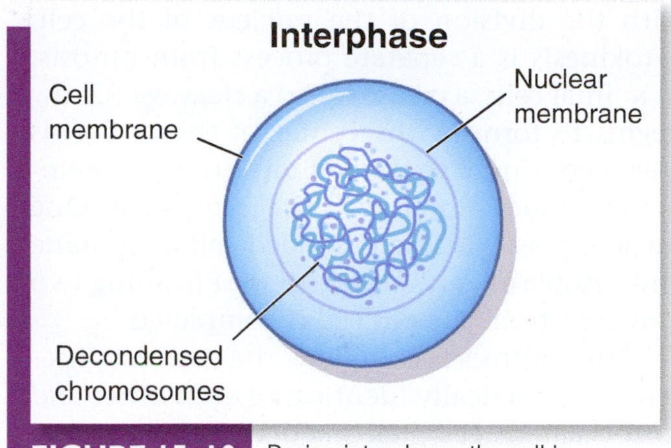

FIGURE 15–10 During interphase, the cell is carrying on processes such as synthesizing of materials.

Prophase

The first actual phase of mitosis is called the prophase. During this phase the chromatin appears in the form of distinct, shortened, rod-like structures. The chromosomes are formed with two strands called chromatids that are attached at the center by a structure known as a centromere. As this formation takes place, the nuclear membrane begins to dissolve and the entire nucleus begins to disperse (**Figure 15–11**). In place of the nucleus, a new structure called the spindle is formed. The spindle is a structure shaped somewhat like a football and composed of microtubules. In animal cells, the centrioles move to opposite sides of the cell.

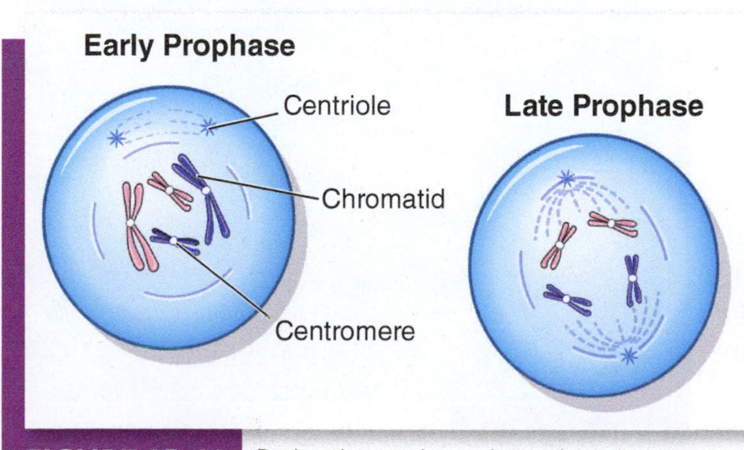

FIGURE 15–11 During the prophase, the nucleus disperses and a new structure called the spindle is formed.

Metaphase

During the next phase, called the metaphase, the nucleus disappears and the chromatids move toward the center of the spindle (**Figure 15–12**). The center of the spindle is referred to as the equator. When they reach the center, the centromeres of the chromatids connect themselves to the fibers of the spindle.

Anaphase

During the third stage of the process of mitosis, the anaphase, the pairs of chromatids separate into an equal number of chromosomes and the centromeres duplicate (**Figure 15–13**). After separating, the chromosomes move to opposite ends of the cell.

Telophase

In the final phase of mitosis, the telophase, the chromosomes continue to migrate to opposite sides of the cell (called poles). When they reach

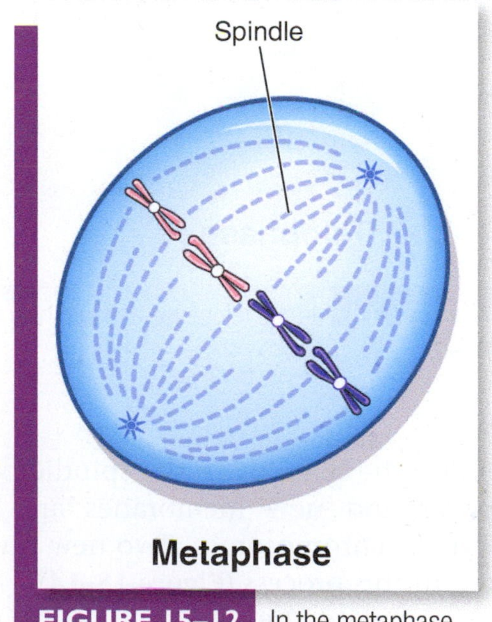

FIGURE 15–12 In the metaphase, the nucleus disappears and the chromatids connect with the spindle.

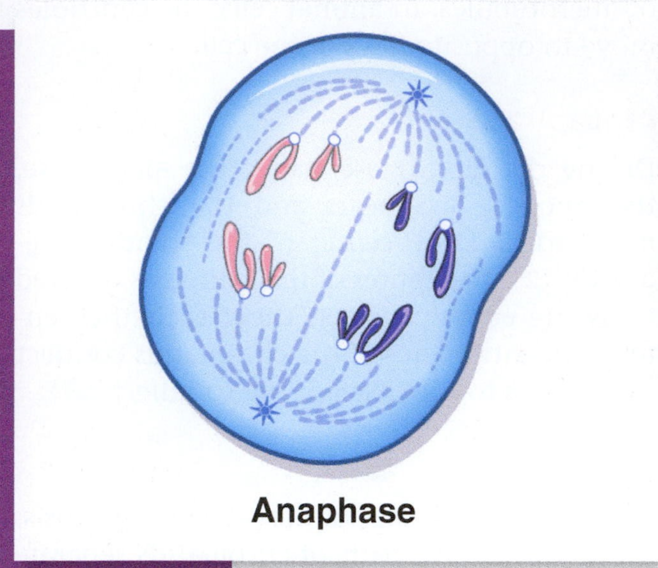

FIGURE 15–13 In the anaphase, the chromatids separate into an equal number of chromosomes and the centromeres duplicate.

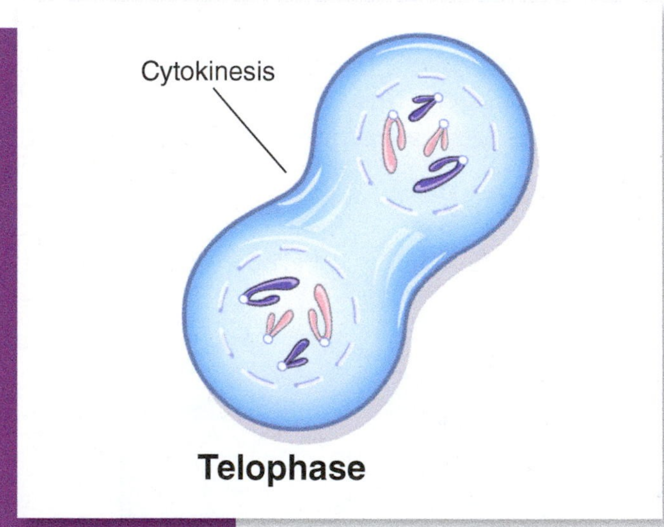

FIGURE 15–14 In the telophase, the two new nuclei are formed.

the poles, the remains of the spindle begin to disappear and new membranes are formed around the chromosomes. Two new nuclei are formed in this process (**Figure 15–14**).

To complete the cell division, a process known as **cytokinesis** occurs that divides the cytoplasm in the cell. Since mitosis is involved with the division of the nucleus of the cell, cytokinesis is a separate process from mitosis. In animal cells, a crease called a **cleavage** furrow begins to form in the center of the cell. This crease continues to deepen until the cell membrane divides along with the cytoplasm. One nucleus goes with each divided cell membrane and cytoplasm, and the process of forming two new cells from the old cell is completed.

After mitosis is complete, the new daughter cells are genetically identical to each other and to the parent cell that divided to form them. After formation, the daughter cells go into interphase and the whole process of mitosis starts over. Through this continuous process, an organism grows and maintains its structure through the replacement of worn out and injured cells.

ANIMAL STEM CELLS

Once fertilization has occurred, a complete cell has formed with all the genetic material necessary for developing into a complete organism. This cell, called a stem cell, is said to be a totipotent cell, meaning that the cell is capable of developing into any type of cells (**Figure 15–15**). Within a few hours, this cell divides

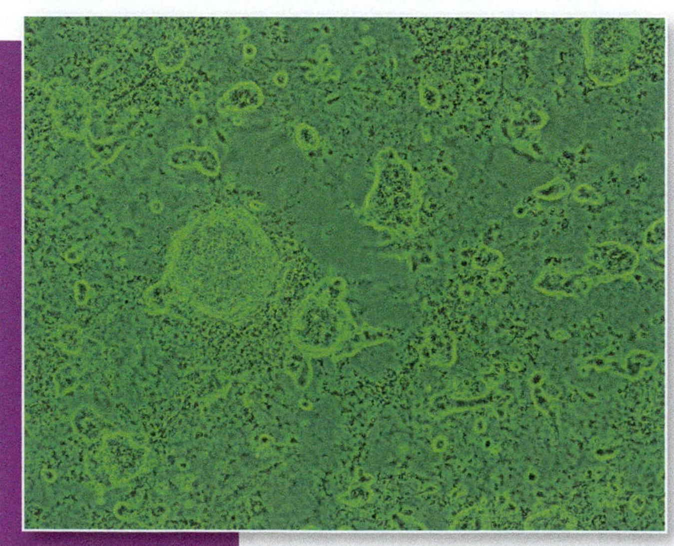

FIGURE 15–15 Stem cells can develop into different types of cells such as bone, muscle, nerves, and so forth.
© iStockphoto/dra_schwartz.

into two totipotent cells, either of which could be implanted into a uterus and develop into a complete animal. Sometimes this happens naturally, resulting in identical twins.

Soon the cells divide and group together to form a ball-shaped mass called the morula, where the cells divide and clump into a mass in the cleavage process. As the cells of the morula begin to increase, they form a spherical shape called a **blastula**, with an outer layer and inner mass of cells. The outer layer of the blastula develops into the **placenta**, which attaches to the uterus and provides nutrients and other support for the fetus. The inner masses of cells form all of the different types of tissues in the body.

As the blastula begins to grow and develop, the cells begin to change and take on different characteristics. They begin to form different layers that later develop into the organs of the body.

Like cells group together to form tissue; the tissue that develops bone is different from the tissue that develops blood; the tissue from which muscles develop has its own unique characteristics, and so on. This process is called cell differentiation, and the cells that begin the differentiation process are called stem cells. Scientists still don't fully understand what causes the cells and tissue to begin to differentiate. Apparently, some mechanism triggers the differentiation of cells at the proper stage of development.

Scientists envision using stem cells to create new tissue to replace diseased or damaged human tissue. Growing new tissues for a specific organ could potentially cure many diseases, such as Parkinson's disease, diabetes, heart disease, and other problems. If scientists are ever able to unlock the secret to the cell differentiation process, the possibilities are enormous. The use of human cell research is extremely controversial, and until issues surrounding the use of these cells are settled, the research cannot proceed unhindered.

SUMMARY

Cells are the building blocks of all life. Trillions of different cells work together to make up all of the systems comprising an animal's body. Almost everything that happens in an animal begins at the cellular level, where all processes take place. By studying and understanding the structure, function, and reproduction of cells, you will have a better comprehension of how the systems in animals function.

CHAPTER REVIEW

Review Questions

1. Why is understanding cells important to the producer?
2. How do prokaryotic and eukaryotic cells differ?
3. Why is cytoplasm so important in the cell?
4. What is the difference between osmosis and diffusion?

5. Why is osmosis important?
6. Why is diffusion important?
7. Name four types of organelles and explain the function of each.
8. What is the difference between meiosis and mitosis?
9. List the phases of mitosis.
10. What is a stem cell?

Student Learning Activities

1. Research and write a report on stem cell research. Explain why some scientists think stem cell research can help to cure some diseases. Discuss the controversy surrounding human stem cell research.
2. Research the structure of plant cells and explain how plant cells differ from animal cells. How are they alike?

CHAPTER 16

Animal Genetics

STUDENT OBJECTIVES IN BASIC SCIENCE

As a result of studying this chapter, you should be able to

- explain the basic function of deoxyribonucleic acid (DNA).
- describe the function of ribonucleic acid (RNA).
- define *allele*.
- describe how traits are passed from parents to offspring through genetic transfer.
- explain the concept of dominant genes versus recessive genes.
- discuss the concept of codominant genes.
- explain how computers are used to predict genetic differences in animals.
- explain how the sex of an animal is determined.

STUDENT OBJECTIVES IN AGRICULTURAL SCIENCE

As a result of studying this chapter, you should be able to

- discuss how producers use the laws of genetics to produce the type of livestock they want.
- describe how the concept of heritability is used in the selection of livestock.
- tell how phenotypic and genotypic characteristics differ.
- explain how performance data are used in the selection process.
- describe how computers are used in the modern selection process.

KEY TERMS

fertilized egg
phenotype
genotype
chromosomes
deoxyribonucleic acid (DNA)
genes
helix
nucleotides
ribonucleic acids (RNA)
differentiation
allele
homozygous
heterozygous
recessive
codominant genes
epistasis
mutations
gamete
zygote
heritability
ewe
index
weaning weight
yearling weight
most probable producing ability
growth ability
estimated breeding value
siblings
pedigree record
expected progeny difference
thurl
sickle-hocked

NATIONAL AFNR STANDARD

AS.04.02.01.a
Summarize genetic inheritance in animals.

AS.04.02.01.b
Compare and contrast the use of genetically superior animals in the production of animals and animal products.

AS.04.02.02.a
Identify and summarize inheritance and terms related to inheritance in animal breeding.

AS.04.02.02.c
Select and evaluate breeding animals and determine the probability of a given trait in their offspring.

OF ALL THE BILLIONS OF ANIMALS in the world, no two are exactly alike. Even animals that originate from the same **fertilized egg** are not alike in every aspect. Although they have the same genetic makeup, one may be slightly taller, may be a little heavier, or may grow faster. Differences in animals are brought about by two groups of factors: genetic factors and environmental factors.

One set of differences is the animal's **phenotype**. Phenotypes are the physical appearance of the animal, such as color, size, shape, and other characteristics. The phenotype can be caused by the environmental conditions under which the animal is raised. For instance, the amount and type of feed an animal receives influence the way it looks. The amount of stress, climatic conditions, exposure to parasites, and diseases can all have an impact on the animal's appearance and performance. Obviously, the producer has a lot of control over the animal's environment.

The other contributor to the animal's phenotype is the **genotype**, or actual genetic makeup of the animal. Characteristics of individual animals are controlled by the animal's genes, passed on by its parents (**Figure 16–1**). By controlling the type of animals used as breeding animals, the producer may (to some extent) also control the genotype. The phenotype is what the producer is able to observe and is used in the selection process.

FIGURE 16–1 Characteristics of an animal are passed on from parent to offspring. © eastern light photography/Shutterstock.com.

Through the application of the science of genetics and with the aid of computers, the producer is able to use the genotype in the process. Whatever methods the producer uses to select animals for breeding, the entire process is built around the concept of gene transfer.

GENE TRANSFER

An animal's characteristics are passed on to the animal by its parents. It gets half of its genetic makeup from each of its parents. The information about how the animal will be structured is passed along through the **chromosomes** contributed by each parent. Chromosomes are composed of long strands of molecules called **deoxyribonucleic acid (DNA)**. DNA is a very complex substance composed of large molecules that are capable of being put together in an almost unlimited number of ways. Segments of DNA, called **genes**, are connected and arranged on the chromosomes. Each segment or gene is responsible for developing a particular characteristic of the animal.

The code for how the animal is to be formed (all of its characteristics) is contained in the DNA that makes up the genes. The molecules forming the DNA have a spiral shape called a **helix**, which resembles a corkscrew (**Figure 16–2**). If this corkscrew-shaped helix were to be straightened out, it would resemble a ladder with rungs where the segments fit together. The helix is composed of two halves that separate when the cell divides.

At each point on the helix where the two halves are connected, different substances such as adenine (A), thymine (T), guanine (G), and cytosine (C) are attached to each other. These substances, called **nucleotides**, are shaped so that each one can pair only with one specific nucleotide. Adenine (A) can pair only with thymine (T), and cytosine (C) can pair only with guanine (G). When the cell divides, the strands of DNA separate and each half replicates itself, forming two strands, exactly alike. This process

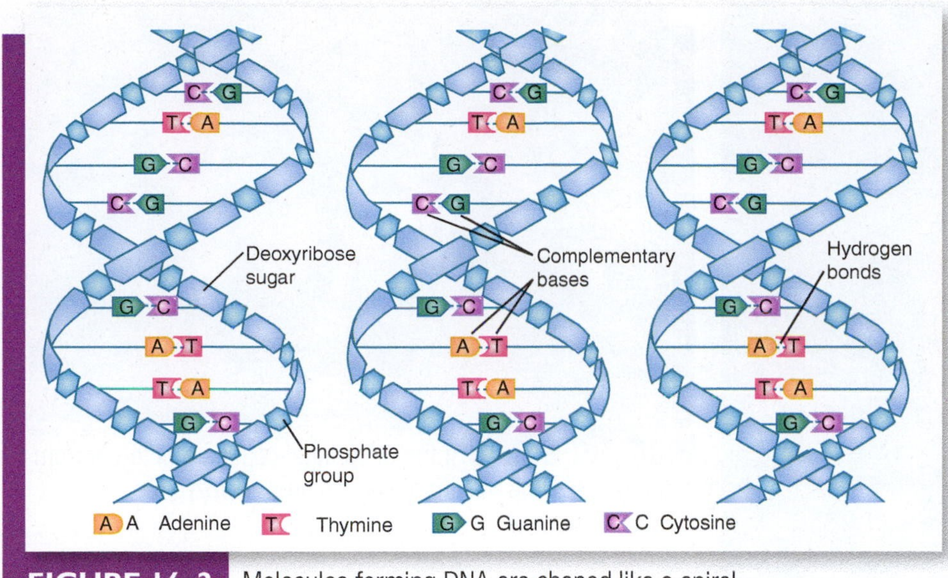

FIGURE 16–2 Molecules forming DNA are shaped like a spiral.

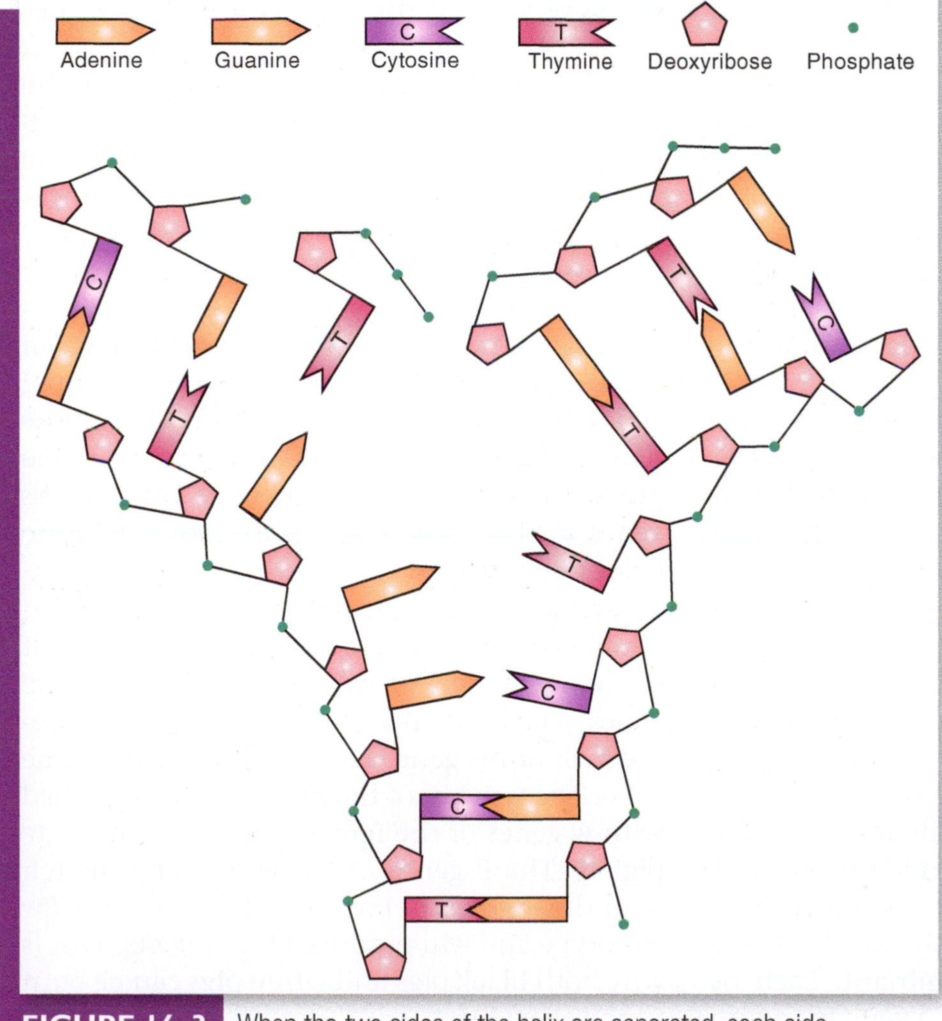

FIGURE 16–3 When the two sides of the helix are separated, each side replicates itself.

is called DNA replication (**Figure 16–3**).

The messages encoded in the DNA are transferred to the rest of the cell by means of a messenger substance known as **ribonucleic acids (RNA)**. RNA uses the model of the DNA molecule to transfer the pattern or blueprint for how the animal is to be constructed. In genetic engineering, segments of DNA are spliced into existing strands of DNA. This places new genetic information into the cell, producing characteristics that are different from what otherwise would be expected.

As the embryo begins to grow and develop, the protein cells start to differentiate. This means that some of the cells begin to form muscle, some hair, some bone, some skin, and some internal organs (**Figure 16–4**). The process by which **differentiation** occurs is not fully understood.

At conception, the chromosome halves from each parent are combined to form fully paired chromosomes, and the chromosomes from each parent are united to form the genetic code for the fertilized egg. The DNA molecules can be arranged in the gene in an almost infinite number of ways. The arrangement of

FIGURE 16–4 As the embryo begins to grow, cells differentiate into various tissues such as hair, muscle, bone, and nerves. © Anneka/Shutterstock.com.

FIGURE 16–5 If the offsprings' coats are different from the parents, the offsprings' genes are heterozygous. © iStockphoto/LAByrne.

molecules and how the molecules are paired at conception determine the makeup of the new animal.

Every gene that comes from the male is paired with a gene of the same type from the female. For example, the gene that controls the color of the animal's coat is made up of a pair of "coat color" genes—one from the father and one from the mother. A pair of genes that controls a specific characteristic is called an **allele**. If both of the genes are the same—that is, both call for a black coat or both call for a white coat—the genes are said to be **homozygous** and the animal will be the color the genes call for.

But what happens if the gene from the father calls for a black coat and the gene from the mother calls for a white coat? In this case, the offspring's genes are said to be **heterozygous** (**Figure 16–5**). The color of the offspring's coat will be determined by the dominant gene; this means that one gene will override the effect of the other gene. If a white sow is mated to a black boar, the piglets probably will be black because the black gene is dominant. Each of the piglets, however, will carry a gene for white color and a gene for black color. If *B* represents the dominant black color and *w* represents the **recessive** white color, the pairing of the genes for the piglets will be *Bw*.

Now suppose that the females from the litter are mated to a purebred black boar (all genes homozygous). Half of the genes from the females that control color will be *B* and half will be *w*. All of the genes from the boar will be *B*. Those genes from the male that match with the females' *B* genes will result in a *BB* genotype. The offspring will be black, and they will possess genes for the black color only. Those genes from the male that match with the *w* female genes will result in a *Bw* genotype. The pigs also will be black, but they will possess genes for both the black color and the white color (**Figure 16–6**).

If the *Bw* females are bred to a purebred white boar, the outcome will be different. Half of the female genes will be *w* and half of the genes will be *B*. Because the male is a purebred, all of his genes for color will be the same (*ww*). The *w* genes of the male that are matched with *w* genes of the female will result in white piglets. The *B* genes of the female that match with the *w* genes of the male will result in a *Bw* genotype and will produce black piglets. This is why both black pigs and white pigs can be born in the same litter.

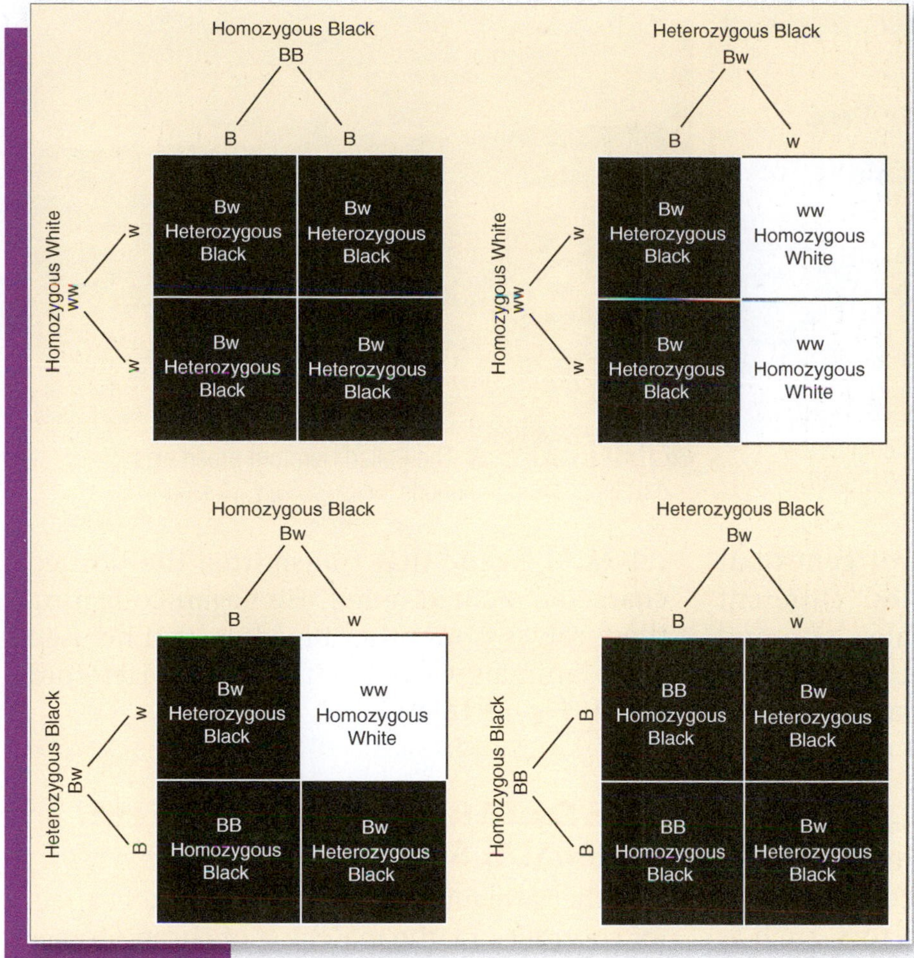

FIGURE 16–6 Black and white parents can produce a variety of genetic combinations.

There are exceptions to the rule of dominance. Some pairs of genes may not have a gene that is dominant over the other. They are of equal power and are said to be **codominant genes**.

A good example is found in Shorthorn cattle. Purebred Shorthorns may be red, white, or roan (**Figure 16–7**). Cattle that are completely red carry genes that call for red color only (*RR*); cattle that are completely white carry genes for white color only (*WW*); and cattle that are roan or spotted carry one gene for red and one for white (*RW*). In this case, neither the red (*R*) nor the (*W*) gene is dominant. The color of the animal will be a combination of red and white; the animal will be spotted or roan in color. These cattle can have a variety of color combinations and still be registered as purebred Shorthorn cattle.

Color coding is a good example of how genes transfer traits. The same general principles can be applied to other traits, such as horned or polled, tall or short, and so on. However, the entire process of defining the characteristics of the animal by genetic makeup is much more complicated. For instance, genes that are not alleles (matched pairs that control a characteristic) may interact to cause an expression that is different from the coding on the genes. This interaction is called **epistasis**.

Another factor in genetic transfer of characteristics is that of the additive expression of genes. This means that a number of different genes may be added together to produce a certain trait in an animal. For instance, the amount of milk the female produces is

FIGURE 16–7 Purebred Shorthorn cattle may be all white, all red, or roan. *© oliverrees/Shutterstock.com.*

FIGURE 16–8 A wide variety of genes control the amount of milk a mother produces. © iStockphoto/JackiNix.

FIGURE 16–9 The Polled Hereford breed was developed from mutations. © iStockphoto/Lynn_Bystrom.

not controlled by a single pair of genes but rather by several pairs of genes. Different pairs of genes control the female's size and body capacity, ability to produce the proper amounts of hormones, and mammary size and functioning ability. Yet, all of these factors contribute to the female's overall ability to produce milk (**Figure 16–8**). The same thing may be said about the animal's rate of gain or its ability to reproduce efficiently. Several genetically controlled factors, such as body size and structure, can affect the animal's ability to grow rapidly and efficiently. A heifer's pelvic size, shape of the genital tract, and output of sex hormones are all controlled by different genes and are all factors in reproduction efficiency.

Occasionally an accident will happen within the genetic material, and traits are not passed on as intended. In this case, the animal will take on characteristics unlike the parents. For example, animals sometimes are born with extra legs or two heads. These are referred to as **mutations**. Sometimes mutations can be used to introduce new kinds of animals. A good example is the Polled Hereford breed of cattle. Hereford cattle are naturally horned. Around the turn of the century, an Iowa Hereford breeder named Warren Gammon noticed that Hereford calves sometimes were born without horns. In these calves, the gene that transmitted the horned characteristic had failed. He began collecting these calves from other breeders, and he used these animals to begin the Polled Hereford breed (**Figure 16–9**).

THE DETERMINATION OF THE ANIMAL'S SEX

Whether a mammal is male or female is a factor controlled by the matching of chromosomes from the mother and father and is determined at conception. Each body cell contains one pair of chromosomes called the sex chromosomes. Each **gamete** (sex cell from the parent) contains one half of the sex chromosome from the parent.

The female chromosome usually is referred to as *XX*. When the chromosome divides and half goes into the gamete (egg), both of the chromosome halves are the same (*X*) (**Figure 16–10**). The male chromosome is referred to as *XY*, and when divided into the gametes (sperm), contains either *X* or *Y* chromosome halves.

When the two halves (in the sperm and the egg) are united at conception, the **zygote** (fertilized egg) will be either *XX* or *XY* chromosomes. The zygotes containing the *XX* sex chromosome develop into females, and the zygotes containing the *XY* sex chromosome develop into males. For this reason, the number of male or female offspring is dependent on the male gamete.

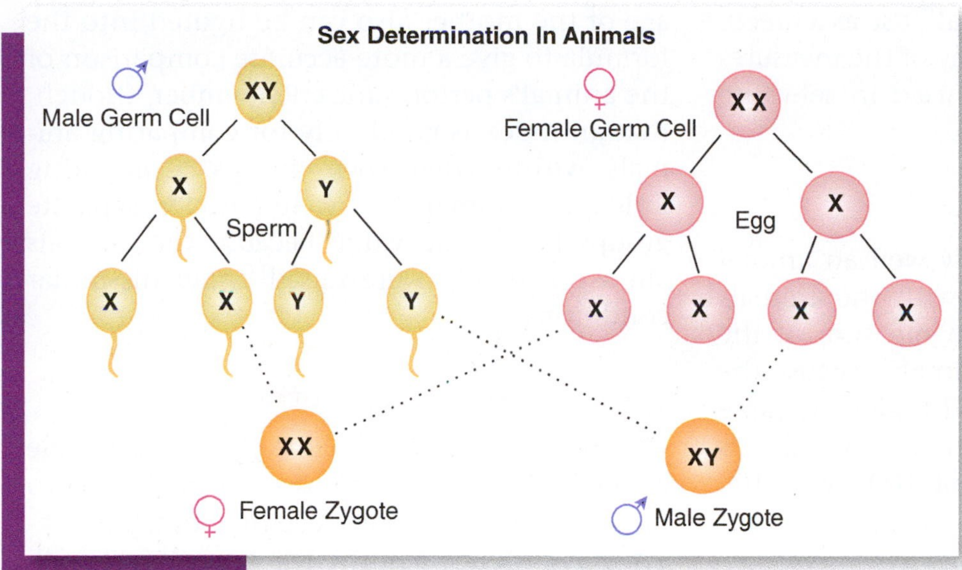

FIGURE 16–10 The sex of an animal is determined by the combinations of *X* and *Y* chromosomes.

USING GENETICS IN THE SELECTION PROCESS

The selection of livestock by physical appearance is discussed in Chapter 17. The modern producer has a number of genetically based tools to use in the selection of livestock. One such measure is known as **heritability**. Heritability is the measure of how much of a trait was passed on to the offspring by genes—how much of a trait is inherited and how much is attributable to the environment in which the animal lives. Heritability measurement varies from zero to one. The higher the number, the stronger is the degree of heritability.

Table 16–1 lists the heritability estimates for traits in agricultural animals. Notice that the heritability for milking ability in sheep is .25. This means that the variation among different sheep in milking ability is caused by factors other than genetics. These factors might include environmental factors such as quantity and quality of feed available to the **ewe** and climatic conditions. Obviously, traits that are more highly heritable are the traits that the producer can use in selecting breeding stock.

PERFORMANCE DATA

Through data collection and computer analysis, the records of how an animal has performed and the analysis of how the animal's ancestry and progeny have performed can be a valuable

BEEF		SWINE		SHEEP	
Trait	Heritability Estimates	Trait	Heritability Estimates	Trait	Heritability Estimates
Birth weight	.40	Litter size (weaned)	.12	Number born	.13
Weaning weight	.25	Weaning weight (3 weeks)	.15	Weaning weight (90 days)	.30
Yearling weight	.40	Number of nipples	.60	Yearling weight	.40
Feedlot gain	.45	Age at 220 lb.	.30	Fleece weight	.40
Efficiency of gain	.40	Length at 220 lb.	.60	Milking ability	.25
Fat thickness	.45	Backfat thickness at 220 lb.	.50	Rib eye area	.53
Rib eye area	.70	Percentage carcass muscle	.45		

TABLE 16–1 Heritability Estimates for Beef, Swine, and Sheep

tool in determining the animal's use as a breeding animal. Following are some of the measures of performance that can be used in selecting breeding animals.

Indexes

An **index** is a measure of how well an animal has performed as compared with the animals raised with it. This can be a measure of the genetic differences in the animals because the animals presumably are raised under the same conditions and are treated alike. An index is measured based on a scale of 100, with 100 being the group average. For example, the formula for calculating an index is

$$\text{Index} = \frac{\text{Individual Animal Weight}}{\text{Average Group Weight}} \times 100$$

If a group of calves are weaned when they are 205 days old and the average weight of the group is 583 pounds, an index of 100 is equal to 583 pounds. If one of the calves weighs 614 pounds, the calf is said to have a 205-day **weaning weight** index of 105.3 (**Figure 16–11**). This means that the animal has outperformed its peers. If a calf has a 205-day weaning index of 82, this means that the calf has performed only 82 percent as well as the other calves with which it was raised. Other indexes are used for comparing animals on **yearling weight**, birth weight, and other measures. Factors such as the age of the mother also can be figured into the formula to give a more accurate comparison of the animal's performance. Remember, though, that an index is good only for comparing animals within their own groups. Comparing indexes of animals that were raised in separate groups is of little value because the animals almost assuredly were raised under dissimilar conditions.

FIGURE 16–11 Weaning weights are used to compare calves within the same herd. © iStockphoto/anthonysp.

Measurements of Mothering Ability

One of the most valuable traits that a female agricultural animal can possess is being able to produce enough milk to feed her young. Mothers that produce a sufficient quantity of milk will wean young that are larger and faster growing than those from mothers that milk poorly. This is true for all agricultural animals, whether sheep, cattle, or swine.

The **most probable producing ability** (MPPA) is a measure of a cow's ability to wean a superior calf. It is figured using the 205-day weaning weight index and the number of calves the cow has produced. MPPA is an indication not only of the cow's ability to pass **growth ability** on to her calf but also her ability to produce enough milk and to care for her calf well enough to wean a large calf.

A similar measurement is used for pigs. This measure, called the *Sow Productivity Index,* takes into account factors such as the number of piglets born alive, the average for the other sows in the herd, and the 21-day litter weight (for both the sow and the herd). The term for this measurement in sheep is the Ewe Index, which takes into account the ewe's ability to wean above-average lambs. The MPPA for cows, the Sow Index for pigs, and the Ewe Index for sheep are all ways of putting the mothering ability of individual animals into numbers so they can be compared.

Estimated Breeding Values

Cattle breed associations use a measure referred to as the **estimated breeding value**, which is an estimate of the breeding value of an animal compared with other animals. It is computed through a complicated formula using the animal's own performance data as well as that of the animal's half-brothers and half-sisters. Through artificial insemination, a given animal can have a large number of **siblings** (brothers and sisters). Data on a large number of siblings can be useful in determining how well the desired characteristics are passed on by the animal's ancestors.

The larger the number of siblings and the larger the amount of data on both siblings and ancestors, the more accurate the estimated breeding value will be. The accuracy also is used to determine how much emphasis a producer will place on an animal's estimated breeding value in the selection process. Measures of performance are summarized on the animal's **pedigree record**, which is a part of the animal's registration papers. Notice that all the types of data that have been discussed are on the record for the producer to use in evaluating the animal.

Expected Progeny Difference

The **expected progeny difference** (EPD) is used to predict the differences that can be expected in the offspring of a given sire over those of other bulls used as a reference. The data for calculating EPDs are obtained from the performance data of the progeny from the bulls. This is another example of how artificial insemination is useful in determining the desirability of a sire because of the tremendous number of offspring made possible through artificial insemination (**Table 16–2**). These data can be hugely helpful to a producer who wishes to increase or improve certain traits in the calves produced. For example, if a bull has

Accelerated Genetics Linnear Profile

		C.E.H		C.EC		Birth Wt.		Wean. Wt.		Yearl Wt.		Mat. C.E.H		Mat. C.E.C		Mat. Wean. Wt.		Mat. Milk	
		EPD	ACC.	EPD	ACC.	EPD	ACC.	EPD	ACC.	EPD	ACC.	EPD	ACC	EPD	ACC.	EPD	ACC.	EPD	ACC.
SIMMENTAL																			
14SM340	AF Redlands 35Y	–6	.07	–2	.07	+3	.16	+18	.15	+31	.14	+1	.06	+0	.06	+11	.08	+2	.08
14SM341	ASR Polled Pacesetter 413Z	–1.0	.16	+0	.16	–1	.21	+16	.20	+32	.18	+0	.15	+0	.15	+25	.17	+17	.17
21SM270	Bold Leader	–4.7	.68	–1.3	.68	+2.8	.92	+26.6*	.91	+47.4*	.88	–2.1	.68	–.5	.68	+24.5*	.78	+11.2*	.77
14SM334	F N Stamina	+4.5	.09	+1.2	.09	–.4	.18	+6.7	.17	+16.8	.17	+1.2	.09	+.4	.09	+1.0	.11	–2.3	.10
14SM338	Hancocks Pineview Regal	–7	.18	–2	.18	+2	.35	+13	.32	+28	.19	+0	.18	+0	.18	+12	.18	+6	.18
14SM326	HCC Prophet	+6.3	.28	+1.6	.28	–.5	.73	+.3	.65	+4.5	.58	+3.6	.19	+1.0	.19	+1.5	.24	+1.4	.23
36SM145	HF Phantom	–6.1	.37	–1.8	.37	+.8	.70	+18.6*	.66	+32.3*	.61	+5.0	.37	+1.3	.37	+4.7	.45	–4.6	.43
14SM327	HMF Gold Bar 304W	+3.4	.16	+1.0	.16	+.7	.53	+1.4	.46	+9.1	.40	+1.2	.14	+.4	.14	+8.5	.18	+7.8	.17
14SM330	HMF Polled Siegfried 230U	–3.2	.17	–.8	.17	+.9	.55	+9.0	.46	+25.0	.40	+2.3	.15	+.7	.15	+6.5	.20	+2.0	.19
14SM336	Keystone	–10.8	.53	–3.4	.53	+5.9	.66	+31.3*	.62	+50.5*	.61	+6.9	.53	+1.7	.53	+11.1	.55	–4.5	.55
14SM339	LCHM Black Barton 235X	+2.2	.17	+.7	.17	–.9	.45	+11.0	.40	+25.3	.35	+4.3	.17	+1.1	.17	+.8	.19	–4.7	.19
14SM337	Mr GF Train	–7	.17	–2	.17	+2	.20	+22	.20	+35	.18	–2	.17	–1	.17	+15	.18	+5	.17
14SM320	Paymaster	–1.5	.19	–.3	.19	+.1	.75	+2.9	.68	+21.4	.66	–5.9	.21	–1.7	.21	+3.8	.41	+2.4	.39
14SM011	Pineview Apache	–7	.12	–2	.12	+1	.35	+32	.32	+57	.18	+1	.12	+0	.12	+23	.14	+7	.13
14SM142	Pineview Jazz	–5.5	.81	–1.5	.81	+.9	.93	+16.1*	.92	+38.8*	.91	–2.2	.80	–.6	.80	+12.7*	.86	+4.7	.85
36SM154	Pineview Presley	–4.8	.34	–1.3	.34	–.6	.77	–7.6	.72	–6.2	.65	+3.1	.23	+.8	.23	–.5	.32	+3.3	.30
14SM342	R&R Magician Z504	–2	.11	+0	.11	–2	.34	+14	.31	+31	.17	+0	.10	+0	.10	+14	.12	+7	.12
14SM332	R&R The Wizard 504X	+4.1	.14	+1.1	.14	–3.6	.42	+15.3	.34	+28.6	.30	–.9	.11	–.2	.11	+10.6	.14	+3.0	.14
14SM318	Royal Can Am	+.4	.16	+.2	.16	+2.2	.46	+9.1	.41	+10.9	.36	+2.6	.15	+.7	.15	+9.2	.19	+4.6	.18
36SM165	S&S Eclipse	–3.3	.27	–.9	.27	+.9	.67	+20.1	.59	+37.3	.57	–2.4	.20	–.6	.20	+9.8	.26	–.3	.24
14SM321	Sunny K Blackjack	+5.4	.18	+1.4	.18	–.4	.57	+2.6	.50	+9.4	.45	+.6	.15	+.2	.15	+1.1	.19	–.2	.18
36SM184	Super Light 19U	+.9	.18	+.3	.18	–.1	.37	+5.1	.32	+9.6	.30	+3.5	.17	+.9	.17	+5.0	.19	+2.5	.18
14SM324	The Greek	–8.7	.21	–2.7	.21	+6.0	.67	+36.7	.55	+58.9	.46	+7.7	.17	+1.9	.17	+3.0	.21	–15.4	.20
14SM357	TNT Mr T	–0.8	.60	–3.4	.60	+2.5	.90	+20.9*	.87	+41.6*	.85	–1.7	.55	–.4	.55	–7.9	.66	–18.3	.65
14SM325	Triumph	–13.8	.65	–4.6	.65	+5.6	.85	+19.9*	.81	+36.3*	.78	+3.7	.64	+1.0	.64	+.2	.68	–9.8	.67
14SM322	WRS Alien	–21.4	.15	–8.3	.15	+7.7	.70	+25.8*	.63	+51.9	.53	+6.9	.10	+1.7	.10	+10.5	.16	–2.4	.15
36SM156	WRS Enterprise	–2.2	.17	–.5	.17	–.2	.34	–.7	.31	+6.9	.29	–.3	.17	+.0	.17	+2.4	.20	+2.8	.19
14SM331	Y1 Yardleys P&B R248	+1.8	.25	+.6	.25	+1.9	.73	+7.1	.66	+24.9	.58	–1.4	.27	–.3	.27	–5.3	.46	–8.9	.45

*Trait Leader

TABLE 16–2 TRI-STATE BREEDERS—Beef Sire EPDs *Courtesy of Accelerated Genetics*

an EPD of 118, his offspring can be expected to have a weaning weight of 18 pounds more than the average bull of the same breed. Of course, the bull may have calves that are below average weight, but it should be expected that most of his offspring will have higher weaning weights than average for the breed.

Linear Classification

A modern tool that dairy producers use in selecting replacement animals is a process known as linear classification. This process combines the use of visual and genetic selection. Cows are evaluated and classified by assigning a value between 1 and 50 for certain traits that are considered to be important to the animal's production ability. The Holstein-Friesian Association uses 17 functional traits for evaluation and assigns such a number to each trait (**Table 16–3**). For instance, a wide pelvis (**thurl**) is related to ease of calving, so a wide pelvis is desirable in an animal that is to be used as a replacement animal. An animal that is appraised as having an extremely narrow pelvis might be given a point score of 1 to 5 points, depending on how severely narrow the thurl is. A cow with an extremely wide pelvis (thurl) may be assigned 45–50 points.

A score of 50 does not necessarily mean that the animal has the most desirable form of the particular trait, though. To be comfortable in walking and standing, an animal must have strong legs. The rear legs must have the proper curve, or set, to provide the most flex and cushion as the animal walks. However, a cow with too much set to the rear legs is not desirable because too much stress will be placed on the muscles and tendons of the legs. A post-legged cow (with straight, "posty" rear legs) will be given 1 to 5 points (**Figure 16–12**); a cow with an intermediate set to the rear legs will be given 25 points; and a cow with an extreme amount of set, called **sickle-hocked**, will be given 45–50 points. In this case, the most desirable condition is given the midpoint score of 25.

FIGURE 16–12 An animal with legs that are too straight will be given a score of 1–5 points. This heifer is too straight on her rear legs. © Rudd Morijn Photography/Shutterstock.com.

The animals are scored periodically by a professional evaluator who has been trained by the breed association. These evaluators come to the dairy producer's farm and evaluate each animal for a set fee. Information from the animals then is compiled, and the producer receives a data sheet on the linear classification of the animals. The producer can use this information to select bulls for artificially inseminating the cows. For instance, if the cows in the herd have problems with the structure of their feet and legs, a bull is selected that is known for siring daughters with strong and structurally correct legs. The offspring from the cows should have stronger feet and legs than their mothers and can be used as replacement heifers.

The Holstein Association lists the following advantages of the linear classification system:

- Provides unbiased and accurate evaluations of cows
- Defines types of trait trends from one generation to another
- Gives the producer a clear understanding of each animal's strengths and weaknesses
- Compares the producer's herd type pattern to breed averages
- Adds trait appraisals to official pedigrees
- Provides a basis for mating services

Page No. 1

Owner Dean Dairyman

Address 1500 Farm Lane

Holstein WI 53535

	ANIMAL NO.	PEDS	MATE	ANIMAL BARN ID / DATE OF BIRTH	LACT. NO.	Stage Code	DATE OF CALVING MO.	DAY	YR.	STATURE	STRENGTH	BODY DEPTH	DAIRY FORM	RUMP ANGLE	THURL WIDTH	REAR LEG SIDE VIEW	FOOT ANGLE	FORE UDDER ATTACHMENT	REAR UDDER HEIGHT	REAR UDDER WIDTH	UDDER CLEFT	UDDER DEPTH	FRONT TEAT PLACEMENT	TEAT LENGTH	REAR LEG REAR VIEW	UDDER TILT	(NOT DEFINED)	(NOT DEFINED)	(NOT DEFINED)	PREVIOUS CLASSIFICATION DATE / SCORES	PERMANENT	MULTI E Date Eligible #E	GENERAL	DAIRY	BODY	MAMMARY	FINAL	NEW E
										LINEARIZED DESCRIPTIVE TRAITS: FORM				RUMP		LEGS/FEET		UDDER					TEATS		RESEARCH TRAITS													
1	10342984	4	0	Regina 10-05	7		12	30	05	34	33	34	41	25	36	25	30	26	32	37	45	28	18	37	40	28				4-12 EVVV	P		E	E	V	E	90	
2	10922314	4	0	Veronica 10-04	7		1	22	06	42	40	37	42	18	34	22	40	10	36	36	34	15	14	36	37	32				4-12 +EVG	P		V	E	E	G	85	
3	11600072		0	Margie 6-10	5		2	02	06	checked cow																				4-12 +EVG	P						✓	
4	11717850			Bon Bon 6-08	4		5	24	05	32	31	32	36	12	27	32	28	32	25	17	36	25	37	26	18	27				4-12 FVVG	P		G	V	V	+	81	
5	11990933	3		Rebecca 5-10	4		1	15	06	42	38	40	42	25	37	25	36	34	36	38	27	28	32	22	32	30				4-12 VVV+			E	E	E	+	88	
6	11990935	3		Lilac 5-10	4		4	07	05	36	38	37	37	18	36	30	12	42	48	40	40	36	34	26	22	29				4-12 +++V			+	V	V	E	86	
7	12052652	4		Suzanne 5-09	3		7	26	05	50	50	43	40	22	45	27	36	32	28	25	20	15	26	37	18	22				4-12 EEEV			E	E	E	V	90	
8	12313697	3		Vera 5-00	3		12	26	05	40	37	40	36	20	38	23	37	22	26	18	32	28	28	32	36	28				4-12 +VV+			V	V	E	+	85	
9	12313698	3		Margaret 5-00	3		8	28	06	37	30	34	37	22	30	34	28	40	34	37	36	34	36	23	27	30				4-12 ++V+			V	V	V	E	87	
10	12495037			Regal 4-07	2		8	18	05	36	36	24	22	27	34	22	32	34	34	36	33	36	32	30	32	30				4-12 +++G			+	+	V	V	83	
11	12615110	3		Bonnita 4-06	2		4	24	05	34	29	34	44	15	28	36	17	43	26	30	37	40	36	14	18	28				4-12 +VVV			+	E	V	E	86	
12	12616706		0	Carmen 4-04	2		7	20	05	24	21	24	34	27	15	21	24	17	34	24	25	19	17	15	35	18				4-12 +++G			G	+	+	G	78	
13	12798352	3		Vanessa 3-09	2		2	26	06	39	22	25	41	34	18	32	27	28	18	20	30	32	28	26	20	21				4-12 ++++			+	E	+	G	82	
14	12857939	3		Margo 3-08	2		12	29	05	34	19	25	40	35	18	28	16	32	34	34	36	34	30	23	25	30				4-12 G+V+			G	V	V	G	81	
15	12932831	3		Romance 3-05	1		4	25	05	35	32	32	18	33	27	27	32	36	30	30	38	34	35	22	18	32							G	G	V	V	81	
16	12932834			Bobbie 3-05	1		5	05	05	27	20	25	28	31	23	28	17	17	27	17	16	28	25	27	16	40							G	+	+	F	75	
17	13225840	3		Sunshine 2-09	1		9	04	05	40	27	28	20	28	27	27	26	27	18	17	32	38	25	28	27	30							V	G	+	G	81	
18	13225841			Carmel 2-09	1		10	29	05	32	30	32	31	32	26	18	26	26	18	18	26	39	25	27	26	40							G	+	+	G	79	
19	13225857	3		Ramona 2-04	1		2	13	06	30	22	21	24	25	27	15	30	31	26	32	38	38	27	25	32	35							G	+	+	+	80	
20	13288385			Mermaid 2-03	1		3	21	06	33	31	31	28	26	27	22	18	12	23	22	25	27	25	22	21	27							+	+	+	F	78	
21	13387133		0	Suzette 2-01	1		5	28	06	not in condition																											N/C	
22																																						
23																																						
24																																						
25																																						

TABLE 16–3 The Holstein-Friesian Association considers 17 functional traits in evaluating animals. *Courtesy of Holstein Association USA, Inc.*

The selection process that modern producers use is truly based on science. Principles of genetics are used as well as the findings of research studies that have discovered physical characteristics of animals that indicate the potential of the animal. Computer technology has greatly enhanced the producer's ability to predict how well an animal will perform, as well as the performance of the animal's progeny.

SUMMARY

Through an understanding of genetics, producers can better choose animals to use in a breeding program. Traits that are desirable are passed from generation to generation through gene transfer. By knowing which traits are dominant and which are recessive, producers can develop a program that will produce the type of animals that will pass along desired characteristics to their offspring. Although we know a lot about how genes are transferred, we still are a long way from understanding how the process works. When we unlock the mechanisms of gene transfer, the way we select and produce animals will be revolutionized.

CHAPTER REVIEW

Review Questions

1. What is the difference between an animal's genotype and its phenotype?
2. What is DNA? What purpose does it serve?
3. What is a helix?
4. What part does RNA play in passing traits from parent to offspring?
5. What is meant by a dominant gene and a recessive gene?
6. Explain how a red Shorthorn bull mated with a white Shorthorn cow will produce a spotted or roan-colored calf.
7. If a male rabbit that is black in color (pure gene for the dominant black color *BB*) is bred to a white female rabbit that has the pure recessive gene *WW* for the white color, what percent of the young would you expect to be black?
8. If a black male rabbit with the gene *Bw* is bred to a white female rabbit (*ww*), what percent of the young would you expect to be black? What percent should be white?

9. How is the sex of an animal determined?
10. What does heritability measure?
11. If a calf has a weaning weight of 478 pounds and the other calves that were raised with it have an average weaning weight of 634 pounds, what will be the weaning weight index for the calf?
12. What measure of mothering ability is used for cows? For sheep? For swine?
13. What do the letters EBV stand for? How is this measure used?
14. What is meant by the expected progeny difference?
15. What is meant by linear evaluation?
16. Explain how the use of artificial insemination and the use of computers have greatly aided in the selection process.

Student Learning Activities

1. Choose a species of agricultural animal such as sheep, cattle, swine, or horses. List the physical characteristics you think are of economic importance to that type of animal. Decide whether these characteristics are influenced more by genetics or by environment.
2. Obtain copies of the pedigree papers of several animals of the same breed. Compare the animals based on their pedigrees and performance records.
3. Write to the American Breeders Society or to Select Sires to obtain a copy of their dairy cattle sire catalog. From a dairy producer, obtain copies of the linear classification data from the dairy's herd. Using the data on the females from the records and the data on the different sires in the catalogs, choose the most desirable sires for the cows in that particular herd.

CHAPTER 17

The Scientific Selection of Agricultural Animals

STUDENT OBJECTIVES IN BASIC SCIENCE

As a result of studying this chapter, you should be able to

- explain the concept of natural selection.
- explain how humans have influenced the development of animals.
- illustrate how scientific research has influenced the development of animals by humans.
- cite examples of how problems have developed in animals because of the selection process controlled by humans.

STUDENT OBJECTIVES IN AGRICULTURAL SCIENCE

As a result of studying this chapter, you should be able to

- justify the selection of different animal traits by the producer.
- trace the stages of development in modern swine.
- describe problems associated with overly muscled pigs.
- interpret the reasoning behind the selection for sex character.
- rationalize the selection of animals for structural soundness.
- describe the physical characteristics associated with growth in animals.
- describe the modern beef and swine animal.

KEY TERMS

natural selection
reproductive efficiency
fertile
growth ability
efficiency
performance data
Porcine Stress Syndrome (PSS)
PSE pork
loin eye
pin nipples
blind nipples
inverted nipples
gilts
infantile vulva
vulva
testicles
viable
sheaths
ligaments
pasterns
splayfooted
pigeon-toed
cow-hocked
cannon bone
backfat
rib eye
frame size
hip height
retail cuts
double muscling
concentrate
pelvic capacity
sex character
brisket
cutability
carcass merit
style
soundness
smoothness
type
balance
muscling
finish
yearling
ram
conformation
vacuum packaging

NATIONAL AFNR STANDARD

AS.04.01.01.c
Select breeding animals based on characteristics of reproductive organs.

AS.04.01.02.c
Evaluate and select animals for reproductive readiness.

AS.04.02.03.b
Evaluate reproductive problems that occur in animals.

AS.04.02.03.c
Treat or cull animals with reproductive problems.

AS.04.03.04.a
Examine the use of quantitative breeding values in the selection of genetically superior breeding stock.

AS.04.03.04.b
Compare and contrast quantitative breeding value differences between genetically superior animals and animals of average genetic value.

AS.04.03.04.c
Select and assess animal performance based on quantitative breeding values for specific characteristics.

AS. 06.03.01.c
Evaluate and select animals to maximize performance based on anatomical and physiological characteristics that affect health, growth, and reproduction.

AS.06.03.02.a
Evaluate an animal against its optimal anatomical and physiological characteristics.

AS.06.03.03.c
Evaluate and select animals to produce superior animal products based on industry standards.

HUMANS HAVE ALWAYS SELECTED the type of animals they want to produce. As was discussed in Chapter 2, breeds were developed because humans chose to select animals with certain characteristics for use in breeding. As breeds developed and animals were bred true for the characteristics of that breed, animals were selected for desirable traits within that breed (**Figure 17–1**). Throughout history, animal producers have selected animals based on the traits they thought would improve the next generation of animals and would be more profitable for the producer. Only in recent history has there been a truly scientific basis for the selection of animals.

In the wild, animals developed traits that would help them survive in their environment. Animals having traits that aided them in survival stood a better chance of living and reproducing than did animals without those traits. For example, wild pigs that had the longest tusks, the thickest hide, and the fiercest nature had a better chance at survival than those with small tusks, a thin hide, and a docile nature (**Figure 17–2**). Wild cattle that could run fast for long distances and had long horns to defend themselves stood a better chance of evading or fighting off predators. Only the strongest animals survive in the wild, and they are the ones that breed and pass along their characteristics to the next generation. This process is known as **natural selection**.

FIGURE 17–2 Wild pigs have long tusks for self-defense. Domestic pigs have no use for tusks.
© JONATHAN PLEDGER/Shutterstock.com.

FIGURE 17–1 As breeds were developed, animals were selected for desirable characteristics. This calf is a Limousin.
© meunierd/Shutterstock.com.

In domestication, animals no longer need many of the characteristics that increase their chance of survival in the wild. To the contrary, many of the traits that were essential for survival in the wild are a great disadvantage to animals in domestication. For instance, the fierce nature of a wild pig is far less desirable than the docile nature of the domesticated pig. Cattle no longer need to be able to run swiftly or to possess long horns for defense. These traits have been bred out of domesticated agricultural animals.

Through selective breeding, humans have attempted to produce those animals that do well in a domesticated state. Obviously, the conditions under which the animals are to be raised (the environment) dictate the characteristics that an animal needs to thrive. For example, during the nineteenth century, the Longhorn breed of cattle was developed in Texas. This breed needed to retain some of the characteristics of wild animals such as long horns for

FIGURE 17–3 Nineteenth-century Longhorns retained some of their wild animal characteristics to be able to survive under harsh conditions. © B Norris/Shutterstock.com.

defense and the toughness to survive under harsh conditions (**Figure 17–3**). During the time when range land was almost limitless and labor was in very short supply, this type of cattle met the requirements of producers very well. These animals could tolerate living on an open range and foraging for food with very little help from people. Indeed, Longhorns still play a part in some cattle operations, but most situations in modern animal production call for a completely different operation. To meet the demands of the modern livestock industry, animals are selected based on what is desired by the people who produce them and by those who buy them.

Of the animals raised for food, there are basically two categories of animals: those that are produced for slaughter and those that produce offspring to be raised for slaughter. Many considerations go into selection of the type of animals for these two categories. Consumers have to be pleased with the type of product they find at the meat counter. To make a profit, meat packers have to approve of the carcasses sent to the meat packing plant (**Figure 17–4**). The buyers want animals that will remain healthy until they reach the slaughterhouse. The growers want animals that can gain weight quickly at an acceptable cost and with a minimum of care. The breeders want an animal that can reproduce efficiently. All of

FIGURE 17–4 Meat packers want carcasses that will make a profit. © racron/Shutterstock.com.

these criteria make the selection of the modern animal a complicated process. There are three basic traits that are desirable in the modern agricultural animal: **reproductive efficiency**, growth ability, and efficiency.

Reproductive efficiency means that breeding animals must be selected that produce offspring at a regular rate. If animals are producing young at a steady rate, producers are more likely to make a profit than they would if the animals produced fewer young (**Figure 17–5**). This means that the males must be **fertile** (produce sufficient numbers of healthy sperm); they must be healthy, aggressive breeders; and they must live a long, productive life. Females must be able to come into estrus regularly, conceive readily, produce an adequate number of healthy offspring, and produce enough milk to ensure that the young are weaned at an adequate size and weight.

Growth ability refers to an animal's ability to grow rapidly. The faster an animal grows, the more likely the producer will be to make

FIGURE 17–5 Animals must produce and raise young at a steady rate. *© Paul Looyen/Shutterstock.com.*

a profit from growing the animal. This trait is inherited from the parents and is influenced greatly by the type of care the producer offers.

Efficiency is the animal's ability to gain on the least amount of feed and other necessities. The producer sees that the animals are well cared for and fed properly. Animals that gain the most on the least amount of feed are the most desirable. If one steer can gain a pound for every 9 pounds of feed it consumes and another steer gains a pound of body weight for every 8.5 pounds of feed it consumes, the steer that requires less feed per pound of gain is said to be more efficient.

These characteristics have always been the important traits that producers have wanted. In years past, producers had a much more difficult time predicting which animals would possess these traits. Now, as a result of modern research, producers are able to predict with much more accuracy which animals will possess the desired traits. For example, research has shown that certain physical characteristics of animals will predict the reproductive capability of an animal. Breed associations have developed a bank of data on the performance of the offspring of a particular sire or dam. This has been brought about through the use of artificial insemination and embryo transfer. As discussed in the previous chapter, **performance data** have been compiled into values that indicate an animal's usefulness as a breeding animal.

THE SELECTION OF SWINE

Prior to the 1950s, swine were raised primarily for lard. People used the lard not only for frying and cooking food but also in the manufacture of cosmetics and lubricants. With the advent of cooking oils made from vegetable oils and cosmetics and lubricants made from petroleum-based synthetics, the demand for lard was reduced dramatically. Efforts then were directed toward developing a hog to produce meat instead of lard. Pigs that were used for breeding were especially selected for their degree of muscling. The idea was to produce an animal with the maximum amount of meat and the minimum amount of fat. These efforts culminated in the late 1960s and early 1970s with what has been referred to as the "super pig" (**Figure 17–6**). Pigs were selected for huge, round, bulging hams and overall thickness of

FIGURE 17–6 Selective breeding has brought many changes to the form of the swine. Bottom: 1960s classic outline; Middle: In the 1970s, a different form was developed; Top: The modern swine is designed to be muscular and carry the proper amount of finish.

muscling. Although producers were successful in producing a very lean pig that had a lot of muscle, some problems arose with the highly muscled, extremely lean pigs. This type of pig failed to be the "ideal" type for three reasons:

1. *Porcine Stress Syndrome (PSS)*. Extremely heavy muscled pigs are associated with a condition known as **Porcine Stress Syndrome (PSS)**. Apparently, this condition is genetic and is passed on to the offspring by the parents. Pigs suffering from PSS have very little tolerance to stress associated with hot weather, moving about, and some management practices. When put under such stress, animals with this condition have muscle tremors and twitching, red splotches develop on their underside, and they suffer sudden death (**Figure 17–7**). Obviously, this condition in hogs is not in the best interests of the pigs or the producer.
2. *PSE pork*. Pigs with extremely heavy muscle tend to produce lower-quality pork. Although they may have a large quantity of muscling, the meat has little or no intermuscle fat (marbling), is pale in color, and is soft and watery (exudative). These characteristics account for the name of the condition known as Pale, Soft, and Exudative **(PSE) pork**. Consumers reject this type of pork because the pale color is not appealing, and when cooked, the meat is dry and lacking in taste.
3. *Reproductive efficiency is lessened*. Heavily muscled, tightly wound boars have problems moving about and mounting females that are in heat. In addition, their sperm count is often very low. These conditions make them less desirable as herd sires. Females that are too heavily muscled are less fertile and have problems conceiving. Those that do become pregnant often have problems farrowing because the birth canal is bound so tightly with muscling that it cannot expand properly. Also, the number of piglets born tends to be fewer than those from less- muscular females.

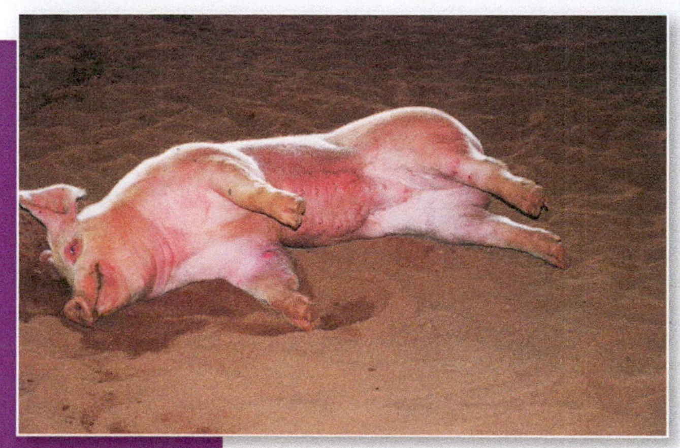

FIGURE 17–7 This pig has Porcine Stress Syndrome. Note the bulging muscles.

In an effort to correct these problems, during the 1970s producers developed long, tall, flat-muscled pigs. The idea was to produce animals that could move freely and reproduce efficiently. The market emphasis was on pigs that were extremely long and tall and could move freely. Pigs with extreme bulge and flare to their muscle pattern were highly discriminated against. The tall, flat-muscled pigs had greater resistance to stress and could reproduce more efficiently. Boars were more efficient breeders, and sows had larger litters. Although this type of pig improved reproductive efficiency, problems were encountered with carcass desirability. The amount of muscle on these animals did not suit the demands of the packers. **Loin eye** areas (an indication of the overall amount of muscle in the carcass) began to be unacceptably small. Consumers demand that loins be of an acceptable size (**Figure 17–8**). In addition, the growth rate and feed efficiency of these pigs were lower than the producers wanted. Market conditions also influenced the type of pig that was needed. Added production costs demanded that producers raise a more efficient pig.

In modern swine operations, the three aspects of the swine production industry that

FIGURE 17–8 Pork loins must be acceptable to consumers. *© Kelvin Wong/Shutterstock.com.*

make money for the producer are (1) reproductive efficiency, (2) growth ability, and (3) carcass quality. Of these, carcass quality is third and least important. With this information in mind, it is easy to understand why the shift is toward the modern type of hog.

Beginning in the 1980s, the emphasis on selection in pigs was on what was termed the "high-volume pig": very wide down the top, especially at the shoulders. Producers wanted pigs that were deep and widely sprung at the ribs and deep in the flank and belly. The reason was that this type of pig had a lot of room in the body cavity for the internal organs such as the heart and lungs. An animal with larger internal organs seems to grow faster and remain healthier.

Also, in a radical change from years past, pigs were selected that had large, loose bellies capable of holding large amounts of feed. These animals are more efficient in their intake of feed and in the conversion of feed to body weight. In addition, pigs with larger bellies will produce more bacon. A wide-topped, deep-sided animal has more capacity and internal volume for the internal organs such as the heart, lung, and digestive tract. This type of animal is a "better doing animal" in that it should have a higher rate of gain than a narrow-topped, shallow-sided animal.

The modern market hog has many traits similar to those of the 1980s pig. In the 1990s, the trend was toward a leaner, more muscular pig. Emphasis was on leanness and pigs that had little wastage in the carcasses. Some of the same problems were encountered in the 1960s. Consumers complained that the sausages were too dry and some fat was needed to flavor the pork. Consequently, the modern trend is toward pigs that are level-topped, have high capacity, and carry some backfat. This type of pig also has greater reproductive potential (**Figure 17–9**).

FIGURE 17–9 The modern trend is toward pigs that are level-topped, have high capacity, and carry some backfat.

THE SELECTION OF BREEDING HOGS

As mentioned, earlier, the factors in swine that increase profitability in raising pigs are reproductive efficiency, growthiness, and carcass merit. Of these, reproductive efficiency and growthiness are by far the two most important. With these factors in mind, breeding hogs are evaluated for characteristics that best combine these factors.

To be reproductively efficient, a female must be feminine—that is, she must look like a female. The same substances (hormones) that control the reproductive cycle also account for the development of sex characteristics. The sex characteristics, therefore, are an indication that the female is producing a large enough quantity of hormones to cause the female to conceive efficiently.

A producer can use several indicators of femininity to select females that are reproductively efficient. The underline should be well defined—that is, the teats should be large and easily seen (**Figure 17–10**). There should be at least six pairs of prominent and evenly spaced teats. If the teats are too close together, there may not be enough surrounding mammary tissue for good milk production.

Pin, blind, or inverted nipples should be avoided. **Pin nipples** are very tiny nipples that are much smaller than the other nipples on the underline. These may not function well enough to feed the young pigs. **Blind nipples** are nipples that fail to mature and have no opening. Obviously, they have little use, as they are nonfunctional. **Inverted nipples** appear to have a crater in the center and are not functional. Usually, a female that is producing enough female hormones will have the proper underline to efficiently feed the young she bears.

Both **gilts,** females that have not had litters, and sows, females who have had at least one litter, should look like females—they should have a head shaped like a female's and not like a boar's. Although all pigs selected for breeding should have a large, broad head, gilts should not have the massive head that is characteristic of the boar.

Another defect that adversely affects reproduction is an **infantile vulva**. This condition is characterized by a tiny **vulva** in a breeding-age gilt. This makes breeding very difficult, and conception rates usually are poor. A gilt with an infantile vulva should be culled from the herd. Gilts should be selected that have large, normally formed vulvas.

Boars should appear massive, rugged, and masculine (**Figure 17–11**). The **testicles** should be large and well developed inside a scrotum that is well attached. Research has shown that the larger the male's testicles, the more **viable** sperm he will produce and the more aggressive he will be in breeding. Large, pendulous, or swollen **sheaths** should be avoided, as these characteristics can lead to breeding problems.

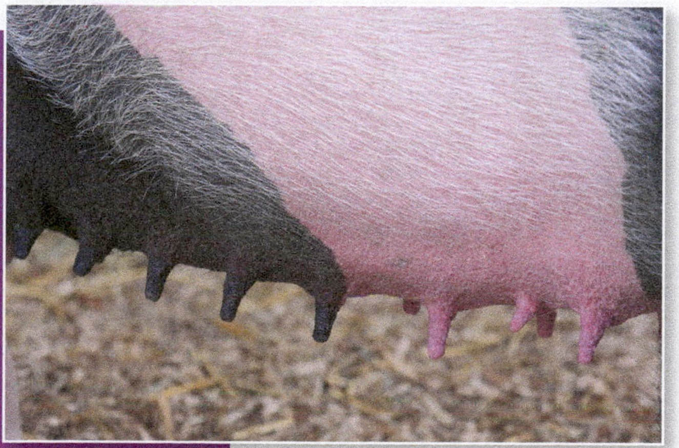

FIGURE 17–10 The teats should be large and easily seen, and there should be at least six on each side. *© bierchen/Shutterstock.com.*

FIGURE 17–11 Boars should appear massive, rugged, and masculine. *Source: USDA, Agricultural Research Service (ARS).*

Structural Soundness

Structural soundness refers to the skeletal system and how well the bones support the animal's body. Bone growth, size, and shape can have quite an effect on the animal's well-being. Animals that are structured well are more comfortable as they move about or stand in one place. Structural soundness also may affect reproductive efficiency. Boars that have problems moving freely are less likely to be interested in breeding than boars that move freely and are comfortable. Also, boars that stand too straight on their legs will have problems mounting females in the mating process.

FIGURE 17–12 Structural defects are amplified by the effect of standing and walking on concrete. This pig is too straight on his rear legs. *© kvasilev/Shutterstock.com.*

The vast majority of today's hogs are raised on concrete and should be selectively bred to be comfortable living on the hard surface. Concrete floors are much easier to keep clean and can be kept more sanitary than wood or dirt floors. Structural defects are amplified by the effect of standing and walking on concrete (**Figure 17–12**). Soreness, stiffness, and pain in moving greatly reduce reproductive ability. In addition, pigs that have problems standing and moving on concrete usually do not live as long and are not as productive as structurally correct pigs that are more comfortable on concrete.

A skeletal structure that allows a pig to be comfortable on concrete will have a topline that is almost level. At one time, pigs were selected for a uniform arch down the topline; however, pigs with a strongly arched back usually have steep rumps and straight shoulders. As **Figure 17–13A** shows, the scapula and humerus (shoulder and front leg bone) are more vertical and provide less flex and cushion than the level-topped, more structurally correct pig in **Figure 17–13B**. Note the vertical position of the aitch bone and femur (hip and rear leg bones) of the pig in Figure 17–13A, as opposed to the same bones that are more nearly parallel to the ground in the pig in

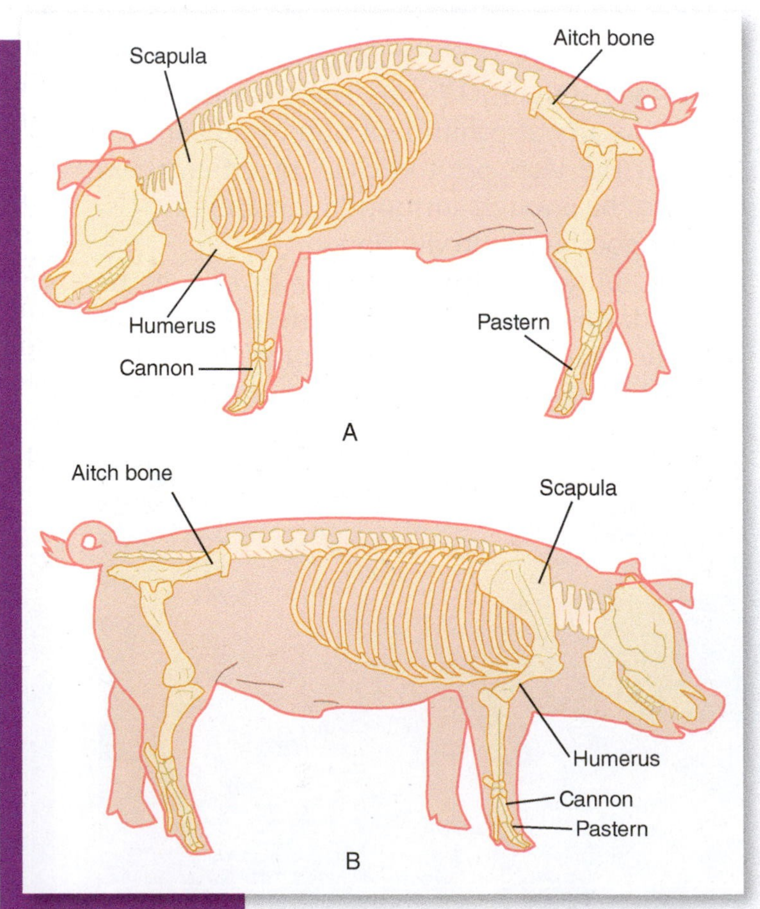

FIGURE 17–13 The skeletal structure of a high-arched pig (A) is more vertical and provides less flex and cushion than the level-topped pig (B).

Figure 17–13B. Again, bones parallel to the ground add more cushion and flex as the animal walks. If these bones are closer to being vertical to the floor, the ends of the bones will jar when the animal walks. This will eventually cause discomfort to the animal. On the other hand, if these bones are closer to being parallel with the floor, they will act more as a hinge. The **ligaments** of the joints will absorb more of the shock of walking, and walking will have less of a jarring effect to the bones. These animals will be more comfortable walking on concrete.

Pasterns (ankle bones) that are too vertical cause too much of a jarring effect to the skeletal system as the animal walks. Conversely, the pasterns should not slope so much that they are weak. Bones should be large in diameter rather than small and refined. Larger-diameter bones are stronger, and research has shown that animals with larger-diameter bones tend to grow faster.

The toes of the animal's feet should be of approximately the same size. Toes that are uneven (most commonly a small inside toe) indicate structural unsoundness. This inherited defect causes misalignment of the feet, weakens the pasterns, and causes an abnormal amount of weight to be placed on the outside toes. This understandably can cause a great deal of discomfort to the animal.

Legs, both front and rear, should be placed squarely under the animal. Front feet that are turned out (**splayfooted**) (**Figure 17-14**) or turned in (**pigeon-toed**) should be avoided, as should rear feet that are splayed out (**cow-hocked**) (**Figure 17–15**) or turned in. The pig should move out with a long, easy stride. As the animal walks, the rear foot should be placed about where the front feet were placed. Pigs that take short, choppy steps (goose-stepping) are either structurally unsound, muscle-bound, or both.

FIGURE 17–14 This pig is splayfooted. Also, note the unevenly shaped toes. *Photo courtesy of National Hog Farmer (www.nationalhogfarmer.com)*

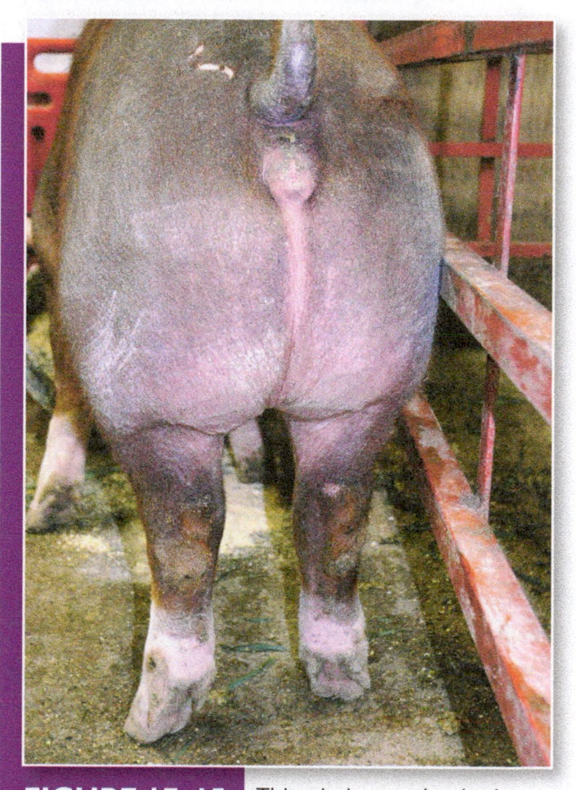

FIGURE 17–15 This pig is cow-hocked. *Photo courtesy of National Hog Farmer (www.nationalhogfarmer.com)*

Growthiness

In selecting the modern animal, a lot of emphasis is placed on capacity. Preference should be given to those animals that are wide down the top and deep in the side. The animal should be wide between the front legs and have a wide chest floor. The reasoning is that those animals that have greater dimensions in the side, down the top, and through the chest have more room for the vital organs such as the heart and lungs. In addition, pigs with a large, loose belly have a greater capacity for holding feed and thus gain more rapidly.

The ribs should be long and well-arched. The rib cage should be rectangular; that is, the fore rib should be about as long as the rear rib. Again, this is an indication that the body cavity has adequate room to house the vital organs.

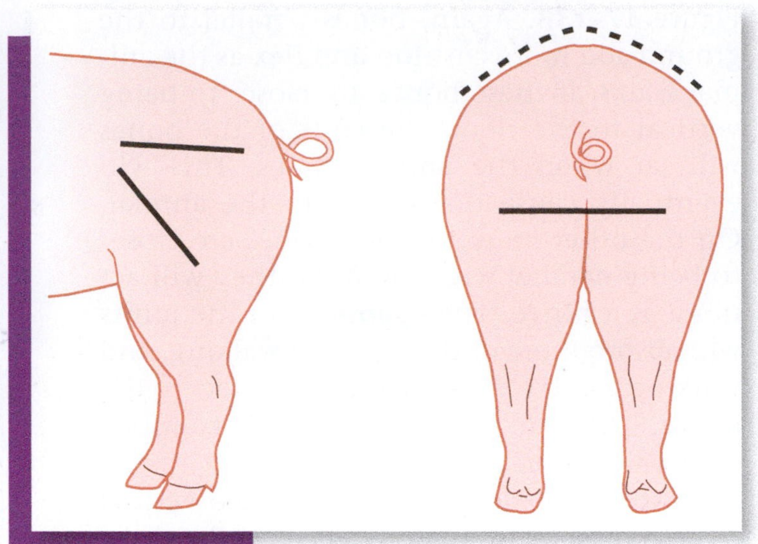

FIGURE 17–16 The best indicator of overall muscling is the amount of muscle in the hams. Total volume of muscle is indicated by length, width, and depth.

Other growth indicators are length of **cannon bone** and length of neck. Research studies have shown that animals having longer cannon bones are later maturing and grow more rapidly. Also, the longer-necked animals have been shown to have more growth potential. Big-headed, broad-skulled pigs that have longer distances between their ears and between their eyes also have been found to be more efficient gainers.

Carcass

As was pointed out earlier, in years past an overemphasis was placed on thickness and degree of muscling in hogs. Muscling in the modern pig should be smooth and loose. The best indicator of overall muscling is the amount of muscle in the hams. Remember—hams are three-dimensional; that is, they have length and width as well as thickness (**Figure 17–16**). Pigs with short, thick, bulging hams should be avoided and preference given to pigs whose hams get their volume in length from tail head to hock and width across the rump. The smoother-muscled pigs move more freely and easily. In addition, gilts with this type of muscling will farrow easier than will females that are so muscle-bound that the pelvis cannot give enough to allow pigs to pass easily through the birth canal.

THE SELECTION OF MARKET BEEF ANIMALS

Consumers play an important role in determining what type of beef animal is raised for slaughter. They usually want beef that is tender, flavorful, and affordable (**Figure 17–17**). To produce this type of product, the right type of animal must be produced. Tenderness and taste are both related to the age of the animal. Most of the beef sold in the supermarket as retail cuts come from a Choice grade of beef. This means that the animal has reached a

FIGURE 17–17 Consumers want beef that is tender, flavorful, and affordable. *© Maria Komar/Shutterstock.com.*

degree of maturity where it begins to deposit fat in the muscle.

As animals grow and mature, fat is deposited differently. In young animals, most of the energy from feed goes into the growth of bone and muscle so the animal can grow larger. When the animal matures, growth of bones and muscles ceases and the animal begins to utilize energy from feed to deposit fat. Fat is first deposited in the body cavity of the animal. This serves to provide energy storage and also to help cushion the internal organs. As the body cavity reaches its peak in terms of fat deposit, the animal begins to deposit fat under the skin. When a certain level of **backfat** is reached, fat is deposited between the muscles and finally inside the muscles. These intermuscular fat deposits, called marbling, are what give the meat its flavor.

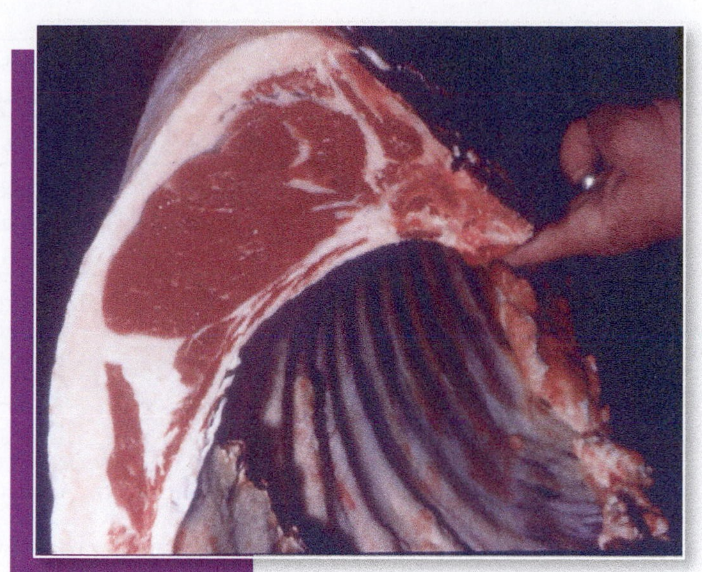

FIGURE 17–18 Rib eyes larger than 15 square inches may be too big for some consumers. *Courtesy of Calvin Alford, Cooperative Extension Service, Univeristy of Georgia*

Carcasses are graded largely on the amount of fat deposited in the muscles. If fed properly, cattle usually reach the proper stage of fatness or finish when they are about 2 years of age. At this age, the animals are generally still young enough to be tender. The consumer does not want cuts of meat that have a lot of excess fat, so the aim is to produce animals that will put marbling into the meat with minimum backfat on the carcass.

In addition, consumers are choosy about the size of the meat cuts. If the animal is too large when slaughtered, the carcass may yield cuts that are too large. Consumers seem to want steaks that one person can consume in one meal. This means that carcasses with a **rib eye** larger than 15 square inches may be too large for the average consumer (**Figure 17–18**).

To be of the proper size at slaughter, animals have to be the proper size when they begin to mature and lay down fat within the muscles. The size of the animal at maturity depends on the **frame size** of the animal. Frame size refers to the skeletal size of an animal at a given age. Small-framed animals mature earlier than do large-framed animals, and they deposit fat in the muscle at a smaller size. A small-framed animal (frame score 1) will probably grade Choice around 750–850 pounds, while a large-framed animal (frame score 7) will usually have to weigh 1,350 pounds or more to grade Choice. The numerical score for frame size is determined by measuring the height of an animal at the hip at a certain age.

Table 17–1 lists the classification of frame scores by **hip height** and age, for males and females. The base point is 45 inches hip height at 12 months of age for a frame score of 3. Allow 2 inches for each frame score at the same age. Allow 1 inch per month from 5 to 12 months of age, 0.50 inch per month from 12 to 18 months, and 0.25 inch up to 2 years.

Commercial packers usually want a carcass that weighs between 600 and 700 pounds. Carcasses within this weight range are more easily managed in the cutting room, and they provide the size of cuts that the consumer wants (**Figure 17–19**). Frame size should be large enough for the animal to grade Choice at about 1,050–1,200 pounds to obtain the desired carcass weight and grade. This means that in selecting slaughter steers, preference should be

FRAME SCORES
MALES

Frame Scores Based on Height (in inches) Measured at Hips

Age in Months	Frame Score 2	Frame Score 3	Frame Score 4	Frame Score 5	Frame Score 6	Frame Score 7
5	36.00	38.00	40.00	42.00	44.00	46.00
6	37.00	39.00	41.00	43.00	45.00	47.00
7	38.00	40.00	42.00	44.00	46.00	48.00
8	39.00	41.00	43.00	45.00	47.00	49.00
9	40.00	42.00	44.00	46.00	48.00	50.00
10	41.00	43.00	45.00	47.00	49.00	51.00
11	42.00	44.00	46.00	48.00	50.00	52.00
12	43.00	45.00	47.00	49.00	51.00	53.00
13	43.50	45.50	47.50	49.50	51.50	53.50
14	44.00	46.00	48.00	50.00	52.00	54.00
15	44.50	46.50	48.50	50.50	52.50	54.50
16	45.00	47.00	49.00	51.00	53.00	55.00
17	45.50	47.50	49.50	51.50	53.50	55.50
18	46.00	48.00	50.00	52.00	54.00	56.00
19	46.25	48.25	50.25	52.25	54.25	56.25
20	46.50	48.50	50.50	52.50	54.50	56.50
21	46.75	48.75	50.75	52.75	54.75	56.75
22	47.00	49.00	51.00	53.00	55.00	57.00
23	47.25	49.25	51.25	53.25	55.25	57.25
24	47.50	49.50	51.50	53.50	55.50	57.50

FEMALES

Frame Scores Based on Height (in inches) Measured at Hips

Age in Months	Frame Score 2	Frame Score 3	Frame Score 4	Frame Score 5	Frame Score 6	Frame Score 7
5	35.75	37.75	39.75	41.75	43.75	45.75
6	36.50	38.50	40.50	42.50	44.50	46.50
7	37.25	39.25	41.25	43.25	45.25	47.25
8	38.00	40.00	42.00	44.00	46.00	48.00
9	38.75	40.75	42.75	44.75	46.75	48.75
10	39.50	41.50	43.50	45.50	47.50	49.50
11	40.25	42.25	44.25	46.25	48.25	50.25
12	41.00	43.00	45.00	47.00	49.00	51.00
13	41.75	43.75	45.75	47.75	49.75	51.75
14	42.25	44.25	46.25	48.25	50.25	52.25
15	42.75	44.75	46.75	48.75	50.75	52.75
16	43.25	45.25	47.25	49.25	51.25	53.25
17	43.75	45.75	47.75	49.75	51.75	53.75
18	44.25	46.25	48.25	50.25	52.25	54.25
19	44.50	46.50	48.50	50.50	52.50	54.50
20	44.75	46.75	48.75	50.75	52.75	54.75
21	45.00	47.00	49.00	51.00	53.00	55.00
22	45.00	47.00	49.00	51.00	53.00	55.00
23	45.25	47.25	49.25	51.25	53.25	55.25
24	45.25	47.25	49.25	51.25	53.25	55.25

The height in inches shown under each frame size is the minimum height for that frame size

TABLE 17–1 Frame size table.

FIGURE 17–19 Commercial packers usually want a beef carcass that weighs around 600–700 pounds. *© iStockphoto/ARSELA.*

FIGURE 17–20 Double-muscled animals can be recognized by the protruding muscles, short tail, and low set head. *© Peter Braakmann. Image from BigStockPhoto.com.*

given to the medium-framed steers that finish at 1,050–1,200 pounds.

The purpose in producing market beef animals is to obtain muscles that can be cut into **retail cuts** of beef for the consumer. It would seem that the more muscle there is on the animal, the more desirable that animal becomes. This is true only up to a point. Just as an animal can have too little muscling, it can also have too much. Selection for extreme muscling leads to the development of cattle with a condition known as **double muscling**. Double muscling is undesirable for the following reasons:

1. These animals are difficult to produce. If the goal of the producer is to select animals with double muscling, then breeding stock of the same type must be selected. Fertility in these animals is very poor, and calving is much more difficult.
2. The meat tends to be coarse and void of intermuscular fat (marbling). Even though the animal may have a sufficient cover of fat, the marbling tends to be less than adequate. Thus, a large percentage of these animals grade Standard.
3. Double-muscled calves are difficult to raise because they are more susceptible to disease.
4. Double-muscled feeder animals must be fed a higher proportion of **concentrate** in the ration to obtain enough marbling to grade Choice.

These cattle can be recognized by their physical appearance. The rump is protruding and round with definite grooves or creases between the thigh muscles. The tail is short and is attached high and far forward on the rump (**Figure 17–20**). The head of the animal is small and long and is carried lower than the top of the body.

Animals that possess an adequate degree of long, smooth muscling should be selected over animals that are lightly muscled or have extremely tight-wound, excessive muscling. Smooth muscling—muscling that does not bulge too much—allows the animal to move freely and smoothly. To determine the amount of muscling in the animal, it should be viewed through the center of the round, as indicated in **Figure 17–21**. Notice that the steer on the left is light muscled and the steer on the right is thick muscled.

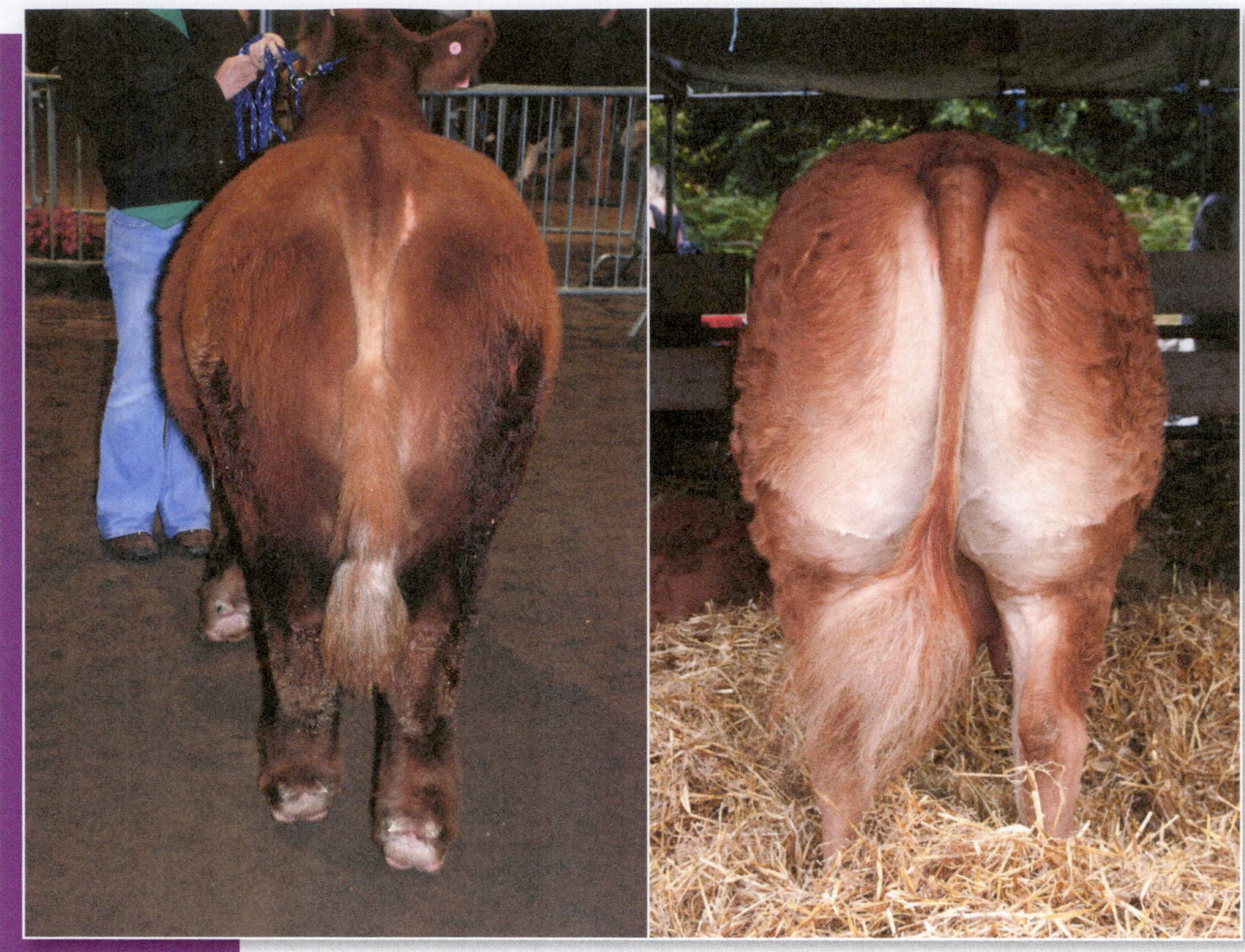

FIGURE 17–21 Notice that the steer on the left is lightly muscled and the steer on the right is thickly muscled.

THE SELECTION OF BREEDING CATTLE

As mentioned earlier, frame size refers to the overall height of the animal at maturity; tall animals are larger-framed than short animals. The frame size can be determined when the animal is a young calf. The leading indicator of frame size is the length of the cannon bone (the bone between the ankle and the knee of the front leg). Research has shown that an animal with a longer length of cannon bone will be a taller animal at maturity than an animal with a shorter cannon bone. In fact, some breed associations request that the producer record the length of cannon at birth of any animals they wish to register.

In heifers, consideration should be given to animals having a longer length between the hooks and pins and those that are wider apart at both hooks and pins. This is an indication of greater **pelvic capacity**. When a female gives birth, the pelvis must open enough to allow the calf to pass through the birth canal. If a heifer has a small pelvis, she probably will have problems delivering a calf (**Figure 17–22**).

Obviously, cattle that grow faster (have a greater average daily gain) are more desirable. Research has shown that there are other physical characteristics of cattle that are associated with growthiness. The length of an animal's face (distance from eye to muzzle) and the length of the neck are both indicators of

FIGURE 17–22 The heifer on the left has a greater pelvic capacity than the other two.

FIGURE 17–24 Bulls should be bold and masculine-appearing. *Source: USDA, Agricultural Research Service (ARS).*

growth. The longer the neck, the faster the animal is likely to grow.

Breeding cattle should have adequate body depth and width to provide adequate room for the internal organs. The larger the internal organs, such as the heart and lungs, the better the animal should do in terms of viability and growth. Indicators of capacity are width through the chest floor; long, well-arched ribs; and depth in the side (**Figure 17–23**).

Sex character simply means that a bull looks like a male and a heifer looks like a female. Because sex hormones control both the physical appearance of animals and their ability to reproduce, it stands to reason that an animal with more sex character should be more reproductively efficient.

FIGURE 17–23 Breeding cattle should have adequate depth and width to provide room for the internal organs. *© MaxPhoto/Shutterstock.com.*

Sex character in a bull is determined by a broad, massive, bull-like head (**Figure 17–24**). The shoulders should be bold and well-muscled, but care should be taken that bulls with coarse, excessively thick shoulders are not selected as a herd bull. Since this characteristic is passed on to the offspring, calving difficulties can be encountered.

One of the most important physical traits is that of testicle shape and size. A 2-year-old bull should have a scrotal circumference of at least 34 cm when measured at the largest part. Research has shown that the larger the testicles, the larger the number of valuable sperm that are produced. In addition, the scrotum should extend to about hock level and have a definite neck. If the testicles are held too close to the body, the temperature will be too high for ideal sperm production. Bull 1 in **Figure 17–25** has a straight-sided scrotum often associated with testicles of only moderate size, and the testicles are held too close to the body. Bull 2 shows the ideal testicle size and scrotal shape. Bull 3 has a tapered or pointed scrotum that usually is associated with undersized testicles that are too close to the body.

The fertile female should be well balanced and present a graceful, feminine appearance.

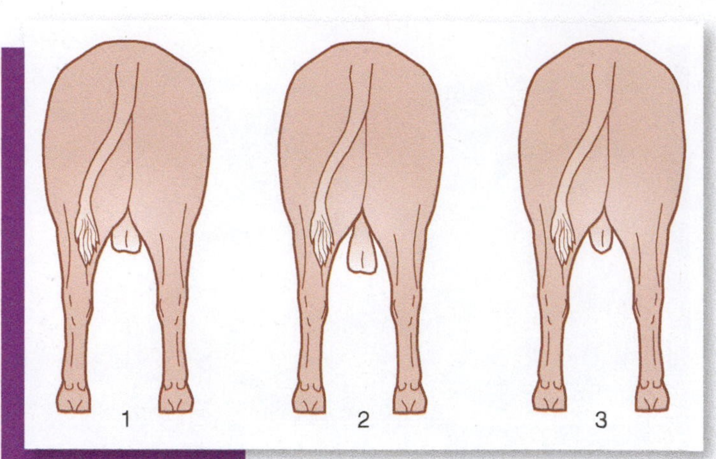

FIGURE 17–25 Testicles should be carried away from the body and have a definite neck. Bull 2 is the best choice.

FIGURE 17–26 A good breeding female should be clean and trim through the brisket and middle, such as this Charolais. *Courtesy of American International Charolais Association*

She should be long and clean in the face and throat. Her neck should be long and blend smoothly into smooth, sharp shoulders. She should be clean and trim through the **brisket** and middle. The pelvic area should be large and wide for easy calving (**Figure 17–26**). Distances should be wide from hook to hook and from pin to pin, and long from hooks to pins.

Structural soundness refers to the correctness of the feet and legs of an animal. The legs should fit squarely on all four corners of the animal. The correct "set" to the back legs of a bull is shown in **Figure 17–27A**. If a plumb line is dropped from the pins to the ground, the line will intersect with the hock. The condition known as post-legged is shown in **Figure 17–27B**. The rear legs are too straight and do not provide enough cushion and flex as the animal walks. In bulls, this condition causes problems in mounting cows. The opposite condition, called sickle-hocked (**Figure 17–27C**), also causes problems in mating. As the animal mounts, undue stress is placed on the stifle muscle, causing the animal to become stifled; that is, the ligament attaching the stifle muscle tears. This results in the animal being worthless as a herd bull. When viewed from the rear, the back legs should be straight.

Figure 17–28 depicts animals that have structural problems in comparison to one with correct form as viewed from the rear. **Figure 17–29** shows animals whose front legs are structurally incorrect as compared with those that are correct. All cattle are *slightly* splayed in the front, but the front feet should not turn out very much. All animals should move out with a free, easy stride. The rear foot should be placed about where the front foot was

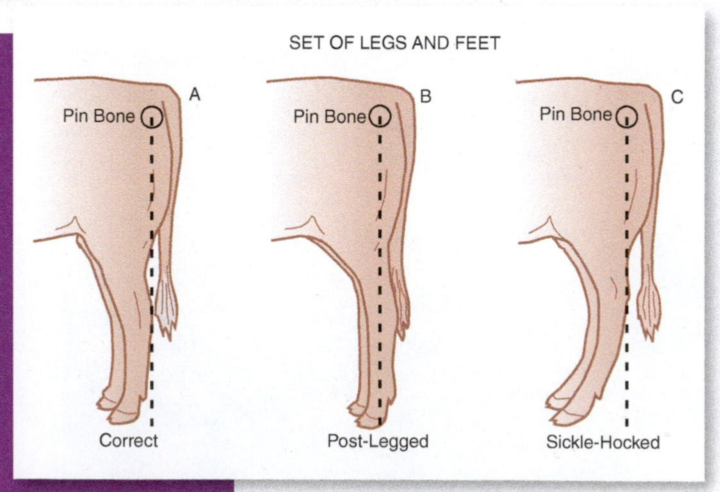

FIGURE 17–27 (A) In a structurally sound animal, the legs fit squarely on all four corners of the animal; (B) Common defect known as post-legged; (C) Common defect known as sickle-hocked. *Courtesy of Vocational Materials Service, Texas A&M University*

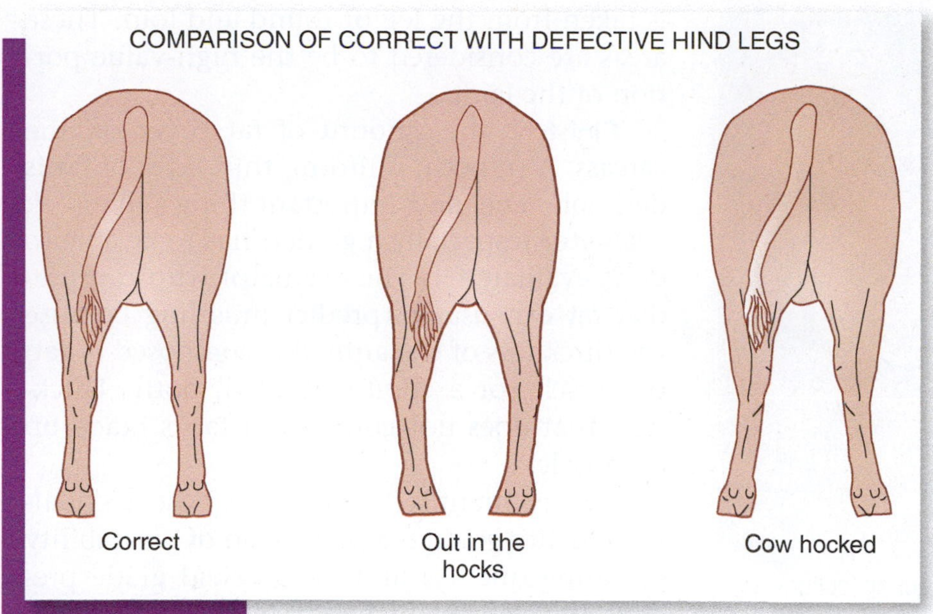

FIGURE 17–28 Comparison of correct with defective hind legs. *Courtesy of Vocational Materials Service, Texas A&M University*

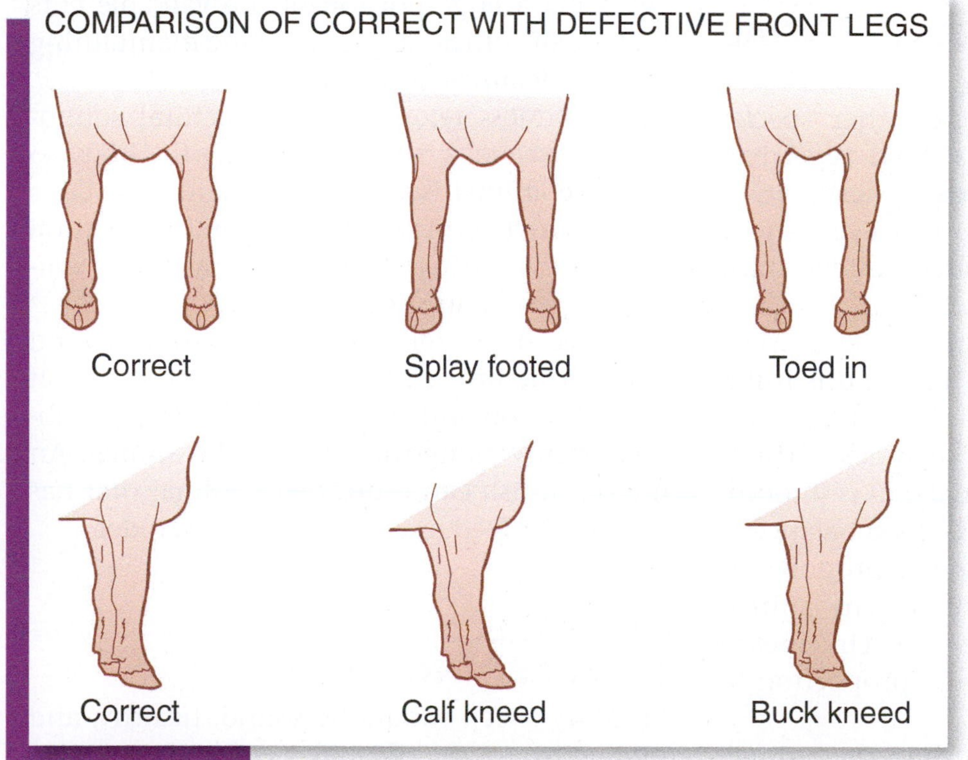

FIGURE 17–29 Comparison of correct with defective front legs. *Courtesy of Vocational Materials Service, Texas A&M University*

picked up. Cattle that take short, choppy steps either have a muscle pattern that is wound too tightly or have problems in their skeletal makeup. Both conditions are objectionable. To feel their best and to grow and do their best, animals must be structurally correct.

THE SELECTION OF SHEEP

The selection of sheep has always been more complicated than the selection of either beef or hogs because two traits have been traditionally considered—meat and wool. Milk sometimes is considered to be a third trait. Shepherds had been selecting more productive sheep long before the principles of inheritance were outlined by Gregory Mendel. Prior to and during the 1950s, as much emphasis was placed on the production of wool as on the production of meat (**Figure 17–30**). When synthetic materials became more popular, the emphasis shifted from wool production to meat production within some breeds. During the 1960s, interest and research were concentrated on **cutability**, the percent of lean cuts that a lamb carcass would produce. Research studies placed more emphasis on the muscling and leanness of lamb carcasses than on the production of wool.

FIGURE 17–30 Unlike the other major agricultural animals, sheep are selected by their fleece quality as well as conformation. *Courtesy of Cooperative Extension Service, University of Georgia*

THE SELECTION OF COMMERCIAL OR WESTERN EWES

Sheep are judged similarly to other species of livestock. Selection is based on type, balance, muscling, finish, **carcass merit**, **style**, **soundness**, and **smoothness**. However, sheep have another trait to consider besides meat production—wool production.

Type is the general build of the animal. Desirable type changes over time and is influenced greatly by trends in the show ring. Desirable type, however, usually includes a thick, moderately deep-bodied animal that is smooth, has straight lines, and exhibits good balance.

Balance refers to the general proportions in the structure of the animal. An animal should appear to fit together well. This means that the front end should be in proportion to the rear end.

Muscling is the natural flesh of the animal, not including the fat. Modern lambs should have thick muscling. Meat is composed of muscling, and the more muscling on the carcass, the more value it has to the packer and the producer. A good portion of meat from the carcass is taken from the leg or round and loin. These areas are considered to be the high-value portion of the lamb.

Finish is the amount of fat cover on the carcass. A smooth, uniform, thin layer of fat is desirable. The most important thing a livestock evaluator (especially a grader) has to be able to do is evaluate fat. The evaluator who can predict fat can usually predict muscling because the thickness of the animal is composed of fat, or muscle, or a combination of both. Thickness that does not come from fat is made up of muscle.

Carcass merit is determined by carcass quality (Quality grade is a prediction of palatability or eating quality) and yield (Yield grade predicts the percent of boneless retail cuts from the leg or round, the loin, the rib or rack, and the chuck or shoulder). A carcass of good merit will yield a large portion of the valuable cuts that are of sufficient quality while maintaining as much leanness as practical.

Soundness refers to the structural soundness or skeletal system supporting the body so that the animal is comfortable and maintains reproductive efficiency. The same structural correctness desired in other agricultural animals is desired in sheep as well.

Smoothness refers to the lack of awkward bone structure and a smooth, even finish along the top and sides. Fat should be distributed evenly along the body of the animal. An uneven finish can indicate an animal that has been off its feed or an animal that is older.

THE SELECTION OF BREEDING EWES

Breeding sheep should be sound, healthy, and productive. Ewes should be vigorous, with normal teeth, feet, legs, eyes, and udders. Ewe lambs or **yearling** ewes should be selected for breeding soundness and should be culled for reproductive problems. However, younger ewes will require more attention during their first lambing. Age is determined by the teeth located on the lower jaw.

The correct type of ewe should exhibit the following traits:

- Purebred ewes should show breed traits.
- Feminine appearance.
- Depth in the fore and rear flanks.
- Good chest capacity.
- Proportional length, depth, and width of body.
- Width and thickness throughout the loin.
- A strong, wide, long, and level rump with a wide dock.
- Straight legs with long, full muscling that provides width between the legs.
- Strong pasterns and feet with medium-size toes.
- Strong, wide back and crop with a tight shoulder that blends into the body.
- Balance and smoothness, with all parts blending well together.
- Wool should be indicative of the breed in fiber diameter, length, and quality.

THE SELECTION OF RAMS

Great care should be given in selecting the **ram** since half the growth and wool-producing ability of every lamb is contributed by the ram. Generally, the same traits listed for consideration in selecting ewes are used in selecting rams. However, rams should be selected with certain goals in mind. Con sideration should be given to the strengths and weaknesses of the ewe flock. If the ewes are lacking in a quality, a ram should be selected that will bring the desired traits into the flock. The rams to be used as terminal sires should be different from those chosen for purebred flocks. Rams should be selected for genetic capability—whether for growth rate, wool production, or increased lambing.

Production records should aid in the selection of rams. Rams should have the traits of muscling and structural correctness that were listed for ewes. In addition, they should be rugged, muscular, and masculine (**Figure 17–31**). Rams should exhibit superior growth, structural soundness, and quality of fleshing. The testicles should be well developed and pliable. A semen test should be obtained or the ram guaranteed as reproductively sound. Finally, the ram's semen should be retested just prior to breeding season.

FIGURE 17–31 Modern rams are large and growthy.
©Ruud Morijn Photographer/Shutterstock.com

JUDGING MARKET LAMBS

The emphasis for judging market lambs and purebred sheep at livestock shows has changed and affected the industry over the years. In the 1950s and 1960s, the ideal lamb was short, fat, and extremely thick. These lambs produced carcasses that were small but thick because of combined muscle and fat. Many of these carcasses had prime **conformation**. Because carcasses were shipped in carcass form, the fat helped to reduce shrinkage and allowed the retailer to trim the carcass in order to produce good looking—but fat—retail cuts.

During the late 1960s and 1970s, the wastefulness of the excess fat was taken into account, and the trend was for producers to select a medium-size lamb that was much

FIGURE 17–32 The ideal modern market lamb is lean and muscular.

leaner. This was due in part to the introduction of **vacuum packaging** and the boxing of lamb carcasses and carcass parts. However, by the late 1970s and until the mid-1980s, show sheep were being selected for height, length, and stylishness, with little emphasis on carcass, except for finish (0.1 inch of backfat was considered ideal). These animals were selected for the hind-saddle (leg, loin, rack) muscling without regard for chest capacity and natural fleshing needed by ewes for livability and reproduction.

From the mid- to late-1980s to the present, selection of show sheep has moderated. The following description reflects the modern type of show sheep (**Figure 17–32**).

Size, muscle, structure, style, and balance are all factors that should be considered when selecting a market lamb at a livestock show. Champion lambs should be heavily muscled, nicely balanced, and correctly finished. Carcass merit should be kept in mind when evaluating the animal. The hind-saddle (leg, loin, and rack or rib area) makes up about 70 percent of the carcass value. However, capacity is also important, especially in the breeding animal, for livability and reproductive volume.

Different breeds have a different frame size related to the correct market weight for each lamb. This makes judging large classes made up of different breeds more difficult. Muscling is shown by the thickness and width of the loin, thickness of muscling in the leg, and prominence of the stifle region. A long hind saddle is desirable, but it must have width and depth as well as length to be of greater total volume. The lamb should have a wide, level rump that blends into a muscular leg. A heavily muscled lamb will stand wide—both front and rear—and exhibit a heavily muscled forearm. Muscling should be firm and smooth, but not too extreme or bulging.

Finish on the carcass is measured 1.5 inches from the middle of the backbone between the twelfth and thirteenth ribs. Lamb carcasses with less than .1-inch fat tend to dry out, and shelf life is reduced. Lamb carcasses with more than .25-inch backfat simply require too much trimming to obtain acceptable retail cuts. The ideal amount of finish is about .15 inch or less. Live lamb finish is determined by handling the lamb over the ribs and backbone. This handling requires experience to accurately feel the amount of finish. A good learning practice is to evaluate live lambs and then evaluate the carcass. Videos are especially useful for learning this practice.

Knowing the amount of finish lets the evaluator know the amount of muscling (flesh − fat = muscling). Other indicators of fat on lambs are that they are heavy-fronted (breast), have fat deposits in the cod or udder region, have fat around the dock, and lack trimness in the middle and flanks. A correctly finished lamb should feel firm and have a trim underline. The judge must be able to compensate for the pelt (skin and fleece) in determining finish. This is especially important in lambs with thick hides and in lambs that are not slick-shorn. Most market lamb shows require that the lambs be slick-shorn.

Bone is important because it relates to frame size and structural correctness. Muscles attach to bones, so sufficient muscle requires a sufficient diameter and length of bone. Structural correctness allows the animal to move in comfort and maintain productivity. It is not as important for market animals to be structurally correct as it is for breeding stock. However, structural correctness is a consideration in determining the overall style and balance of the market animal.

FIGURE 17–33 Young goats should have ample bone, be well muscled, and structurally correct. *© iStockphoto/hidesy.*

THE SELECTION OF GOATS

As discussed in Chapter 8, goats are becoming increasingly important in the animal industry. Like any other agricultural animal, selection is an important tool to produce good animals. Any animals being considered for purchase should be selected from a reputable producer and should have complete medical records available for review. If selecting animals for a breeding program, different characteristics will have to be evaluated than selecting for market animals.

The Selection of Market Goats

Market goats should be selected for potential growth and structure (**Figure 17–33**). A market goat, when purchased young, should have ample bone to accommodate lots of muscle. No animal should be purchased that is not structurally sound and healthy. Lame or sickly animals can take months to become healthy enough to gain the appropriate amount of weight. Structural correctness includes a level topline and legs that are well set. This means that the animal should not stand post-legged or sickle-hocked. As in other animals, the pasterns should be well set and not slope too much. They should have a wide stance between both the front legs and the rear legs. As with heifers, there should be ample length between the hooks and pin bones in the hip area (**Figure 17–34**).

Growth potential is another important factor to consider. The market animal should be selected on its potential for muscle development. As with

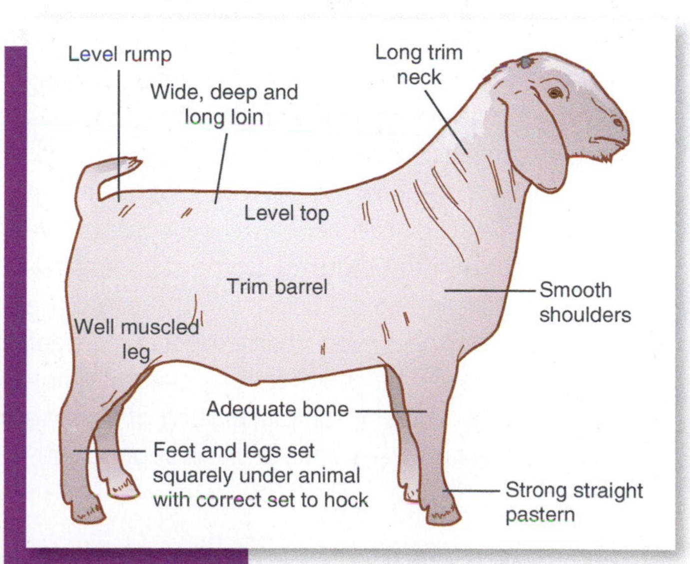

FIGURE 17–34 Ideal goat conformation.

other meat animals, buyers are most interested in the more expensive meat cuts located in the loin and leg. Observe the animal when it is standing broadsided. Examine the length of the side to determine how long the animals is, as well as the frame size. The length of the canon bone and length of the face of the animal are indications of how well the animal will grow.

If a goat is being purchased for show, the show rules should be consulted and followed. Some junior shows have rules that the goat cannot be over a certain age, so the goat should be bought in plenty of time for it to grow sufficiently for the show.

Any animals purchased should be quarantined from other animals for at least 30 days to prevent any new disease from infecting the herd, even if medical records are presented. Goats new to the herd can bring in pathogens that can contaminate a whole herd that has not been previously exposed. Goats are herd animals, so when purchasing them, they should either be purchased in groups or be housed next to other animals during their quarantine period.

The Selection of Breeding Meat Goats

Breeding goats are selected according to a variety of characteristics including bloodlines and reproductive fitness. Nannies should have

PUTTING IT INTO PRACTICE

Livestock Evaluation Career Development Event

The concepts and skill in livestock evaluation or judging you learn in the classroom can be further developed by participating in several FFA career development events. The FFA Livestock Judging Event is for teams consisting of four members. Each member competes individually, and the final team score is calculated by adding the three highest individual scores. The FFA Livestock Judging Event aids classroom instruction by teaching students practical skills in selecting quality livestock. All livestock producers know that the selection of quality livestock is essential to be competitive and profitable in the industry. Some skills and talents you will need to participate in the event are understanding the breed characteristics of beef, swine, and sheep. The ability to rank classes of animals and give reasons for placement of animals is also required. Team members independently evaluate and place each livestock class. After the judging, students are asked to give the reasons they used to place the livestock. Points are awarded for each class placing and correct reasons. This FFA event is usually held at the chapter, area or district, state, and national levels.

Students can develop livestock evaluation skills by participating in the FFA Livestock Judging Career Development Event. *Courtesy of the National FFA Organization*

strong bone structure and be able to produce kids without any problems. Because goats normally produce more than one kid at each birth, some animals do have trouble giving birth. Multiple births increase the chance of something going wrong. The most common problems at birth are that the baby is too large or that the baby goat (or goats) is in an improper position.

Herd sires or billies should not have any structural defects, should be of sufficient size, and should have ample muscling, because most of these features are passed on to their offspring in future generations. Bloodlines are also important for the herd sire if the producer is planning to sell the offspring for purebred replacements. Growth characteristics are taken into consideration, but not as heavily as reproductive capabilities.

The Selection of Dairy Goats

Dairy goats should have a structure different from meat goats. Remember—the goal for meat goats is to produce muscle, and the goal for dairy goats is to produce milk. When viewed over the topline, meat goats should be thicker and exhibit muscle, and dairy goats should be trimmer and have a more refined, wedge shape.

Dairy goats should be selected based on their potential for milk production. The mammary system should be well attached to the body. If the udder sags or becomes pendulous, the productive life of the nanny will become shortened. Major faults with nannies include having more than two teats, having multiple orifices, and being round-boned rather than flat-boned. Meat animals have round bones, which can support more meat development. Dairy animals are flat-boned because muscle development is not the main concern.

Bucks or billies pass on so much of their genetics to the offspring that they should be ideal animals based on breed standards. As with other animals, the male should look like a male. Remember—the reproductive hormones control all of the reproductive process as well as the physical characteristics of the male. The more the animal looks like a male, the more likely he is to produce sufficient amounts of the male reproductive hormones.

SUMMARY

Ever since the beginning of the animal industry, selective breeding has been a cornerstone. Choosing the right animals to mate is essential in order to produce the type of offspring desired. Although the style and type of animal that producers want has changed over the years, some of the basics remain the same. The advent of embryo transplant and artificial insemination has greatly aided the selection process. The computer also has had an impact. Perhaps we are just on the brink of realizing how the computer can revolutionize the process.

CHAPTER REVIEW

Review Questions

1. What is the process of natural selection?
2. Why did animals develop characteristics such as thick hides, long horns, and the ability to run fast?
3. Why are these characteristics not desirable in modern agricultural animals?
4. What is meant by selective breeding?
5. What are two basic categories of agricultural animals involved in the producer's selection process?
6. In the selection of market animals, what does the consumer want? The packer? The producer?
7. Describe the type of hogs raised in the following eras: the period prior to the 1950s; the 1960s; the 1970s.
8. Describe the modern market hog, and tell why these characteristics are important.
9. What is Porcine Stress Syndrome?
10. What is meant by sex character, and why is it so important?
11. Why is it so important that animals be correct on their feet and legs?
12. Why should a hog have a level back rather than an arched back?
13. Define *capacity*. Why is it important?
14. How are beef carcasses graded?
15. In what sequence is fat deposited in a beef animal's body?
16. Of what use is an animal's frame size in the selection process?
17. What are double-muscled cattle? Explain why this type of cattle is undesirable.
18. Describe at least two traits that are desirable in breeding heifers.
19. Why are testicle size and scrotal shape important in selecting bulls?

Student Learning Activities

1. Choose a breed of animal (swine, beef, or sheep). Using a set of pictures or viewing the animals live, list all of the animal's physical traits. From the list, decide which of the characteristics are a result of natural selection and which are the result of selective breeding.
2. Go to the library and research the development of a breed of livestock. Include in your report the place where the animals originated, the traits for which they were selected, and changes in the breed over the years.
3. Visit with a producer in your area, and determine the traits the producer selects for when he or she selects replacement animals.
4. Visit a packing plant and determine the traits the packer likes in the animals that he or she buys for slaughter.

CHAPTER 18

Animal Systems

STUDENT OBJECTIVES IN BASIC SCIENCE

As a result of studying this chapter, you should be able to:

- discuss the interdependency of the systems of an animal's body.
- name the major animal systems.
- discuss the major parts and functions of the skeletal system.
- discuss the major parts and functions of the muscular system.
- discuss the major parts and functions of the digestive system.
- discuss the major parts and functions of the respiratory system.
- discuss the major parts and functions of the circulatory system.
- discuss the major parts and functions of the nervous system.
- discuss the major parts and functions of the endocrine system.

STUDENT OBJECTIVES IN AGRICULTURAL SCIENCE

As a result of studying this chapter, you should be able to:

- explain why a basic understanding of animal systems is important to producers.
- describe the important aspects of the heart and lung system important to producers.
- describe the important aspects of the muscular system important to producers.
- describe the important aspects of the endocrine system important to producers.
- describe the important aspects of the digestive system important to producers.

KEY TERMS

cartilage
periosteum
vertebrae
synovial fluid
myofibrils
striate
tendons
myoglobin
monogastric
ruminant
plasma
hormones

HAVE you ever wondered what makes a car run? After all, it is composed mostly of steel and plastic, and certainly no magic is involved. What is so mysterious about how the machine operates? The secret lies in the different systems of the car that make use of the laws of nature. Within each automobile is an electrical system, a fuel system, a chassis that supports the car, an intake and exhaust system, and a gear system. All of these systems go together to make a complete automobile. Each system serves a distinct purpose, yet each is dependent on the others. If one fails, the vehicle will not function properly or may not function at all. Animals are much like cars in that they are composed of a variety of interdependent systems that perform separate functions. No one system can function entirely on its own, and each depends on the other systems for support. The animal systems include the skeletal, muscular, digestive, respiratory, endocrine, nervous, and reproductive systems. The reproductive system will be discussed in Chapter 19.

Producers of agricultural animals pay close attention to the systems of the animals they grow. In order to keep animals healthy, contented, and growing efficiently, producers have to make sure that all of the systems of their animals' bodies function properly (**Figure 18–1**). This is done through the selection process, the formulation of feed, a medication program, and management practices.

THE SKELETAL SYSTEM

The skeletal system of an animal provides the frame and support for all of the other systems and organs (**Figure 18–2**). This system is made up of bones, connective tissue, and cartilage. **Cartilage** is composed of firm tissue that is not as hard as bone and is somewhat flexible. The skeletal systems of immature animals contain a larger proportion of cartilage than mature animals. As bones grow, the cartilage portions of the bone harden into bone, and the bone increases in length. As an animal matures, cartilage in some parts of the skeleton continue to turn to bone. One method to determine the age of a slaughtered animal is to examine the amount of cartilage that has turned to bone in the ribs and backbone of the carcass. Other cartilage, such as that in the ears and nose, remains as cartilage throughout the animal's life.

FIGURE 18–1 Animal producers have to make sure their animals are healthy so the animals' systems will operate properly. *© Monkey Business Images/Shutterstock.com.*

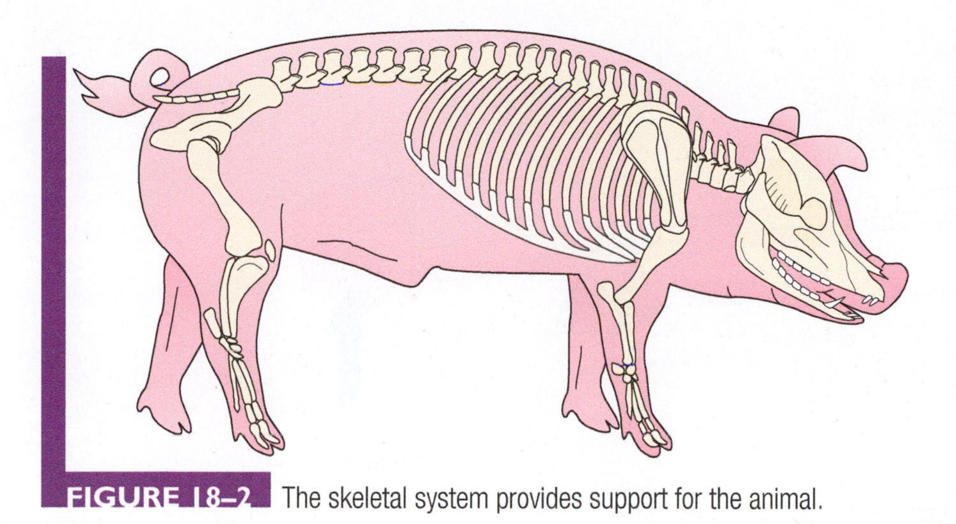

FIGURE 18–2 The skeletal system provides support for the animal.

Bones grow and develop in much the same manner as other tissues in the body. However, they are structured differently. Several types of bones are in an animal's body. These are classified according to the purpose the bone serves.

Long Bones

Long bones are, as the name implies, the longest in the animal's body. They support the body by giving it the rigidity necessary to stand and move. The large bones in the legs provide locomotion for the animal. The long bones act as levers that aid animals in moving about (**Figure 18–4**). A lever is a simple machine that consists of a bar that is used to pry against a load. The use of a lever magnifies the amount of force exerted. The longer the lever, the greater the force that is exerted. Think of a racehorse; an animal with long legs

BONES

The mature size of an animal is determined by the size and length of the bones in its body. Bones provide a place to attach the muscles and a means of movement. They also protect the internal organs of the body and are a key storage area for the body's mineral supply. The hollow portions of the bones serve as sites for blood formation.

Bones are covered with an outer layer known as the **periosteum** that cushions the hard portion of the bone and aids in the repair of a broken bone. Beneath this covering is a layer of hard mineral matter, called the compact bone, that is made of living cells, minerals, and protein (**Figure 18–3**). Most of the hard mineral matter is made up of calcium phosphate and calcium carbonate. This layer gives bones their strength. Inside the hard outer layer is a spongy-appearing structure called the spongy bone. The ends of bones are filled with this material, and the hollow portions of the bones are lined with it. The cavities of the spongy bone are filled with a soft substance called red marrow. This substance is responsible for the formation of blood cells. Inside the hollow portion of the bone is another type of marrow called yellow marrow. This marrow is yellow because it is made up of fat storage cells. This serves as an energy storage area for the animal.

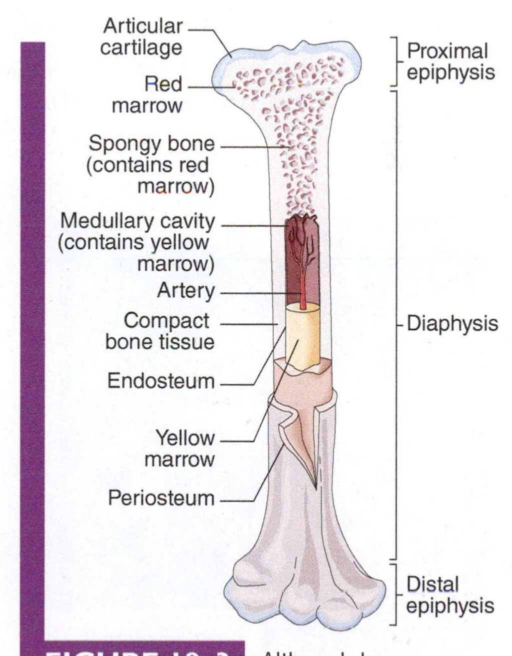

FIGURE 18–3 Although bones are mostly composed of hard mineral material, they are living tissue.

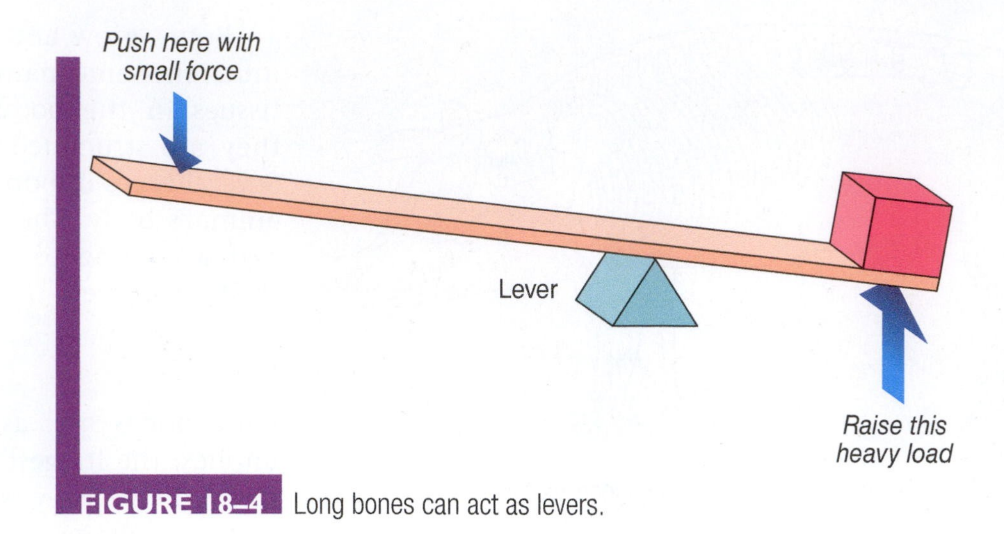

FIGURE 18–4 Long bones can act as levers.

and long bones in the hip region has a distinct advantage over horses with shorter bones. The greater leverage of the longer bones allows the horse to have more power as it propels itself forward (**Figure 18–5**).

Producers often select animals based on the size and length of the long bones. Research has shown that the length of the cannon bone (the bone between the ankle or pastern and the knee) of young animals is a good predictor of how tall the animal will be at maturity. Charts have been developed to help producers determine the mature size of cattle based on the hip height of the animal at a certain age. Obviously the hip height is directly proportional to the length of the long bones in the legs (**Figure 18–6**).

FIGURE 18–5 A greater length of the long bones gives a horse an advantage in power and speed.

Pelvic Bones

The pelvic bones, which are modified long bones, are important to producers in the selection of breeding stock. Producers want females that can give birth easily, and the size and shape of the pelvic bones can be a significant indication. For example, cows that have long pelvic bones (the distance between the hooks and pins) usually have greater calving ease because the birth canal can open wider (**Figure 18–7**).

Ribs

The ribs, which are another group of modified long bones, are an important point of selection. The ribs form the thoracic cavity and protect internal organs such as the heart and lungs. They are long, relatively flat bones that attach to the backbone on one end and to the sternum (the bone on the chest floor) at the other end. Producers know that animals with larger rib cages usually have larger hearts and lungs than animals with smaller rib cages (**Figure 18–8**). A widely sprung rib cage and wide floor to the thoracic cavity indicate an animal that is vigorous and will grow efficiently.

(A) (B)

FIGURE 18–6 (A) In the 1950s, cattle were selected for short hip height. (B) Today, cattle are selected for a taller hip height.

FIGURE 18–7 Cows with long pelvic bones usually have greater calving ease.

© uzuri/Shutterstock.com.

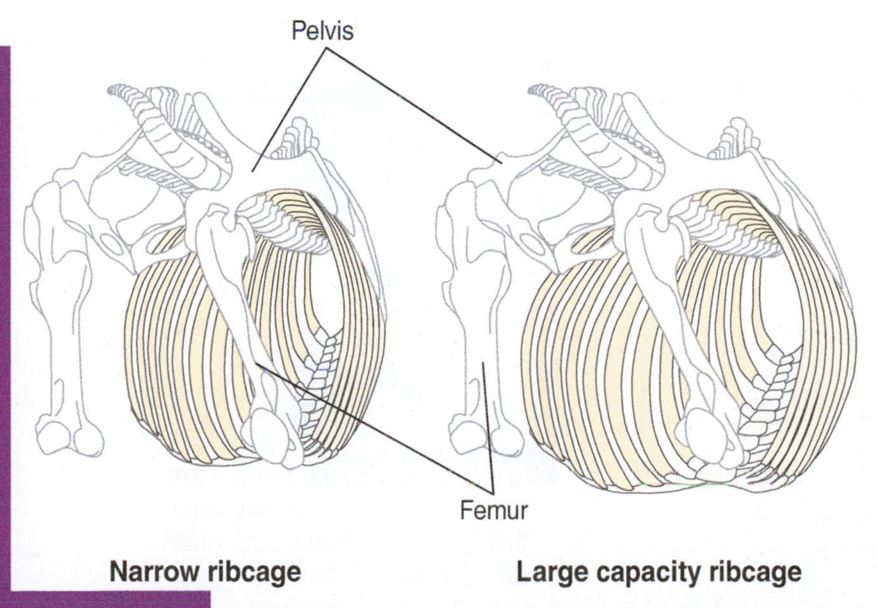

FIGURE 18–8 A widely sprung rib cage provides more room for the internal organs.

Short Bones

The short bones are smaller than the long bones and serve a different purpose. These bones may be about as large around as they are long. Most short bones are found in the joints and serve as hinges that help the joints to be flexible (**Figure 18–9**). Also, they help cushion shock and protect the long bones by being flexible or "giving" before the long bones are injured. Strong joints are important to the producer because the comfort and mobility of animals depend on the joints that hold the bones together. A thorough discussion of joints will be provided later in the chapter.

Irregular Bones

Irregular bones have an irregular shape. Their function is support and protection. Most agricultural animals are vertebrates. This means that they have a vertebra or backbone running the length of their bodies. Despite its name, the backbone is not a single bone, but rather a series of irregular bones called **vertebrae**. The bones begin behind the head and continue like a chain to the end of the animal's tail. These bones provide an attachment either directly or indirectly for all of the limbs and other bones. They flex and bend to give the animal movement.

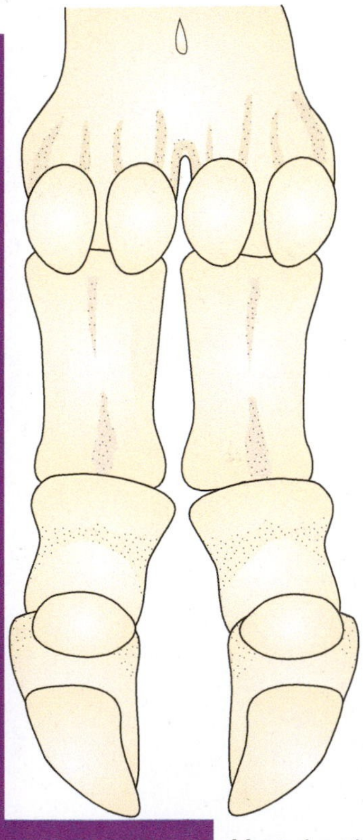

FIGURE 18–9 Most short bones are found in joints and help the joint to be flexible.

In the center of the vertebrae is a channel through which the spinal cord runs (**Figure 18–10**). This channel contains soft tissue and fluid that protects the cord from injury.

The vertebrae are divided into several different areas (**Figure 18–11**). The region of the neck from the skull to the first rib is called the cervical region. Almost all agricultural animals (mammals) have seven cervical vertebrae. These bones support and allow movement of the head and neck.

The thoracic region extends along the rib cage. Each vertebra has a rib attached to each side, although sometimes ribs at the front or rear of the rib cage have no attachments to the backbone. Cattle have 13 thoracic vertebrae, horses have 18, and pigs have either 13 or 14 depending on the breed.

The area of the spinal column from the last rib to the pelvis is called the lumbar region or the loin. If you have judged sheep, you

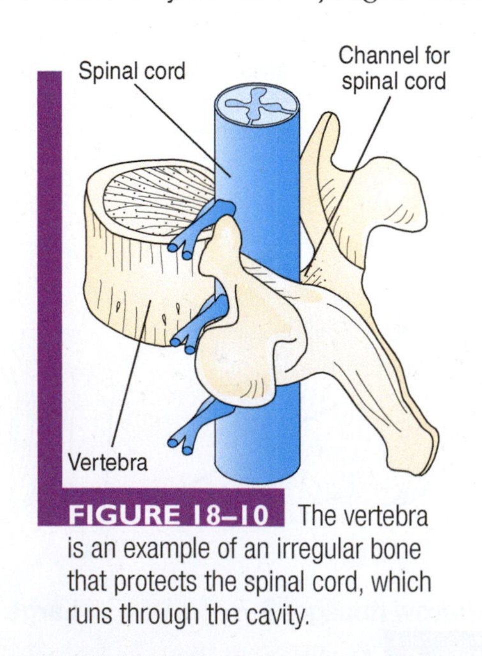

FIGURE 18–10 The vertebra is an example of an irregular bone that protects the spinal cord, which runs through the cavity.

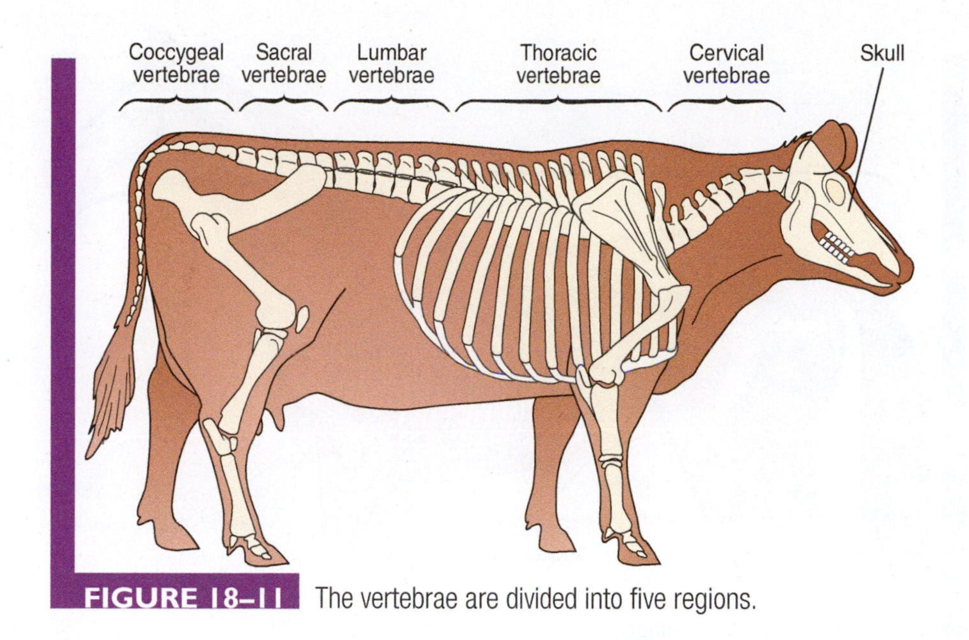

FIGURE 18–11 The vertebrae are divided into five regions.

probably have measured this area with your hand. Animals that have long lumbar vertebrae are more desirable because this is where some of the most expensive cuts of meat come from.

The sacral region of the vertebrae extends through the pelvic area. The bones in this region are usually fused together to form a rigid section called the sacrum. This structure provides an attachment for the pelvis.

The vertebrae continue to the end of the tail of the animal. This section is called the coccygeal region. In most animals the bones at the top of the tail are larger than the bones at the end. The size tapers to the end where the bones have no channel.

Flat Bones

The flat bones are relatively thin and flat and protect organs. The bones of the head are flat bones (**Figure 18–12**). They encase the brain and protect this delicate organ. The flat bones of the skull have openings through which the animal takes in air and nourishment. The openings or passageways are called sinuses. The ears, nose, eyes, and mouth all have openings in the skull.

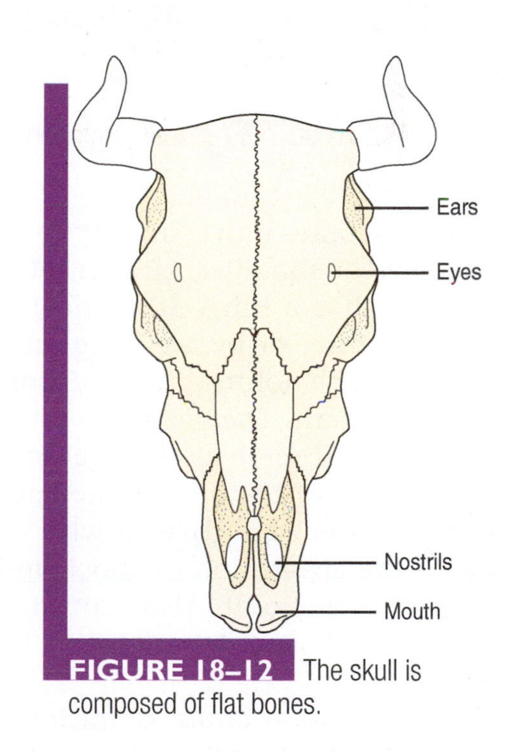

FIGURE 18–12 The skull is composed of flat bones.

Joints

All of the bones in an animal's body are connected to make up the skeletal system (**Figure 18–13**). They are held together by bands of tough tissue called ligaments that bind the bones at the joints. Some joints, like the bones of the skull, do not move. Others,

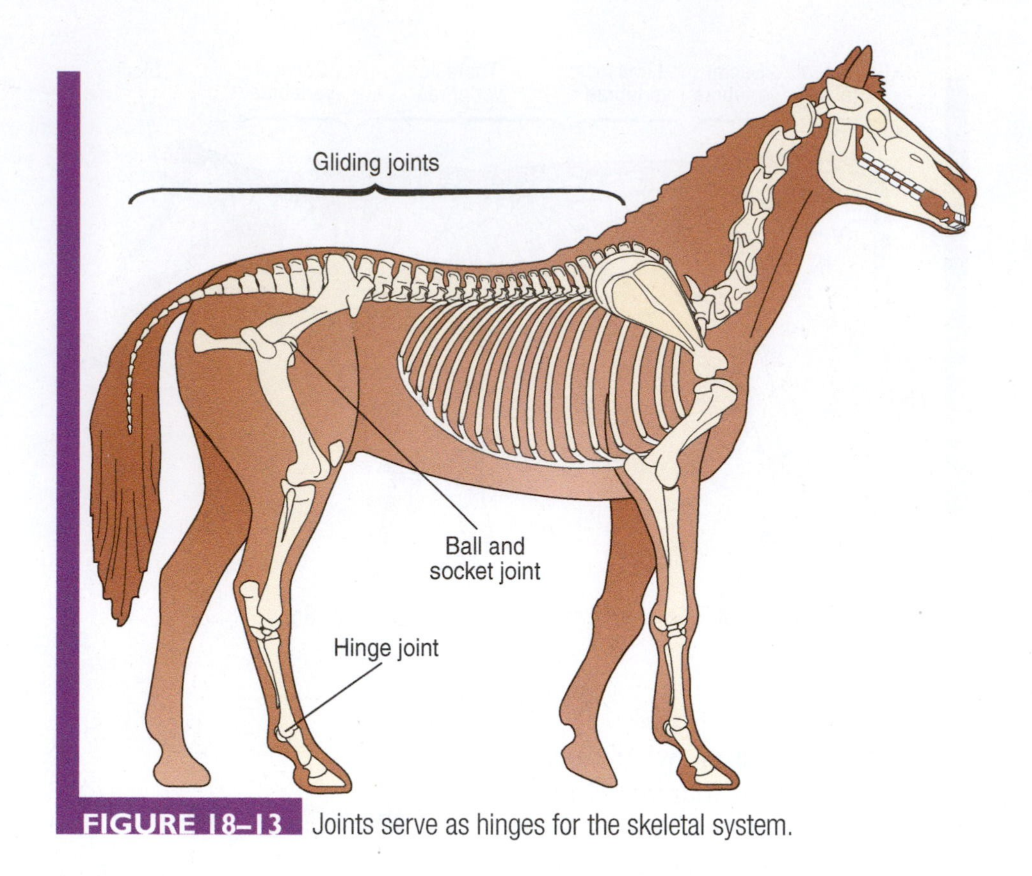

FIGURE 18–13 Joints serve as hinges for the skeletal system.

such as the vertebrae, move slightly; others, such as the knee or shoulder, move a lot. The joints can move like a hinge (the knee), as a ball and socket (hip), or by a gliding motion (the vertebrae). The joints give the animal freedom of movement. The ligaments give the joints flexibility and serve as shock absorbers to protect the ends of the bones. The ends of the bones in the joint are covered with cartilage to aid in the absorption of shock and in the lubrication of the joint. Also, within each joint is a pocket of **synovial fluid** that lubricates the joints.

Producers pay close attention to the joints of animals they select for breeding (**Figure 18–14**). Strong joints can help ensure that the animal moves about freely and is capable of mating. This is particularly true of animals raised on concrete floors. The legs must be set properly under the animal for the animal to be able to move freely and efficiently. If the joints are set at an improper angle, the animal will have problems walking and will not be as productive as animals with proper feet, legs, and joints.

THE MUSCULAR SYSTEM

The skeletal system supports the animal and allows movement. However, the skeleton cannot move without the muscular system. Throughout the animal's body, a complex system of muscles provides the means for the animal to move about and for the proper functioning of the organs.

Muscles are a major component of an animal's body. For example, muscle may make up 35 to 40 percent of a beef animal's total body weight. Producers of beef and other meat animals are in the business of producing muscles. It is the muscles of animals that are processed

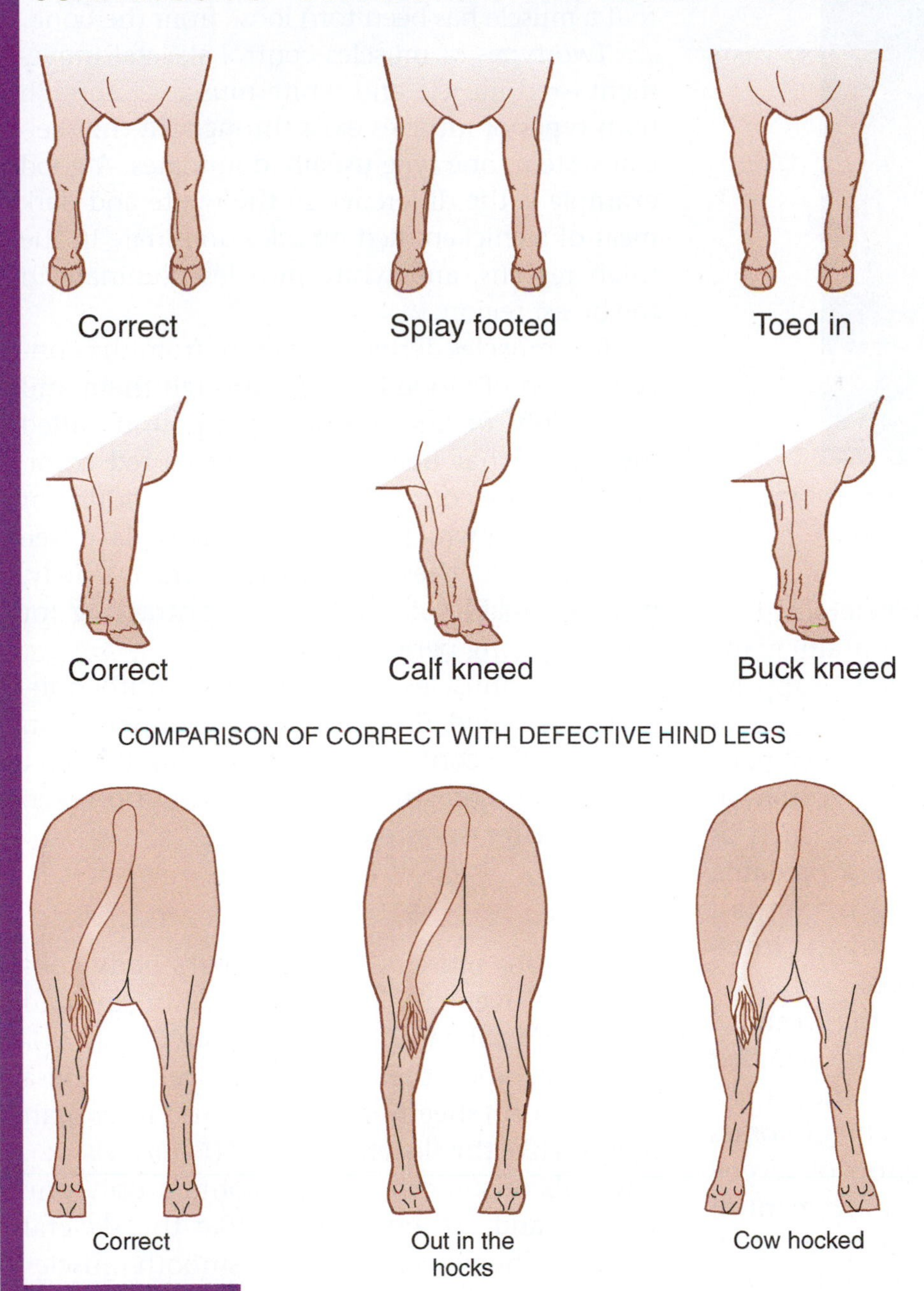

FIGURE 18–14 For breeding animals, producers select animals without irregularities in the joints.

into the meat that occupies such an important part of our diet (**Figure 18–15**).

There are basically three types of muscles: skeletal, smooth, and cardiac.

Skeletal Muscle

Skeletal muscles make up the largest portion of the muscles. These muscles are long bundles of fibrous tissue called **myofibrils**. (*Myos* means

FIGURE 18–15 Muscles are processed into meat. © Alaettin YILDIRIM/Shutterstock.com.

muscle, and *fibrilla* means fibers.) Skeletal muscle cells are long and narrow and contain many nuclei. The cells are **striate**, or striped, in appearance. Their function is to provide movement for the bones of the skeleton and other parts of the body. These muscles cause movement in response to particular circumstances, such as moving toward food or away from a predator. They are called voluntary muscles, in contrast to involuntary muscles, like the heart and intestines, that maintain basic functions.

Movement comes about by the contraction of muscles. In other words, the muscle that causes movement becomes shorter. A muscle seldom works by itself. Usually movement comes about through the coordinated effort of several muscles. As an animal walks, the movement of a leg is not controlled by a single muscle but by a group of muscles that surround the bones of the leg. These muscles work opposite each other: When one muscle contracts, another relaxes, and vice versa. For example, when the leg is extended forward, the muscles on the top of the leg contract, and the muscles on the bottom of the leg relax. When the leg is drawn back, the opposite occurs—the muscles on the bottom of the leg contract, and the muscles on the top relax.

The muscles move the bones through connective tissue called **tendons**. This tough tissue binds the muscle to the bones. A torn tendon means that a muscle has been torn loose from the bone.

Two types of muscles control skeletal movement—red muscle and white muscle. Although both types of muscles exist throughout the skeletal system, one type usually dominates. A good example is the difference in the white and dark meat of a chicken. Red muscles dominate in the thigh regions, and white muscles dominate in the breast region.

Red muscles derive their color from the concentration of blood flowing through them and the supply of an iron-rich compound called **myoglobin** that helps give blood its red color. Red muscles contain many mitochondria, or organelles in which respiration takes place (see Chapter 15). These muscles contract slowly, but are capable of continually contracting for relatively long periods of time.

White muscles contain fewer mitochondria, less blood flow, and a lower myoglobin content. By contrast, these muscles contract faster and are stronger. They do, however, fatigue faster than red muscles.

Smooth Muscle

The smooth muscles of an animal's body control the movements of the internal organs and are found in the walls of the digestive tract, urinary tract, and other organs. The cells of these muscles form sheets of muscle tissue rather than bundles like the skeletal muscles (**Figure 18–16**). The cells of smooth muscles contain only one nucleus and are not striated like the skeletal muscles. The movement of the smooth muscles is called involuntary because the processes of the internal organs occur automatically.

Cardiac Muscle

The third type of muscle is cardiac muscle. As the name implies, these muscles control the heart. In fact, most of the heart is made of this muscle. Cardiac muscle has some characteristics of both skeletal and smooth muscle. Like the skeletal muscle, its cells are striated and are arranged in

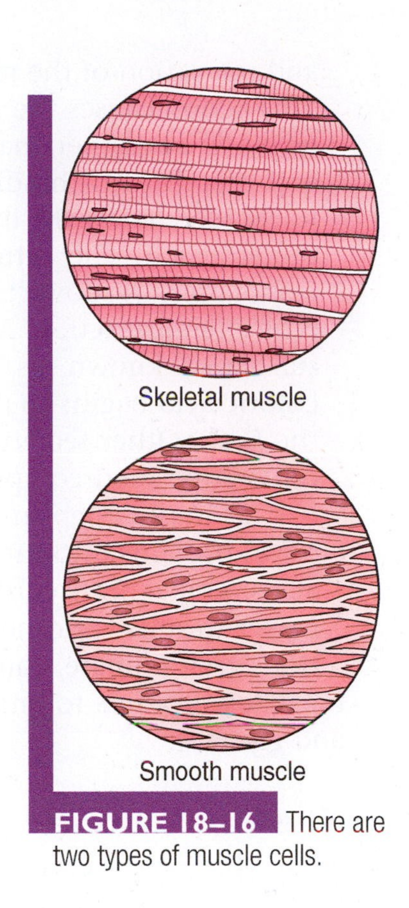

FIGURE 18–16 There are two types of muscle cells.

bands. Like smooth muscle, cardiac muscle has only one nucleus and is operated automatically. These muscles have amazing stamina; they must act almost continuously from before the animal is born until it dies.

THE DIGESTIVE SYSTEM

For the other systems to function, energy must be supplied. Energy comes from the food taken in or ingested by the animal. The digestive system takes the food ingested by the animal and converts it into a form that the animal can use (Figure 18–17). This conversion involves breaking down the food into components that can be absorbed and used by the cells of the animal. Basically there are two types of digestive systems in agricultural animals, **monogastric** and **ruminant** systems.

FIGURE 18–17 The digestive system takes food ingested by the animal and converts it to a form the animal can use. *© daseaford/Shutterstock.com.*

Monogastric Systems

Monogastric systems are often referred to as simple stomach systems. This means that animals with this system have only one compartment in their stomachs. These include the pig, horse, dog, cat, and birds (Figure 18–18). The horse is different from the other monogastrics in that it has an enlargement known as a cecum that enables it to utilize high-fiber feeds by means of microbial fermentation, much as do ruminants (Figure 18–19). Simple stomach animals cannot digest large amounts of fiber like ruminants. Feed for monogastrics is referred to as a "concentrate" because of the relatively high concentration of nutrients and the low level of fiber.

The Mouth

The mouth is the organ that begins the digestive process. The tongue is used for grasping the food, mixing, and swallowing. The teeth tear and chew the feed into smaller particles that can be swallowed and further broken down. The mouth contains salivary glands, three pairs

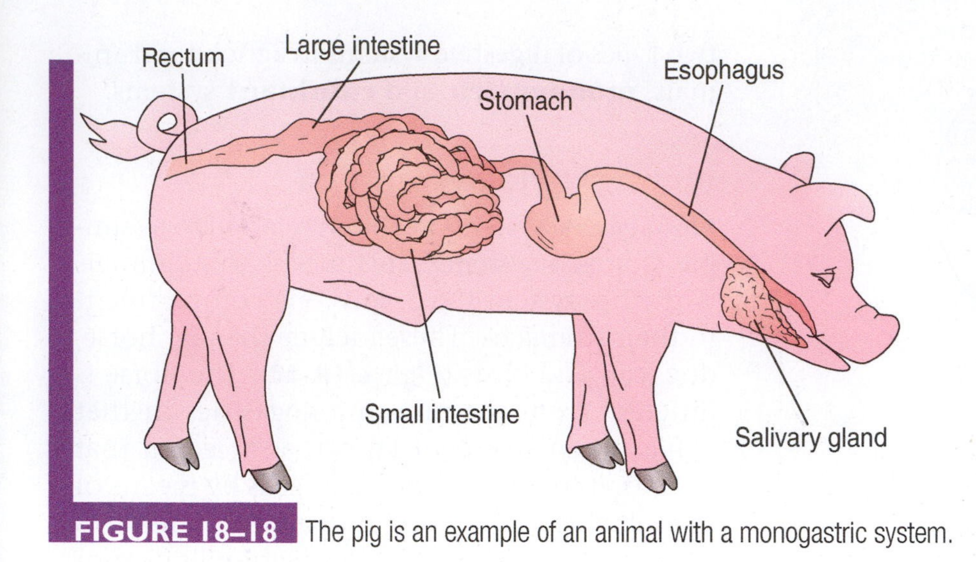

FIGURE 18–18 The pig is an example of an animal with a monogastric system.

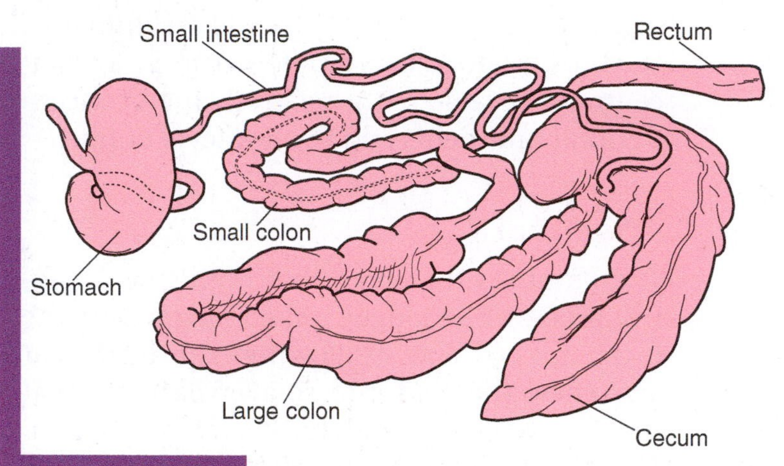

FIGURE 18–19 The horse has a pouch known as the cecum that allows it to digest roughage.

of glands that excrete saliva. Saliva contains several substances and serves several purposes: water to moisten, mucin to lubricate, bicarbonates to buffer acids in the feeds, and the enzyme amylase to initiate carbohydrate breakdown.

The hollow muscular tube that leads from the mouth to the opening of the stomach is the esophagus. It moves food from the mouth to the stomach. This is accomplished by muscles within the esophagus that contract and push the food down.

The Stomach

The stomach is a hollow muscle that causes further breakdown of foods by the contraction and relaxation of the muscles. This action causes the food to be pressed together, massaged, and mixed with the digestive juices secreted by the lining in the stomach. The actual process of breaking down the food is by chemical action. A strong substance known as hydrochloric acid begins to dissolve the food. Other secretions act on specific food components. For example, pepsin breaks down proteins into the amino acids, and rennin curdles the casein in milk. Another secretion, gastric lipase, causes the breakdown of fats to fatty acids and glycerol.

The Small Intestine

After the food is sufficiently broken down by muscular contractions and chemical reactions in the stomach, it moves into the next organ in the system, the small intestine. The entrance to the small intestine is controlled by a sphincter muscle that helps move food into and through the tract. This digestive organ consists of a long hollow tube that leads from the stomach to the large intestine. The small intestine is made up of several segments—the duodenum, the jejunum, and the ileum. Food passes through each of these segments and processes occur in all parts.

The duodenum is the first segment of the small intestine. This section uses secretions from the pancreas to break down proteins, starches, and fats. The intestinal walls also secrete intestinal juices that contain enzymes that continue the process of breaking down the food.

After the food leaves the duodenum it enters the next segments of the small intestine, known as the jejunum and the ileum.

These are the areas where nutrient absorption takes place. Absorption is the process by which the nutrients are passed from the intestine into the bloodstream. The walls of these sections are lined with small fingerlike projections called villi that absorb the food nutrients into the bloodstream and lymph system through membranes that surround the villi (**Figure 18–20**). These are what is known as semipermeable membranes. This means that the membranes allow particles to pass through in a process called diffusion.

The Large Intestine

The large intestine is the last organ of the digestive tract. The first section of this organ is a blind pouch called the cecum. Although the cecum has little purpose in most monogastric animals, it serves a very important function in animals such as the horse. It is in this area that fibrous food, such as hay and grass, is broken down into usable nutrients.

The second segment of the large intestine, called the colon, provides a storage space for the waste from the digestive process and is the largest part of the organ. Water is removed from waste, and the process of decomposition of fibrous materials begins through microbial action.

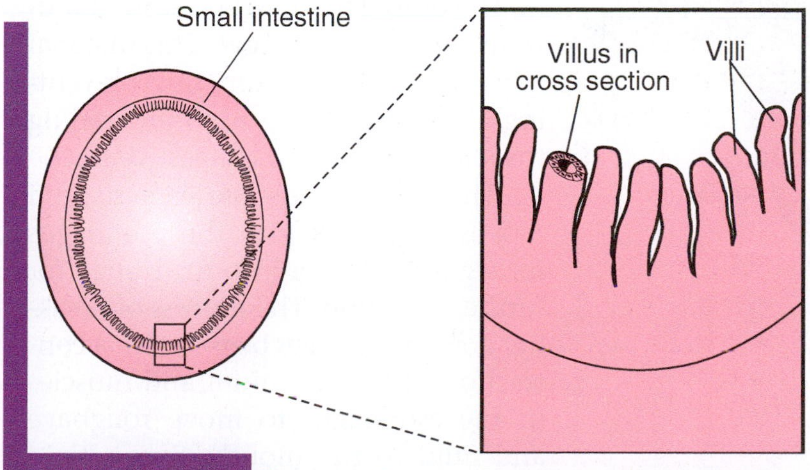

FIGURE 18–20 The walls of the small intestine are lined with villi that absorb food nutrients.

The terminal end of the large intestine and the entire digestive system is the rectum. It serves to pass waste material through to the anus, where it is finally eliminated from the body.

RUMINANT SYSTEMS

A large group of animals including cattle, goats, and sheep eat large quantities of fibrous material such as grass and hay (**Figure 18–21**). This type of feed is called roughage. The reason these animals are able to digest all the fiber is that they have multicompartment stomachs that break these materials down into forms that are usable by the body. These animals are often called "cud chewers" because they regurgitate chunks of feed called boluses, masticate (chew), and reswallow. If you observe cows lying in a pasture you will often see them chewing on their "cud." The digestive systems of **ruminants** differ from monogastric systems in several ways.

The beginning of the system, the mouth, is different in ruminants because there are no upper front teeth. Instead a dental pad works in concert with the lower front teeth (incisors) in tearing off forages and other foodstuffs (**Figure 18–22**). The forage is then chewed between the upper and lower jaw teeth (molars). It is necessary that the mouths of ruminants produce large quantities of saliva to begin the process. This saliva is highly buffered (has a high pH) and contains phosphorus and sodium that aid the microorganisms that live in the rumen. Unlike most monogastric animals, there are no enzymes in the saliva, although some urea is released. This substance has a high level of nitrogen that is used by the bacteria in the rumen.

Ruminants are often said to have four stomachs. This is incorrect: They have only one stomach,

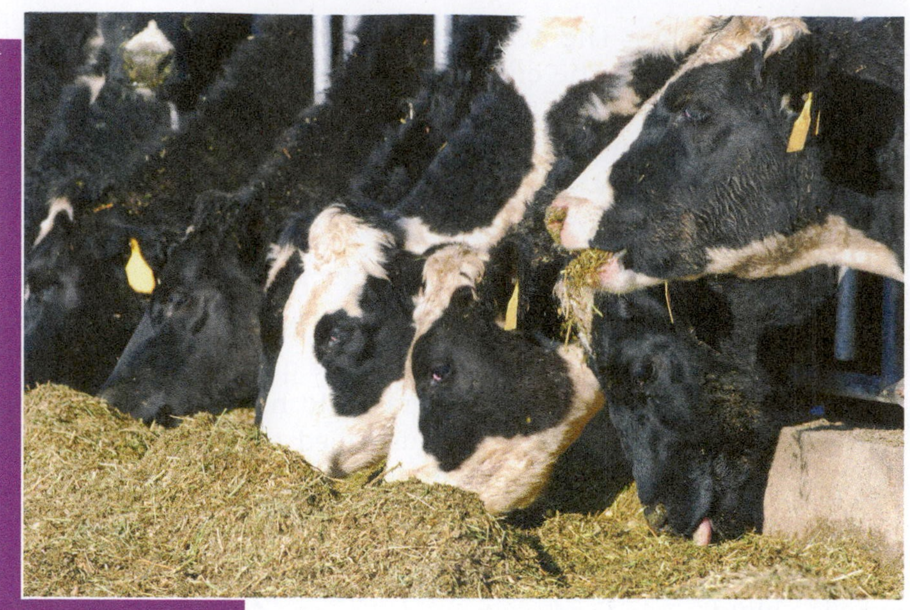

FIGURE 18–21 Ruminants are able to digest large amounts of fibrous materials.
© Diane Garcia/Shutterstock.com.

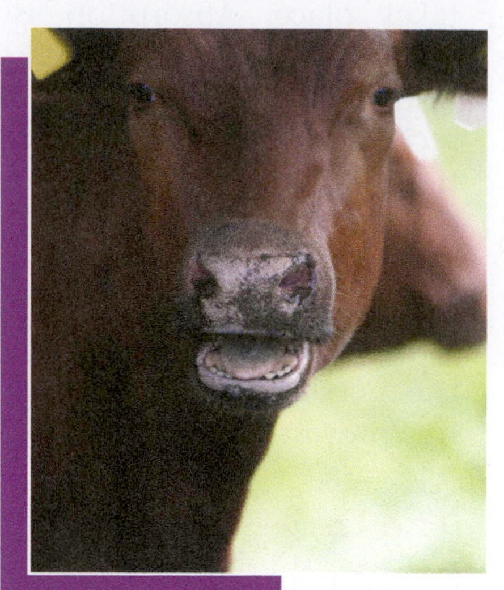

FIGURE 18–22 Ruminants have a dental pad that helps tear off forage.
© Gemenacom/Shutterstock.com.

but it is divided into four separate and distinct compartments (**Figure 18–23**). The four compartments are the reticulum (honeycomb), rumen (paunch), omasum (many plies), and the abomasum (true or glandular stomach). In addition, the young ruminant animal has a structure called an esophageal groove or heavy muscular folds that allows milk from the mother to bypass the rumen and reticulum to go directly to the abomasum.

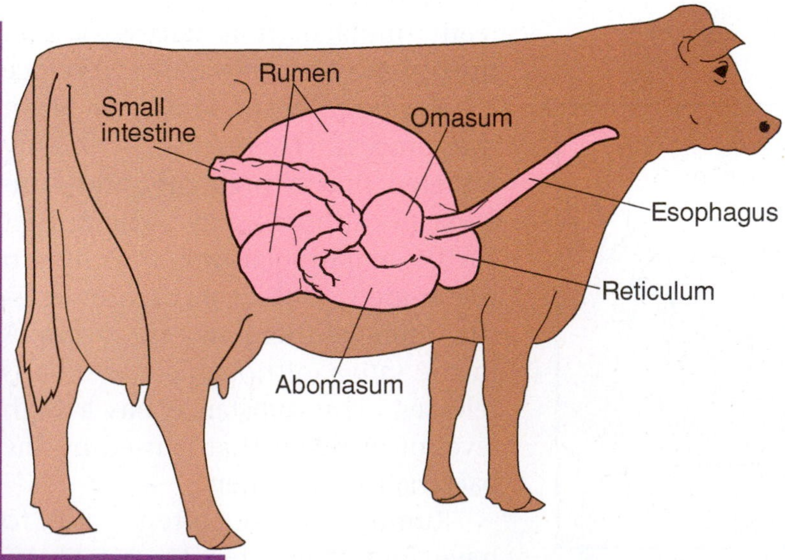

FIGURE 18–23 The ruminant digestive system has several compartments that perform specific purposes.

The Reticulum

From the mouth, the esophagus leads to both the reticulum and the rumen. Ruminants sometimes pick up objects such as nails and small stones that are not digestible and that might harm the digestive system. These materials are heavier than most of the material swallowed, and they fall into the reticulum. The inside of the reticulum is lined with tissues called mucous membranes. Mucous membranes secrete a viscous, watery substance called mucus. These membranes form subcompartments that have the appearance of honeycomb. They trap and store "hardware" that does not float. This material remains in the reticulum, thus preventing dangerous objects from proceeding through the rest of the digestive tract.

The reticulum also stores, sorts, and moves feed back into the esophagus for regurgitation or into the rumen for further digestion. The process of breaking down roughages begins with a contraction of the reticulum and muscles in the esophagus to move roughage and fluid to the mouth. Excess fluid is squeezed out, and the material is reswallowed.

The food next moves to the rumen, which serves as a storage vat where food is soaked, mixed, and fermented by bacteria. These bacteria live in the rumen in a symbiotic relationship with the ruminant. This means that one organism lives in another, and both of the organisms benefit from the relationship. Bacteria thrive in the rumen environment and break down fibrous feeds for the ruminant.

The Rumen

The rumen is a hollow muscular paunch that occupies the left side of the abdominal cavity and contains two sacs, each lined with papillae (fingerlike projections) that aid in the absorption of nutrients. Here, carbohydrates are broken into starches and sugars. Volatile fatty acids are released as the carbohydrates are broken down, and these fatty acids are absorbed through the rumen wall to provide body energy.

Bacteria in the rumen also use nitrogen to form amino acids, and the amino acids form proteins. The bacteria can also synthesize water-soluble vitamins and vitamin K.

The Omasum

From the rumen, the food material passes through to the omasum. The omasum is a round organ located on the right side of the animal and to the right of the rumen and reticulum. The walls of the omasum contain many folds that are lined with blunt muscular papillae that grind roughage.

The Abomasum

After leaving the omasum, the food moves into the last compartment of the ruminant's stomach, the abomasum. The abomasum is the only glandular (true stomach) stomach of the ruminant. The abomasum is located below the omasum and extends to the rear and to the right of the rumen. By the time food materials reach the abomasum, the fibers of the roughages have been broken down to the extent that they can be handled by the abomasum. The abomasum and the small and large intestines of the ruminant animal function much the same way as they do in monogastric animals.

THE RESPIRATORY SYSTEM

For bodily processes to take place and for the animal to live, oxygen must be taken into the systems for their use. The respiratory system takes oxygen from the air and places it into the bloodstream for distribution to the cells of the animal's body.

The respiratory system begins with the nostrils, which are openings on the face of the animal. Large amounts of air are brought in through these openings. One of the selection criteria in racehorses is that they have large, broad nostrils that can take in sufficient air when the animal is running (**Figure 18–24**). Air is also brought in through the mouth. The nostrils open into the

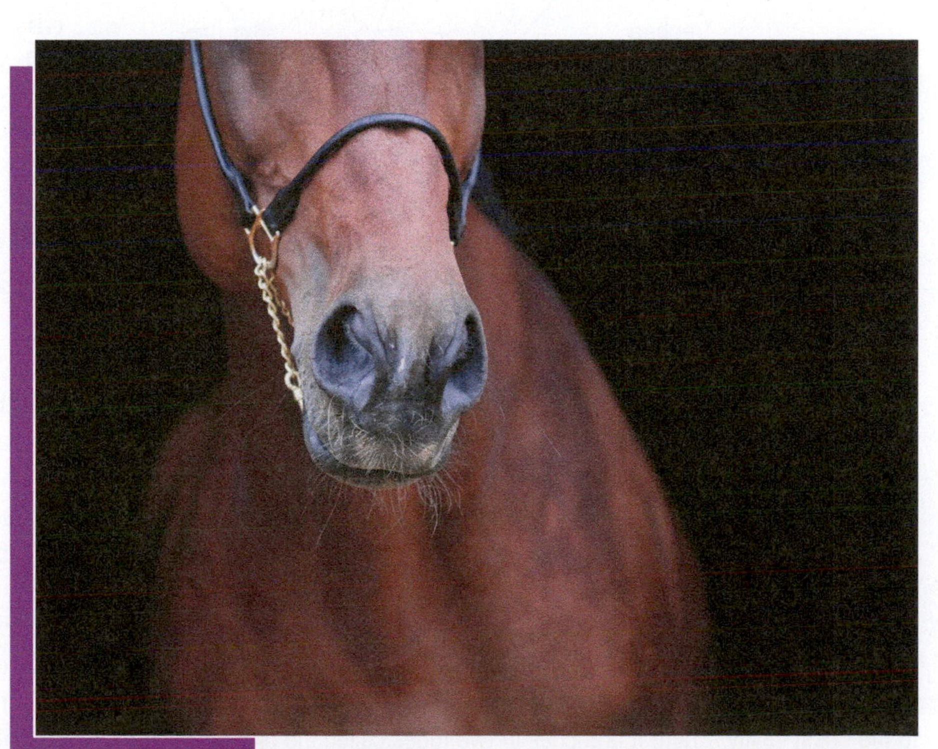

FIGURE 18–24 One selection criterion for racehorses is large nostrils that can take in sufficient amounts of air when the animal is running. *© Melory/Shutterstock.com.*

nasal cavities inside the skull, and these cavities lead into the pharynx, which serves as a common passageway for the food, water, and air that the animal ingests. The opening is controlled by a valvelike structure called the epiglottis that closes when the animal swallows. This prevents water and food from entering the respiratory system. If these materials accidentally enter the airway, the animal coughs and expels it back into the pharynx. The pharynx also provides a junction for the esophagus and the larynx. The larynx, also called the voice box or Adam's apple, controls the voice of the animal and aids in preventing material other than air from entering the lungs. A large tube known as the trachea leads from the larynx to the chest cavity, where it branches out into two tubes called the bronchi (**Figure 18–25**). Both the trachea and the bronchi are made up of ridged rings of cartilage that prevent the tubes from collapsing during breathing.

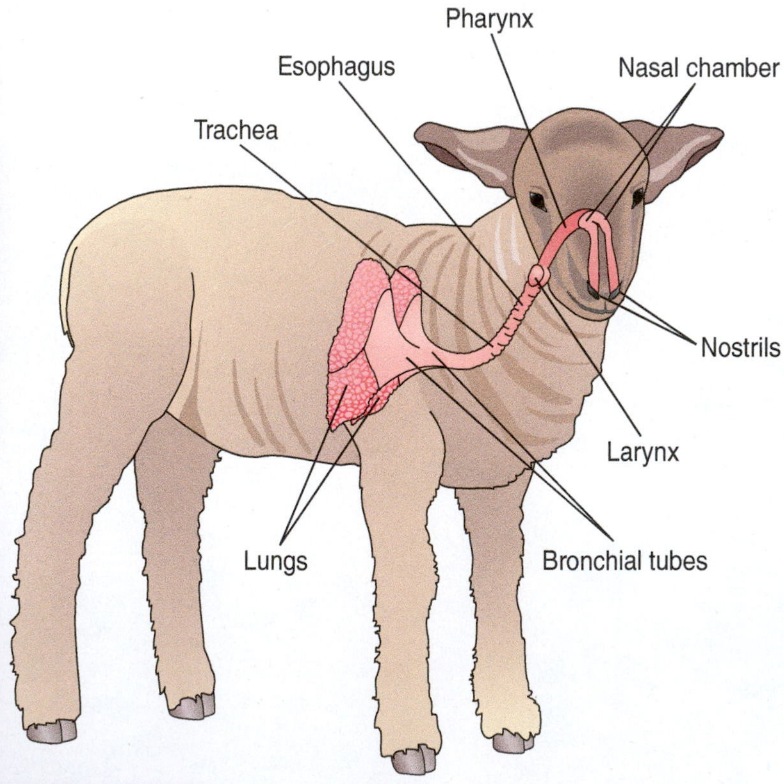

The bronchi branch out, and each branch divides further with each succeeding branch becoming smaller and smaller until they terminate in small sacs called alveoli. These tiny structures are so numerous that their combined surface area would be several times that of the total skin area of the animal.

The alveoli exchange gases with the bloodstream (**Figure 18–26**). The alveoli are surrounded by blood vessels that absorb oxygen from the ducts in the alveoli. Carbon dioxide is taken from the blood through the alveoli and expelled from the lungs. This process takes place through diffusion.

Breathing takes place by muscular contractions of the rib cage and the diaphragm. The diaphragm is a muscular structure that separates the chest cavity from the abdominal cavity. The inside of the chest cavity is a partial vacuum. When the muscles of the diaphragm contract and are drawn downward, the rib cage expands and air is drawn in. When the muscles of the diaphragm relax and return to their normal upward arch, the ribs move inward, and the animal exhales. The process of exhaling expels carbon dioxide out of the lungs.

FIGURE 18–25 A large tube called the trachea leads from the larynx to the chest cavity, where it branches to two tubes called the bronchi.

THE CIRCULATORY SYSTEM

The transportation of food nutrients, water, and oxygen is accomplished through the circulation of blood through the animal's body. Additionally, the bloodstream cleanses the body by carrying toxic materials to the kidneys and sweat glands for excretion. Body temperature is regulated by the circulating blood, and disease agents are removed by cells in the blood.

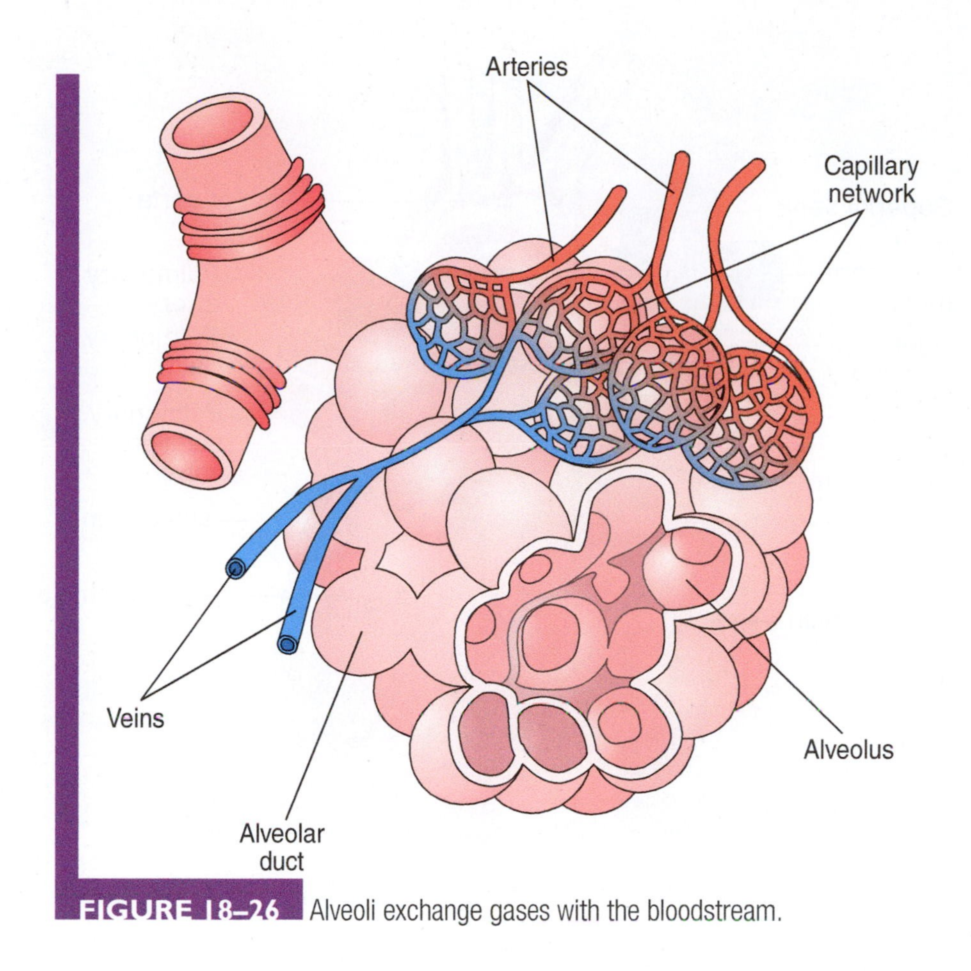

FIGURE 18–26 Alveoli exchange gases with the bloodstream.

The Heart

The center of the circulatory system is the heart. This muscular organ operates continuously to pump blood throughout the animal's body. In fact, several thousands of gallons of blood are pumped each day through the body of a large animal. The heart is divided into four chambers (**Figure 18–27**). The heart is divided lengthwise by a thick wall of muscle called the septum, making right and left chambers. Each of these is further divided into smaller chambers, the atrium and the ventricle. Blood returning from the body collects in the atriums and is pushed into the ventricles through valves that open and close. The sound of the heartbeat is the sound of these valves opening and closing. The walls of the ventricle are composed of thick, strong muscles that contract and force the blood through to the lungs and the other parts of the body. The right side of the heart pumps blood to the lungs through a large vessel called the pulmonary artery, and the left side pumps blood to the body through another large vessel, the aorta. These vessels branch out until they reach all areas of the body.

Arteries

The blood vessels that take the blood from the heart are known as the arteries. Those that return blood to the heart are called veins. Veins and arteries are connected together by tiny, thin-walled vessels called capillaries that deliver the nutrients to the cells from the arteries and take away waste material through the veins (**Figure 18–28**). As the blood passes through the kidneys, the waste material is filtered out and is passed out in the urine. If all of the blood vessels of a large animal's body were laid out end to end they would reach around the world!

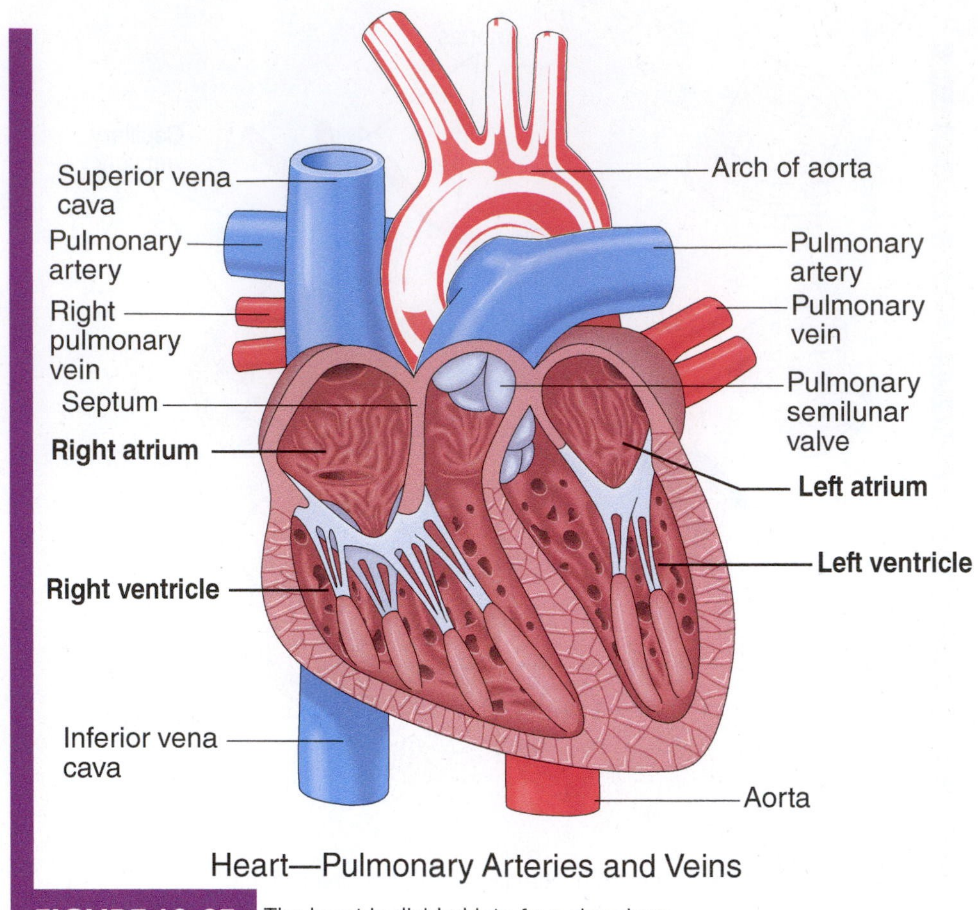

FIGURE 18–27 The heart is divided into four chambers.

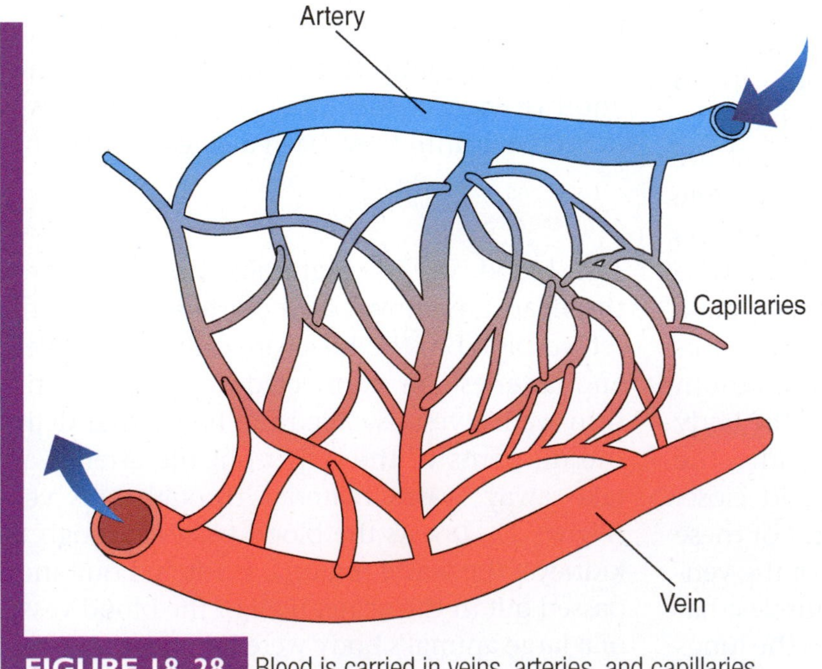

FIGURE 18–28 Blood is carried in veins, arteries, and capillaries.

Blood

Blood is an amazing fluid. More than half of its volume is composed of a tan-colored fluid called **plasma** that suspends several substances that help sustain life. The rest of the blood's volume is made up of three types of blood cells. Red blood cells give the blood its color and are shaped like donuts with holes that are not quite clear through (**Figure 18–29**). These cells are filled with a substance called hemoglobin that is composed of an iron-rich protein. The red blood cells bind with oxygen as they pass through the lungs.

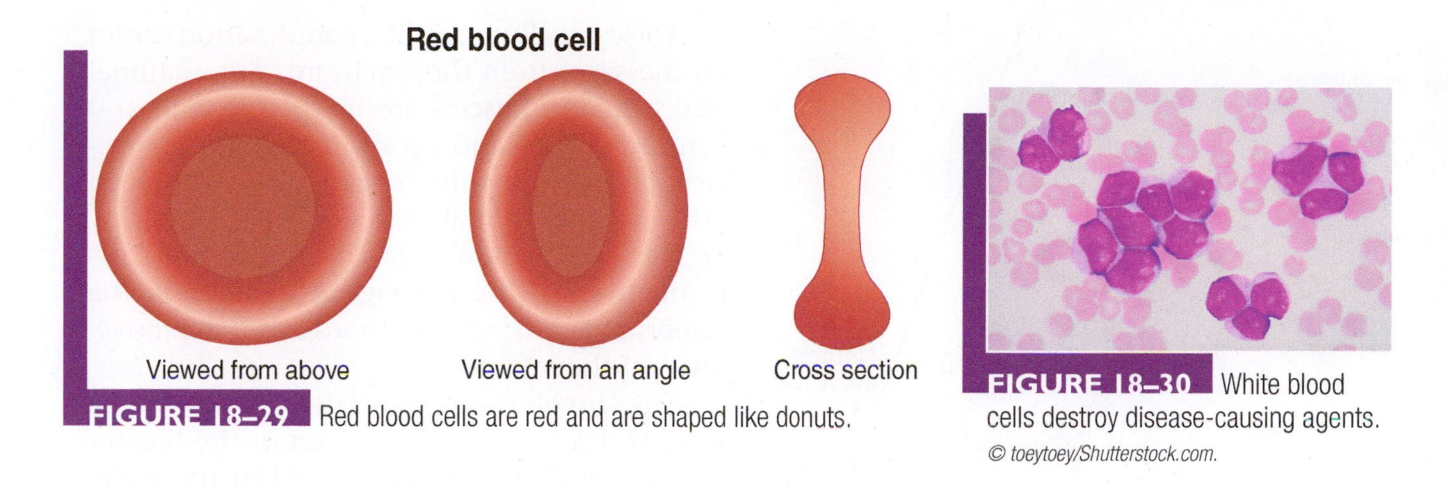

FIGURE 18–29 Red blood cells are red and are shaped like donuts.

FIGURE 18–30 White blood cells destroy disease-causing agents.

White blood cells are actually clear in color and destroy disease-causing agents (**Figure 18–30**). (More will be discussed about this process in Chapter 25.) The third type of cell is the platelets. These are actually only fragments of cells and contain bits of cytoplasm. Their function is to prevent blood loss when a vessel is injured. The platelets release chemicals called clotting factors that "patch" breaks in blood vessels.

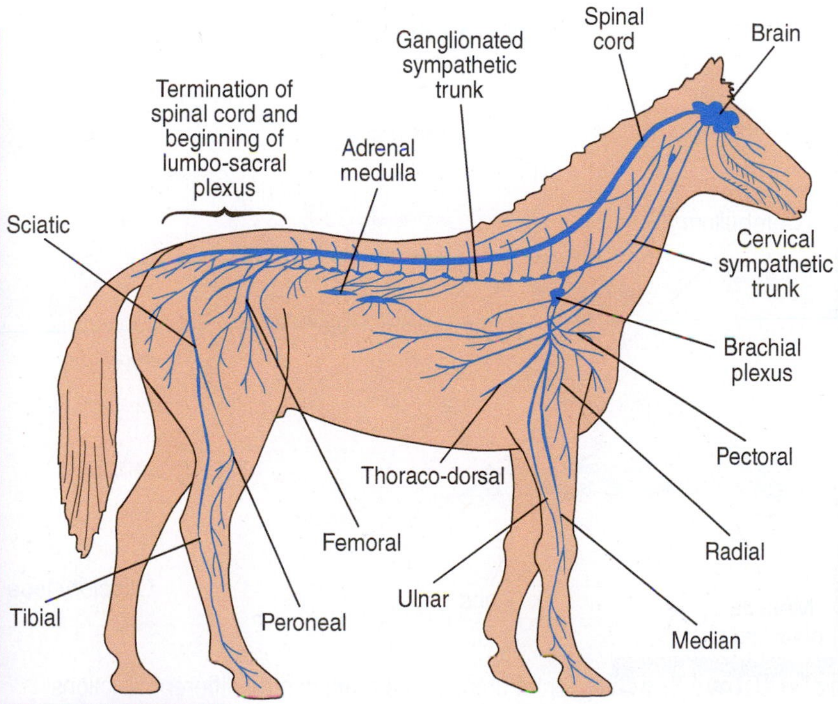

FIGURE 18–31 The nervous system is the control center of the animal.

THE NERVOUS SYSTEM

For all of the systems to function properly, the movements and processes have to be controlled by a central system. This is the function of the nervous system (**Figure 18–31**). The control center for the nervous system, the brain, is an almost incomprehensibly complex organ. The most advanced computers in the world are simple compared to the brain of a higher-order animal. Messages are sent to all parts of the body from the brain through long fiberlike structures called nerves. The nerves that conduct impulses from the brain to other parts of the animal's body are called motor or efferent neurons (**Figure 18–32**). Nerves that send impulses from the body back to the brain are known as sensory or afferent neurons. All nerves are connected directly or indirectly to the spinal cord, which runs through the center of the backbone. Different sections of the spinal cord receive and deliver messages to and from certain parts of the body.

The upper portion of the spinal cord is attached to the brain.

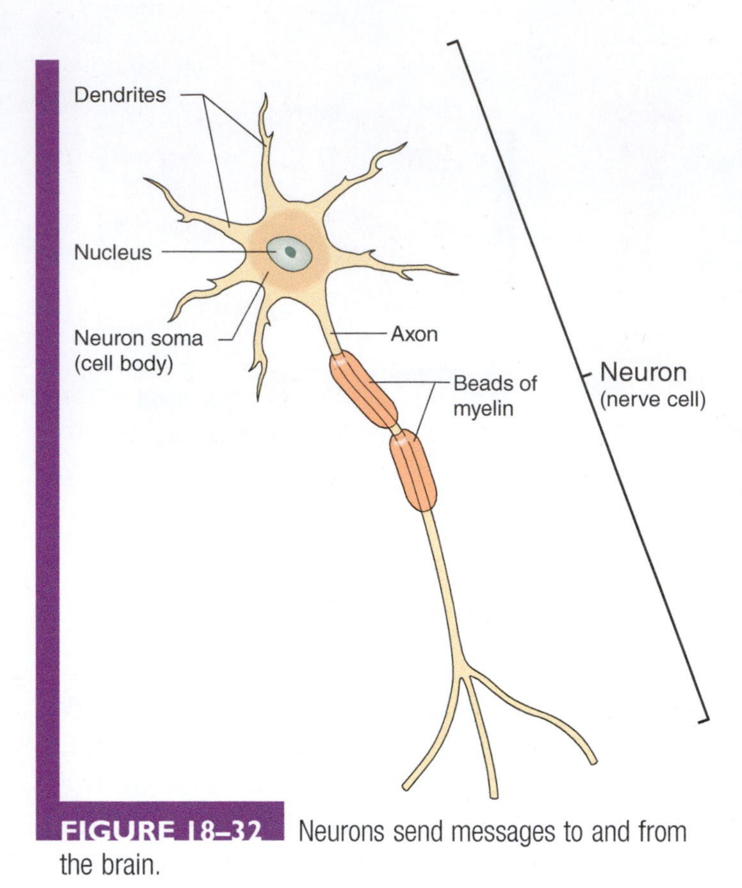

FIGURE 18–32 Neurons send messages to and from the brain.

Impulses transmitted through the spinal cord are sent to or received from the brain and carried to or from the various parts of the body. The brain is divided into several sections, each of which controls certain body functions (**Figure 18–33**). The largest part of the brain is the cerebrum. This wrinkled, folded portion controls the thought processes of the animal. If an animal decides to move a foot, the impulses must come from the cerebrum. If the foot is placed on a sharp stone, the impulse is sent to the brain from the foot, and the brain sends a message back to the foot to lift off the stone.

The cerebellum acts as a coordination center for messages from the cerebrum. As an animal walks, many muscles are used to extend and contract the leg and back muscles. These muscles have to move in coordination with each other or the animal will not walk smoothly or may not walk at all. The cerebellum sifts through all of the messages to and from the cerebrum and times and coordinates the movement of the muscles.

The involuntary activities of the body such as the beating of the heart or the breathing of the animal are controlled in the lower part of the brain called the medulla oblongata. This part of the brain responds to the need for changes in heartbeat and blood pressure. If an animal begins to work harder (running, for instance), the breathing rate increases along with the heartbeat. This area of the brain also controls such bodily functions as the body temperature, the movement of food through the digestive system, and feelings such as fear or thirst.

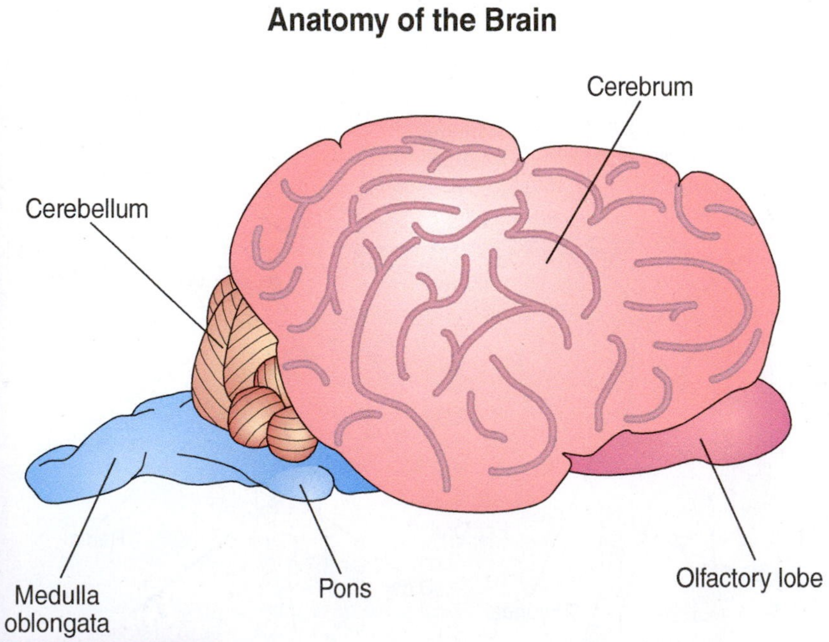

Anatomy of the Brain

Cerebrum
Cerebellum
Medulla oblongata
Pons
Olfactory lobe

FIGURE 18–33 Different areas of the brain control different functions.

THE ENDOCRINE SYSTEM

The endocrine system is composed of glands that secrete substances called **hormones** (**Figure 18–34**). Hormones are chemical agents that are sent to specific areas of the animal's body to stimulate or inhibit a response. Because hormones control such vital bodily functions as growth and reproduction, they have been the subject of a lot of scientific research. Producers of agricultural animals frequently use hormones or synthetic hormones to stimulate responses in animals. (Chapters 19 and 27 will discuss the uses producers make of hormones.)

Pituitary Gland

The pituitary gland is located near the base of the brain. This gland is often referred to as the "master gland" because it controls all of the functions of the other endocrine glands through the hormones it secretes (**Table 18–1**). It also manufactures and secretes hormones that control such processes as growth and reproduction.

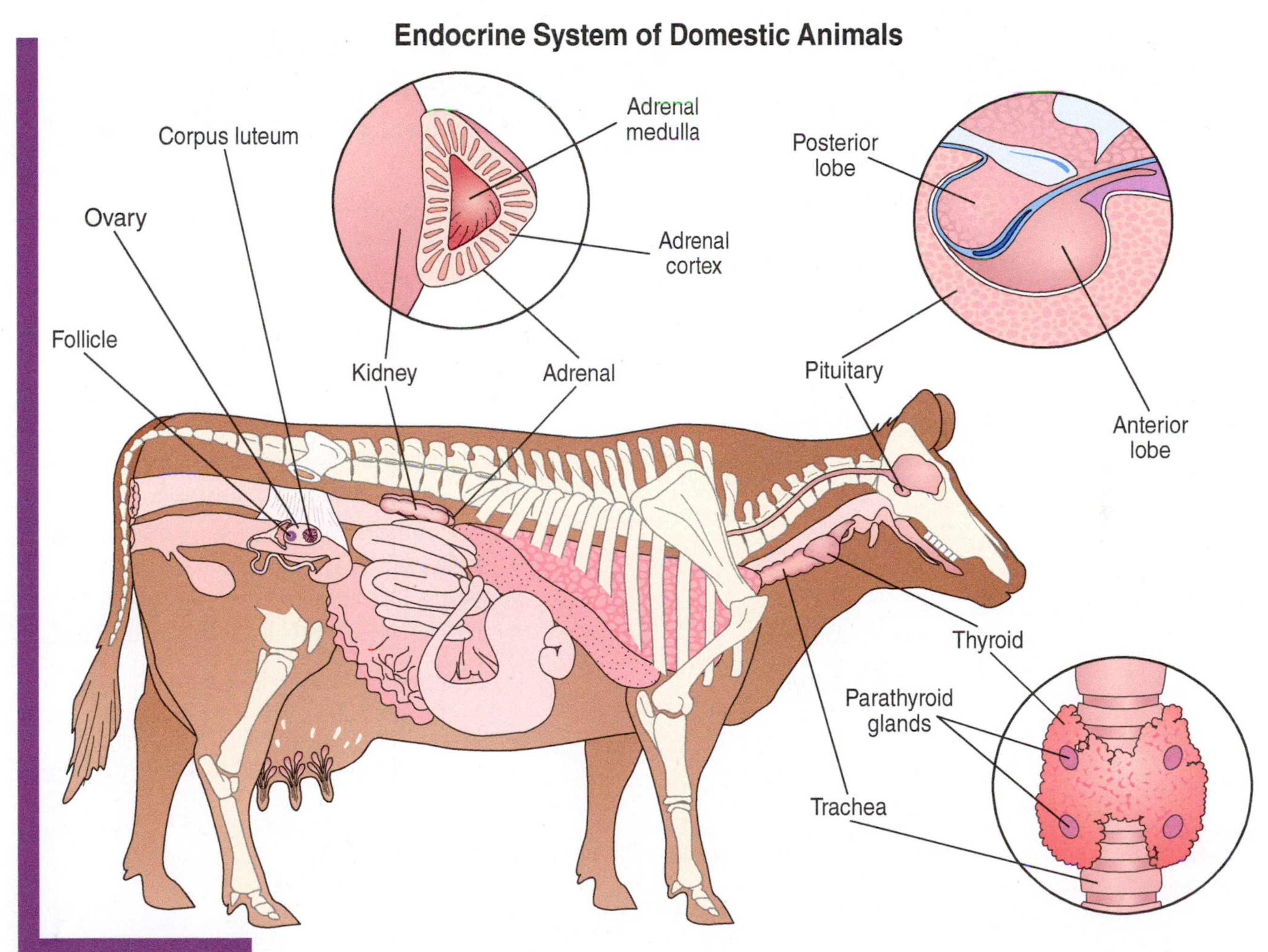

FIGURE 18–34 The endocrine system is composed of glands that secrete hormones.

Functions of Endocrine Glands and Hormones		
Endocrine Gland	**Hormone Secreted**	**Primary Physiological Function**
Hypothalamus	Gonadotropin-releasing hormone (GnRH)	Stimulates release of LH and FSH
	Corticotropin-releasing hormone (CRH)	Stimulates release of ACTH
	Thyrotropin-releasing hormone (TRH)	Stimulates release of TSH
	Growth-hormone-releasing hormone (GHRH)	Stimulates release of growth hormone
	Growth-hormone-inhibiting hormone (somatostatin)	Inhibits release of growth hormone
	Prolactin-releasing hormone (PRH)	Stimulates release of prolactin
	Prolactin-inhibiting hormone (PIH)	Inhibits release of prolactin
	Oxytocin	Causes ejection of milk in mammals, expulsion of eggs in hens, and uterine contractions
	Vasopressin (antidiuretic)	Causes constriction of the peripheral blood vessels and water reabsorption in the kidney tubules
Anterior pituitary	Growth hormone (GH or somatotropin)	Promotes growth of tissues and bone in the body
	Adrenocorticotropin (ACTH)	Stimulates secretion of steroids (especially glucocorticoids) from the adrenal cortex
	Thyrotropin or thyroid-stimulating hormone (TSH)	Stimulates thyroid gland to secrete thyroxine
	Prolactin (Prl)	Initiates lactation and promotes maternal behavior
	Gonadotropic hormones	
	Follicle-stimulating hormone (FSH)	Stimulates follicle development in the female and sperm production in the male
	Luteinizing hormone (LH)	Causes maturation of follicles, ovulation, and maintenance of the corpus luteum in the female. Causes testosterone production by the interstitial cells of the testes in the male
Thyroid	Thyroxine, triiodothyronine	Increases metabolic rate
	Calcitonin	Lowers the concentration of calcium in the blood and promotes incorporation of calcium into bone

TABLE 18–1 Hormones of the endocrine system stimulate or inhibit responses in the animal

Hypothalamus

The hypothalamus is located under the lower front part of the brain. It secretes hormones that control body temperature, hunger, sleep, and certain functions of the digestive process. Another major function is to manufacture hormones that stimulate the pituitary into producing hormones.

Adrenal Glands

The adrenal glands are positioned on top of each kidney. They produce a hormone called adrenaline that stimulates physiological responses in the animal in times of stress or danger. The heart rate and breathing rate increase to give the animal added oxygen and energy to meet a crisis.

The Thyroid Gland

The thyroid gland is located on the front of the trachea (windpipe) and produces a hormone called thyroxine that aids the body in the use of energy from digested food. This substance controls the rate at which the body cells break down carbohydrates and sugars. If the body needs more energy, the thyroid releases more thyroxine, and the rate of breakdown increases. When the need is lessened, the production of thyroxine is slowed. The thyroid also secretes calcitonin, which causes the storage of calcium in the bones. The parathyroids are situated inside the thyroid glands and regulate the amount of calcium and phosphorus in the blood. If the calcium level of the blood is too low, the parathyroid hormone causes some of the bone tissue to break down and restores the proper level of calcium in the blood.

The Pancreas

The pancreas is a gland situated below the stomach. It secretes two hormones, insulin and glucagon, that regulate the amount of glucose in the blood. Humans sometimes suffer from a disorder called diabetes that results from a lack of the proper amount of insulin. The condition is corrected through the daily intake of insulin.

Other organs of the body, such as the reproductive organs, also produce hormones. (Chapter 19 will deal with the hormones produced by the reproductive system.) Scientists often discover hormones they did not previously know existed. Animals are complex organisms composed of interdependent systems that continue to baffle our most knowledgeable scientists.

SUMMARY

Higher-ordered animals are complicated organisms. Many systems have to function properly, and at the same time function in coordination with each other. By understanding and using the naturally occurring processes of animal systems, agriculturalists are better able to produce the meat and other animal products so vital to our survival.

CHAPTER REVIEW

Review Questions

1. How might producers ensure that all the systems of their animals' bodies function properly?
2. Name the types of bones found in the skeletal system. Describe their functions and give an example of each.
3. Describe the differences between white muscle and red muscle.
4. Name each compartment of the ruminant stomach and briefly describe the function of each.
5. Describe the pathway that food travels in the monogastric digestive system.
6. Identify the three types of cells found in blood and the function of each.
7. Name the parts of the brain and briefly describe their functions.
8. Which system is responsible for the production of hormones? Why are hormones so important to bodily functions? Why might producers use hormones or synthetic hormones in the production of animals?

Student Learning Activities

1. Make a chart of the different animal systems and their uses. Be sure to include food, pharmaceuticals, and cosmetics.
2. Create a chart contrasting plant and animal systems. List all of the similarities and differences in the systems.
3. Using the analogy of a car and its systems, compare it to the systems of an animal. For example, the nervous system could be compared to the automobile electrical system.
4. Obtain an organ, such as a heart, from a slaughterhouse. Dissect it and label the parts. Be sure to wear protective gloves and a lab coat.

CHAPTER 19

The Reproduction Process

STUDENT OBJECTIVES IN BASIC SCIENCE

As a result of studying this chapter, you should be able to

- distinguish between asexual and sexual reproduction.
- explain the process by which gametes are produced in both the male and the female.
- describe the steps involved in meiosis.
- list and describe the parts and function of the male reproductive system.
- list and describe the parts and function of the female reproductive system.
- describe the functions of the hormones that control reproduction.
- describe the phases of the female reproductive cycle.
- explain the process by which fertilization takes place.

STUDENT OBJECTIVES IN AGRICULTURAL SCIENCE

As a result of studying this chapter, you should be able to

- list the reasons why artificial insemination is valuable to livestock producers.
- explain the procedures used in artificial insemination.
- explain the importance of embryo transfer.
- list and explain the steps used in embryo transfer.
- describe the advantages of estrus synchronization.
- explain the process of estrus synchronization.
- describe new scientific technology that will be of benefit to livestock producers.

KEY TERMS

asexual reproduction
sexual reproduction
zygote
mitosis
sterile
meiosis
spermatogenesis
spermatogonia
spermatozoa
chromatids
synapsis
oogenesis
polar bodies
cytoplasm
nucleus
testosterone
libido
epididymis
vas deferens
seminal vesicles
urethra
Cowper's gland
prostate gland
penis
prepuce
estrus
estrogen
progesterone
fallopian tubes
cervix
vagina
vulva
clitoris
endocrine system
ovulation
copulation
corpus luteum
conception
fertilization
estrus cycle
ejaculation
motile
fertilization membrane
differentiate
protectant
quarantine
artificial vagina
extenders
straws
estrus synchronization
artificial hormones
embryo transfer
progeny testing
genetic base
donor cows
recipient cows
superovulation
prostaglandin
catheter
clone

NATIONAL AFNR STANDARD

AS.04.01.01.a
Identify and categorize the male and female reproductive organs of the major animal species.

AS.04.01.01.b
Analyze the functions of major organs in the male and female reproductive systems.

AS.04.01.02.b
Assess and describe factors that lead to reproductive maturity.

AS.04.03.01.a
Identify and categorize natural and artificial breeding methods.

AS.04.03.02a
Analyze the materials, methods, and processes of artificial insemination.

AS.04.03.02.b
Demonstrate artificial insemination techniques.

AS.04.03.03.a
Identify and summarize the advantages and disadvantages of major reproductive practices, including estrous synchronization, superovulation, flushing, and embryo transfer.

AS.04.03.03.b
Analyze the processes of major reproductive management practices, including estrous synchronization, superovulation, flushing, and embryo transfer.

REPRODUCTION IN ANIMALS

In order for living things to remain on the earth, they must reproduce. If an organism does not reproduce, that type of organism would disappear as soon as death occurs. By reproducing, animals ensure that their type of animal will continue to exist. There are two basic means of reproducing. One-celled organisms and some plants reproduce by means of **asexual reproduction**—they produce another organism from only one parent. All higher-order animals reproduce by means of **sexual reproduction**. This means that animals come from two parents, a male and a female.

In mammals, reproduction is achieved by each of two parents' contributing genetic material to the young. Half of the characteristics of the young come from the father and half come from the mother. Each of the two parents creates reproductive cells called gametes. The male gamete is known as a sperm cell and the female gamete is known as an egg cell. The uniting of the two cells results in the beginning of a new animal that is similar to the parents. The new cell produced by the uniting of the sperm and the egg is called a **zygote**. The zygote divides by a process called **mitosis**. Both the egg and the sperm contain the material that dictates what the young animal will look like.

Even though the gametes are so small that they cannot be seen without the aid of a microscope, they contain all of the material necessary to determine all of the characteristics of the new animal. This material is called deoxyribonucleic acid (DNA). DNA is the matter that carries the code that determines exactly how the animal will develop and how it will look. Specific segments or units of DNA that are grouped together are called genes. Each gene has a unique structure that controls a trait of an animal. For example, a gene may determine that Angus cattle are hornless, or may determine that Watusi cattle will have horns that grow to a length of 6 feet (**Figure 19–1**).

Groups of genes combine to form thread-like structures called chromosomes. The name comes from *chromo*, which means "colored," and *soma*, which means "body." Each cell in an animal's body contains a number of chromosomes. In body cells, the chromosomes always come in pairs. Different species have different numbers of pairs. **Table 19–1** provides a listing of the number of numbers of chromosomes for several domestic animals. Each gamete (the sperm and the egg) contains one of each pair or half of the chromosomes. For example, each horse cell contains 32 pairs, or 64 chromosomes. The sperm or egg from a horse would

FIGURE 19–1 Watusi cattle have specific genes that determine that their horns will be long. *© Oleg Znamenskiy /Shutterstock.com.*

Animals	Chromosome Number (2n)
Donkey	62
Horse	64
Mule	63
Swine	38
Sheep	54
Cattle	60
Human	46
Dog	78
Domestic cat	38
Chicken	78

TABLE 19–1 Characteristic numbers of chromosomes in selected animals.

each contain 32 chromosomes. When the egg and the sperm unite at the time of conception, they each contribute 32 chromosomes that go together to form a full set of 64 chromosomes. Each chromosome from the father is matched with a chromosome from the mother.

Notice that the donkey has 31 pairs or 62 chromosomes. Horses and donkeys can be mated successfully to produce an offspring known as a mule. However, mules cannot reproduce because the donkey's 31 pairs of chromosomes combined with the horse's 32 chromosomes will not divide into an even pairing of chromosomes. As a result, the gamete produced by a male or female mule will not successfully unite to form a zygote. For this reason, mules are almost always **sterile**, which means that mules are not generally capable of reproducing.

Production of Gametes

The formation of the sperm cell takes place in the testicles of the male; the formation of the egg takes place in the ovaries of the female. Within these organs, a process known as **meiosis** takes place. This process differs from mitosis in that meiosis results in a cell that contains only half the number of chromosomes of the original cell (**Figure 19–2**). In mitosis, the dividing cells contain the same number of chromosomes as the parent cells. Meiosis is necessary in order to allow the contribution of half of the chromosomes by each parent.

The production of the male gamete or sperm is called **spermatogenesis**. In the testes of the male, cells called **spermatogonia** are produced. Through a four-step process, these cells develop into **spermatozoa**. During the first step, the chromosomes replicate (make an exact copy of themselves) and remain attached. The replicated chromosomes are called **chromatids**. In the next step, the chromatids come together and are matched in pairs in a process called **synapsis**. In the third step, the cell divides,

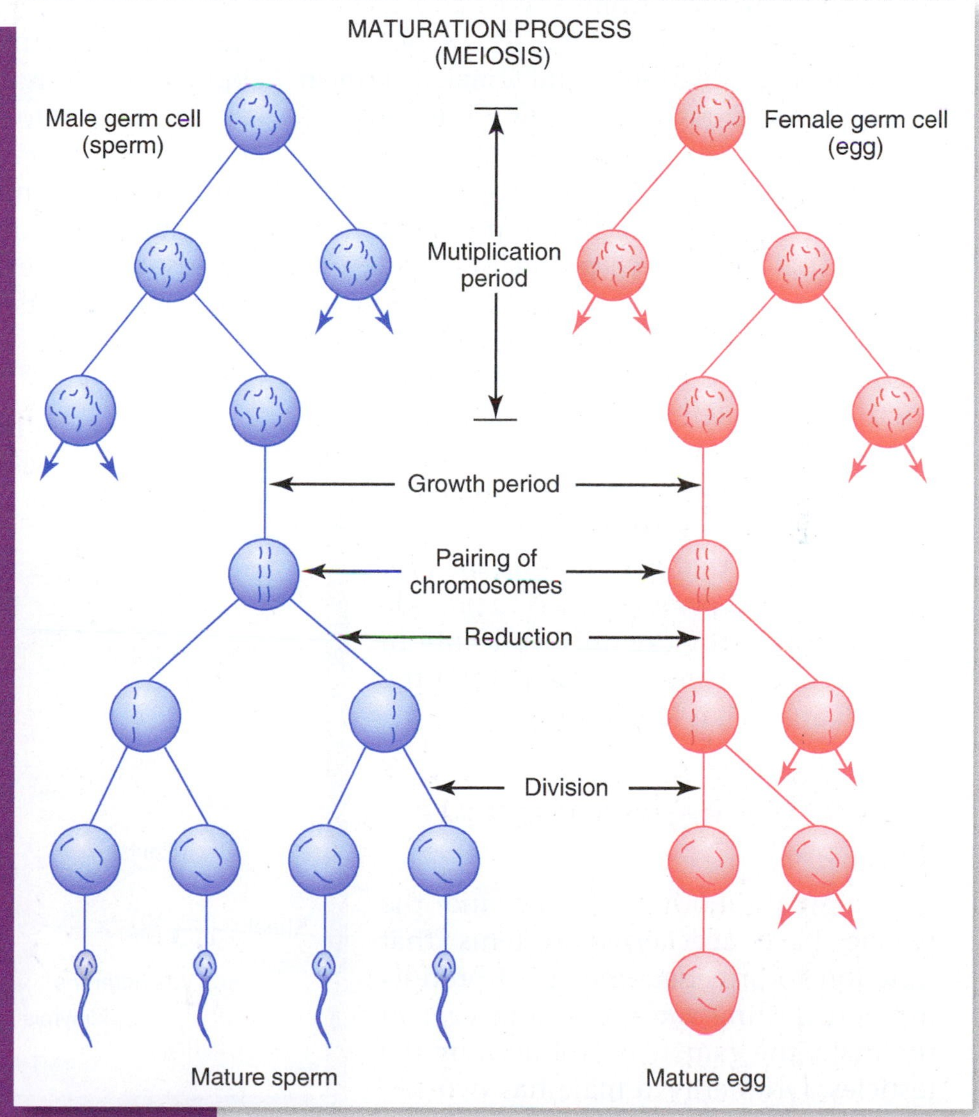

FIGURE 19–2 Meiosis is the process by which mature sperm and egg cells develop.

the chromosomes are separated, and each cell receives one of each chromosome from each pair. However, remember, that each chromosome replicated itself (a chromatid) and is still attached to another chromosome. In the final step, the cells separate again and the chromatids separate and become chromosomes.

Remember that these cells (the new sperm cells) each contain only half of the chromosomes that the original cell contained. The end result of this process is that four new sperm cells are produced from the original cell. When the sperm is united with the egg at conception, the original number of chromosomes is restored because the sperm furnishes half of the chromosomes and the egg furnishes half.

Gamete production in the female is known as **oogenesis**. The stages of egg production are similar to those in sperm production. The one important exception is that in sperm production, four new sperm cells are produced from the original cell. In egg production, only one egg cell is produced. Instead of producing four new eggs, three of the newly divided cells become what are known as **polar bodies** and only one cell becomes a viable egg. Polar bodies are produced as a result of most of the **cytoplasm**—cell material outside the **nucleus**—from the cells going to the one cell that will become the egg. The function of polar bodies is to provide sustenance for the egg until conception. The sperm cell is much smaller than the egg cell and needs less to subsist.

THE MALE REPRODUCTIVE SYSTEM

In mammals, both the male and the female have specialized systems that function to provide a means of producing and uniting the sperm and egg. In the male, the gamete is produced by the testicles. Ordinarily, a male has two testicles that are suspended away from the body. An exception is poultry, in which the testes are on the inside of the rooster's body.

In mammals, the testicles are enclosed in a saclike structure called the scrotum. The scrotum functions not only to encase the testicles but also to act as a means of regulating the temperature of the testicles. For sperm production to occur, the testicles must have a temperature lower than the animal's body. This is why the testicles are suspended away from the body. The skin of the scrotum is thin, relatively hairless, and contains no subcutaneous (under the skin) fat. This aids greatly in the dissipation of heat in the summer. In the winter, the scrotum is retracted by small muscles that draw the testicles toward the body to keep them warm. The reproductive tract of a bull is shown in **Figure 19–3**.

In addition to producing sperm, the testicles produce the hormone **testosterone**. This hormone controls the animal's **libido**, or sex drive, and stimulates the development of sex characteristics. For example, the development of large forequarters in a bull and the growth of

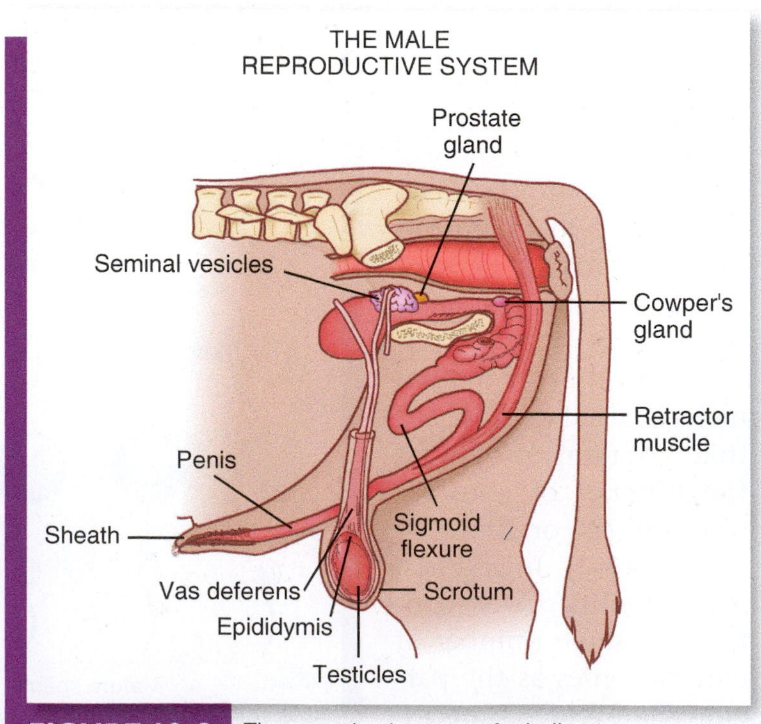

FIGURE 19–3 The reproductive tract of a bull.

large tusks and the unpleasant odor in boars are induced by testosterone.

Along the outside of the testicles is a tuberous structure called the **epididymis** that provides a place for the storage and maturation of the sperm produced by the testicles. A long tube called the **vas deferens** leads from the epididymis to the **seminal vesicles**, located at the upper end of the **urethra** (the tube through which urine is passed from the bladder). The vas deferens serves as a transportation route for the sperm.

The male reproductive tract has three accessory glands. The first, the seminal vesicles, function to secrete a fluid that is mixed with the sperm to protect the sperm and provide a mechanism by which the sperm can be transported. The two other glands also secrete fluid. The mixture of fluids is referred to as semen. The seminal vesicles also act as a holding place for the sperm. The **Cowper's gland** secretes fluid that helps to cleanse the urethra before the sperm is passed along the tube. This secretion also acts to coagulate or to thicken the semen. The **prostate gland** also secretes fluid that is added to the semen mixture. Its purpose is to provide nutrients for the sperm and to expel the semen during the mating process.

The male organ that deposits sperm in the female tract is the **penis**. This organ also serves as the means of expelling urine from the body. The penis of the boar, bull, and ram is composed of a high concentration of connective tissue. The upper end of the penis is S-shaped and flexes to extend the penis outward during mating. The penis of the stallion is made up of a high concentration of vascular tissue that allows the organ to become engorged with blood. This causes the penis to become extended until it is said to be erect. This allows penetration of the female. The external covering of the penis is called the sheath, or **prepuce**; its purpose is to protect the penis from injury or infection.

THE FEMALE REPRODUCTIVE SYSTEM

The female reproductive system is much more complex than that of the male (**Figure 19–4**). Sperm production in the male is constant, but production in the female comes about only in carefully controlled cycles. The cycle produces the egg, places the egg in the proper place, causes the female to accept the male for mating—called **estrus** or *heat*—and ensures that the fertilized egg remains in place throughout the gestation period. This cycle from egg production to fertilization occurs at different intervals in different animals.

The female reproductive system consists of several organs that make a contribution to the process. The ovaries are two small organs supported in the abdominal cavity by strong ligaments. Inside these ligaments are the arteries and vessels that supply blood to them. The main function of the ovaries is to produce the egg, or ovum. This is where the oogenesis, or production of the egg, takes place.

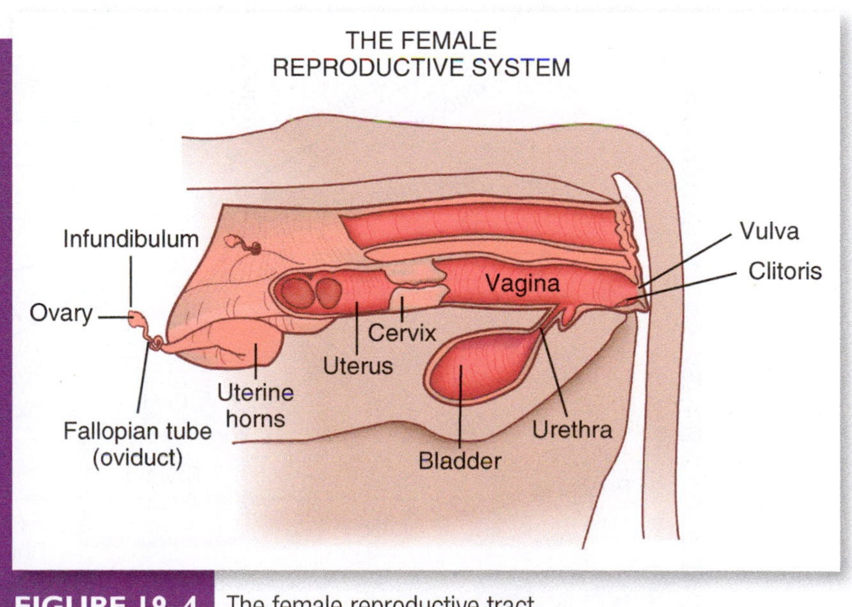

FIGURE 19–4 The female reproductive tract.

Another important role of the ovaries is to produce the hormones **estrogen** and **progesterone**. (These two hormones play essential roles in the reproductive cycle of the female.) Leading from the ovaries are two tubes known as the **fallopian tubes,** which serve to transport the egg from the ovaries to the uterus. It is within the fallopian tubes that the egg is united with the sperm and conception takes place. The fallopian tubes open into a muscular saclike organ known as the uterus (sometimes called the womb). The uterus serves as the chamber in which the fertilized egg (zygote) develops into an embryo, then into a fetus, and finally expels the newborn animal.

The uterus is sealed by a thick group of circular-shaped muscles called the **cervix**. The cervix acts as a valve that keeps foreign matter from entering the uterus. It contains glands that secrete a waxlike material that serves as a seal to the uterus. When the animal comes into estrus, the cervix opens to allow passage of the sperm. The cervix opens into the **vagina**, which is a sheathlike organ that accepts the male's penis during mating. The semen is deposited here. When the fetus has matured, the vagina serves as the birth canal through which the young animal leaves the uterus. The exterior part of the female reproductive system is the **vulva**. The vulva provides a closing for the vagina and serves as the end of the urinary tract that expels the urine. Within the vulva is a small sensitive organ called the **clitoris** that provides stimulation during the mating process.

As mentioned earlier, the entire process of the female reproductive cycle is controlled by hormones (**Figure 19–5**). Hormones are produced by the **endocrine system** and serve to stimulate or inhibit the development or operation of body functions such as reproduction. This system includes the pituitary gland, which is located near the base of the brain in mammals and acts as a type of master control for most of the other glands in the endocrine system.

The reproductive cycle of the female begins with a hormone secreted from the

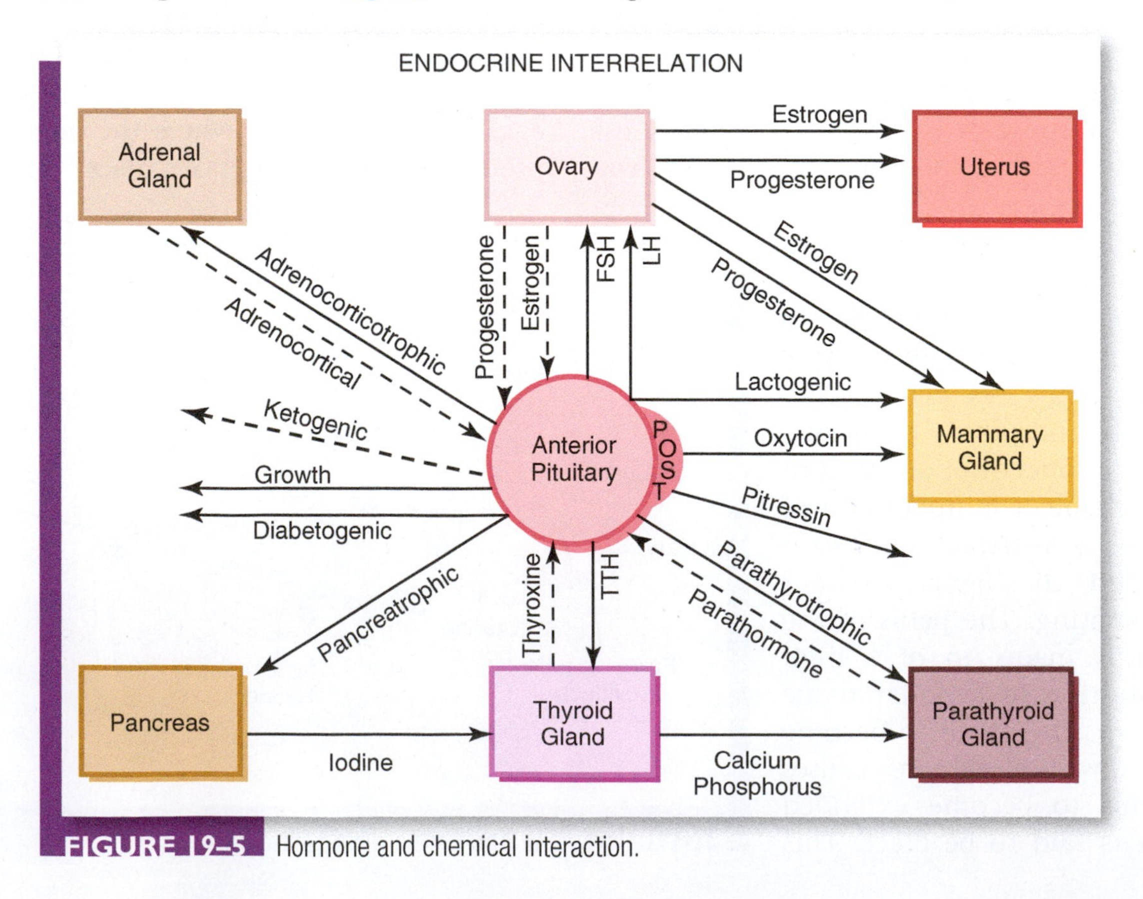

FIGURE 19–5 Hormone and chemical interaction.

pituitary gland that stimulates the ovary to produce a blisterlike structure called a follicle. The hormone, therefore, is called the follicle-stimulating hormone (FSH). The follicle, which appears as a clear blister on the surface of the ovary, secretes the hormone called estrogen (**Figure 19–6**). Estrogen acts as a messenger that stimulates the rest of the reproductive system to prepare to receive the egg. The follicle continues to produce estrogen and provides a place for the ovum to grow and mature. Oogenesis occurs in the follicle. When the gamete (the egg or ovum) is matured, the follicle becomes soft and expels the egg into the fallopian tube. (Some animals such as pigs have more than one young. The follicles of these animals release several eggs instead of one. Occasionally, even cattle may release two eggs instead of only one, resulting in the birth of twins.)

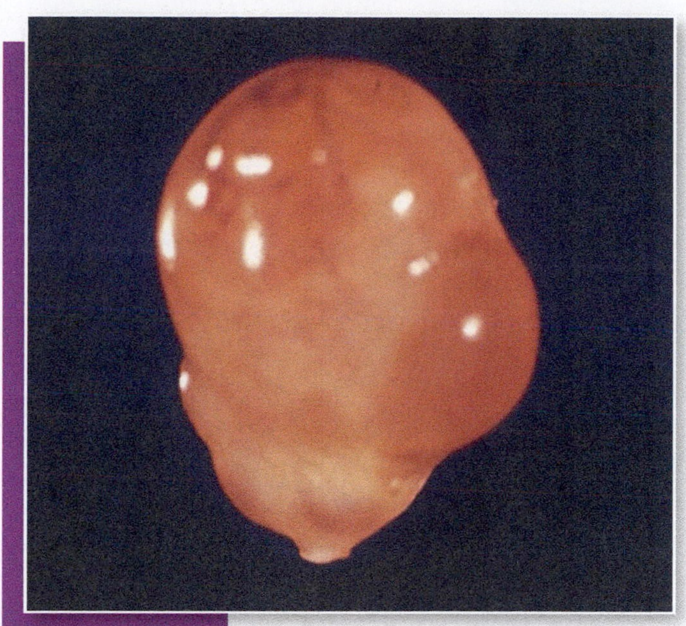

FIGURE 19–6 The follicle, which appears as a clear blister on the surface of the ovary, secretes a hormone called estrogen. *Courtesy of NAL Image Gallery. Photo by Harold Hafs.*

This process is known as **ovulation**, **Figure 19–7**. As ovulation occurs, the estrogen produced by the follicle causes the animal to go into the condition known as estrus or heat.

During this time, which may last from a few hours to two or more days, the female allows the male to mate with her. In this process, called **copulation**, the male's sperm are deposited into the female's vaginal tract. The expulsion of the egg from the follicle leaves a rupture that is filled with yellow cells that develop into a body called the **corpus luteum**. The development of the corpus luteum is caused by a hormone from the pituitary gland known as the luteinizing hormone (LH). The corpus luteum that develops where the follicle ruptured secretes a hormone called progesterone. This hormone causes the walls of the uterus to thicken in preparation for receiving the fertilized egg. After **conception** (the uniting of the sperm and the egg) occurs, the corpus luteum continues to produce progesterone and the female remains pregnant. If conception does

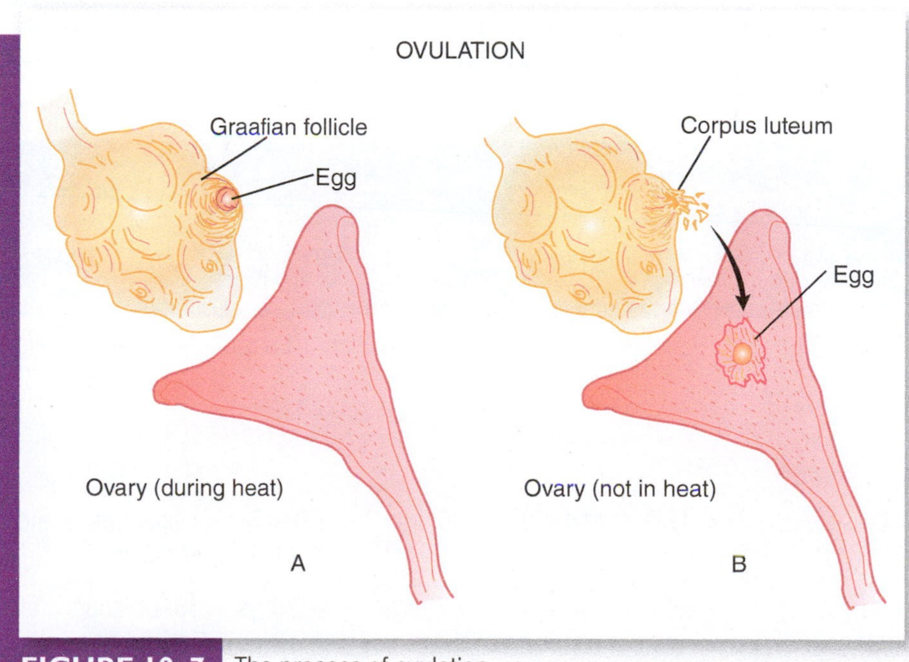

FIGURE 19–7 The process of ovulation. *Courtesy of Vocational Materials Service, Texas A&M University*

not occur, the corpus luteum recedes, the ovary returns to normal, and the cycle begins again.

When the sperm from the male are deposited into the female's vaginal tract, the sperm cells make their way up through the cervix, through the uterus, and into the fallopian tubes. If a mature ovum (egg) is present, then conception may occur.

FERTILIZATION

Fertilization is the process by which the sperm is joined with the egg. Following this process, the embryo begins to develop. Because the sperm may only live for 20 to 30 hours, mating must take place at a time when the egg has matured and is released from the ovary. The entire process of gamete production is controlled by hormones. For example, the hormone estrogen, secreted by the follicle that develops on the ovary, causes the animal to enter the estrus phase. During this time, which may last from a few hours to a few days, depending on the species of animal, the female allows the male to approach her and to mate. Estrus is timed to occur as the follicle releases the egg. If fertilization does not occur and the female does not become pregnant, the whole process of egg production begins again. This cycle is referred to as the **estrus cycle**. It normally occurs every 21 days in hogs, cattle, and horses, and every 17 days in sheep (**Table 19–2**).

During mating, a combination of sperm and fluid (semen) is deposited into the vagina of the female. The process is called **ejaculation,** and the semen is sometimes referred to as ejaculate. As stated earlier, the fluid in the semen serves two purposes: to provide nourishment for the sperm and to provide a means for the sperm to move. In each ejaculation, millions of sperm are deposited. Each sperm is shaped like a tadpole and has a tail that causes the sperm to move in a whip-like action (**Figure 19–8**). Sperm that are able to

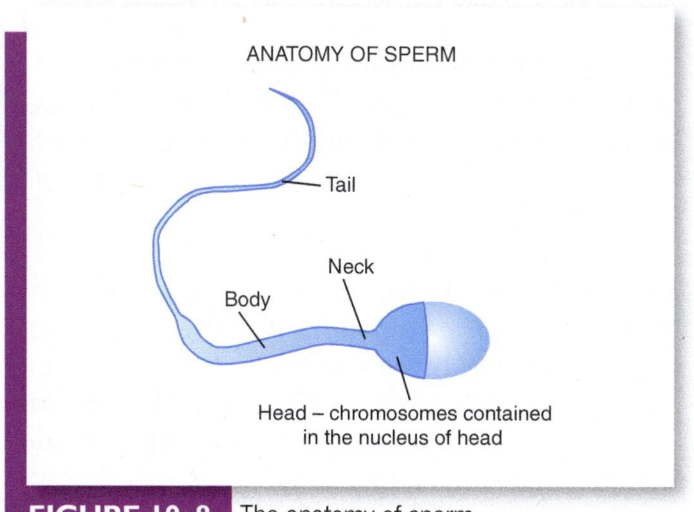

FIGURE 19–8 The anatomy of sperm.
Courtesy of Vocational Materials Service, Texas A&M University

Species	Length of Estrus Cycle (days)		Length of Estrus		Usual Time of Age at Ovulation	Length of Gestation (days)		Age at Puberty (months)
	average	range	average	range		average	range	
Mare	21	10–37	5–6 days	1–14 days	24–48 hours before end of estrus	336	310–350	10–12
Cow	19–21	16–24	16–20 hrs.	8–30 hrs.	10–14 hours after end of estrus	281	274–291	8–12
Ewe	16	14–20	30 hrs.	20–42 hrs.	1 hour before end of estrus	150	140–160	4–8
Sow	21	18–24	1–2 days	1–2 days	18–60 hours after estrus begins	112	111–115	5–7

TABLE 19–2 The reproduction cycle in farm animals. *Adapted from Vocational Materials Service, Texas A&M University.*

move about freely are said to be **motile**. Through this means, the sperm begin a journey through the cervix, into the uterus, and into the fallopian tubes (oviduct), where fertilization occurs. The sperm are attracted to the egg by a chemical that is secreted by the egg. An obvious question is: If only one sperm is needed to fertilize the egg, why are millions of sperm deposited? This large number of sperm is needed for two reasons. First, not all sperm are hardy enough to make the long trip to the oviduct where the egg is. Therefore, a large number of sperm are necessary to ensure that some sperm reach the right destination. Second, even though only one sperm penetrates and fertilizes the egg, many sperm are needed for the process. The sperm swarm around the egg and secrete an enzyme that loosens the cells surrounding the egg.

The outer membrane of the egg is covered with a protective layer of a jellylike substance that must be dissolved before the sperm can enter the egg (**Figure 19–9**). The sperm release a chemical that works to dissolve the coating. One of the sperm forms a tubelike connection with the membrane of the egg. The nuclear material of the sperm then enters the egg and fertilization occurs. Only the nuclear material actually enters the egg; the tail of the sperm is left outside of the egg. When the nucleus of the sperm enters the egg, the egg releases carbohydrates and protein to form a layer around the egg that will prevent any more sperm from entering. This new layer is called the **fertilization membrane**.

Fertilization is completed when the nucleus of the sperm and the nucleus of the egg fuse together and the correct number of chromosomes for that particular species is restored. Remember that during meiosis, only half of the number of chromosomes from the original cell are transmitted in the egg and half are transmitted in the sperm. At the completion of fertilization, the original number is reestablished. The resulting fertilized cell is referred to as a zygote (**Figure 19–10**).

Very soon after fertilization, the zygote begins to divide. This type of cell division is called mitosis, as compared with meiosis, which is the process by which the gametes are formed. As mentioned previously, the division of the fertilized egg by mitosis is called cleavage. The cells continue to divide rapidly, and as they do, they make their way out of the oviduct into the uterus. As the cells enlarge and begin to differentiate, the mass of cells is called an embryo. **Differentiate** means that the cells begin to specialize; some begin to form cells that will become skin, some bones, some internal organs, and so on. No one completely understands what causes differentiation, but we do know that all parts of the animal's body come from the original fertilized egg.

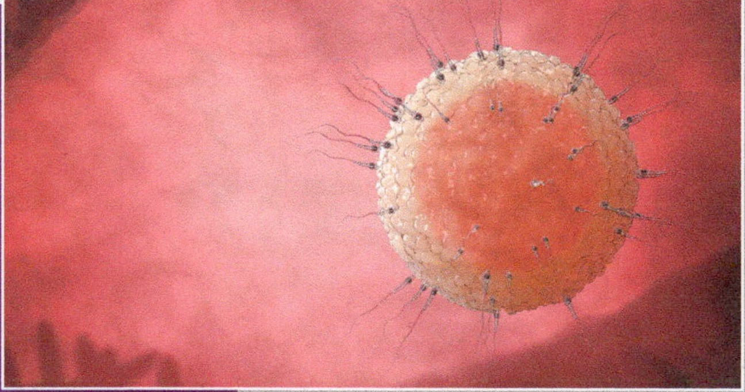

FIGURE 19–9 The outer membrane of the egg must be dissolved before the sperm can enter the egg. Note the sperm that has penetrated the egg. *© Pavel Chagochkin/Shutterstock.com.*

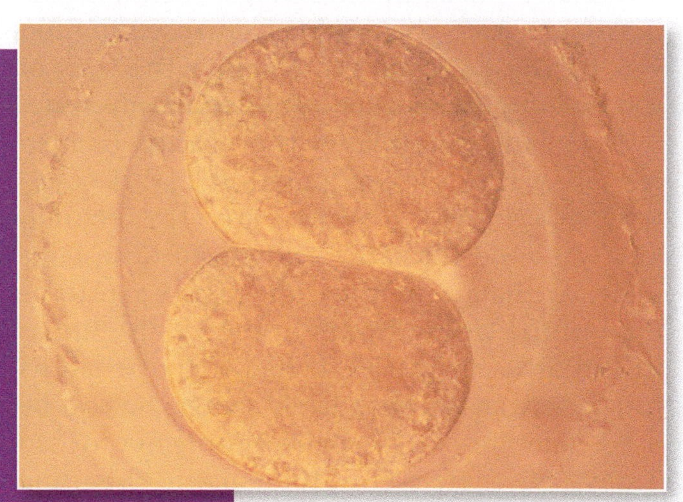

FIGURE 19–10 Fertilization occurs when the nucleus of the sperm and the nucleus of the egg fuse. The cell multiplies and begins to differentiate. *Courtesy of Richard Fayrer-Hosken, University of Georgia*

Once the embryo reaches the uterus, it attaches to the wall of the uterus and begins to develop. In the embryo development stage, the corpus luteum maintains a level of the hormone progesterone that causes the uterus to implant and nourish the embryo. As a result, the lining of the uterus remains intact, and the embryo can continue to develop, meaning the female maintains pregnancy.

ARTIFICIAL INSEMINATION

As humans began to understand how reproduction works in animals, they began to use techniques that aided in natural reproduction. Healthier, faster-growing, more efficient animals could be produced with help from the humans who cared for the animals. One of the production practices that has been a real asset to animal producers is artificial insemination. Artificial insemination is not a new technology. Some say that the process goes all the way back to the Middle Ages, when Arabs used this method to collect semen from stallions that belonged to their enemies and bred their own mares in order to produce superior foals.

The first recorded use of artificial insemination was in 1780, when Lazarro Spallanzani, an Italian scientist, was successful in artificially inseminating dogs. Perhaps the first large-scale use of artificial insemination was by the Russians shortly after the turn of the twentieth century. A Russian physiologist named Ivanoff used the process to help replenish the horse population in his country following World War I. Later, the technology was used for cattle and sheep on a large scale. Artificial insemination first began to be used in the United States in the 1930s. But, as in the other countries, it had not reached its full potential because fresh semen had to be used. The lifespan of sperm is only about 2 to 3 days, so there were problems in obtaining semen when it was needed.

In the 1950s, the technique of freezing semen was perfected. A **protectant** such as glycerin is added, and the semen is frozen at a specific steady rate until the temperature reaches –320°F (**Figure 19–11**). If the semen is kept at this temperature, it can remain viable for years. In fact, bull semen has been stored successfully for as long as 30 years. Semen from bulls, stallions, and rams can be frozen, stored, and thawed successfully. Semen from boars, however, usually is shipped immediately and is used fresh because of problems with sperm viability when it is frozen.

Artificial insemination is used most widely in the dairy industry (**Figure 19–12**) but also is used extensively for beef and, to a lesser extent, with horses, sheep, and swine.

FIGURE 19–11 A protectant such as glycerin is added to semen, and the semen is frozen at a specific steady rate until the temperature reaches –320°F. *Courtesy of Dr. Frank Flanders*

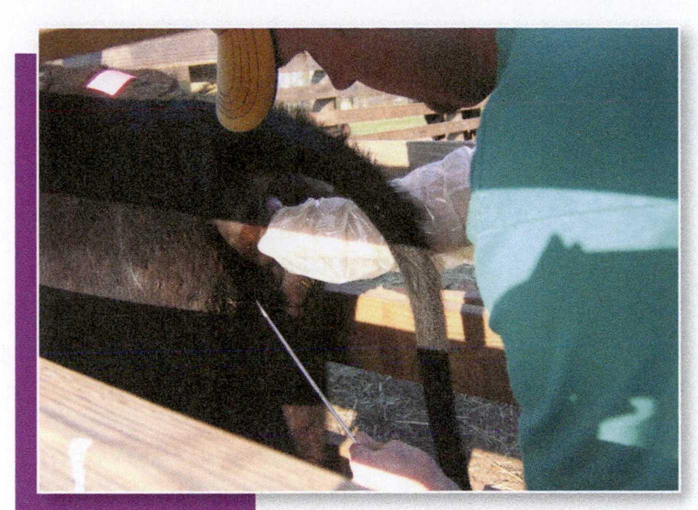

FIGURE 19–12 Artificial insemination is widely used in the dairy industry. *Courtesy of Dr. Frank Flanders*

The advantages of using artificial insemination are numerous:

1. Producers may use sires of higher quality than they otherwise could afford. A high-quality sire from any species is expensive; the cost of semen from artificial insemination is much lower.
2. Data from the progeny of sires used in artificial insemination are available to aid the producer in determining the quality of the sire. One bull may produce many thousands of offspring. In fact, a Holstein bull named Toystory lived for 13 years and produced over 2.4 million units of semen. These units were used in more than 50 countries. Obviously, if production data for thousands of offspring can be compiled, the producer can get a clear idea of the type of animals to expect from the sire.
3. Artificial insemination allows the producer to select the type of sire needed for a particular group of females. For instance, a hog producer may need to increase the size of bone in the herd, or a beef producer needs a bull that will sire smaller calves at birth for calving ease. Through the use of sire data, the producer can select sires that are known for these characteristics.
4. Producers do not have to keep male animals. This can be an advantage not only because of expense but also from a safety standpoint. Mature male animals are by their nature aggressive and often dangerous. A large boar or bull can kill or seriously injure those who care for them.
5. The likelihood of disease is lessened. Many diseases are transmitted through direct contact with other animals. By using artificial insemination, contact between animals is avoided.
6. Sires from all over the world can be used. One of the largest problems associated with importing animals from other countries is strict **quarantine** laws that require the animals to be kept in isolation for a period of time to make sure that they do not bring disease into the United States. By using frozen semen, new genes can be brought into the country with less risk of importing disease, and they can be brought in at much less cost.
7. Sires can be easily replaced. If producers own their own sires, the expense of changing sires is substantial. If a producer is not pleased with the offspring of the sire, the old sire has to be sold and a new one bought. By using artificial insemination, all the producer has to do to change sires is to order semen from a different sire.

Semen Collection and Processing

Semen is collected through the use of an **artificial vagina**. The artificial vagina consists of a rigid tube that is lined with a smooth surface water jacket that is filled with warm water. At the end is attached a receptacle for collecting the semen. As the male approaches and mounts the dummy or live animal, the penis is guided into the artificial vagina, where ejaculation occurs. The amount of the ejaculate or semen varies with different species (**Table 19–3**).

Once the semen is collected, it is examined in the laboratory under a microscope

Animal	Volume per Ejaculate (milliliter)	Sperm per milliliter (one thousands)	Total Sperm per Ejaculate (billions)	Number of Females per Ejaculate
Boar	150–250	100	15–25	10–12
Bull	5–15	1,000	3–5	100–600
Rooster	0.6–0.8	3,000	1.8–2.4	
Ram	0.8–1.0	1,000	0.8–1.0	40–100
Stallion	70–100	100	7–10	8–12

TABLE 19–3 Semen volume and numbers for farm animal species.
Courtesy of Vocational Materials Service, Texas A&M University

FIGURE 19–13 Once the semen is collected, it is examined in the laboratory under a microscope. *© diplomedia /Shutterstock.com.*

(**Figure 19–13**), to check for foreign material and for quality. Quality is determined by the number of sperm in a milliliter of semen, how active the sperm are (motility), and the shape of the sperm (morphology). Very active sperm are desirable because of the distance they must travel to reach the oviduct of the female. Sperm of different species are shaped differently (**Figure 19–14**). The sperm are checked for the normal shape; a large number of sperm with an unusual shape is not desirable (**Figure 19–15**).

Once an ejaculate has been checked and determined to be of an acceptable quality, it is processed. Processing involves adding **extenders** such as milk, egg yolk, glycerin, and/or antibiotics. One purpose of the extenders is to provide a means of diluting the semen. The semen from one bull ejaculation may be divided into several units, depending on the number of sperm in the ejaculate. Another purpose of extenders is to provide protection to the sperm during the freezing procedure. Extenders also provide nourishment for the sperm.

After adding the extenders, the semen is checked again to make sure that the sperm are still motile. The semen then is packaged in small, hollow tubes called **straws**, sealed, and labeled with the name of the company, the date, and the name of the sire (**Figure 19–16**). The straws containing the semen are frozen at a specific rate to 320°F and are stored and transported in liquid nitrogen tanks. Boar semen does not freeze as well as bull semen. Although frozen boar semen is used, fresh semen is preferred because its use results in greater conception rates.

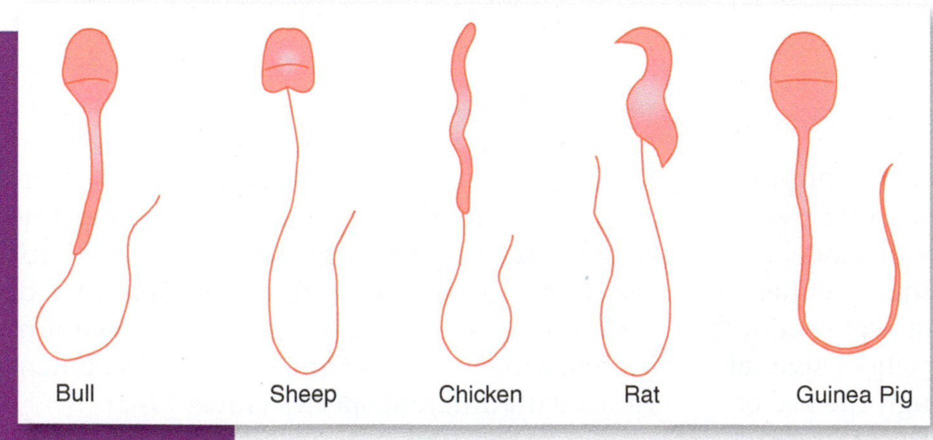

FIGURE 19–14 Species differ in sperm anatomy. *Courtesy of Vocational Materials Service, Texas A&M University*

When the technician is ready to artificially inseminate an animal, the straws are carefully removed from the liquid nitrogen tank. Precautions have to be taken because liquid nitrogen can

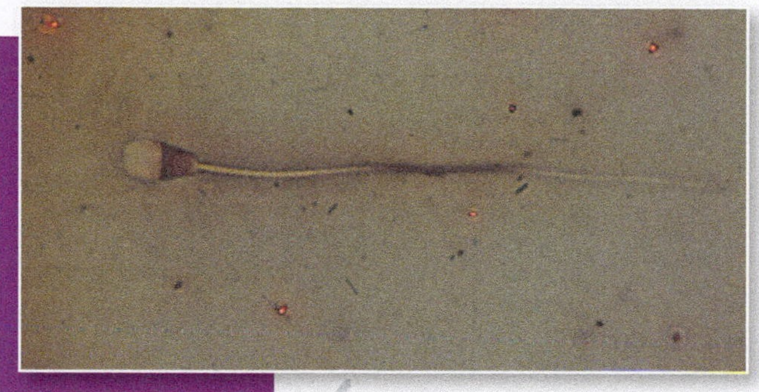

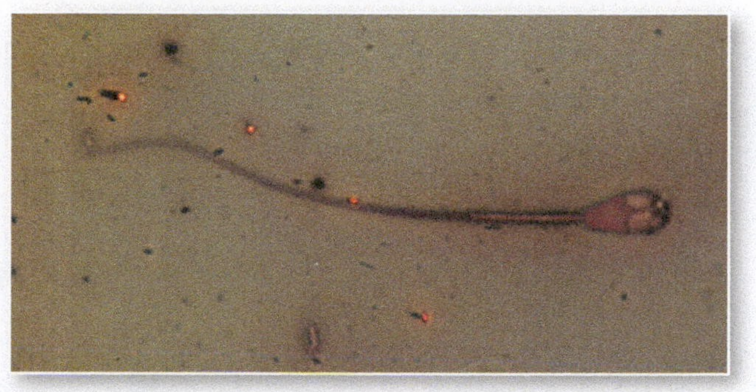

FIGURE 19–15 Sperm with abnormal shapes are not desirable. (A) normal. (B) defective. *Courtesy of Dr. Ben Bracket, Department of Physiology School of Veterinary Sciences, University of Georgia*

cause a frostbite-like injury if it contacts the skin. The straws of semen have to be thawed at the proper temperature and speed. Thawing may be accomplished through the use of a special apparatus that heats water to a certain temperature or through the use of a water-filled thermos bottle. The straw is placed into the water and left for not less than 30 seconds and not more than 15 minutes. Proper thawing ensures that the thawed sperm will be healthy and motile. Once the semen is properly thawed, the straw containing the semen is placed in a tubelike instrument that will be used to place the semen in the tract of the female (**Figure 19–17**). After the semen is placed in the female tract, the process of fertilization takes place just as in natural mating. The people who do the actual insemination must undergo special training before they can develop the skills necessary to properly thaw the semen and place it correctly.

Control of the Estrus Cycle

Another scientific advance that has aided greatly in improving the reproductive efficiency of agricultural animals is the use of **estrus synchronization**. Recall that estrus (the time the female allows breeding) is controlled by the production and secretion of hormones at the proper time. The estrus cycle is a chain of events that occur as certain hormones are released. If **artificial hormones** (from an outside source) are introduced into the female, the hormones will cause the same reaction as a naturally produced hormone.

FIGURE 19–16 The semen is placed in straws, labeled, and stored in tanks cooled with liquid nitrogen. *Courtesy of Dr. Frank Flanders*

For example, remember that a hormone causes the follicle to develop on the ovary, and the egg is developed from the follicle. By injecting females with a hormone that stimulates the

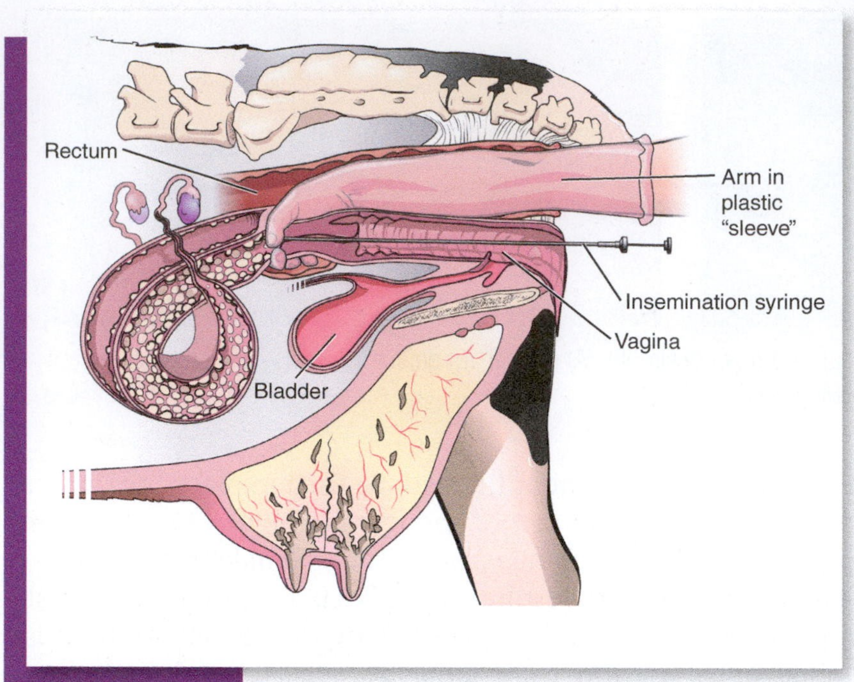

FIGURE 19–17 Once the semen is placed in the female tract, fertilization takes place just as in natural mating.

follicle, the female will begin the cycle at that point and come into estrus. The advantage to the producer in inducing estrus is that by injecting all of the females in the herd at the same time, the cycle can be synchronized so that they all come into heat at about the same time. The obvious advantage is that a producer can have all of the animals artificially inseminated at the same time. Not only does this save time and resources at breeding time, but it will also save time later because the females will all calve or farrow at about the same time. The crop of young animals will be of about the same age, so they can be managed alike as they are grown out.

EMBRYO TRANSFER

Among the newest of the reproductive technologies is that of transferring embryos from one female to another. Just as artificial insemination has allowed genetic improvement from a single sire to be greatly increased, **embryo transfer** has increased the reproductive capacity of superior females. If a producer breeds a superior female, the animal will produce one offspring per year. Although a female is capable of producing many thousands of eggs during her lifetime, only a relatively few will develop into offspring. If the eggs are collected from a superior female and implanted in an inferior animal, the superior female has the capacity to produce many offspring in a year.

Embryo transfer has many benefits:

1. The use of embryo transfer allows the rapid advancement of genetics from the dam. Just as artificial insemination allows the production of many offspring from a superior male, embryo transfer allows the production of many offspring from a superior female (**Figure 19–18**).
2. Embryo transfer allows the **progeny testing** of females. This involves gathering data

FIGURE 19–18 Embryo transfer allows the production of many offspring from a superior female. *Courtesy of James Strawser, Photography, Inc.*

from the offspring of a particular animal. The data are analyzed to determine how valuable the animal is as a parent. Through the use of artificial insemination, a male can be progeny-tested in a short time because of the tremendous number of offspring that can be born and raised at the same time. The problem with progeny testing with females is that their offspring are limited in number, and this method does not allow sufficient numbers of offspring from which data can be collected. Through the use of embryo transfer, one female can produce many offspring in a short period of time, allowing for the testing of her progeny.

3. As in artificial insemination, embryo transfer permits the import and export of quality animals without the quarantine measures required of animals that are already born.
4. Embryo transfer allows the use of a dual production system. For example, by using embryo transfer, dairy cattle can produce calves that are pure beef animals. Dairy cows are bred so they will continue to produce milk. But if they are bred naturally or by using artificial insemination, the calves will still be half dairy animals. When beef calves are preferred, embryo transfer is the method used.
5. Implanting two embryos into a recipient female can produce twin offspring.
6. Producers can rapidly convert their herds from grade animals to purebred herds. By implanting a female of mixed breeding with purebred embryos, they can raise replacement animals that are both purebred and high quality.

Some argue that the use of embryo transplant has a big disadvantage. They say that if producers use embryo transfer and artificial insemination over a period of many years, the **genetic base** of the various breeds of animals will narrow. This means that in time there will be only a relative few animals that will eventually provide genetic material (egg and sperm) for the perpetuation of the breed. The fear is that producers, by demanding only embryos from the best animals, will eventually cause the loss of animals the producers feel are inferior. If the only animals of a breed that exist are related, there is reason to believe that this will lead to weakening rather than strengthening the breed. Others contend that through the use of embryo transfer, the importation of genetic strains from all over the world will prevent this problem from occurring. They contend that there are enough different strains in the world to make a narrowing of the genetic base highly improbable. Who is right? Only time will tell.

The Process of Embryo Transfer

The process of embryo transfer begins with the selection of **donor cows** and **recipient cows**. Cows selected as donors are usually animals that are of unusual value as breeding animals. They possess characteristics that are highly desirable to pass on to offspring (**Figure 19–19**). These characteristics might include high milking ability, growth ability, or reproductive capacity. Or, the characteristics might be the type in demand for the show ring. In any case,

FIGURE 19–19 Only high-quality females such as this one are used as donor cows. © LandFox/Shutterstock.com.

the donor animals are too valuable to produce an offspring only one time a year.

Producers may purchase frozen embryos from one of many companies that specialize in the sale of genetically superior embryos. The producer selects the embryos he or she wishes to order by analyzing data that have been compiled about the donor and the sire. These data usually consist of production data about the animal's ancestors and their progeny. In this way, the producer can select for those traits that will be of most use in the producer's herd. By contrast, recipient cows (those into which the embryo will be transferred) usually are cows of ordinary value, but these animals also are selected carefully. They must be healthy animals that are able to reproduce efficiently. They must be able to maintain pregnancy and to deliver a healthy, growing calf at the end of the gestation period. Some producers like to use recipient cows that have at least some dairy breeding so they will produce adequate milk for the calf.

After the donor and the recipient animals have been selected, both groups must be synchronized so they are at the same phase in their estrus cycle. This allows for the proper transfer of the embryo from one reproductive system to another. This synchronization is accomplished using the procedures discussed earlier. The only difference is that the donor animals undergo a process known as **superovulation**, which causes them to release several eggs instead of just one. In this way, as many as 12 to 15 eggs can be collected from one ovulation.

Superovulation is accomplished by injecting the donor with a follicle-stimulating hormone (FSH). This hormone causes the ovaries to produce several follicles instead of just one (the follicles provide a place for the growing and maturing of the egg). During the process, the female is injected with **prostaglandin** to cause her to come into estrus. About 48 hours later, the female should be in estrus or heat. At this time, the cows are artificially inseminated or are bred naturally. Because there are multiple eggs to fertilize, more semen must be used than would be used in regular artificial insemination.

Once fertilization occurs, the fertilized eggs (embryos) are allowed to grow for about a week before they are collected. In the earlier days of embryo transfer, the eggs were collected by removing the embryos surgically. This caused problems because of the scarring of tissue in the reproductive tract of the donor female. Today, the embryos are removed by a process called flushing. In this procedure a long, thin rubber tube called a **catheter** (**Figure 19–20**) is passed through the cervix and into the uterine horn. The catheter has an inflatable bulb about 2 inches from the end that fills like a balloon and seals the entrance to the uterus. Then a solution is then injected through the catheter into the uterus. When the fallopian tubes and the uterus are filled with solution, the flow

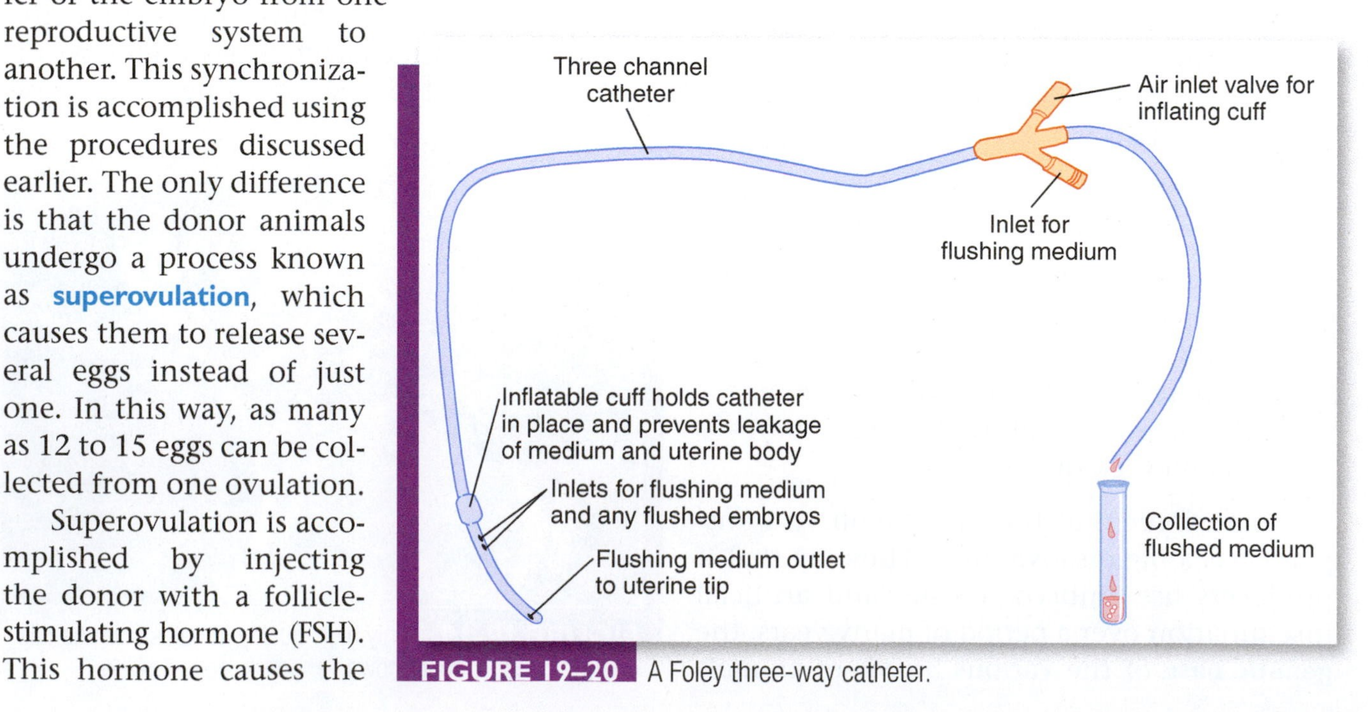

FIGURE 19–20 A Foley three-way catheter.

of the solution is stopped and the solution is drained off into a collection cylinder. The fertilized eggs (embryos) are carried out of the uterus with the solution. An average of about six embryos are collected with each flush. After the embryos have been flushed out, the uterus is flooded with another type of solution that kills any embryos that were missed. This solution also helps to prevent infection.

Once the embryos are collected in the solution, they are strained from the solution and examined under a microscope to determine their quality. Only embryos that are in the proper stage of maturity and appear normal and undamaged are used for transferring (**Figure 19–21**). The embryos may be transferred directly to a recipient female or may be frozen and stored for implantation at a later date.

The recipient cow is induced to come into estrus using injections of prostaglandins. When the corpus luteum reaches the proper stage, the embryo is placed in the uterus of the recipient cow (**Figure 19–22**). The pregnancy is allowed to progress as it would in a normal conception. Research has shown that pregnancies from embryo transfer are as likely to go full-term and deliver a normal calf as are pregnancies from natural conception.

FIGURE 19–22 Here, technicians are placing embryos in the recipient female.

New Technology in Embryo Transfer

Exciting possibilities for the use of embryo transfer exist in the not-too-distant future. In fact, much of the knowledge required to make these practices a reality is already known, if not feasible economically. For example, one of the techniques that producers would like to use is to be able to determine the sex of an embryo. Being able to transplant an embryo that they would be sure would produce a ram, a heifer, a boar, or a filly would benefit livestock producers tremendously. For example, dairy producers might like most of the calf crop to be female, to be used as replacements in the herd (**Figure 19–23**). If the embryos could be separated before implantation, all female embryos could be used. Or if a purebred breeder wants to produce mostly males to sell as herd sires, the process could be

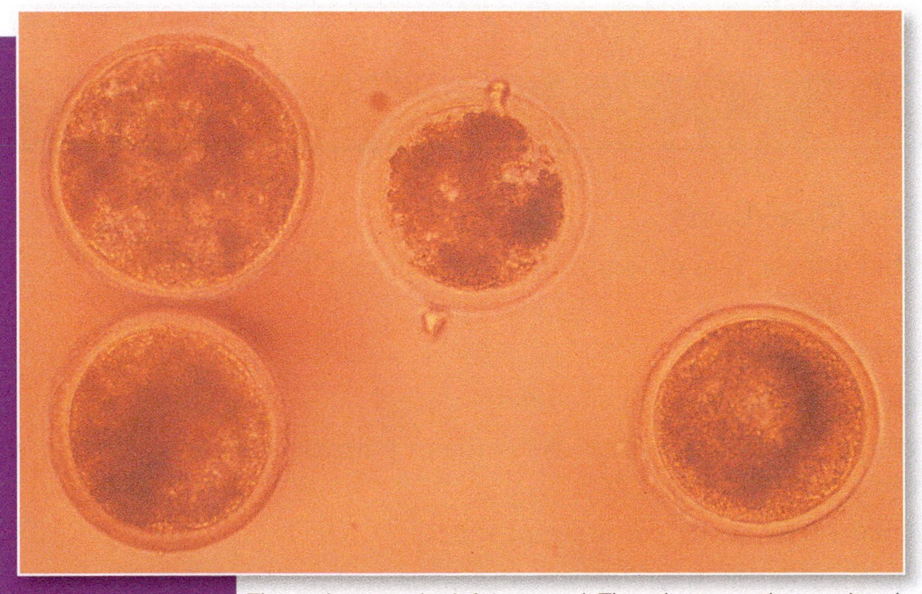

FIGURE 19–21 The embryo on the left is normal. The others are abnormal and not suitable for implant. *Courtesy of Ben Bracket, DVM, University of Georgia*

FIGURE 19–23 If embryos can be separated by sex, dairy producers could plan for a majority of calves born to be female. © iStockphoto/Oktay Ortakcioglu.

beneficial. Of course, scientists already are able to determine the sex of an embryo before it is implanted in the recipient female. The goal now is to make the process inexpensive enough to make sex determination profitable.

Another technology that holds promise is that of cloning. A **clone** is an animal that is genetically identical to another animal. Cloning involves the splitting of embryos into two or more parts that will produce genetically identical offspring. The ability to clone exists already and is rapidly becoming economically feasible. In fact, cloning now is done routinely on a commercial scale by several embryo transfer companies. This process will be discussed in the next chapter.

SUMMARY

The most important aspect of animal agriculture is reproduction. Through this process, new animals are brought into the world. For many years, the natural process of breeding has been controlled by humans with increasingly more dramatic results. The use of artificial insemination and embryo transplantation has revolutionized the entire industry of animal agriculture. Selective breeding has brought about gains in efficiency and the type of animal better suited for production. The exciting new world of gene manipulation and cloning will bring about changes we cannot yet comprehend.

CHAPTER REVIEW

Review Questions

1. What are the two different methods by which organisms reproduce?
2. What is a gamete?
3. What function does a gene serve?
4. Why do chromosomes always come in pairs (in higher-ordered animals)?
5. Explain why mules cannot reproduce.
6. What is the difference between meiosis and mitosis?
7. List and explain the steps in the production of spermatozoa.

8. Name and explain the purpose of the hormone produced by the male's testicles.
9. What purposes do the accessory glands of the male reproductive tract serve?
10. Name and give the functions of three hormones that assist in control of the female reproductive cycle.
11. What function does the cervix serve?
12. Explain how the female gamete (the egg) is produced in the ovary.
13. Where does fertilization take place?
14. Why is a large number of sperm needed to ensure that fertilization takes place?
15. What is the difference between a zygote and an embryo?
16. What are the advantages of using artificial insemination?
17. How is semen stored in order to last a long period of time?
18. What is estrus synchronization?
19. Why would producers want to synchronize the estrus cycles of their herds?
20. What are the benefits of embryo transfer?
21. How are embryos collected from the female?
22. What potential problem do some people see in the continued use of embryo transfer?

Student Learning Activities

1. From a local slaughterhouse, obtain reproductive tracts of beef and pork animals. In the laboratory, dissect and examine the tracts. To protect yourself from any disease organisms that might be present in the tracts, be sure to wear rubber gloves at all times.
2. Obtain samples of frozen semen and embryos from a local veterinarian. Following the veterinarian's instructions, carefully thaw the semen and embryos and examine them under the microscope. Look for sperm or embryos that may be damaged or of low quality.
3. Using the catalogs of breeder companies, compare the costs of semen and embryos of different animals. List some reasons why there would be a price difference. Explain why a producer might want to buy the cheaper or the more expensive semen or embryos.
4. Conduct a survey of livestock producers in the area. Determine how many of the producers use artificial insemination and/or embryo transfer. Ask for their reasons for or against using the techniques.

CHAPTER 20

Cloning Animals

STUDENT OBJECTIVES IN BASIC SCIENCE

When you have finished studying this chapter, you should be able to

- describe animals that have been successfully cloned.
- explain the difference in cloning derived from embryos and cloning derived from differentiated cells.
- explain the process of nuclear transfer.
- discuss why there can be differences in animal clones.

STUDENT OBJECTIVES IN AGRICULTURAL SCIENCE

When you have finished studying this chapter, you should be able to

- discuss the benefits of cloning animals.
- list ways in which cloning has already been used.
- explain how cloning and genetic altering can be used together to produce useful animals.
- discuss how the cloning of cattle has been made more efficient.

KEY TERMS

clone
surrogate mother
genetic engineering
genetic improvement
endangered species
genetic code
oocyte
zygote
differentiation
morula
nuclear transfer
enucleated oocyte
quiescent cells
genetically altered clone
genotype
phenotype

Cloning is the process of producing offspring that are genetically identical from a parent. Notice that the term "parent" is used instead of "parents." **Clones** are derived from a single parent without the usual mating process. In nature, there are animals such as starfish and sponges that reproduce asexually (reproduction from a single parent), but almost all higher-ordered animals reproduce through the uniting of a sperm and an egg. Remember from Chapter 16 that an offspring gets half of its genetic makeup from each parent and receives traits from both parents. In this way, no two animals are identical genetically. There is, however, an exception: Sometimes the fertilized egg divides and two or three embryos develop. When this happens, the two (twins) or three (triplets) offspring are identical in genetic makeup because the egg was already fertilized when it divided, so the genetic coding was already determined. In a sense, these twins or triplets are clones because they have the same genetic scheme (**Figure 20–1**).

Not all twins and triplets are identical. Sometimes the mother releases two or more eggs instead of one. Different sperm fertilize the different eggs, and twins or triplets occur that are not identical. In some animals such as pigs, dogs, and cats, this is common, and they have litters of offspring. In other animals, such as cattle and horses, twins and triplets do happen, but this is not common. Sheep may have twins or a single lamb.

FIGURE 20–1 Twins and triplets can be considered clones if they come from the same zygote. *© Calek/Shutterstock.com.*

The purpose of cloning is to reproduce animals from a single parent that will be genetically identical to the parent. This is different from identical twins because the egg divides after fertilization and the twins are genetically identical but are genetically different from either parent. Today, scientists can produce clones that are identical to a single parent. The process is extremely complicated and took many years of research to achieve.

Many types of animals have been cloned. Cattle, pigs, dogs, rabbits, cats, rats, and sheep have all been successfully cloned. However, the process is very expensive and time-consuming. If cloning is so expensive, the obvious question is: Why, with all of the modern techniques such as selective breeding, artificial insemination, and embryo transplant, do we want to clone animals? The following section discusses some of the rationale for conducting further research on the cloning process.

REASONS FOR CLONING

Cloning is controversial as well as expensive. Almost all producers have to make a profit to remain in business, so any aspect of production has to be cost-effective. Cloning is a reproduction method that cannot be done by producers or the average veterinarian. Labs are expensive, and they take a lot of time to produce cloned animals. Nevertheless, there are some good reasons for cloning.

Genetic Superiority

Perhaps the most important reason for cloning is to take advantage of genetic superiority. Even with the technology of artificial insemination and embryo transplant, the process of genetic improvement in animals can be slow. Also, keep in mind that with these technologies, the resulting offspring are still a combination of the genetics of both parents. Through the use of cloning, each cell in the body of a superior animal could theoretically produce

FIGURE 20–2 Through the use of cloning, each cell in the body of a superior animal theoretically could produce a new animal with the same gene pool. © Roger Hall/Shutterstock.com.

a new animal with the same gene pool (**Figure 20–2**). If these cells can be harvested, grown into embryos in the lab, and transplanted into **surrogate mothers**, there could be a tremendous increase in the efficiency of **genetic improvement**. A vastly superior animal could be used to produce a large number of offspring that theoretically would have the same characteristics as the parent.

One of the problems with selective breeding is that unwanted traits sometimes are passed along with the desirable traits in superior animals. Scientists are able to insert DNA into animal cells that control certain characteristics. This process, called **genetic engineering**, allows scientists to produce characteristics in animals that might not be possible through the normal selective breeding process. However, the process of inserting DNA into animal cells is a very complicated, time-consuming process and is expensive. If the cells from a genetically engineered animal could be cloned, the process could be made a lot more efficient. By carefully selecting the genes that carry only desirable traits, this problem could be eliminated. Theoretically, through the process of genetic engineering, a "superclone" could be developed that would be an ideal agricultural animal. This animal could then be reproduced through multiple clones.

Animal and Product Uniformity

Although genetic diversity has many advantages, a drawback is that genetically diverse animals raised together are given care and nourishment aimed at the *average* animal in the flock or herd. This means that there will be animals that need less nourishment or medication and there will be animals that will need more nourishment and medication. If the animals were all genetically identical, the environmental care, medications, feed, and other management techniques could be tailor-made for the entire group of animals in the flock or herd.

Also, animal products from cloned animals could be more uniform. If an entire flock of cloned chickens were raised in the same environment with the same management practices, the resulting chicken carcasses should be uniform. Drumsticks, as well as all other parts of the chicken, would all be close to the same size. The same would be true for any agriculture animal. If a particular market wanted small T-bone steaks, then cloned cattle that matured early and produced relatively small T-bone steaks could be raised in the same pen, managed in the same way, and sold at the same time through the same market (**Figure 20–3**). Both retailers and consumers could order steaks of a uniform size of their choice. This could make

FIGURE 20–3 Through cloning, cuts of meat such as T-bone steak could be made uniform and the proper size. © Brent Hofacker/Shutterstock.com.

the process of raising meat more efficient and ultimately benefit the consumer by lowering prices and increasing the quality of meat products.

Endangered Species

Most likely you have seen the movie *Jurassic Park,* where scientists use DNA harvested from the digestive tract of mosquitoes preserved in amber. The idea was that the mosquitoes were trapped in the amber during the Jurassic period, when dinosaurs roamed the earth. The blood in the mosquitoes' system was blood they took from dinosaurs, and thus the DNA of the dinosaurs was preserved. Because all cells of an animal carry the animal's entire genetic code, the DNA sequencing of the dinosaurs could be extracted and duplicated.

Although most scientists think this story line can never be brought to reality, many scientists think that cloning may be a solution to the extinction of animal species that are on the verge of dying out. Throughout the world, literally hundreds of different animal species are in danger of becoming extinct. Even with the best conservation efforts, whole species of animals continue to disappear. If new animals could be reproduced from the tissue of the few remaining animals, many **endangered species** might be saved (**Figure 20–4**). In fact, scientists have dreamed of bringing back long-extinct species through the cloning of preserved DNA.

FIGURE 20–4 If new animals could be reproduced from the tissue of the few remaining animals, many endangered species might be saved. *© Hung Chung Chih/Shutterstock.com.*

For example, wooly mammoths and mastodons have been extinct for thousands of years, yet frozen remains of these animals are periodically discovered in the arctic regions. The thought is that if enough intact DNA is preserved in the remains, an embryo of the animal might be cloned, using elephants as surrogate mothers. So far there has been no success with this goal, primarily because no DNA has been found that is complete and sound enough to be used for cloning.

Some success in the cloning of endangered species has already been achieved. In 2000, an endangered species was successfully cloned. The Asian gaur is an oxlike animal native to India and Burma that has been endangered for several years. The adults can reach a mature weight of over a ton and have been a favorite game animal for many generations of hunters. Overhunting and the destruction of their habitat have caused the populations of these animals to decrease drastically to the point of extinction.

Using cow eggs with the nucleus removed, DNA from the skin cells of a dead gaur was implanted into the eggs, and the resulting embryos were placed in the reproductive tract of cows. Altogether, 42 embryos were implanted into 32 cows; however, only one live birth of a real cloned gaur resulted. Even though the cloned calf, named Noah, died only 2 months after birth, scientists proved that the technology was possible. In the future, genetic scientists and wildlife biologists will attempt to clone many other endangered species.

Research

Perhaps the greatest advantage of cloned animals is their use as research animals. One of the biggest problems facing researchers is that of controlling all the differences among animals within a group or between groups. A group of animals selected for research will have different genetics even though they may be closely

FIGURE 20–5 Because they are genetically the same, cloned animals could greatly reduce the number of animals needed for a research study. *© Jason Benz Bennee/Shutterstock.com.*

related. Often, this interferes with the findings of a research study. For example, differences may show up that are related more to genetics than to the treatment given to the experimental group. Placing a large number of animals in both the experimental and the control groups usually controls this genetic difference (**Figure 20–5**).

Having to use large numbers of animals and having to repeat experiments many times greatly increase the cost and amount of time involved in conducting animal research. However, if genetically identical animals could be produced through cloning, the number of animals needed for the study would be much fewer and the results also might be more meaningful. The differences among animals in an experimental group because of genetic differences could be eliminated.

DEVELOPMENT OF THE CLONING PROCESS

For decades, scientists have known that all the cells in an animal's body contain the **genetic code** for the entire animal. This code is created when half the code is passed from the father and half is passed from the mother when the sperm and the egg unite during fertilization. After the egg, called an **oocyte**, is fertilized, it becomes a **zygote** with the complete genetic code intact. As discussed in Chapter 18, the zygote begins to divide into identical cells, and the process of cell division continues for 10 to 12 days until a ball-shaped mass of cells called a **morula** is formed. At this point, cells begin to change and to develop into different types of cells that will divide and grow to form bones, muscle, skin, and so on.

The cells from which the **differentiation** begins are called stem cells. Once the cells begin to differentiate, the genetic coding is locked into place, and bone cells produce nothing but bone cells and muscle cells produce muscles, and so on, even though the entire genetic code is contained in each cell. For many years, scientists thought that once the cells differentiated, the process could not be reversed into producing undifferentiated cells.

Scientists developed the first clones by dividing a zygote to produce two or more animals. The scientists stimulated each half of a zygote to continue the division process, and clones were born in much the same manner as ordinary identical twins in which the zygote divided naturally (**Figure 20–6**). It was long thought that this was the only means of creating clones because the process of differentiation could not be reversed.

Animal clones derived from differentiated cells first became a reality back in 1962 when a scientist named John Gurdon of Oxford University developed a procedure called **nuclear transfer**. Gurdon was able to take the DNA from a cell in the intestine of an adult frog

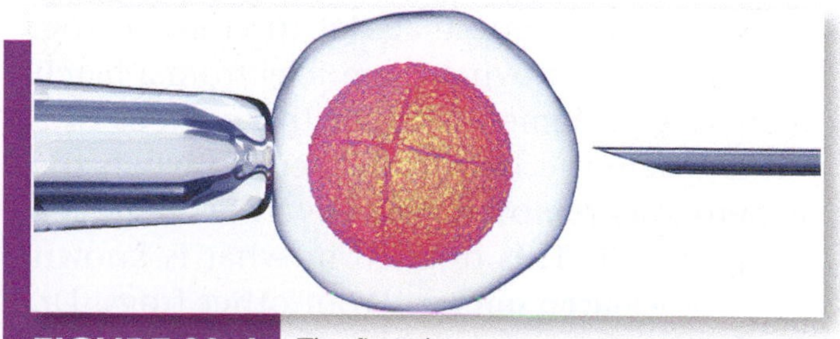

FIGURE 20–6 The first clones were created using microscopic tools to divide a zygote. *© the_lightwriter. Image from BigStockPhoto.com.*

NUCLEAR TRANSFER PROCEDURE

Holding Pipette
DNA
Polar Body
Enucleation Pipette

STEP 1: Remove DNA from unfertilized egg (enucleation).
Why: We do not want the genetic information from the egg (unknown genetics).

Enucleated Oocyte
Embryonic Cell

STEP 3: Place embryonic cell next to egg from Step 1 (enucleated egg).

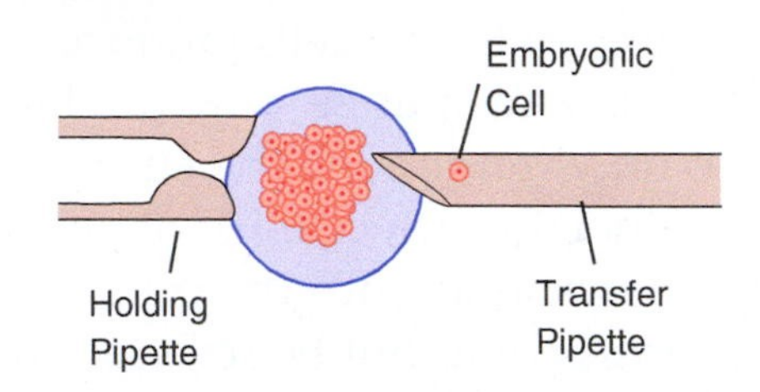

STEP 2: Pick up one cell from a good embryo.
Why: We want to use the good genetics of the embryo.

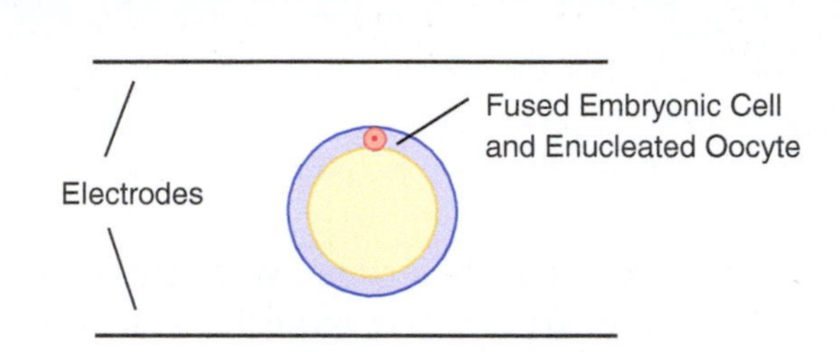

STEP 4: Fuse egg and cell by brief electrical pulse.
Why: Embryonic cell will supply genetic information and egg will supply cytoplasm to form new embryo (copy).

STEP 5: Repeat Steps 1-4 until each cell of the embryo has been used.
Why: Each cell has potential to form a copy of the original embryo. A 32 cell embryo could produce up to 32 copy embryos (clones) during one cycle of the cloning process.

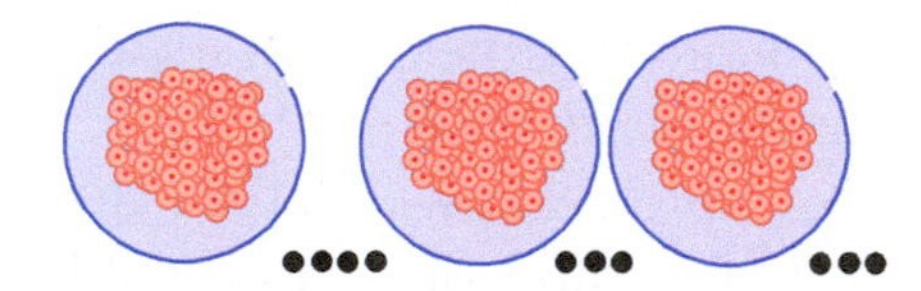

STEP 6: Culture clones for 5 to 6 days until they reach the morula or blastocyst stages. Transfer or freeze clones.
Why: The best stages of embryo development for embryo transfer or freezing are that of morula and blastocyst.

FIGURE 20–7 The nuclear transfer process. *Adapted from Dr. Carol L Keefer, ABS Specialty Genetics*

and use this genetic material to clone a frog. He began by removing the nuclei from a batch of frog eggs. Remember that the nucleus of a cell is where the DNA is located, so when the nucleus was removed, the DNA was removed along with it. This resulted in what is known as an **enucleated oocyte**. From other frogs, Dr. Gurdon took cells from the intestines, removed from the nuclei of the cells, and placed them into the enucleated eggs. Through this process of nuclear transfer, he created many cells with new nuclei. Although most of the cells died, some of the nuclear transfer cells began to behave in much the same way as a fertilized egg. These cells began to divide, and after a while a morula formed (**Figure 20–7**). From this structure, the cells began to differentiate and develop into tadpoles.

For the first time, a scientist had demonstrated that cell differentiation could be reversed. However, a large problem arose from the experiments: Even though the tadpoles appeared to be normal, they never developed into frogs. Other scientists duplicated Gurdon's research, but no one could get the cloned tadpoles to develop into grown frogs. The reason still remains a mystery.

Gurdon's procedure was tried many times with other animals. Scientists were particularly interested in cloning mammals such as pigs and cattle because of the usefulness and economic value of these animals. However, no one was able to get nuclear transferred mammal cells to grow and divide beyond a few cells. The scientific thought at the time (1970s and 1980s) was that for some reason the process would not work in mammals as it had in tadpoles. However, some scientists refused to give up and continued to research methods of cloning mammals. They were successful in the late 1980s when sheep, cattle, and rabbits were successfully cloned. However, these clones were accomplished by dividing embryos and were not the result of the nuclear transfer of DNA from an adult cell.

Cloning Mammals

A theory was developed in the 1980s when scientists began to study how cells divide and differentiate. For many years they had known that some cells divide more rapidly than others. Cells such as those that make up skin, hair, or inner organs go through their division cycle more rapidly than other body cells. This is because repairs have to be made as cells are injured or wear out on such exposed areas as skin. Cells go through periods of inactivity and are said to be **quiescent**. During these periods, cells do not divide because there is no need for rapid cell division. Then when the cells are needed, some mechanism that is still not very well understood triggers the cells into action.

In the 1990s, scientists at Roslin Institute in Scotland began to experiment with quiescent cells as a way of cloning mammals. They began by taking rapidly dividing cells from the mammary gland of a pregnant white-faced sheep and culturing the cells in the laboratory. As a treatment, they deprived the cells of nutrients to stop the cells from growing. Using previously developed technology, they removed the nucleus of an oocyte from a black-faced ewe, and the DNA from the white-faced sheep's mammary gland was placed in the cell. A small current of electricity caused the foreign DNA to enter and fuse with the cytoplasm in the enucleated oocyte. Nutrients then were added to cause the new cells to begin to divide. The dividing embryo then was placed in the reproductive tract of a black-faced sheep.

Of course, this was not a one-time process that achieved immediate results. The scientists created 277 new embryos from this process, and of these, they recovered only 29 that were good enough to transplant into ewes. Thirteen ewes were given either one or two of the embryos, and of these 13, only one grew to full term and produced a normal, healthy lamb. The lamb had a white face and was born to a black-faced surrogate mother, so the scientists could be almost sure that the lamb was a product of their cloning.

All doubt was removed when DNA fingerprinting proved that the DNA from the cloned sheep, named Dolly, matched the cells from the tissue taken from the mammary gland. Dolly created quite a phenomenon in the media. TV and radio newscasts all over the world carried stories about the birth of Dolly. The Roslin Institute estimated that in the first week after the story broke, the scientists there received more than 2,000 phone calls about the clone. In addition, during that same week, they talked at length with nearly 100 reporters, and Dolly was filmed by 16 film crews and photographed by more than 50 media photographers.

Many of the news stories raised ethical concerns about cloning animals. Questions were asked about whether Dolly was a "normal" animal and the type of creature she might develop into. Also, concerns were voiced about the likelihood that the research might lead to the cloning of humans. Since that time, Dolly has matured and produced normal lambs of her own, and fears about her have subsided. Several stories were published about abnormalities with Dolly, but almost all of these have been proven to be unfounded. For example, it was once thought that Dolly was aging more rapidly than normal. However, a closer examination revealed that while the arthritis in her joints was unusual, it is known to occur in sheep of her age. Speculation is that the problems were not related to cloning.

Genetically Altered Clones

Later, the Roslin Institute produced two cloned lambs, named Molly and Polly, that had been genetically altered. The animals were given a human gene responsible for the production of a protein that aids in the clotting of blood after an injury. The protein, Factor IX, could be very useful as a pharmaceutical in treating human patients who suffer from hemophilia. The idea is that if genetically altered animals can be mass-produced, the substance can be commercially produced at a much lower cost than by conventional methods.

The first **genetically altered clones** of calves were produced in 1998. Dr. Steven Stice and Dr. James Robl of the University of Georgia developed two genetically altered cloned calves named George and Charlie (**Figure 20–8**). These researchers placed two genes, a genetic marker and a gene that makes cells resistant to antibiotics, into Holstein cattle DNA that was used to clone the two calves. Although the main benefit gained from this procedure was research, the process proved that genetically altered calves could be produced.

Later, Dr. Stice led efforts to clone calves possessing a gene to enable cows to produce milk with the human serum albumin. Every year, around 440 tons of human serum albumin are used in hospitals to treat patients. If this process can be made commercially feasible, a tremendous boon to the medical establishment will be achieved. Also, it will open the door for many more beneficial products that can be produced in this way.

FIGURE 20–8 These two calves were the first genetically altered cloned calves. *Courtesy of Dr. Steven Stice, University of Georgia*

Another breakthrough occurred in 2002 at the University of Georgia when Dr. Stice succeeded in cloning a calf from cells taken from a cow that had been dead for 48 hours. The female calf grew normally, matured, and had a normal healthy calf by the natural breeding process. The cells were taken from a kidney of the slaughtered animal because the kidneys usually remain with the carcass until the carcass is divided into retail cuts. The importance of this technology is that the possibility may exist to take cells from a superior carcass and clone animals to produce similar carcasses. If a truly superior carcass is found, the animal is already dead, and reproducing the animal would be possible only through cloning (**Figure 20–9**).

PERFECTING THE PROCESS

Several times, research has shown that mammals can be cloned successfully. The challenge now is to make the process easier and more efficient. Remember that when Dolly was cloned, 277 cloned embryos resulted in only one lamb, and of the 13 ewes that received 29 healthy embryos, only Dolly was carried to full-term.

FIGURE 20–9 KC was the first animal cloned from an animal that had been dead for 48 hours. *Courtesy of Gary Farmer*

FIGURE 20–10 All eight of these calves were cloned from the same cow. *Courtesy of Dr. Steven Stice, University of Georgia*

Obviously, this success rate is far too inefficient to be of any practical use. If cloning is ever to achieve the potential outlined by futurists, the process will have to be greatly refined and perfected.

In 2001, a team of scientists led by Dr. Stice at the University of Georgia reached a milestone in this effort when eight calves cloned from the same adult cow were born (**Figure 20–10**). All of the calves were born from different surrogate mothers over a period of about 2 months. The significance of this achievement is that in the past, the best viability rate of cloned cattle embryos was about 1 in 20. Dr. Stices's team reduced that rate to 1 in 7.

The team used eggs harvested from the cattle ovaries obtained from the slaughter plant. Using the eggs as host cells, the nuclei were removed in much the same way as described for the previous research. The process differed in that a chemical inhibitor was applied to the cells from the donor cow. This chemical makes the cell more uniform in preparation for cloning, meaning that the DNA material used for cloning was made more uniform through the use of the chemical inhibitor.

Today, cloning has become more commonplace. Several companies in the United States provide cloning services for horses, livestock, and even pets. A good example is an outstanding Holstein bull named Starbuck that produced semen units that amounted to close to $25 million. His highly productive offspring were born in countries all over the world. When he died at age 19 years, he was cloned and his offspring Starbuck II was born with the same genetics as Starbuck.

DIFFERENCES IN CLONES

An interesting phenomenon about cloning is that there can be observable differences in clones. Take a close look at the calves in **Figure 20–11**. Can you detect any differences in the calves? Obviously, the size difference is due to the differences in the ages of the calves. But what about the different color patterns on the animals? Since they all have an identical genetic makeup, how can the patterns be so different?

FIGURE 20–11 The color pattern on cloned calves may be different. *Courtesy of Dr. Steven Stice, University of Georgia*

Remember from the discussion of genetics in Chapter 16 that the **genotype** of an animal is the actual genetic makeup and the **phenotype** is how the genes are expressed or how the animal actually looks. Environmental factors can have a large impact on an animal's phenotype. Remember that all the calves were born to different surrogate mothers. Differences in nutrition, condition of the placenta, or the degree of heat absorbed by the fetus can all affect the color patterns of individual calves. Also notice that the color pattern on the heads of all the calves appears to be the same. Head markings on calves do not seem to migrate during gestation as do markings on other regions of the body. With all the knowledge we have accumulated about cloning animals, there is a tremendous amount yet to learn!

SUMMARY

Until recently, the cloning of animals has been a fantasy of science fiction writers. In recent years, several breakthroughs in our understanding of DNA and the gene transfer process have allowed scientists to make this concept a reality. Several animals including rabbits, cattle, and sheep have been successfully cloned using DNA from adult animals. Many benefits can be gained by the cloning of animals if the process can be made commercially feasible. Although the technology to make this happen is still just around the corner, most geneticists believe that the techniques can be developed to the point where the cloning of agricultural animals is a common occurrence. Given concerns raised over cloning, it remains to be seen whether or not the public will accept cloning as part of our society.

Despite the controversy surrounding animal cloning, research and development of this technology have continued at an ever-increasing pace. Predictions are that in the near future, cloned animals will be commercially feasible and a common occurrence. In fact, large mammals such as cattle and sheep have already been cloned and have received a lot of media attention. Several large companies are investing a huge amount of time and resources into the development of animal cloning. A discussion of these concerns will be addressed in another chapter.

CHAPTER REVIEW

Discussion Questions

1. List three benefits of cloning animals.
2. Explain the benefit of making all the animals in a herd uniform.
3. How can the use of cloned animals reduce the cost of conducting animal research?

4. List and explain the steps involved in nuclear transfer.
5. What is the difference between a stem cell and a differentiated cell?
6. What were some of the concerns over the cloning of Dolly the sheep?
7. What is the advantage of using genetically altered embryos for cloning?
8. Explain how genetically identical cloned calves may have different color patterns.
9. What is the difference between a phenotype and a genotype?
10. Discuss how cloned animals may be used for pharmaceuticals.

Student Learning Activities

1. Search the Internet and locate information on cloning pets. Prepare a position paper on your thoughts concerning the use of cloning techniques to re-create a beloved pet that has died. Should time and funding be invested in this technology? Share your findings and conclusions with the class.
2. Read a story or watch a movie that depicts cloning as science fiction. Was the plot feasible and realistic? Record your thoughts, and share them with the class.
3. Conduct a survey of teachers and students in your school to determine attitudes toward animal cloning. Ask if they can give examples of animals that have been successfully cloned. Find out whether they think scientists should develop and perfect the cloning process. Do they think animal cloning will lead to human cloning? Choose a species of agricultural animal, and write a report on the benefits of cloning that species. If the animal could be genetically altered before it is cloned, what beneficial traits should be added? Share the results with the class.

CHAPTER 21

Animal Growth and Development

STUDENT OBJECTIVES IN BASIC SCIENCE

As a result of studying this chapter, you should be able to

- describe how an animal grows.
- distinguish between prenatal and postnatal growth.
- define mitosis.
- explain the three phases of prenatal growth.
- define cleavage.
- list the layers of the blastula and the organs that are derived from each layer.
- describe the function of the placenta.
- explain how muscle cells are different from most body cells.
- discuss the sequence in which fat tissue is deposited in an animal's body.
- define the role of fat cells.
- explain the effects of hormones in the growth process.
- describe the aging process in animals.
- distinguish between chronological and physiological age.

STUDENT OBJECTIVES IN AGRICULTURAL SCIENCE

As a result of studying this chapter, you should be able to

- explain why animal growth is so important to producers of agricultural animals.
- explain why selection for muscling in breeding cattle is important.
- explain the phases of an animal's life in which the most rapid growth occurs.
- discuss the effect of castration on the growth of animals.
- define the lean-to-fat ratio.
- explain the effects of aging on the productivity of animals.

KEY TERMS

prenatal
postnatal
hyperplasia
hypertrophy
cytoplasm
placenta
amniotic fluid
ectoderm
mesoderm
endoderm
morphogenesis
cartilage
ossification
organic matter
artificial hormones
castration
lean-to-fat ratio
barrows
aging
chronological age
physiological age
vertebrae
collagen

ALL ANIMALS MUST GO through the process of growth in order to develop from a fertilized egg into a mature adult. During their lives, animals go through several stages of development in which their growth takes place in slightly different ways. A vast amount of both basic and applied research has been completed to better understand how growth occurs and how to make the growth process more efficient. Animal producers earn their living based on the amount of growth that occurs in animals. This is true whether the producers raise cattle for beef, sheep for wool, or horses for pleasure riding. Without this growth, animal products would be nonexistent (**Figure 21–1**).

Growth is generally defined as an increase in the size or volume of living matter. An animal begins as a microscopic speck and grows within its mother until a certain mass of body weight and degree of maturity are achieved. A newborn calf may weigh 85 pounds at birth, but by the time this animal is mature, it may weigh more than a ton (**Figure 21–2**).

The two major phases of growth in the animal's life are **prenatal** and **postnatal**. Prenatal refers to occurrences that take place before the animal is born; postnatal refers to occurrences after the animal is born. During prenatal growth, all of the organs of the animal's body will be formed. When the animal is born, the organs begin to function in order for the animal to live and grow outside the mother's womb. During postnatal growth, the animal increases in size and the body systems develop and mature. During both prenatal and postnatal growth, the increase in size is a result of the cells increasing in size or in number. The increase in the number of cells is called **hyperplasia**, and the increase in the size of the cells is called **hypertrophy**. Growth in the size of cells usually is a result of the accumulation of materials such as protein or calcium in the cytoplasm of a cell. **Cytoplasm** is the material within the cell wall that does not include the nucleus.

FIGURE 21–1 Producers depend on the growth of animals to make a living. *Courtesy of North American Limousin Association*

FIGURE 21–2 Calves begin small but may grow to weigh more than a ton. *© Ramona Heim/Shutterstock.com.*

PRENATAL GROWTH

Prenatal growth may be divided into three phases: ovum, embryonic, and fetal.

The Ovum Stage

The ovum phase lasts from fertilization of the ovum by the sperm until the mass of cells attaches to the wall of the uterus. This phase lasts about 10 days in sheep and 11 days in cattle. During this period, little change takes place, and the shape of the cell mass remains approximately spherical.

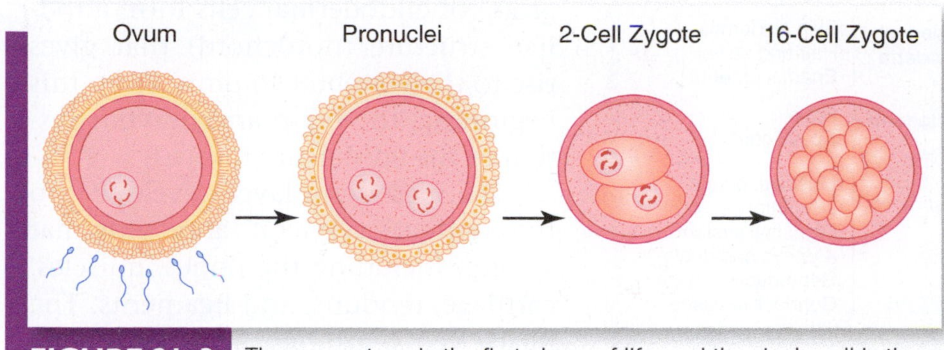

FIGURE 21–3 The ovum stage is the first phase of life, and the single cell is the beginning of animal life.

The single cell (fertilized egg) is the beginning of animal life (**Figure 21–3**). Remember from Chapter 15 that the sperm cell and the egg cell develop in a process called *meiosis*. Each of these cells contains half the number of chromosomes needed to create a new animal. When the two gametes (the egg and the sperm) unite at fertilization, they combine to supply the correct number of chromosomes for the new animal.

After fertilization, the fertilized cell must reproduce itself so it can begin to grow. This process of cell division is called *mitosis*. Through mitosis, the nucleus of a cell divides and the DNA is replicated so each cell contains exactly the same genetic information. As the number of these cells increases, the fertilized egg develops into a tight ball containing a large number of cells. This process is called *cleavage*.

As the cells multiply and divide, they form a spherical mass called the **morula**. The cells of the morula are arranged to form an outer layer and a central core (**Figure 21–4**). As the mass of cells continues to grow, a fluid-filled cavity develops in the center of the group of cells. This mass of cells with a cavity in the center is called a blastula. From the blastula, the cells begin to differentiate.

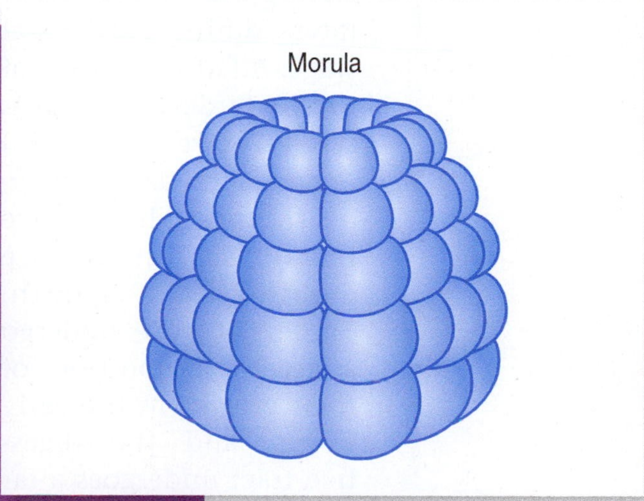

FIGURE 21–4 As the cells multiply, they form a spherical mass called the *morula*.

The Embryonic Phase

During the embryonic phase, the major tissues, organs, and their major systems are differentiated. The embryo is attached to the uterus, and a saclike pouch develops around the embryo. This enclosure is called the **placenta**, and it contains fluid known as the **amniotic fluid**. The purpose of the placenta is to give the embryo (fetus) nutrition and oxygen from the mother. Also, the placenta absorbs waste materials, and the mother's lungs and kidneys dispose of these wastes. The amniotic fluid in the placenta serves as a mechanism to absorb shock and to protect the fetus. It also provides a lubricant when the fetus is moving through the birth canal.

The embryonic phase lasts from the 10th day to the 34th day in sheep and from the 11th day to the 45th day in cattle. During the embryonic phase, the body undergoes a series of successive changes without much weight gain.

The blastula goes through a process that begins the development of all the organs and tissues in the animal's body. This process begins with the cells dividing into three layers (**Figure 21–5**). The outer layer is called the **ectoderm**, the middle layer is called the **mesoderm**, and the inner layer is called the **endoderm**. All parts of the animal's body develop from these layers. This process of cell development into different tissues and organs is known as **morphogenesis**.

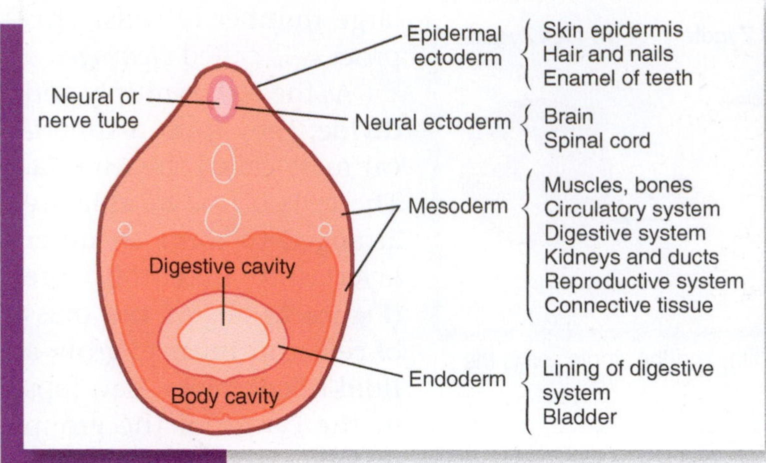

FIGURE 21–5 From the blastula, the cells begin to differentiate into distinct types of tissue.

After the layers are formed, the outer layer (the ectoderm) develops into the tissues that are on the outside or near the surface of the animal's body. These include the skin, hair, hooves, and certain endocrine glands (ductless glands that secrete hormones into the bloodstream). In addition, this layer forms the central nervous system—the brain, the spinal cord, and all the nerve branches. The development of the brain and spinal cord begins with formation of the neural tube. The neural tube is formed by a thickened strip of ectodermal cells that fold together to form a tube. Another group of endodermal cells form a rodlike structure (notochord) that gives rise to the vertebral column. From this beginning, the brain and nervous system of the animal are formed.

The mesoderm layer develops into the animal's skeletal and muscular system, including the bones, muscles, cartilage, tendons, and ligaments. The heart, veins, arteries, and other parts of the circulatory system also develop from this layer. In addition, the reproductive system is formed from the mesoderm layer. The inner organs, such as the liver and the digestive system, develop from the endoderm layer. The endoderm forms the digestive system, lungs, liver, and other endocrine glands.

The animal begins to take on a recognizable form in the late embryonic stage (**Figure 21–6**). Scientists do not fully understand what triggers cells into taking on different forms in the differentiation process. They believe that the answer is in the genetic code of the DNA. As more and more genetic research is completed, scientists are realizing that this code is much more complicated than they once thought.

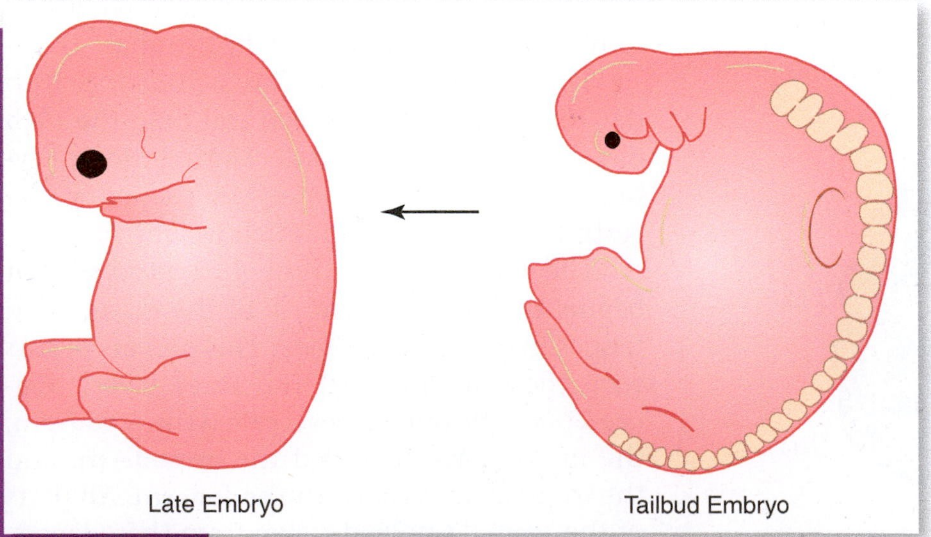

FIGURE 21–6 The animal begins to take on a recognizable form in the late embryonic stage.

The Fetal Stage

Animals of different species grow at different rates during the fetal stage. All of the organs of the body develop at this stage (**Figure 21–7**). Vital organs such as the liver, heart, and kidneys have functional importance during fetal growth. These organs undergo a greater proportion of their growth in the early stages, and the digestive tract undergoes a far greater proportion of its growth in the later stages.

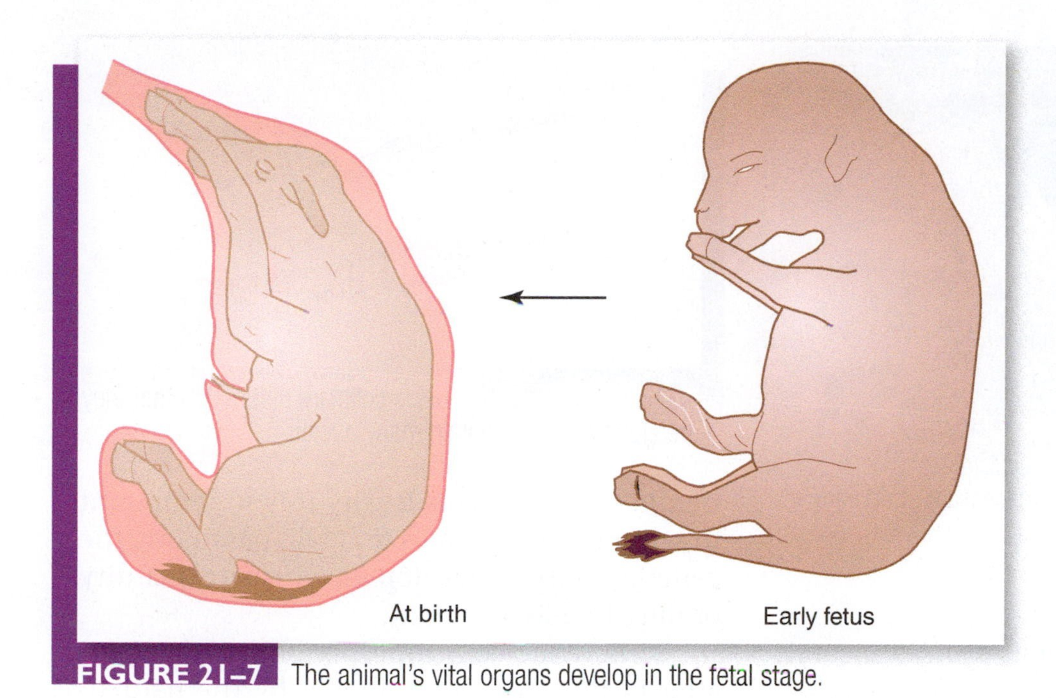

FIGURE 21–7 The animal's vital organs develop in the fetal stage.

Continuous changes in the fetus occur during the fetal stage. As the fetus matures, it becomes less and less dependent on the mother. When the fetus is able to live on its own, a hormone called *oxytocin* stimulates the muscles of the uterus into contracting. The birth canal relaxes, and the new animal is expelled from the uterus. As the animal hits the ground, the lungs are stimulated into functioning and the animal begins to live on its own (**Figure 21–8**).

The reason is that the animal must have fully functioning organs, such as the heart and lungs, before it is born. Because nutrients are passed from the mother's bloodstream, the digestive system of the fetus is not as vital.

Growth of the various parts of the fetus and the organs is characterized by changing rates of growth. This means that various organs grow faster at times than other organs.

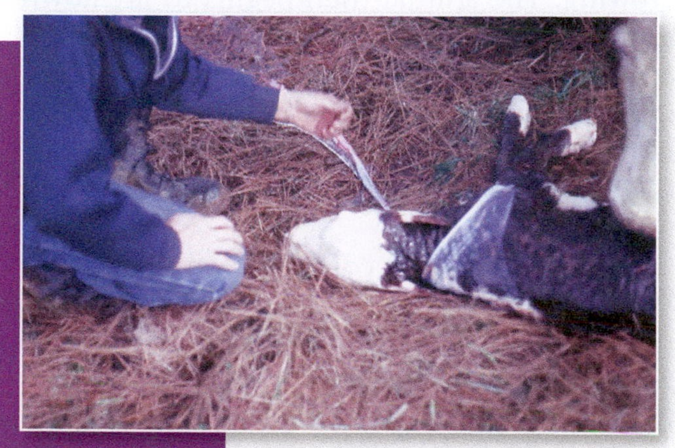

FIGURE 21–8 At birth, the animal is expelled from the uterus. When the animal hits the ground, its lungs are stimulated into operating. *Courtey of Cooperative Extension Service, Univeristy of Georgia*

POSTNATAL GROWTH AND DEVELOPMENT

After an animal is born, the parts of its body do not grow and develop at the same rate, nor is the development rate of different species the same. However, the order in which the animal's parts and systems develop is much the same in all species. Generally, tissues develop in the order of importance to the animal's survival.

In most species of animals, the head comprises a larger portion of the body at birth than at any time during the animal's postnatal development. The head contains the brain, and this organ directs not only the growth but also the functions of all the other systems and organs. The animal's legs tend to comprise a larger portion of the body at birth than at later stages. Well-developed legs are necessary because young animals must be able to stand and nurse (**Figure 21–9**) or be able to get away from predators. The brain, central nervous system, heart, and circulatory system are all well developed at birth. These organs are essential because when

FIGURE 21–9 Strong legs are necessary at an early stage so animals can stand and nurse. *© iStockphoto/cgbaldaug.*

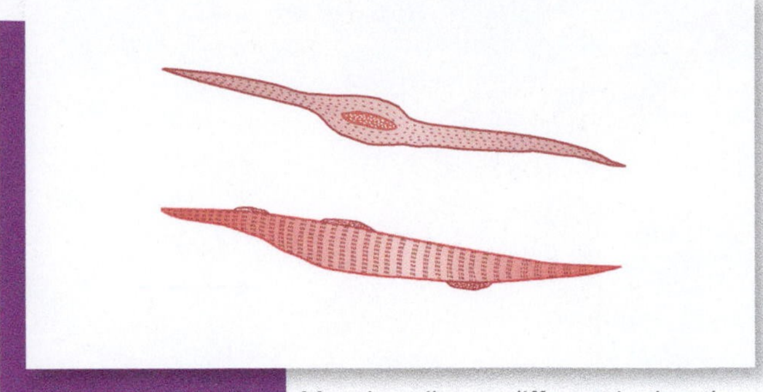

FIGURE 21–10 Muscle cells are different in that they are elongated and contain many nuclei.

the animal is expelled from its mother's womb, it must survive on its own.

The respiratory system and the digestive system of animals usually develop soon after birth. The first milk from the mammary glands of the mother is called *colostrum*. This milk is rich in nutrients and passes immunity from the mother to the newborn. Also, the colostrum cleanses the digestive tract and stimulates it into functioning.

Animals grow rapidly from the time they are born until they reach sexual maturity. In the brief period just after birth, the animal grows relatively slowly as the organs adjust to functioning outside the mother's placenta. After that period, the animal begins a stage of rapid growth when bone and muscle tissue grow steadily. This period continues until the animal reaches sexual maturity. During this phase, the animal achieves the fastest growth rate and the greatest feed efficiency.

The size of an animal depends for the most part on the size and amount of bones and muscles in the animal's body. All of the animal's muscle cells are in place by birth. Muscle cells are different from most other cells in that they are long and relatively thin and contain many nuclei (**Figure 21–10**). After the animal is born, growth in the animal's muscle system results from increases in the size of, but not in the number of, cells. This is why it is so important to select parent animals that will pass along the genetic ability to develop an adequate quantity of muscle cells.

Bone tissue cells multiply both before and after birth. Bones grow longer by the hardening of **cartilage** tissue at the end of the bones. Cartilage tissue is softer than bone and solidifies as the bone matures. Once the cartilage solidifies all the way to the end of the bone, the bone ceases to grow. This process is called **ossification**. When bones mature, they are made up of about half minerals; the two main minerals are calcium and phosphorus. The other half is made up of **organic matter** such as protein.

After the animal reaches sexual maturity, it continues to grow, although it grows at a slower pace until the muscle and bone stop growing. The animal then begins to lay down layers of fat. Fat is present in all phases of the animal's growth if there is adequate nutrition in the diet to allow for the storage of fat. Fat is nature's way of storing energy for the animal to use when it is not getting as much feed as it normally does.

An animal's body contains two types of fat: White fat stores energy; and brown fat functions to maintain the animal's body heat. The animal first deposits fat in the abdominal cavity, where the fatty tissue serves to cushion the internal organs. After a sufficient amount is deposited in the abdominal cavity, the fat is then placed between the muscles; this tends to give the animals a sleek, smooth appearance.

The last place for storage of fat is inside the muscles. In retail cuts of meat, these deposits of fat are called marbling.

THE EFFECTS OF HORMONES ON GROWTH

Hormones are chemical substances formed by the endocrine glands and secreted into the bloodstream. They are carried to other parts of the body where they stimulate organs into action or control bodily processes. The growth of an animal is regulated by hormones. The more important of the glands that secrete hormones affecting the growth of animals are the pituitary, thyroid, testicles, ovaries, and adrenal glands. **Table 21–1** shows the hormones that are responsible for the stimulation and regulation of animal growth. These hormones regulate the ultimate size of the animal. If there is an excessive amount, the animal will be abnormally large or malformed. An absence of the hormones can cause the animal to be a dwarf.

Producers make use of **artificial hormones** to cause animals to grow more rapidly or more efficiently. These hormones or hormonelike substances are implanted in the animal (usually in the ear). The implants slowly release a very small amount of the growth stimulant over a period of time. The Food and Drug Administration (FDA) closely regulates the amounts of these substances that are allowed to be present in the meat from implanted animals (see Chapter 26).

Castration of males slows growth and increases fat deposition in cattle, sheep, and swine. As indicated in Table 21-1, the testicles produce testosterone that stimulates growth to a certain degree. When the testicles are removed, the production of this hormone stops and growth is somewhat affected. The degree of this effect varies within the species. It also affects the amount of lean and fat in the animal's body. This is referred to as the **lean-to-fat ratio**. In cattle and sheep, the female has the lowest lean-to-fat ratio (**Table 21–2**); a larger portion of the female body contains fat than contains lean. Uncastrated males have the highest lean-to-fat ratio, and castrated males fall somewhere in between. However, **barrows** (castrated male pigs) have a lower lean-to-fat ratio than gilts (females that have not had a litter), because barrows reach maturity earlier and have more fat deposition than gilts. Animals with the highest lean-to-fat ratio (intact males) usually have the least marbling. This is especially true for bulls.

Gland	Hormone	Effects on Skeleton	Effects on Protein Metabolism
Pituitary	Somatotropin (STH)	Stimulates growth and closure of long bones	Increases nitrogen retention and protein synthesis
Thyroid	Thyroxin	Stimulates growth of long bones; essential for STH effect	Controls body metabolism by increasing energy production and oxygen consumption by tissues
Testicles	Testosterone	High dose is weak stimulator of epiphyseal closure and inhibits STH; low dose increases the width of epiphysis and helps STH effect	Stimulate s nitrogen retention; promotes muscle growth and development of sex characteristics
Ovaries	Estrogens	Inhibits skeletal growth; promotes epiphyseal closure	Increase nitrogen retention and protein synthesis in ruminants
Adrenal	Glucocorticoids	Decreases growth of epiphysis; decreases stimulation of epiphysis by STH	Increases protein and amino acid degradation; inhibits protein synthesis and increases fat deposition

TABLE 21–1 Hormones have quite an effect on the way animals grow.

Lean-to-Fat Ratio	Cattle	Sheep	Swine
	Bull	Ram	Boar
Intermediate	Steer	Wether	Gilt
Lowest	Heifer	Ewe	Barrow

TABLE 21–2 Lean to Fat Ratios

THE AGING PROCESS IN ANIMALS

Aging, from the time of birth to the time of death, involves a series of changes that lead to deterioration of the animal and eventually to death. There are two different terms for aging in animals. **Chronological age** refers to the actual age of the animal. **Physiological age** refers to the stage of maturity the animal has reached.

In determining the physiological age of an animal after it has been slaughtered, the most common method is that of examining the bones. As mentioned earlier, as bones mature, they harden. Soft cartilage tissue, such as that on the rib cage and the **vertebrae**, continues to solidify throughout the animal's life. Physiological age can be determined by the degree to which these cartilage tissues have solidified.

It is said that as soon as an animal is born, it begins to die. This is true from a physiological sense because after formation of the embryo, many tissue cells stop dividing and only the cells essential to life (skin, blood, etc.) continue to divide. The animal's lifespan is directly related to the rate of growth or the length of time to reach maturity. Sheep mature in about a year and have a life expectancy of about 10 years. Cattle require 2 to 3 years to reach maturity and have a life expectancy of 20 to 25 years (**Figure 21–11**).

The physiological functions of animals decrease with age. Muscular strength and speed decline, reproductive organs secrete lower levels of hormones, and the time of recovery from substance imbalances increases with age. There is a gradual breakdown of separate nerves, the nervous system, and glandular control involved in aging. Wrinkles form as **collagen**, and protein become less elastic.

Production by agricultural animals is often limited by age. This is because animals tend to lose their teeth as they age and become inefficient. Mothers produce less milk as they age, so their young do not grow as rapidly. Some animals, such as the sow, reach excessive size, causing management problems. Sows are culled at 2 to 3 years of age. Cows usually are culled by age 10 to 11 years, ewes are culled at about age 7 to 8 years, and sows are culled at 2 to 3 years of age. The factors affecting how long an animal may live or be productive may be genetic or environmental. How the animals are cared for determines to a large extent how well they do at an advanced age.

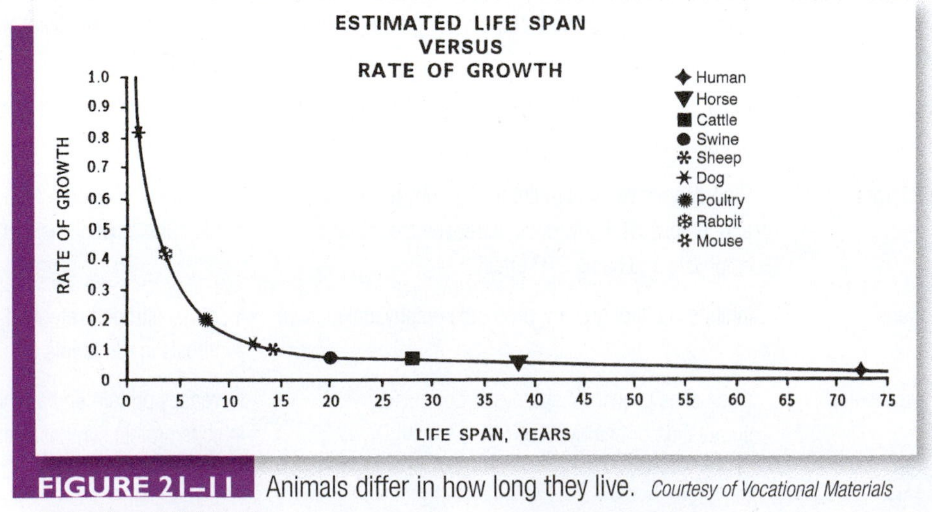

FIGURE 21–11 Animals differ in how long they live. *Courtesy of Vocational Materials Service, Texas A&M University*

SUMMARY

Animals begin to grow at birth and in some ways continue this process until they die. All the body's systems must develop, grow to the proper size, function, and replace cells as they are lost or wear out. Growth and development are what the animal industry is all about. Ensuring that animals grow properly is the job of all those connected with the industry, from conception to the slaughter process.

CHAPTER REVIEW

Review Questions

1. Why is animal growth essential to producers?
2. By what two methods does growth occur?
3. What is the difference between the morula and the blastula?
4. What is the purpose of the placenta?
5. Name three layers of the blastula.
6. What organs arise from the ectoderm?
7. What is meant by differentiation?
8. What is oxytocin?
9. Why is it essential that newborn animals have well-developed legs?
10. At what stage in an animal's life does the most rapid growth occur?
11. What is meant by ossification?
12. List the sequence in which an animal's body deposits fat.
13. List five glands that secrete hormones that affect growth.
14. What is meant by the lean-to-fat ratio?
15. Tell the difference between chronological and physiological age.

Student Learning Activities

1. Obtain from a slaughterhouse the fetuses from animals that have been slaughtered. Try to determine the stage of maturity for each. Be sure to wear latex gloves when handling the material.
2. Visit with a producer and determine how he or she determines when animals have reached the level of maturity desired for market. Try your hand at making the determination of animals in the producer's herd.

CHAPTER 22

Animal Nutrition

STUDENT OBJECTIVES IN BASIC SCIENCE

As a result of studying this chapter, you should be able to

- explain why animals must have nutrients.
- list the six nutrients that are essential to life.
- discuss the role of water in supporting life.
- discuss the relationship between proteins and amino acids.
- distinguish between a carnivore, an omnivore, and an herbivore.
- discuss the importance of protein in animals' diet.
- discuss the importance of carbohydrates.
- list the types of common sugars.
- distinguish between a starch and a sugar.
- discuss the importance of fats in the diet of animals.
- discuss the role minerals play in sustaining life.
- list the vitamins that are important in an animal's diet.
- discuss the functions of vitamins.
- distinguish between a monogastric digestive system and a ruminant digestive system.
- list and define the function of the organs of the monogastric digestive system.
- list and define the function of the organs of the ruminant digestive system.

STUDENT OBJECTIVES IN AGRICULTURAL SCIENCE

As a result of studying this chapter, you should be able to

- name the sources from which protein is obtained for feeds.
- list the common grains that are used as a source of carbohydrates.
- distinguish between a concentrate and roughage.
- list the sources of fats in animal rations.
- list the sources of minerals in animal feeds.
- list sources for the various vitamins that are of use to animals.
- explain the differences in the feed used by monogastrics and the feed used by ruminants.

KEY TERMS

maintenance ration
feedstuff
anabolism
catabolism
amino acids
essential amino acids
nonessential amino acids
crude protein content
carnivores
herbivores
tankage
cellulose
monosaccharides
disaccharides
glucose
fructose
galactose
sucrose
lactose
lipids
inorganic
macrominerals
microminerals
trace minerals
free choice
carotene
monogastric
gastrointestinal tract
alimentary canal
ruminant
cecum
esophagus
pepsin
duodenum
jejunum
ileum
semipermeable membrane
diffusion
boluses (*sing.*, bolus)
rumen
reticulum
omasum
abomasum
mucous membranes
bloat

NATIONAL AFNR STANDARD

AS.03.01.01.a
Identify and summarize essential nutrients required for animal health and analyze each nutrient's role in growth and performance.

AS.03.01.01.b
Differentiate between nutritional needs of animals in different growth stages and production systems.

AS.03.01.01.c
Assess nutritional needs for an individual animal based on its growth stage and production system.

AS.03.01.02.a
Differentiate between nutritional needs of animal species.

AS.03.02.01.a
Compare and contrast common types of feedstuffs and the roles they play in the diets of animals.

AS.03.02.02.a
Examine the importance of a balanced ration for animals based on the animal's growth stage.

AS.03.02.03.a
Examine the purpose, impact, and mode of action of feed additives and growth promotants in animal production.

FOR AN ANIMAL to continue living, growing, reproducing, and performing all of the bodily functions, it must have nourishment. All of its movement and body processes require the use of energy. An animal can obtain energy from only two places: from the food it ingests and from the energy stored in its fat cells. Obviously, even for the animal to store energy in the fat cells, it must have an intake of food.

In the wild, animals must spend most of their time in search of food to sustain themselves. In contrast, most agricultural animals are given their food every day. The nutrients obtained from their feed can go into growing and producing the products desired by the producers. Therefore, producers carefully balance the diets to fit the needs of their animals.

A certain level of nutritional needs, known as the **maintenance ration**, must be met first. This is the level of nourishment needed by the animal to maintain its body weight and not lose or gain weight. Nourishment over that amount can be used for growing, gestating, and producing milk or other products (**Figure 22–1**).

In the wild, most animals eat a variety of foods. This variety gives the animals the nutrients they need to support their bodily functions. In agricultural operations, producers balance the feeds of their animals to ensure that the proper nutrients are consumed. In confinement, the animals have to eat what the producer gives them. A lot of research has gone into the development of feeds that give animals exactly what they need to remain healthy and to perform at their peak (**Figure 23–2**). One type of feed may supply several of the needed nutrients, but usually a certain **feedstuff** contains a certain concentration of a particular nutrient. A feedstuff is generally a feed component that producers would not normally give by itself, but combined with other types of feedstuffs, it helps comprise the animal's feed.

FIGURE 22–2 A lot of research has gone into developing feeds that provide animals with the proper nutrients. *Courtesy of Dr. Frank Flanders*

FIGURE 22–1 Animals must have nourishment above the maintenance level to grow, gestate, and/or produce milk. *Source: USDA, Agricultural Research Service (ARS). Photo by Scott Bauer.*

Animals must have nutrients in each of six major classes: water, protein, carbohydrates, fats, vitamins, and minerals. Each of these classes of nutrients serves a specific function in the metabolism of the animal. Metabolism refers to all the chemical and physical processes that take place in the body. These processes provide energy for all of the functions and activities of the animal's body. Metabolism that builds up tissues is called **anabolism**. Examples of anabolism are the maintenance of the body, growth, and tissue repair. Metabolism that breaks down materials is called **catabolism**. An example is the breakdown of food within the digestive system.

WATER

Water is the most abundant compound in the world. More than two-thirds of the world's

FIGURE 22–3 Water is an essential nutrient needed to provide all the bodily fluids. © smereka/Shutterstock.com.

FIGURE 22–4 A sow nursing a litter of piglets needs a lot of water. © Anthony Gaudio/Shutterstock.com.

surface is covered with water. Because this nutrient is essential for sustaining life, animals take in water frequently to remain alive (**Figure 22–3**). Even animals such as llamas and camels, which can go for long periods of time without drinking, have to have water. This nutrient provides the basis for all the fluid in the animal's body. The bloodstream must be a liquid in order for circulation to occur. Digestion requires moisture to break down the nutrients and move the feed through the digestive tract. Water is needed to produce milk. It is needed to provide fluid for manufacturing all the bodily fluids. It provides the cells with pressure that allows them to maintain their shape. It helps the body maintain a constant temperature. Another vital function of water is that of flushing the animal's body of wastes and toxic materials. This nutrient is so vital that over half the animal's body is composed of water. A loss of 20 percent of this water will result in the death of the animal.

Using ballpark figures, animals generally need about 3 pounds of water for every pound of solid feed they consume. Some of this water comes in the feed itself. For instance, animals that graze obtain water from the succulent green forages they eat. Some water can be obtained in feeds such as silage that have a relatively high water content. However, most of the water an animal needs comes from the water it drinks. Because water is so essential, producers make sure that animals are given a constant supply of clean water.

Animals may require more water at some times than others. A horse that is working hard in hot weather will sweat profusely and will need more water intake to replenish the fluid lost from its body. Likewise, a sow that is nursing a litter of 12 piglets requires a lot of water to produce milk for the young (**Figure 22–4**).

PROTEIN

Protein can make up to around 15 to 16 percent of an animal's ration and may be the most costly part of the ration. Proteins are composed of compounds known as **amino acids**. Amino acids are often said to be the building blocks of life because they go into the formation of tissues that provide growth for the animal. Muscle production in particular depends on the amino acids found in proteins. All of the enzymes and many hormones in the animals' bodies are composed of protein. To a certain extent, protein also is used to provide energy.

FIGURE 22–5 Animals use protein to grow muscle, hair, and other tissues. *Source: Photo by Tim McCabe, USDA Natural Resources Conservation Service.*

Digestible protein is the protein in a feed that the animal can digest and use. Digestible protein ranges from about 50–80 percent of the crude protein.

Although protein can be found in most feedstuffs, some have a much lower content than others. For example, yellow corn has a protein content of around 8 percent. A growing pig may need a ration that consists of 16 percent protein. Being fed corn alone will not give the pig an adequate amount of protein to provide for building the body cells to sustain growth. This means that a feedstuff that is higher in protein content will have to be added.

Protein can come from basically two sources: animal and plant. **Carnivores** (animals that eat other animals), such as dogs, cats, and foxes, get almost all of their protein from meat. After all, the muscles in an animal's body are composed primarily of protein and can serve as food for another animal. Omnivores

As with water, some animals need more protein than others do. Young, rapidly growing animals need more protein than mature animals because the amino acids in the protein are required to build muscles, skin, hair, bones, and all of the other cells that go into the growth process (**Figure 22–5**). A cow that is giving large amounts of milk needs more protein than an animal that is not lactating.

In all, there are over 20 different types of amino acids that an animal's body uses. Of these, there are 10 essential amino acids that the animal must obtain from its feed. The other amino acids can be synthesized in the animal's digestive tract. This means that the nonessential amino acids can be made from the 10 that the animal consumes. In this sense, the 10 are essential in that they cannot be manufactured by the animal and must be consumed. **Table 22–1** lists the **essential amino acids** and the **nonessential amino acids**.

Animals may not be able to digest all the protein in a given feed. The total amount of protein in a feed is called the **crude protein content**. The amount of crude protein in a feed is calculated by analyzing the nitrogen content and multiplying that percentage by 6.25.

Essential Amino Acids	Nonessential Amino Acids
Arginine	Alanine
Histidine	Aspartic acid
Isoleucine	Citrulline
Leucine	Cystine
Lysine	Glutamic acid
Methionine	Glycine
Phenylalanine	Hydroxyproline
Threonine	Proline
Tryptophan	Serine
Valine	Tyrosine

TABLE 22–1 Amino Acids

(animals that eat both plants and animals), such as humans and pigs, can get protein from both plants and animals. Animals that eat only plants are called **herbivores**, and they must get protein exclusively from plants.

Most feedstuffs that are rich in protein come from plant sources. Pigs once were fed slaughterhouse by-products such as **tankage** and blood meal. Tankage is a combination of animal material left from rendering meat (cooking at a high temperature) and products not suitable for human consumption. Recent research has shown that these protein sources are inferior to plant sources in terms of protein that is usable to the animal. Cattle also were once fed animal by-products, but such feeding is banned now. In fact, almost all feeding of animal by-products has been banned because of concerns about transmitting diseases to humans (**Figure 22–6**).

Some dried fish meal is fed to hogs as a supplement. Much of plant protein that goes into the feed of animals comes from the vegetable oil industry. Cooking oil usually is pressed from cottonseed, soybeans, peanuts, or corn (**Figure 22–7**). These seeds are run through huge presses, where the oil is squeezed out. The material that is left is in the form of a cake composed of the seeds minus the oil. It is dried and ground into a meal for feed. This material is usually 40–45 percent crude protein and can greatly increase the percentage of protein in a feed. This feedstuff then is mixed with the other feedstuffs in the proper ratio to give the desired protein content for the feed.

FIGURE 22–6 Animal by-products have been banned from almost all animal feed. © IrinaK/Shutterstock.com.

FIGURE 22–7 Soybeans provide protein for animal rations. © Fotokostic/Shutterstock.com.

The protein source that is used is often determined by the animal that is being fed. For instance, pigs are not fed cottonseed meal because this feedstuff contains a substance known as gossypol that is toxic to them. Other needs of the animal also determine the protein source that is used. Modern livestock operations no longer just balance a feed ration based on the percentage of protein. Now the feed formulas are based on the types and amounts of amino acids that are needed by a particular group of animals. The process of balancing feed rations based on amino acid content is so complicated that it is done by computers. Large, modern feed mills have computers that control the formulating and blending of the different types of feed (**Figure 22–8**). Two feedstuffs may have the same percentage of protein and have different percentages of the essential amino acids. Different amino acids are needed for different body functions. For example, a different amino acid is needed for growth than is needed for milk production. Growing animals need a different type of protein supplement than a lactating dam needs. Feeds balanced on the type of amino acids needed by the animal are more cost-efficient (**Figure 22–9**).

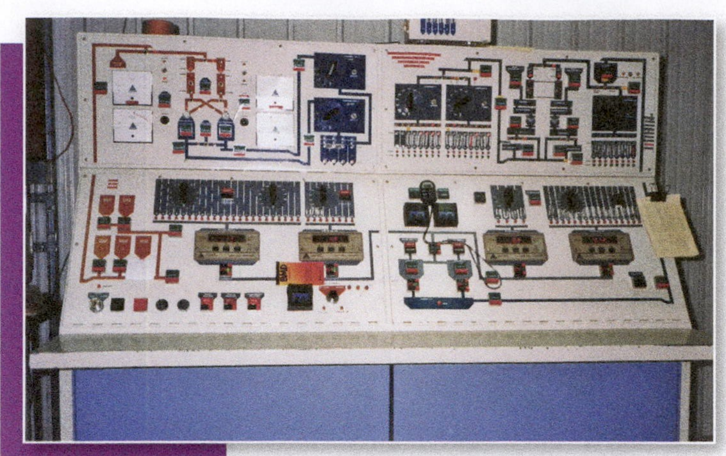

FIGURE 22–8 Feeds are balanced using computers. This is a computer control panel of a large, modern feed mill. *Courtesy of Dan Rollins*

FIGURE 22–9 Feed samples are constantly monitored to ensure the proper balance and quality of the feed. *Courtesy of Dan Rollins*

CARBOHYDRATES

The main source of energy from animals comes from carbohydrates. Carbohydrates are compounds made up of carbon, hydrogen, and oxygen. They include sugars, starches, and **cellulose** and are the major organic compounds in plants. Almost all carbohydrates come from plants and are developed by photosynthesis. By weight, plants are composed of about 75 percent carbohydrates. As will be discussed later, some animals are more efficient than others at making use of these carbohydrates.

Starch is generally found in grain. The plants use starch as energy storage for the seed. Grains such as wheat and corn contain a lot of starch and therefore a lot of energy for the animal to use. Starches are composed of sugars, and in the process of digestion, the starch is broken down into the component sugars.

There are several different types of sugars. Two broad groups are **monosaccharides** (the simple sugars) and **disaccharides** (the more complex sugars). "Simple" and "complex" refer to the chemical composition of these sugars and the different ways the molecules are formed. There are several common simple sugars (monosaccharides); among these are **glucose**, **fructose**, and **galactose**. Glucose is the simplest of all the sugars and is found in a low concentration in plant materials. It is also the major energy source found in an animal's blood. The animal's body breaks down some of the other sugars into glucose.

Fructose is found in fruits and honey and is the sweetest of all the sugars. Common table sugar, **sucrose**, is a disaccharide composed of fructose and glucose. Galactose is obtained from the breakdown of the disaccharide **lactose** (milk sugar).

Cellulose is the portion of cell walls that gives the plant its rigid structure. The enzymes in an animal's digestive system cannot break down cellulose. However, some animals have microorganisms in their digestive system that break down the cellulose fiber so the enzymes can digest the material.

The most important source of carbohydrates for agricultural animals is grains. Most of the millions of tons of corn grown in the United States each year go into the production of livestock feed (**Figure 22–10**). Other grains, such as wheat, oats, and barley, also are used. Feeds that are high in grain content are called concentrates because of their high concentration of carbohydrates.

For horses and ruminant animals, forages grown for grazing and for hay are valuable sources of feed. These food sources are referred to as roughages because of the amount of fiber

FIGURE 22–10 Most of the millions of tons of corn grown each year go into the production of animal feeds. *Source: Photo by Tim McCabe, USDA Natural Resources Conservation Service.*

FIGURE 22–11 Corn is chopped to create silage, which is a combination of roughages and concentrates. *© Alisa24/Shutterstock.com.*

in the diet. Sometimes a combination of grain and forage is used in the form of silage or similar types of feed (**Figure 22–11**).

FATS

Fats are part of a group of organic compounds known as **lipids**. These compounds will not dissolve in water but will dissolve in certain organic solvents. Besides fats and oils, lipids also include cholesterol. Fats are found in both plants and animals. They serve as concentrated storage places for energy. An example of plant fats is the oil within seeds such as corn and soybeans (**Figure 22–12**).

Fats serve the purposes of providing energy for the animal and of storing excess energy. When an animal consumes more energy (especially in the form of fats) than it needs to provide for all the needed bodily functions, the excess is stored in the form of fat. When the body does not take in enough energy to perform the normal bodily functions, these reserves of fat are used.

Certain acids, referred to as the essential fatty acids, are also derived from fats. These acids are necessary in some animals for the production of certain hormones and hormonelike substances.

The most important sources of fats in feeds for agricultural animals are the grains that contain oil. Corn and most other feed grains contain oil that the animals use as a fat source. Some types of animals, pigs for example, may have problems if they are fed too much oil. Hogs fattened on oily

FIGURE 22–12 Oil within seeds such as corn (A) and soybeans (B) is an example of plant fats. *© JIANG HONGYAN /Shutterstock.com.; Source: Photo by Lynn Betts, USDA Natural Resources Conservation Service.*

feeds such as whole peanuts may produce soft, oily pork that is not acceptable to consumers.

MINERALS

Minerals are the only group of nutrients besides water that are **inorganic,** meaning that a compound does not contain carbon. Although they provide only a small portion of the total feed intake, they are vitally important. Animals must have a sufficient intake of these inorganic materials to provide the building materials for their body structure. Bones are formed by a combination of calcium and phosphorus. In addition to building bones, minerals aid in the construction of muscles, blood cells, internal organs, and enzymes. This group serves the important role of providing structural support for the animal. Bones are formed by a combination of calcium and phosphorus. Another example is eggshells, which are mainly composed of calcium. Eggshells are ground up and added to chicken feed as a source of calcium. Animals must have a sufficient intake of these inorganic materials to provide the building materials for their body structure.

In addition to building bones, minerals provide other essential needs. They aid in the construction of muscles, blood cells, internal organs, and enzymes. Animals with a deficiency in minerals never develop properly and are more susceptible to disease.

The mineral elements required by animals include seven **macrominerals** (required in relatively large amounts in the diet) and nine **microminerals** or **trace minerals** (required in very small amounts in the diet). The macrominerals are calcium, chlorine, magnesium, phosphorus, potassium, sodium, and sulfur. The microminerals are cobalt, copper, fluorine, iron, iodine, manganese, molybdenum, selenium, and zinc. These inorganic, crystalline, solid elements make up 3–5 percent of the body on a dry-weight basis, with calcium (approximately one-half the body mineral) and phosphorus (approximately one-fourth the body mineral) accounting for the largest portion of the total mineral content.

FIGURE 22–13 The shells of eggs are composed mostly of the mineral calcium. *© Surachan/Shutterstock.com.*

Minerals usually are added to animal feed in their chemical form. Calcium sometimes is added from other animal sources. For example, ground-up oyster shells and eggshells are fed to laying hens to provide materials for their bodies to create strong eggshells (**Figure 22–13**).

Minerals are often fed **free choice**. This means that the animals are given free access to the minerals and are allowed to eat all they wish. For cattle, this is done by a mineral box or trough or by the use of a salt block (**Figure 22–14**). Essential

FIGURE 22–14 Cattle are fed minerals "free choice." This mineral box turns with the wind to help keep out the rain.

minerals are in the block, and the animals get them as they lick the block for salt.

FIGURE 22–15 Sun-dried forages can provide the basis for the synthesis of vitamin D. *Source: Photo by Keith McCall, USDA Natural Resources Conservation Service.*

VITAMINS

Vitamins are considered to be micronutrients. This means that the body needs them in very small amounts. Even though only small amounts are required, vitamins are essential for life. They are essential for the development of normal body processes of growth, production, and reproduction. They are also vitally important in providing the animal with the ability to fight stress and disease and to maintain good health.

Some animals are able to synthesize certain vitamins in their body tissues. Other vitamins cannot be created by the animal from other nutrients and must be obtained from the diet or by microbial synthesis in the digestive system.

There are 16 known vitamins. The B vitamins and vitamin C are water-soluble. The fat-soluble vitamins are A, D, E, and K.

Vitamin A

Vitamin A is not found in feeds, but it is converted by the animal's body from the provitamin **carotene**, which is found in green, leafy forages from pastures, hay, silage, and dehydrated legumes (alfalfa). Other sources include yellow corn, fish liver oils, and whole milk. Vitamin A can be stored in fats and the liver for several months, to be used when forage quality is low or when stress conditions increase the body's demand for vitamin A. Supplementation is usual for ruminants and swine.

Vitamin D

Vitamin D is sometimes referred to as the "sunshine vitamin" because animal and plant sources both depend on ultraviolet light to make a form of vitamin D. The liver and the kidneys convert this form of the vitamin to usable forms. Animals make their own vitamin D, and diets of sun-cured forages, yeast, and certain fish oils provide the basis for synthesis (**Figure 22–15**). Commercial vitamin D is available, generally made from irradiated yeast. Excessive amounts of vitamin D can reduce an animal's efficiency and is toxic in some incidences. Animals in total confinement often receive supplements of vitamin D.

Vitamin E

The cereal grains, germ oils, and green forage or hay supply vitamin E. This vitamin is found in several forms of a complex organic compound called tocopherol. Commercially produced vitamin E is available for supplemental feeding. There are no known toxic effects from excessive levels in the diet.

Vitamin K

Vitamin K is utilized to form the enzyme prothrombin, which in turn helps to form blood clots. Deficiencies rarely occur because vitamin K is synthesized in the rumen and in the

monogastric intestinal tract. Green forages, good-quality hays, fish meal, and synthetic forms of vitamin K can be used to increase the level of vitamin K in the diet.

The B Vitamins

Thiamine is essential as a coenzyme in energy metabolism. Dietary sources of thiamine include green forage and well-cured hays, cereal grains (especially seed coat or bran), and brewer's yeast. Heat-processing of grains reduces the amount of available thiamine. Thiamine usually is synthesized in the rumen. The diet of monogastric animals usually provides enough thiamine. However, thiamine is commercially available in vitamin premixes.

Riboflavin is important as a part of two coenzymes that function in energy and protein metabolism. Sources include green forages, leafy hays, or silage; milk and milk products; meat; fish meal; and distiller's or brewer's by-products (**Figure 22–16**). Commercially available riboflavin typically is added to swine rations and may be needed in ruminant rations.

Pantothenic acid is a component of coenzyme A, which is important in fatty acid and carbohydrate metabolism. Sources include brewer's yeast, liver meal, dehydrated alfalfa meal, molasses, fish solubles, and most feedstuffs. Commercial sources should be included in the diets of confined animals.

FIGURE 22–16 Green forages are a good source of vitamins such as vitamin K, thiamine, and riboflavin. *Source: Photo by Lynn Betts, USDA Natural Resources Conservation Service.*

Niacin is part of an enzyme system that is essential in the metabolism of fat, carbohydrates, and proteins. Niacin is found in animal by-products, brewer's yeast, and green alfalfa. Most feeds contain some niacin, but the niacin in grains is largely unavailable to nonruminants, so supplementation often is needed.

Pyridoxine is a coenzyme component necessary for fatty acid and amino acid metabolism. Most feedstuffs are fair-to-good sources of pyridoxine, including cereal grains and their by-products, rice and rice bran, green forages and alfalfa hay, and yeast. Supplementation in animal diets usually is not needed.

Biotin is widely distributed. It is found in large quantities in egg yolk, liver, kidney, milk, and yeast. Biotin is a part of an enzyme involved in the synthesis of fatty acids. Animals can readily synthesize biotin, and it is not deficient in normal farm animals.

Folic acid is needed in body cell metabolism. It is found in green forages, such as alfalfa meal, and in some animal proteins. The animal body synthesizes some folic acid, and although it is available in synthetic forms, supplements are not greatly needed in farm animals.

Choline is a component of fats and nerve tissues and is needed at greater levels than other vitamins. Most commonly used feeds are good sources of choline. Choline is synthesized in the animal body when other vitamins such as B_{12} are abundant.

B_{12} functions as a coenzyme in several metabolic reactions and is an essential part of red blood cell maturation. Synthesis of vitamin B_{12} requires cobalt. Sources of B_{12} include protein feeds of animal origin and fermentation products. Most swine rations are supplemented with vitamin B_{12}.

Inositol is found in all feeds and synthesized in the intestinal tract, so it is not generally needed as a supplement.

Although the function is not well known, para-aminobenzoic is synthesized in the intestine. It usually is not deficient in livestock rations.

Vitamin C

Vitamin C is essential in forming the protein collagen. Vitamin C is found in citrus fruits; green, leafy forages; and well-cured hays. Animals normally can synthesize sufficient quantities of vitamin C to meet their needs.

DETERMINING FEED RATIONS FOR AGRICULTURAL ANIMALS

A balanced ration is a key to keeping animals healthy, disease free, and producing. Most large-scale producers use computers to calculate the ingredients that go into animal feed. By using computers, micronutrients (nutrients that are needed in very small quantities) and precise proportions of macronutrients (nutrients that compose the largest percentages of the ration) can be blended in accurate amounts that fit the animals' specific needs. Specific rations are needed for the growth or reproductive stage of the animal. Different rations are needed for the following stages:

- Weaning
- Growth
- Finishing (fattening)
- Gestation
- Lactation
- Breeding
- Maintenance

Each of these stages requires a different ration and computers are an efficient way of balancing the rations. For smaller scale producers or people who raise only a few animals, feed rations can be calculated without the use of computers. The following is a method that can be used.

Calculating Balanced Feed Rations for Agricultural Animals

A. To balance a ration the following information must be found; the animal's nutrient requirements and the nutrient composition of the feedstuffs to be used. The nutrient composition would also include the dry matter percentage and the dry matter value verses the as-fed value.

B. 100% Dry Matter verses As-Fed Basis.

- 100% Dry matter is the value of the nutrients in a feedstuff without any moisture present.
- As-fed Basis is the value of the nutrients in a feedstuff based on the average amount of moisture in the feed.
- It is easier to compare feeds that have different moisture contents when using the 100% dry matter basis. But, the values must be changed to an as-fed basis to determine the amount of actual feed needed.

To convert from as-fed (air-dry) basis to 100% dry matter basis: $\mathbf{a = b \times c}$

When: a = pounds (kilograms) of feed on 100% dry matter basis
b = pounds (kilograms) of feed on an as-fed (air-dry) basis
c = the percent of dry matter in the feed (refer to table)

To convert from 100% dry matter basis to as-fed (air-dry) basis: $\mathbf{b = a \div c}$

C. Steps to Balancing a Ration

1. Identify the animals – species, age, weight, and its function.
2. Determine the animal's nutrient needs using a nutrient requirement table (feeding standards).
3. Select the feeds to be in the ration and determine the nutrient content from the feed composition table.
4. Calculate the amount of each feed to be used in the ration. (Pearson Square)
5. Compare the ration formulated against the requirements of the animal. Make sure it meets the mineral and vitamin content and recalculate if any of the nutrients are in excessive amounts.

D. Using the Pearson Square

The Pearson Square is used to show the proportions or percentages of two feeds to be mixed together to give a percent of the needed nutrient. The Pearson Square may also be used to find out how much of two different grains can be mixed together with a supplement. It can be used to mix two protein supplements with one grain source. The Pearson Square Method is not restricted to rations, it can be used for solving for any ratios and proportions.

E. Algebraic Equations to Balance Rations

Equation 1: X + Y = total pounds (kilograms) of mix needed

Equation 2: (percent nutrient in grain) × (X) + (percent nutrient in supplement) × (Y) = pounds (or kilograms) of nutrient desired in mix

Either X or Y must be canceled by the multiplication of equation 1 by the percentage of nutrient for either X or Y. The resulting equation 3 is then subtracted from equation 2.

F. Using Computers to Balance Rations

Computer ration balancing software is available from numerous sources. Computers can do it quickly and can calculate a least cost ration.

Source: *Georgia Agricultural Education Curriculum*

THE DIGESTION PROCESS

Animals use feed nutrients on a cellular basis—all of the different nutrients that an animal takes in must be converted to a form that the cells in the body can use. Once this conversion (digestion) is completed, the nutrients must be transported to the cells where they are needed. The system that performs this task is referred to as the digestive system. The organs that make up this system are known as the **gastrointestinal tract** (GI tract).

The gastrointestinal tract is also referred to as the **alimentary canal**. This is the tract reaching from the mouth to the anus, through which feed passes following consumption and where it is exposed to the various digestive processes. The digestive system consists of the various structures, organs, and glands involved in procuring, chewing, swallowing, digestion, absorption, and excretion of feedstuffs. Although there are similarities in the digestive systems, farm animals are often classified according to the nature of their digestive system. There are basically two types of digestive systems. One is known as **monogastric** (single-compartment stomach) (**Figure 22–17**); and the other is known as **ruminant** (multicompartment stomach). Following are descriptions of processes in the two digestive systems in the order of occurrence.

Monogastric Digestive Systems

Monogastric systems are those that have only one-compartment stomachs. These include the pig, horse, dog, cat, and birds. The horse has an enlargement, known as a **cecum**, that enables it to utilize high-fiber feeds by means of microbial fermentation, much as do ruminants (**Figure 22–18**). This means that the horse is not typical of monogastric animals. Simple-stomach (monogastric) animals are not capable of digesting large amounts of fiber and are usually fed concentrate feeds.

The digestive process begins in the mouth, which is the first organ of the digestive tract. Within the mouth, the tongue is used for grasping the food, mixing, and swallowing. The teeth are used for tearing and chewing the feed. This is the first step in the process of breaking down the feed into fine particles.

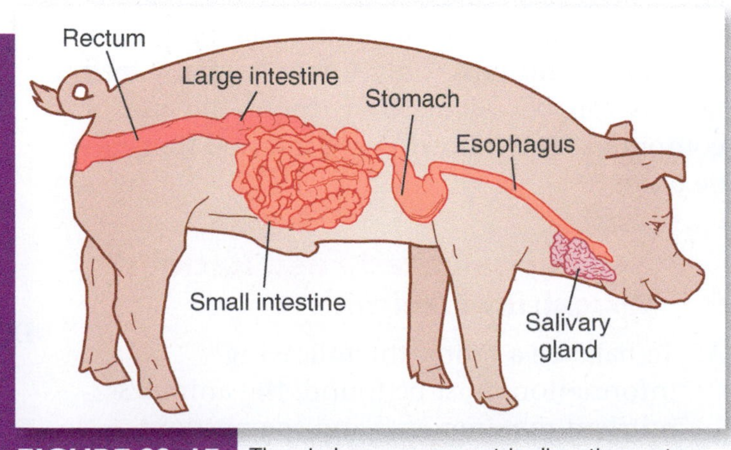

FIGURE 22–17 The pig has a monogastric digestive system.

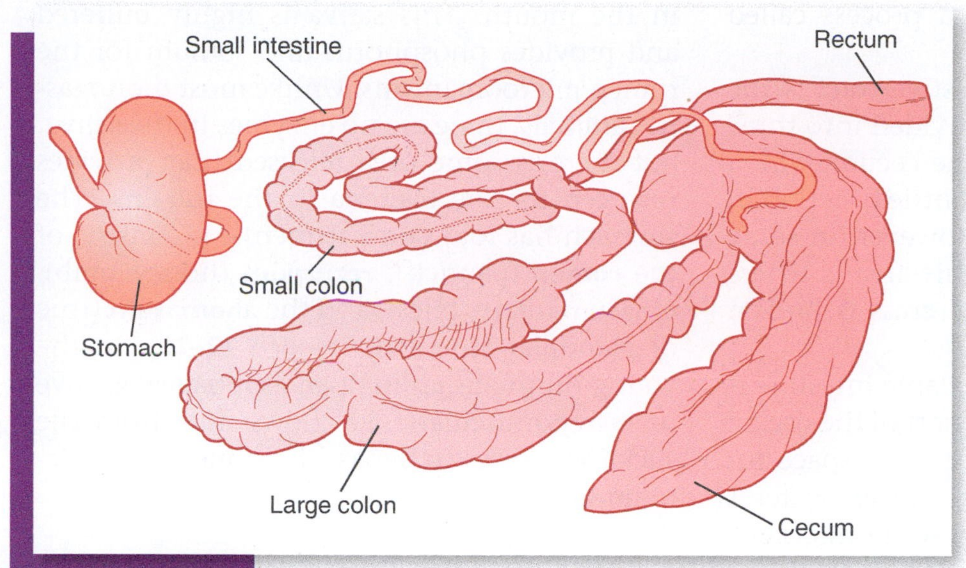

FIGURE 22–18 Horses have a large pouch called a *cecum* that helps digest high-fiber feeds such as hay.

The mouth also contains salivary glands, which consist of three pairs of glands that excrete saliva. Saliva contains several substances: water to moisten, mucin to lubricate, bicarbonates to buffer acids in the feeds, and the enzyme amylase to initiate carbohydrate breakdown.

The **esophagus** is a hollow, muscular tube that moves food from the mouth to the stomach. This is accomplished by muscular contractions that push the food along.

The stomach is a hollow muscle that causes further breakdown of foods by physical muscular movement. The food is pressed together and massaged by the movement of muscles in this area. In the stomach, food is also broken down by chemical action. The walls of the stomach secrete hydrochloric acid that begins to dissolve the food. Another secretion, **pepsin**, begins to break down proteins into the amino acids. The secretion rennin acts to curdle the casein in milk. Gastric lipase causes the breakdown of fats to fatty acids and glycerol.

The small intestine is a long, hollow tube that leads from the stomach to the large intestine. The small intestine is made up of three main parts: the **duodenum**, the **jejunum**, and the **ileum**. The entrance to the small intestine is controlled by a sphincter muscle that helps to move food into and through the tract.

The duodenum is the first segment of the small intestine. It receives secretions from the pancreas, which break down proteins, starches, and fats. Here, the intestinal walls secrete intestinal juices containing enzymes that further break down the food. The next segments of the small intestine, the jejunum and the ileum, are the areas of nutrient absorption. Absorption is the process by which the nutrients are passed into the bloodstream. The villi (small fingerlike projections) in these areas facilitate absorption into the bloodstream and/or the lymph system through membranes that surround the villi (**Figure 22–19**). This type of membrane is called a **semipermeable membrane**. This means that the membrane allows

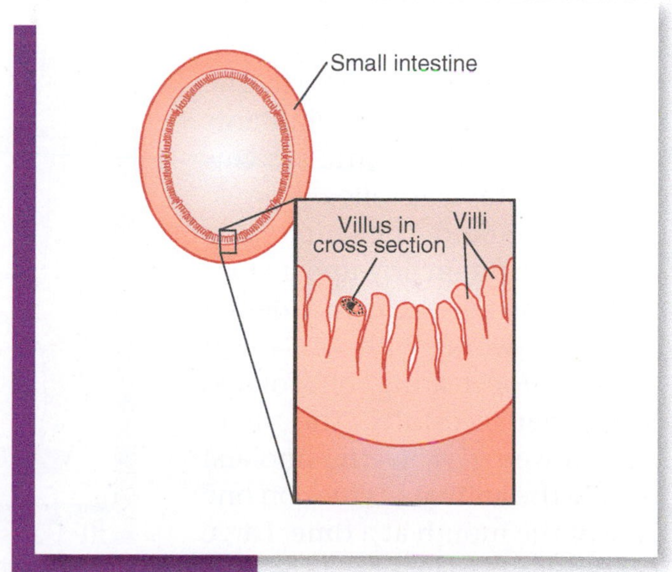

FIGURE 22–19 The small intestine is lined with fingerlike projections, called *villi*, that absorb nutrients into the bloodstream.

particles to pass through in a process called diffusion.

The last organ of the digestive tract is the large intestine. This organ is divided into three sections. The first section is the cecum, which is a blind pouch. A cecum is of little function in most monogastric animals. However, in some animals, such as the horse, this area is where fibrous food such as hay and grass is broken down into usable nutrients.

The second segment of the large intestine is the colon, which is the largest part of the organ. Its function is to provide a storage space for wastes from the digestive process. Here water is removed from the wastes and some microbial action begins on fibrous materials.

The rectum is the final segment of the large intestine and the final part of the digestive system. It serves to pass waste material through to the anus, where it is eliminated.

Ruminant Digestive System

Animals such as cows and sheep have multi-compartment stomachs that allow them to use high-fiber feeds such as grasses and hays. These animals are often called "cud chewers" because they regurgitate **boluses** of feed that they consumed earlier and remasticate (chew) and reswallow it. The digestive systems of ruminants differ from monogastric systems in several ways, as follows.

The mouth of ruminants does not contain any upper front teeth. Instead, there is a dental pad that works with the lower incisors for tearing off forages and other feedstuffs. The upper and lower jaw teeth (molars) enable the animal to chew on one side of the mouth at a time. Large quantities of saliva are produced in the mouth. This saliva is highly buffered and provides phosphorus and sodium for the rumen microorganisms. Unlike most monogastric animals, there are no enzymes in the saliva, but there is some urea released that provides nitrogen for the bacteria in the rumen. The stomach has four compartments. It consists of the **rumen** (paunch), **reticulum** (honeycomb), **omasum** (many piles), and the **abomasum** (true or glandular stomach) (**Figure 22–20**). In the young ruminant animal, an esophageal groove or heavy muscular fold allows milk from the suckling animal to bypass the rumen and reticulum to the omasum.

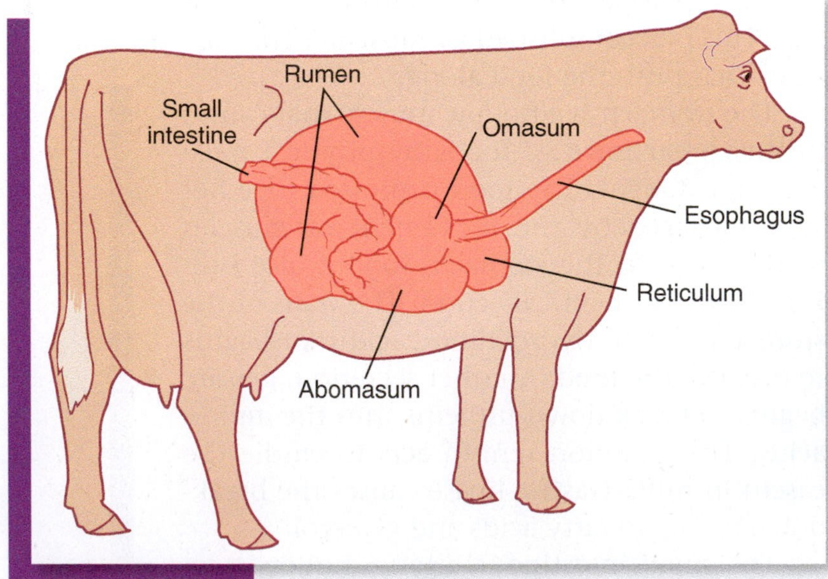

FIGURE 22–20 The stomach of cattle has four compartments, which allows them to use large amounts of roughages.

Compartments of the Ruminant Stomach

The esophagus leads to both the reticulum and the rumen. As ruminants graze, they tend to pick up hard, indigestible objects such as small stones, nails, and bits of wire. These heavy materials fall into the reticulum. The walls of the reticulum are made up of **mucous membranes**, which form subcompartments with the appearance of a honeycomb. These small compartments trap and provide a storage place for "hardware" that does not float.

The reticulum also functions to store, sort, and move feed back into the esophagus for regurgitation, or into the rumen for further digestion. The process of breaking down roughages begins with contraction of the reticulum and muscles in the esophagus to move roughage and fluid to the mouth. Excess fluid is squeezed out, and the material is reswallowed.

After the material is reswallowed, it moves to the rumen. The rumen functions as a storage vat where food is soaked, mixed, and fermented by the action of bacteria. The hollow, muscular paunch fills the left side of the abdominal cavity and contains two sacks, each lined with papillae (nipplelike projections) that aid in the absorption of nutrients. Bacteria and other microorganisms such as protozoa thrive in the rumen environment and break down fibrous feeds (**Figure 22–21**). Carbohydrates are broken down into starches and sugars. Volatile fatty acids are released as the carbohydrates are broken down, and these fatty acids are absorbed through the rumen wall to provide energy for the body.

Bacteria also use nitrogen to form amino acids, and eventually proteins. The bacteria also can synthesize water-soluble vitamins and vitamin K. By-products of the microbial activity include methane and carbon dioxide. The blood absorbs a small portion of these gases, but much of the gas is eliminated by belching. Belching occurs when the upper sacs of the rumen force gases forward and downward so the esophagus can dilate and allow gases to pass. If gases are not eliminated due to froth or foam blocking the esophagus, a condition called **bloat** (an inflation of the rumen) will sometimes occur.

After leaving the rumen, the food material passes through to the omasum. The omasum is a round organ on the right side of the animal and to the right of the rumen and reticulum. The omasum grinds roughage using blunt muscular papillae that extend from many folds of the omasum walls.

The last compartment of the ruminant's stomach is the abomasum. The abomasum is the only glandular (true stomach) compartment of the ruminant. The abomasum is located below the omasum and extends to the rear and to the right of the rumen. This compartment functions similarly to the stomach of monogastric animals. By the time food materials reach the abomasum, the fibers of the roughages have been broken down to the extent that they can be handled by the abomasum. The small and large intestines of the ruminant animal function much the same way as they do in the monogastric animal.

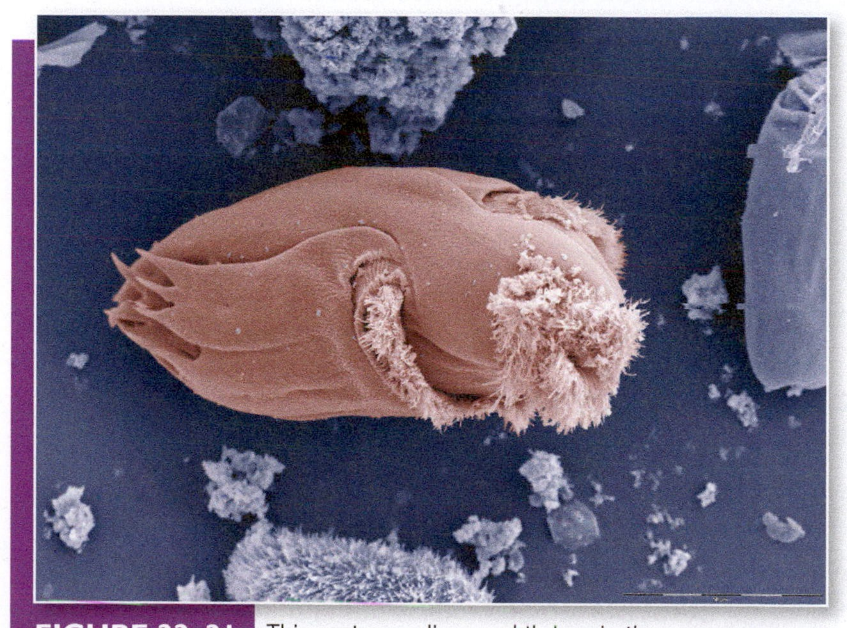

FIGURE 22–21 This protozoan lives and thrives in the rumen. *Source: USDA, Agricultural Research Service (ARS). Photo by Sharon Franklin. Colorization by Stephen Ausmus.*

SUMMARY

In order for animals to grow and thrive, they must have the proper nutrition. An understanding of how the digestive process works in different types of animals is essential in order to formulize the proper diet for animals. Factors such as species, age, sex, and environmental conditions all must be taken into account in formulating rations. Today's feeds are scientifically balanced to give animals exactly what they need in each particular circumstance. Research scientists are continuing to find new and better ways of providing nutrients to animals.

CHAPTER REVIEW

Review Questions

1. What is the difference between a feed and a feedstuff?
2. Define metabolism.
3. List at least four functions water plays in sustaining life.
4. Explain why a young animal needs more protein than a mature animal does.
5. List the sources of protein for livestock feeds.
6. From what source do all carbohydrates come?
7. What purpose do fats serve?
8. What essential needs do minerals provide?
9. List the vitamins that are needed by animals.
10. List the parts of the monogastric digestive system and briefly describe the function of each.
11. List the compartments of the ruminant stomach and describe the function of each.

Student Learning Activities

1. Locate and report on an article telling about research that has been completed on livestock feeds. Explain what practical differences you feel this research will make.
2. Obtain the tag from a bag of feed. Make a list of the ingredients and tell which nutrients are derived from each component.
3. From a slaughterhouse, obtain the digestive tract of a monogastric animal (pig) and a ruminant (sheep or cow). Dissect the tract and identify all of the parts. Be sure to wear latex gloves when the organs are handled.

CHAPTER 23

Meat Science

STUDENT OBJECTIVES IN BASIC SCIENCE

As a result of studying this chapter, you should be able to

- describe physiological processes that take place in an animal's body at death.
- describe the process of ossification.
- list the different types of tissue that make up muscles.
- explain the factors that affect the sensation of taste.
- explain why meat is so highly perishable.
- discuss the types of microbes that cause spoilage.
- list the factors that favor the growth of microbes.
- discuss the scientific principles behind the preservation of meats.

STUDENT OBJECTIVES IN AGRICULTURAL SCIENCE

As a result of studying this chapter, you should be able to

- explain the steps in the slaughter of meat animals.
- distinguish between quality grading and yield grading of carcasses.
- list the wholesale cuts of beef, pork, and lamb.
- discuss the factors that affect the palatability of meats.
- discuss the various methods used to preserve meats.

KEY TERMS

wholesale cuts
immobilization
kosher
exsanguination
chine
shroud
rigor mortis
aged
primal cuts
adipose
mastication
elastin
oxidation
rancid
microbes
psychrophiles
thermophiles
mesophiles
aerobic organisms
anaerobic organisms
facultative
dry curing
injection curing
combination curing
blast freezing
radiation
irradiation
E. coli

NATIONAL AFNR STANDARD

AS.01.02.01.b
Analyze the impact of animal production methods on end product qualities.

AS.06.03.03.b
Evaluate and select products from animals based on industry standards.

MEAT INDUSTRY

Americans are a nation of meat eaters. Each year the average person in this country consumes 48 pounds of beef and veal, 29 pounds of pork, and 24 pounds of poultry. Very few nations in the world even come close to us in the per-capita consumption of meat. In addition, when compared with the rest of the world, Americans spend a small percentage of their annual income for food. This means that they can afford to buy the type of food they prefer, and, obviously, they prefer meat.

Lean meat is dense in nutrients. A pound of meat may equal or surpass the nutritive content of the feed that it took to produce the meat. Meat is among the most nutritionally complete foods that humans consume. Foods from animals supply about 88 percent of the vitamin B12 in our diet because this nutrient is very difficult to obtain from plant sources. In addition, meats and animal products provide 67 percent of the riboflavin, 65 percent of the protein and phosphorus, 57 percent of the vitamin B6, 48 percent of the fat, 43 percent of the niacin, 42 percent of the vitamin A, 37 percent of the iron, 36 percent of the thiamin, and 35 percent of the magnesium in our diet.

In recent years, the trend in meat consumption has changed rapidly. Modern consumers want meat products that are already cooked or are ready to place in the oven or microwave. Meat products such as chicken wings, that once were considered to be almost by-products, are now packaged fully cooked, frozen, and ready to place in the microwave oven. Individually wrapped chicken parts, chops, and steaks that are ready to go on the grill are all gaining in popularity with consumers. The meat industry is changing rapidly to accommodate the wishes of the consumer.

The vast majority of agricultural animals raised in the United States are produced for meat. Meat is defined as the edible flesh of animals and may include the muscles, certain internal organs such as the liver, and other edible parts. A huge industry has developed around processing animals into edible products (**Figure 23–1**).

FIGURE 23–1 Animals are processed into meat in large, automated plants. *© Alf Riberio/Shutterstock.com.*

The United States has over 2,700 processing plants that slaughter animals. Of this number, less than 900 are under federal inspection. The trend in recent years is for these plants to become larger. Of the animals slaughtered under federal inspection, 80 to 90 percent are processed in fewer than 200 plants. These plants may slaughter more than 100,000 animals in a single year. In the larger plants, the slaughter process is highly organized and automated. Like an assembly line, workers perform only a few tasks with the animal or carcass, and the product goes on down the line to the next person.

When the animals are brought in, they must be inspected and slaughtered. The head, entrails, hair or hide, and feet must be removed. What is left is referred to as the carcass, **Figure 23–2**. The carcass is then cut into **wholesale cuts** and sold to grocery stores and other outlets, where meat cutters divide the wholesale cuts into retail cuts. The consumer buys the retail cuts that are ready for cooking.

THE SLAUGHTER PROCESS

After the live animals are inspected they are brought down a chute onto the slaughter floor. The first of several steps in the slaughter

FIGURE 23–2 The carcass is the part of the animal left after the head, entrails, hide (or hair), and feet are removed. © ASA studio/Shutterstock.com.

process is **immobilization**. The purpose is to render the animal unconscious so it feels no pain. This must be done in such a way as to allow the heart to continue to pump, to drain the animal's body of blood. Regulations for the immobilization process are set by the Federal Humane Slaughter Act. Immobilization may be accomplished in several ways. The animal may be placed in a chamber of carbon dioxide until it goes to sleep from lack of oxygen. Some slaughter plants use electric shock to render the animal unconscious. Others use a cartridge or a mechanical bolt that is driven by compressed air to stun the animal.

Livestock slaughtered for **kosher** markets are exempt from the requirement of stunning, under the Humane Slaughter Act. Kosher means that the animal is slaughtered under the regulations of Jewish religious laws. In any case, humane slaughter must be done quickly and with as little stress as possible to the animal.

After immobilization, the animal must be bled quickly to keep it from regaining consciousness and to prevent hemorrhaging (escape of blood from a ruptured blood vessel) resulting from a rise in blood pressure. This process, **exsanguination**, usually is done by severing the jugular vein with a sharp knife. Unless bleeding is accomplished within a few seconds of immobilization, the blood pressure may cause hemorrhaging, resulting in blood spots in the meat. As soon as the blood pressure begins to drop, the heart speeds up to maintain pressure. This action fills many of the organs with blood, and, thus, blood loss at exsanguination is only about 50 percent of the total volume of blood.

The animal is hoisted on a rail, and the hide and entrails are removed (**Figure 23–3**). Pigs that have been slaughtered may be dipped into scalding water and placed on a machine that scrapes the hair from the hide. Other slaughter plants skin the pigs. When the internal organs are removed, care is taken to preserve those that are to be used for food. The liver is the most commonly eaten of the internal organs (**Figure 23–4**). The brains, the pancreas (sweetbread), intestines (chitterlings or tripe), and the heart are all used as human food. In some parts of the world, the kidneys also are used as food.

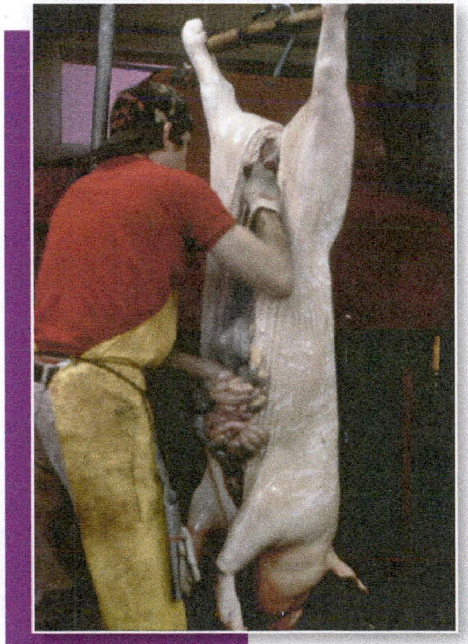

FIGURE 23–3 The animals are hoisted onto a rail, and the entrails are removed. *Courtesy of Dr. Estes Reynolds, Cooperative Extension Service, University of Georgia.*

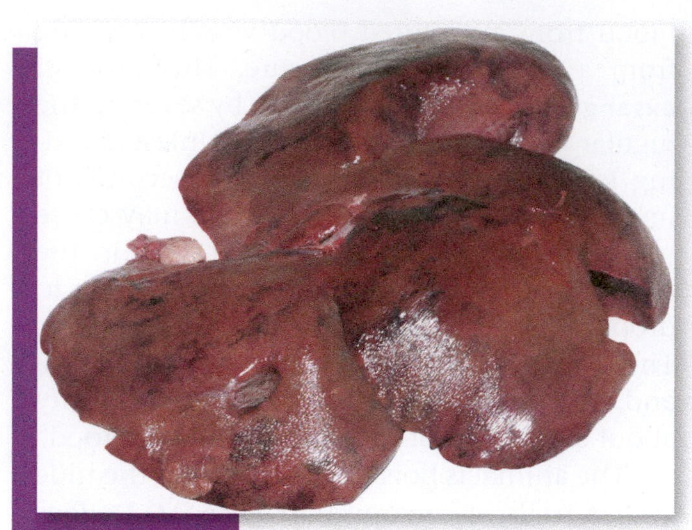

FIGURE 23–4 The liver is an example of an internal organ used for food. © schankz/Shutterstock.com.

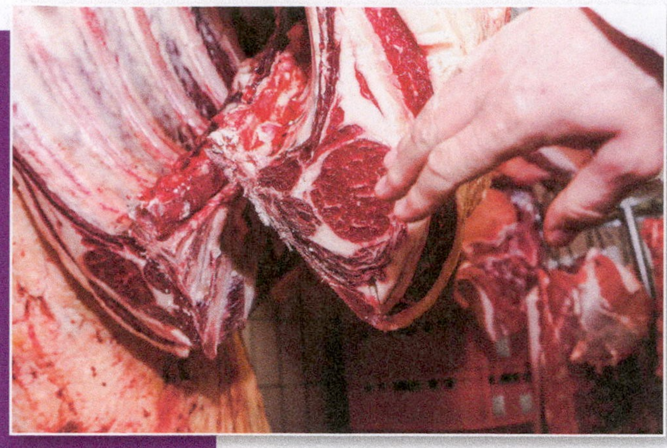

FIGURE 23–6 Beef carcasses are divided down the backbone and between the 12th and the 13th ribs. © iStockphoto/AndiGrieger.

During the slaughter process, federal inspectors are on hand to inspect the internal organs and the carcass to detect any health concern about the meat from the animal (**Figure 23–5**). If they find a problem, the carcass may be condemned as unfit for human consumption. Each carcass that is to be sold for human consumption must be inspected.

Beef carcasses are generally sent into the cooler in halves called sides of beef. After slaughter, the carcass is sawed down the backbone, called the **chine**, and the carcass is divided into two pieces. In very large carcasses, each side may be divided into two quarters. To do this, the side is divided between the 12th and 13th rib into the fore quarter and the hind quarter (**Figure 23–6**). Lamb carcasses usually are sent to the cooler whole. Hog carcasses usually are divided in half down the backbone. The carcasses of beef and pigs that have been skinned are covered in a heavy cloth soaked in salt water. The purpose of this covering, called a **shroud**, is to prevent the carcass from drying out.

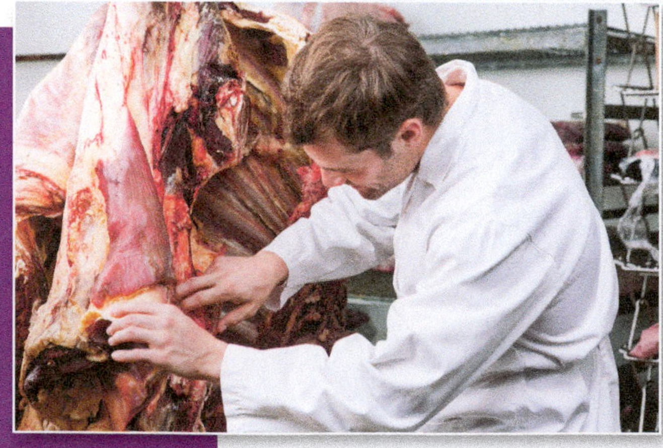

FIGURE 23–5 Federal inspectors examine the carcass and internal organs of each carcass. © leaf. Image from BigStockPhoto.com.

After slaughter, the carcasses need to be cooled down rapidly. Here, the carcasses undergo a process known as **rigor mortis**, in which the muscles lock into place and the carcass becomes stiff. The physiology of rigor mortis is similar to muscle contractions in a live animal, except that in the carcass, the muscles do not relax. The onset of rigor mortis usually takes from 6 to 12 hours for beef and lamb, and 30 minutes to 3 hours for pork. As enzymes and microorganisms begin to break down the muscle tissue, rigor mortis is partially relaxed.

Usually it is desirable to reduce muscle temperature as quickly as possible after death to minimize protein degradation and inhibit growth of microorganisms. Pork and lamb carcasses are usually cooled for 18 to 24 hours before they are cut into wholesale cuts. Larger

FIGURE 23–7 After slaughter, the carcasses are chilled by hanging in the cooler. *© Mark Agnor/Shutterstock.com.*

beef carcasses may require 30 or more hours of cooling before they are ready to be cut (**Figure 23–7**).

Higher-quality beef carcasses may be **aged** in the cooler for as long as a week. The carcasses undergo a period of aging to allow enzymes and microorganisms to begin the process of breaking down the tissue. This improves tenderness and flavor but adds to the expense of processing the meat.

An alternative to aging is the use of electrical stimulation of the muscles. A current of 600 volts is sent through the carcass immediately after slaughter and before the hide is removed. The stimulation speeds up natural processes that occur after death, like the depletion of energy stores from the body. Although this process affects the tenderness of the meat only slightly, there are other benefits such as improved color, texture, and firmness. This process also makes the hide easier to remove.

GRADING

After the carcasses are cooled, they are graded according to USDA standards. Federal Meat Grading was officially established in 1925 and is administered by the Agricultural Marketing Service (AMS) of the United States Department of Agriculture (USDA). The meat grade certifies the class, quality, and condition of the agricultural product examined to conform to uniform standards. Quality grades are a prediction of the eating quality (palatability) of properly prepared meats. Yield grades indicate expected yield of edible meat from a carcass and the subsequent wholesale cuts from that carcass. Grading is voluntary and is paid for by the meat packer.

Quality grades of beef are as follows: Prime, Choice, Select, Standard, Commercial, Utility, Cutter, and Canner. Grades are determined by the age of the animal at slaughter and the amount of fat intermingled with the muscle fibers. Age is determined by the maturity of the cartilage and bones in the carcass. Remember that as an animal ages, the cartilage hardens and turns to bone. Graders inspect the rib cage and vertebrae of the carcass for the degree of bone and cartilage hardening, called ossification (**Figure 23–8**).

As the animal ages, vertebrae in the lower end of the backbone tend to fuse, or grow

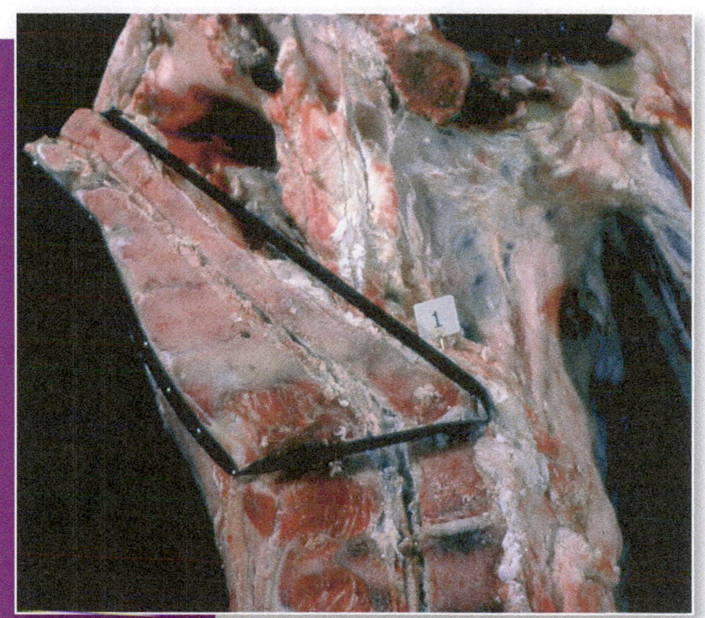

FIGURE 23–8 Graders determine the age of an animal by examining the vertebrae. Note how the bones of the vertebrae are fused together. This indicates the carcass of an older animal. *Courtesy of Dr. Estes Reynolds, Cooperative Extension Service, University of Georgia.*

together. By determining the degree of ossification, graders are able to classify the animal according to its maturity. Animals that appear to be older than about 42 months cannot receive the highest two grades (Prime and Choice). Younger animals tend to be more tender than older animals.

The amount of fat among the muscle fibers is determined by examining the longissimus dorsi muscle that runs the length of the vertebrae on each side. The carcass is separated between the 12th and 13th ribs to expose this cross section. The fat, known as marbling, shows up as specks of white across the rib eye (**Figure 23–9**). The more specks of fat that are visible, the higher is the grade (assuming maturity is the same).

Beef that grades Prime has the highest degree of fat in the muscle. Fat is what gives meat its flavor and juiciness. Fat is expensive to put on animals, so the leaner grades usually are less expensive. Most feedlot owners want their animals to grade a low Choice at slaughter. Those who feed animals to grade Prime cater to the restaurant trade. Most beef bought in the grocery store is Choice grade, although a few market chains are selling the leaner Select grade as a low-fat meat.

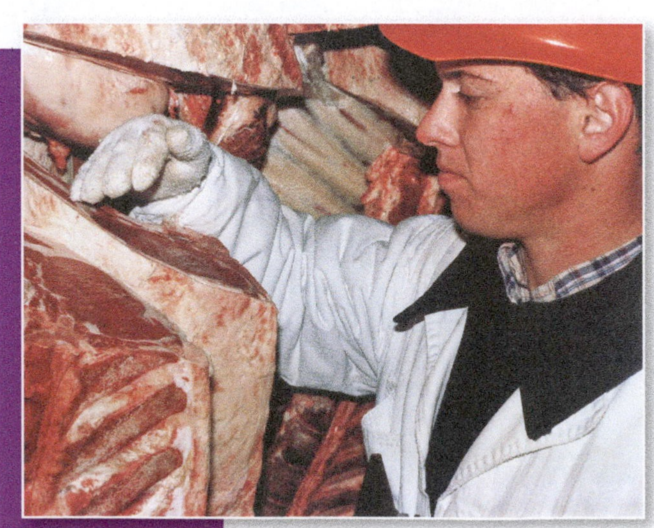

FIGURE 23–9 A USDA grader inspects the amount of marbling in the rib eye of a beef carcass. *Courtesy of North American Limousin Association.*

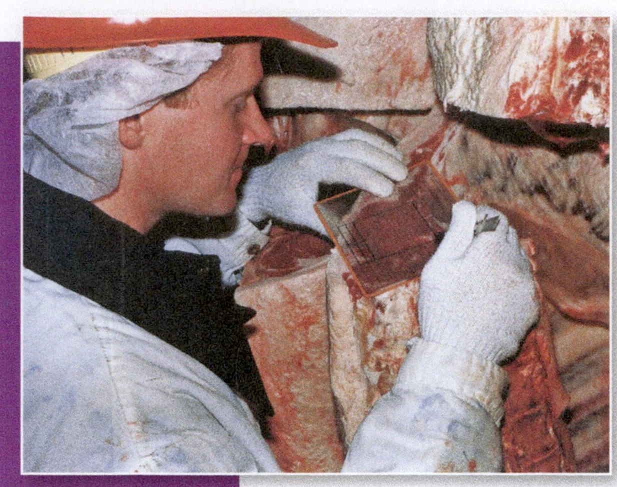

FIGURE 23–10 A USDA grader uses a rib eye grid to measure the area of the rib eye. *Courtesy of North American Limousin Association.*

Yield grade is an estimate of the percentage of boneless, closely trimmed retail cuts that come from the major lean **primal cuts** (round, loin, rib, and chuck). USDA yield grades for beef are as follows:

1. more than 52.3 percent lean primal cuts
2. 50.0–52.3 percent lean primal cuts
3. 47.7–50.0 percent lean primal cuts
4. 45.4–47.7 percent lean primal cuts
5. less than 45.4 percent lean primal cuts

The Yield grade is determined by a formula used by the grader. Factors in the formula are the chilled carcass weight, amount of internal (kidney, pelvic, and heart) fat, size of the rib eye area, and amount of backfat on the carcass (**Figure 23–10** and **Figure 23–11**).

THE WHOLESALE CUTS

After the carcasses have been chilled for the proper time, they are taken out of the cooler and cut into pieces that are sold to retail outlets. These cuts, known as primal cuts, are packaged in vacuum packs, placed in boxes, and shipped to the retailers. One advantage of this is that around 30 percent of the carcass weight is trimmed away with the excess fat and bone.

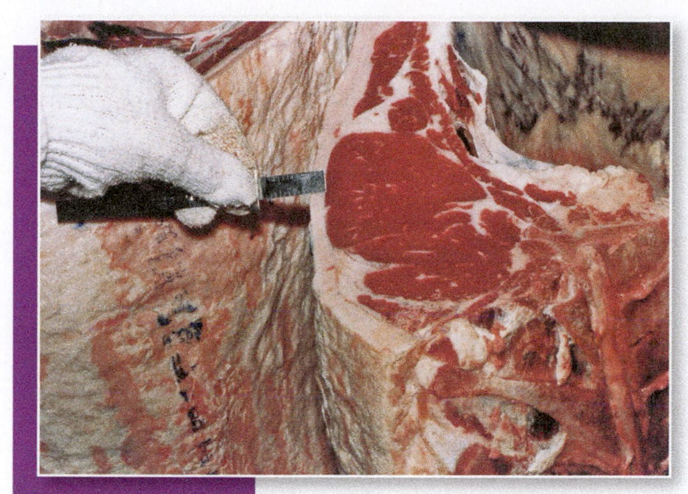

FIGURE 23–11 The thickness of the backfat is measured at the rib eye. *Courtesy of North American Limousin Association.*

This saves in the shipping costs. Boxed primal cuts are also easier to handle than are sides of beef or pork. Retailers also see this as an advantage because they can order the cuts of meat they need without getting parts they have no market for (**Figure 23–12**).

Wholesale or primal cuts of beef usually consist of the chuck, loin, rib, and round. Pork cuts include shoulder, loin, sides, and ham. Lamb cuts include shoulder, rib, loin, and leg. Primal cuts sometimes are divided into smaller units called subprimal cuts.

The wholesale cuts are divided into the retail cuts. These are the cuts of meat that the consumer buys at the grocery store. They are sized into portions that can be easily cooked and eaten without further cutting or trimming (**Figure 23–13**). The most expensive retail cuts usually come from the loin area (**Figure 23–14**). This muscle group is usually the most tender of the muscle groups. This is the area from which chops and steaks such as T-bones come.

As the retail cuts are made, there is always muscle that is trimmed or portions that do not make good retail cuts of meat. These portions or trimmings are made into sausage or ground meat (**Figure 23–15**). Often, the sausage is spiced and preserved by drying or smoking. Ground beef becomes hamburger meat.

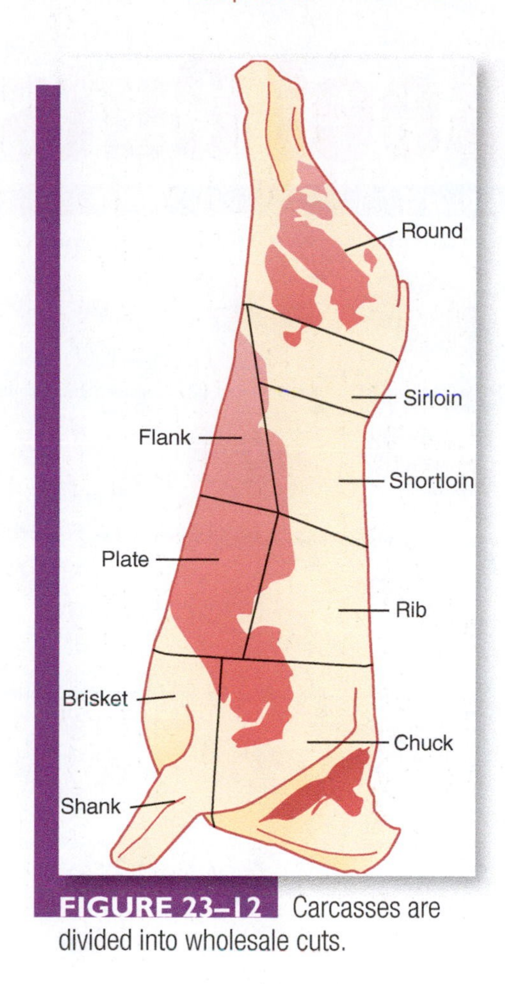

FIGURE 23–12 Carcasses are divided into wholesale cuts.

FACTORS AFFECTING PALATABILITY

Palatability refers to how a food appeals to the palate (a portion of the roof of the mouth). Meat palatability depends on qualities such as appearance, aroma, flavor, tenderness, and juiciness. The palatability of a piece of meat depends on a combination of these listed qualities and the way in which it was cooked. However, consumers buy most retail meat in the uncooked form.

Processed meats are an exception. These meats are composed of those parts of the carcass that do not make good retail cuts (much the same as sausages and ground meat). These meats undergo a processing that may include pressing, forming, and slicing. These products are usually fully cooked and are used as cold cuts and sandwich material. Bologna, hot dogs,

FIGURE 23–13A Retail cuts of beef. *Courtesy of the Beef Checkoff Program*

Pork's Most Popular Cuts

SHOULDER

Shoulder Steak; bone-in

Shoulder Roast; bone-in

Shoulder Country-Style Ribs; bone-in

LOIN

New York Chop

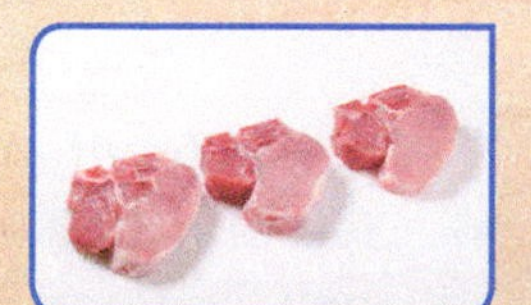
Porterhouse Chop

Ribeye Chop

Sirloin Chop; boneless

Loin Back Ribs

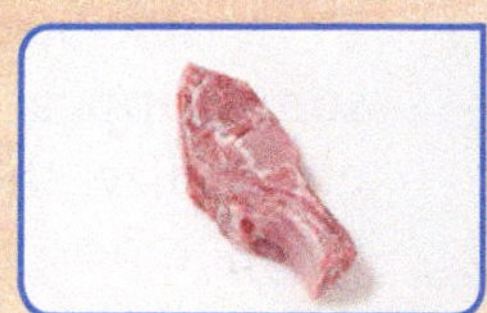
Loin Country-Style Ribs; bone-in

Loin Country Style Ribs; boneless

New York Roast

Tenderloin

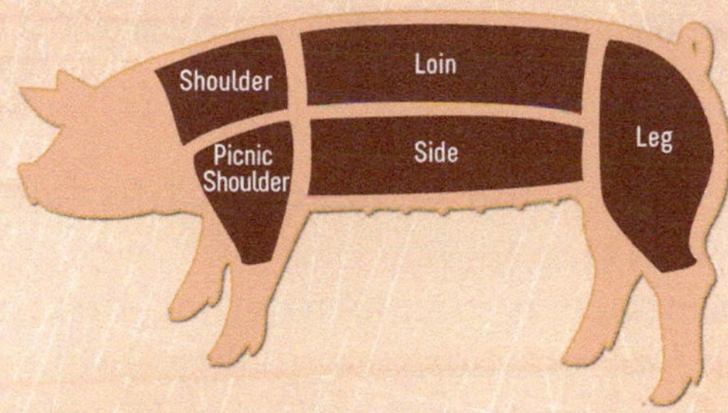

SIDE

Spareribs

St. Louis-Style Ribs

For recipe ideas visit: **PorkBeInspired.com**

pork checkoff.

FIGURE 23–13B Retail cuts of pork. *Courtesy of National Pork Board*

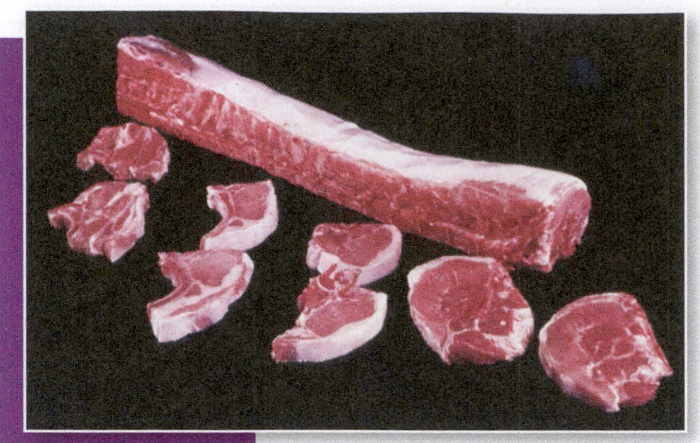

FIGURE 23–14 This is a wholesale cut (the loin) and the retail cuts that come from it. *Courtesy of Dr. Estes Reynolds, Cooperative Extension Service, University of Georgia.*

FIGURE 23–16 Many types of processed meat are made from various parts of the carcass. *© gresei/Shutterstock.com.*

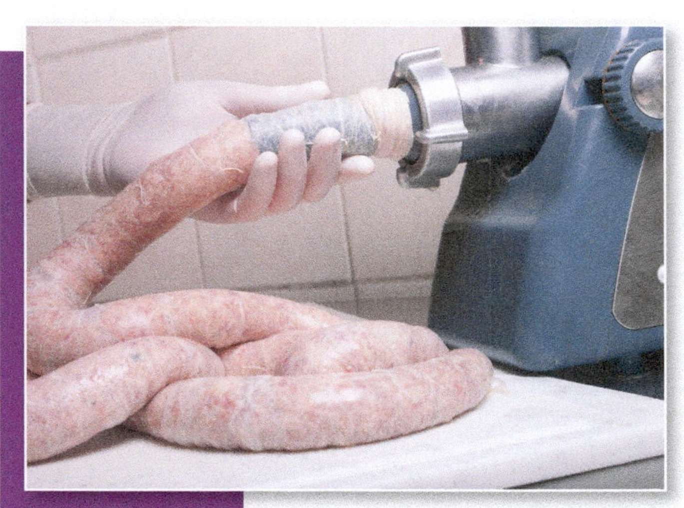

FIGURE 23–15 Sausages are made from ground meat. Here, tubes made from intestines are being filled with sausage meat. *© aboikis/Shutterstock.com.*

processed ham, and salami all fit this category (**Figure 23–16**).

Appearance is the factor that first influences the expectations of quality. Beef, pork, and lamb all vary in the shades of red color. Darker meat tends to be associated with either a lack of freshness or meat from older animals. Bright red meat gives the appearance of being fresh and wholesome.

Fat that is yellow instead of creamy white is less appealing to consumers. Yellow fat is generally associated with certain breeds that are unable to convert carotene (yellow to red pigment found in vegetables, body fat, and egg yolks) to vitamin A. Cattle that have been grass-fed and have consumed an excess of carotene may have yellow fat. Grain-fed beef generally is considered to taste better than grass-fed beef.

The amount of fat and bone in proportion to muscle also affects the appearance and consumer appeal of meat (**Figure 23–17**). Consumers realize that fatter meat with bones has a smaller portion of edible meat or more plate waste (meat scraps, bone, and fat that is not consumed).

Tenderness is a sensation that has several components and has been the object of considerable study. Tenderness is difficult to

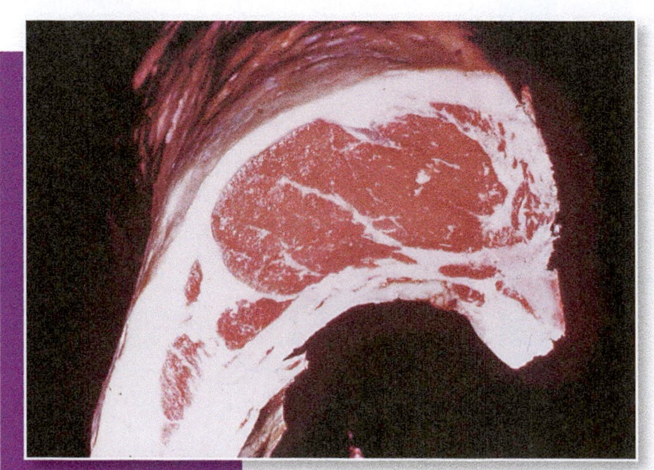

FIGURE 23–17 The amount of fat in proportion to the muscle affects consumer approval. This steak has too much fat. *Courtesy of Cooperative Extension Service, University of Georgia*

measure. Terms that have been ascribed to tenderness during chewing include resistance to tooth pressure, softness to tongue and cheek, ease of fragmentation, mealiness, adhesion or stickiness, and residue after chewing. Not only is it difficult to describe eating tender versus tough meat, but it also has been difficult to use mechanical devices that measure the many aspects of chewing and taste.

Components of muscle that contribute to tenderness are connective tissue, state of the muscle fibers (what degree of contraction), and the amounts of **adipose** (fat) tissue. However, the amount of adipose tissue or fat, particularly intramuscular (within the muscle) fat known as marbling, probably does not influence tenderness to any great degree. Research has not been able to relate marbling to tenderness except that marbling may tend to act as a lubricant during **mastication** (chewing) and swallowing, and that marbling is related to high-energy feed and production systems.

Connective tissue connects various parts of the body and is distributed throughout the body. Sheaths of connective tissue surround muscle bundles, nerve trunks, blood vessels, tendons, and fat cells (adipose tissue). Connective tissue consists of a structureless mass (called ground substance), embedded cells, and extracellular fibers (fibers outside the cell). Extracellular fibers include collagen and elastin. Collagen is the most abundant protein in the animal and is found in all tissues and organs. The presence of collagen in skeletal muscle is generally proportionate to the physical activity of the muscle—the more the activity, the more collagen. Muscles of the limbs contain more collagen than the muscle along the spinal column. As the animal grows older, the collagen becomes less soluble. An example of collagen solubility is the gelatin material in the pan after cooking a pot roast in moist heat. Therefore, the muscles of younger animals and the muscles responsible for less physical activity, such as those along the spinal column, are more tender.

Elastin is an elastic-like protein found throughout the ligaments, arterial walls, and organ structures. Elastin fibers are easily stretched, and they return to their natural state when the tension is released. Cooking has no appreciable effect on elastin fibers. Cuts of meat with high elastin content tend to appear tougher.

When selecting fresh cuts of meat in the retail case, the consumer should avoid extremes in apparent juiciness. A dark, dry appearance usually indicates either age in the retail case or age of the animal from which the cut was obtained. Extremely moist-appearing meat (moisture oozing out of the retail cut) indicates the pale, soft, and exudative (PSE) condition associated with some meats.

The juiciness of cooked meat is important in the perception of palatability to consumers. The juice is made up of water and melted intramuscular fats. As the meat is chewed, juices are released that stimulate the flow of saliva, thereby further increasing the apparent juiciness. The juices contain flavor components, and they assist in lubricating, softening, and fragmenting the meat during chewing.

The flavor of meat often changes after extended storage periods. The chemical breakdown of nucleotides (flavor compounds) imparts a desirable aged flavor, whereas **oxidation** of fatty acids (oxidative rancidity) results in a **rancid** flavor and a sharp, unpleasant aroma.

Aroma is detected from numerous gaseous aspects of meat that stimulate nerve endings in the linings of the nasal passages. The total sensation is a combination of taste (gustatory) and smell (olfactory). The meaty flavor and aroma stimulate the flow of gastric juices and saliva that aid in digestion and increase the apparent juiciness of the meat.

POULTRY PROCESSING

As mentioned in a previous chapter, poultry is becoming an increasingly larger part of our diet. Each day, millions of chickens are slaughtered. Just a few years ago, this was a relatively

FIGURE 23–18 Chicken is now processed into products such as chicken strips. © Moolkum/Shutterstock.com.

simple process because most poultry was marketed as whole carcasses or carcasses cut into breasts, legs, thighs and backs. Now there is a wide array of different products such as wings, nuggets, chicken strips, and other products (**Figure 23-18**).

Most of the newer poultry products are cut from the carcass using machinery designed specifically for cutting strips, nuggets, or other pieces (**Figure 23-19**). One of the newest technology is the use of a laser to guide the machine. The laser "reads" the size and shape of the carcass and makes the most efficient cuts.

Another challenge of poultry processing is that of *E. coli* bacteria that comes from residual fecal matter from the slaughter process. These bacteria present a serious health risk if the meat is not properly cooked. The meat has always been examined for fecal material, but it is sometimes difficult to detect. A new innovation is the use of a laser-induced fluorescent light that can detect small residues of the matter.

FIGURE 23–19 Poultry processing is now highly mechanized. © Alf Riberio/Shutterstock.com.

PRESERVATION AND STORAGE OF MEAT

Meat is a highly perishable product that can spoil in a very short time. Meat is preserved by creating conditions that are unfavorable for the growth of spoilage organisms and the development of off-flavors, which usually are a result of chemical oxidation of fatty acids or proteins. Historically, methods of preservation have included drying, smoking, salting, refrigeration, freezing, canning, and freeze-drying.

Degradation of the animal tissue begins upon slaughter as a result of chemical, biological, and physical reactions. Microorganisms are an integral part of this process, as they thrive on the nutrients supplied by meat. Meat provides an ideal medium for the growth of many **microbes** (microorganisms). Molds, yeast, and bacteria are all found on or in meat. Molds are multicellular, multicolored organisms that have a fuzzy or mildew-like appearance. They spread by producing spores that float in the air or are transported by contact with objects. Yeast consists of large, unicellular buds and spore forms that are spread by contact or in air currents. Most yeast colonies are white to creamy in color and usually are moist or slimy in appearance or to the touch.

Among the factors affecting the growth of microbes are temperature, moisture, oxygen, pH, and the physical form of the meat. Temperature can influence the rate and kind of microbial growth. Some microbes grow well in cooler temperatures of 0°–20°C (32°–68°F). These are known as **psychrophiles**. Microbes that grow at higher temperatures, between 45°C and 65°C (113°–150°F), are called **thermophiles**. Microorganisms with an optimum growth temperature between the psychrophiles and the thermophiles are called **mesophiles**. Temperatures below 5°C (40°F) greatly retard the growth of spoilage microorganisms and prevent the growth of most pathogens.

Moisture and relative humidity greatly affect the growth of certain microorganisms. Most microbes must have moisture to reproduce and grow.

Oxygen availability determines the type of microorganism that grows. Microorganisms requiring free oxygen in order to grow are called **aerobic organisms**. Those organisms growing in the absence of oxygen are called **anaerobic organisms**. Microbes that can grow with or without free oxygen (reduced oxygen availability) are termed **facultative**. Molds, yeast, and many of the bacteria commonly associated with meat are aerobic. Many of the bacteria found in meats, such as lactobacillus, are anaerobic or facultative and will, over time, cause spoilage in vacuum-packaged meat products. In vacuum packaging, the meat is placed in a plastic bag and all the air is removed. This creates a package that does not let air or moisture pass through. The vacuum packaging of meat and meat products increases the storage time of these products by inhibiting the growth of aerobic organisms.

Most microorganisms have an optimum pH near neutrality (pH 7.0). Molds survive in a wide range of conditions (pH 2.0 to 8.0). Yeasts favor a slightly acidic condition of pH 4.0 to 4.5. Most bacteria favor a range of pH 5.2 to 7.0. Meat and meat products generally range from pH 4.8 to 6.8, with the norm being approximately pH 5.4 to 5.6. Therefore, meat conditions favor the growth of molds, yeast, and the acidophilic (acid-loving) bacteria.

The physical form of meat affects the growth of microorganisms. As carcasses are separated into primal, subprimal, and retail cuts, more surface area is exposed. This surface area provides nutrients, moisture, and oxygen for the microorganisms to grow. When meat is ground for products such as sausage and hamburger, the maximum surface area is exposed and the microorganisms are spread throughout the product. Therefore, sanitation and keeping the temperature as low as possible are the major considerations in maintaining acceptable numbers of microbes.

Meat that has been frozen and thawed is more susceptible to microbial growth because of ruptured cells and increased surface moisture. Refreezing meat that has been thawed does not cause serious deterioration or breakdown in itself, but refreezing will not reverse the deterioration caused by microbes. Refreezing in less than optimum conditions, such as often exist in home refrigerator/freezer units, will cause decreases in juiciness and flavor due to the formation of large ice crystals that rupture the cells. This causes drying and the oxidation of fats.

Curing and Smoking

Meat processing developed soon after people became hunters in prehistoric times. The salting and smoking of meat has been documented as far back as 850 B.C. by Homer, and as far back as the thirteenth century B.C. by the Chinese. In these early times, the smoking (drying) and salting (curing) of meats were the only known methods of preservation. Today, the curing and smoking of meat is a method of imparting a particular flavor to the meat (**Figure 23–20**). Very few people in this country still rely on curing

FIGURE 23–20 These sausages are hung on a rack and will be placed in a smoker. Smoke imparts flavor to meat. *© mashurov/Shutterstock.com.*

and smoking as methods of preservation. There are just about as many cured products today as there are regions of the world, many of these being descendants of the ancient curing methods.

The two main ingredients used to cure meats are salt and nitrite. Some of the most frequently used are sugar, ascorbate, erythorbate, phosphates, and delta gluconolactone. Today, salt is used at levels that generally impart flavor to the product (1 to 3 percent) instead of the amounts used to preserve the meat item being cured (9 to 11 percent). Nitrates (saltpeter) and nitrites are used to impart the "cured meat" color and flavor and to inhibit bacteria action. The use of nitrites or nitrates is not permitted to result in more than 120 parts per million (ppm) of nitrite in the finished product.

The oldest of these methods is known as **dry curing**. In this method, the cure ingredients are rubbed onto the surface of the product and allowed to move into the product by osmosis. This method takes the longest amount of time to infiltrate the meat. Over time, technology has improved on this simple procedure.

A more modern method of curing meat is **injection curing** (**Figure 23–21**). This method involves pumping the curing solution (brine) into the meat product. This shortens the curing process since the curing ingredients are dispersed directly into the meat. There are three methods of placing the cure into the product. The first of these uses the artery system in the meat (primarily used in hams) to disperse the curing adjuncts. The second of these methods involves placing a hollow needle (stitch) into the major muscle masses of the product to inject the curing solution. The third method of curing, **combination curing**, is simply a combination of dry curing and injection curing.

FIGURE 23–21 These hams have been injected with curing solution and are about to be placed in the smoker.
© Milos4U/Shutterstock.com.

Refrigerator Storage of Meat

Carcass temperatures upon exiting the slaughter floor generally range between 30°C and 35°C (85°–95°F). Chill coolers generally operate at about 23°C to 1°C (27°–34°F), and carcasses need to be chilled to less than 5°C (40°F). The time required for the chilling process is affected by the carcass size, amount of fat on the carcass (fat reduces heat dissipation), and the initial heat of the carcass. Rapid air movement can reduce the chilling time up to 25 percent. After 12 to 24 hours in the chill cooler (often called a hot box), beef carcasses are moved to aging or holding coolers at 0° to 3°C (32°–37°F) until they are fabricated or shipped.

The storage life of carcasses or meat products in refrigeration depends on factors such as initial numbers of microbes on the meat, temperature and humidity conditions during storage, use of protective coverings or packaging, animal species, and the type of product being stored. As a rule of thumb, fresh meat under good home refrigeration conditions should be consumed within four days of purchase.

Freezer Storage of Meat

Freezing acts as a preservation method because microbial and enzymatic activity is stopped at about 10°C (14°F) (**Figure 23–22**). However, some changes do still take place, such as the development of rancidity and surface discoloration due to dehydration. Factors that affect the quality of frozen meat include freezing rate, length in freezer storage, packaging

FIGURE 23–22 Meat can be preserved by freezing.
© Strakhov Sergei/Shutterstock.com.

materials used, and the variability of the freezer temperatures.

The most common method of commercial freezing is to utilize high-velocity air and temperatures of −10°C to −40°C (14°–40°F). This method is commonly termed **blast freezing**. Methods of freezing utilizing condensed gases in direct contact with the product are termed cryogenic.

Problems associated with relatively slow freezing include the formation of large ice crystals and the loss of moisture during thawing. Temperature fluctuations during frozen storage also may cause these problems.

The length of time that meat can be stored in a freezer varies with freezer temperature, temperature fluctuations, species, type of product, and type of wrapping material. The product must be packaged using vapor-proof materials to keep oxygen out and to keep moisture in the package. Oxygen causes oxidative reactions such as rancidity. Moisture loss causes dehydration and a condition known as freezer burn.

Storage time may be extended by lowering storage temperatures. Although it is not economically feasible, maintaining a temperature of −80°C (−112°F) stops most chemical changes. Most commercial and home freezers are maintained at approximately −18°C (0°F) or lower. Even at this temperature, fluctuations in temperature may cause migration of water and increased moisture loss upon thawing.

Meats from different species differ in the time that freezer storage will maintain acceptable quality. The difference is primarily due to differences in fat composition. Softer fats, such as those found in pork, are more susceptible to oxidative changes and subsequent loss of flavor. Because of the differences in fat composition, recommendations for length of freezer storage at −18°C (0°F) or lower are as follows:

beef—6 to 12 months

lamb—6 to 9 months

pork—4 to 6 months

cured meats—1 to 2 months.

This can vary with packaging material, cuts of meat, and fluctuation in freezer temperature.

Processing of the meat into sliced, ground, or cured products influences the acceptable freezer storage time. Exposure to oxygen or the addition of salts enhances the development of rancidity and reduces flavor acceptability.

Preservation of Meat by Drying

Because moisture is critical to microbial growth, removing moisture from meat is an effective means of preservation. Low-moisture foods are those that contain less than 25 percent moisture. Beef jerky is an example of a low-moisture food.

Intermediate-moisture foods have less than 50 percent moisture. Dry salami is an example of an intermediate dry-meat product that is shelf-stable (requires no refrigeration) but still subject to mold growth unless it is treated with a mold inhibitor. Most meat products that are dried also contain some salt, which assists in lowering the moisture of the product.

Irradiated Meat

In 1997, the U.S. Food and Drug Administration (FDA) granted approval to use **radiation** to help preserve red meat. This technology has been around for several years and has been used on a wide variety of foods, ranging from wheat to onions. In 1990, the technology was approved for use on the poultry to control salmonella and on pork to stop trichina, but approval for beef was delayed. Currently, **irradiation** is used in more than 40 countries to preserve food.

The technology makes use of low levels of radiation to kill pathogens in food products. Of course, additional preservation techniques such as refrigeration are required because the food is not permanently preserved. However, the rate of spoilage is greatly reduced, and dangerous bacteria such as salmonella and ***E. coli*** are killed. This makes meat much safer for the consumer.

The greatest hindrance to the use of irradiation on meats and other foods is consumer acceptance. There seems to be an aversion to eating anything that was submitted to radiation. However, the FDA has declared the process to be safe and effective. A symbol indicating that the product has been treated by irradiation is required on all packages. After treatment, the food is no more radioactive than your teeth are after a dental x-ray. Perhaps after the irradiated foods have been sold for a while, consumers will accept them more readily.

PUTTING IT INTO PRACTICE

The Meats Career Development Event

Another FFA team event related to the livestock industry is the Meats Evaluation Career Development Event. Like the Livestock Judging Event, teams are made up of four FFA members. As a participant in this event, you will gain valuable knowledge of meat identification and grading. This knowledge will benefit you in the future if you are seeking a job in the meatpacking industry or if you are a consumer of meat. Preparation for this event combines team practice and individual study. The team should practice as often as possible. In most career development events, students identify primal cuts of beef and pork; retail cuts of pork, beef, and lamb; and quality or yield grades of carcasses. By visiting meat-processing plants and local grocery meat departments, the identification of wholesale and retail cuts can be reviewed. Each contestant will need to study for the written test relating to meat selection, storage, cookery, nutrition, and safety. Situational problems involving the least-cost formation of a batch of meat products such as hamburger or bologna are given. The student works through procedural questions and the actual determination of the least-cost price. The FFA Meats Evaluation Career Development Event develops employment knowledge with practical skill applications preparing students for entry-level employment in the meatpacking industry.

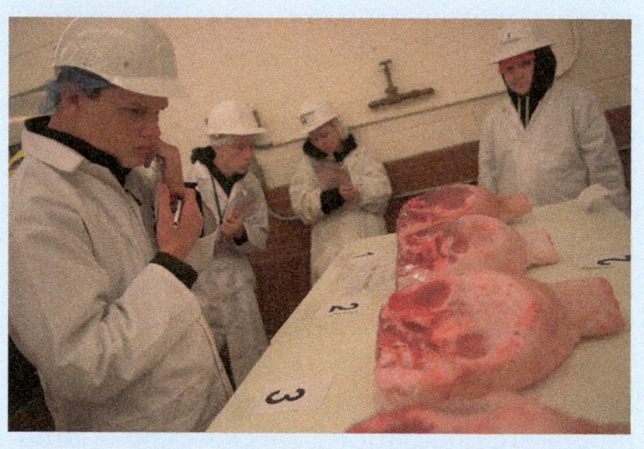

The Meats Evaluation Career Development Event helps students develop skills in identifying and evaluating cuts of meat. *Courtesy of the National FFA Organization*

SUMMARY

Americans eat a lot of meat, and this has spawned a tremendous industry that is constantly changing in an attempt to meet the needs of the consumer. We want products that are safe, tasty, and affordable. The government, through the USDA and the FDA, regulate the growing, processing, grading, packaging, and sale of all meat products. These products require a wide variety of preservation methods to keep the meat from spoiling. As a result of our high-tech agricultural food systems, we enjoy the best, cheapest, and safest food supply in the world.

CHAPTER REVIEW

Review Questions

1. Define the term *meat*.
2. What methods are used to render an animal immobile during the slaughter process?
3. What parts of the animal are edible besides the muscles?
4. What is rigor mortis?
5. What is the difference between Quality grading and Yield grading?
6. List, in order, the Quality grades of beef.
7. List three factors that affect the palatability of meat.
8. List six methods of preserving meats.
9. Name three types of microbes that cause meat spoilage.
10. What problems are encountered when meat is frozen slowly?
11. Discuss the greatest hindrance to preserving meat with irradiation.

Student Learning Activities

1. Visit a local grocery store and list all of the cuts of meat in the meat counter. Develop a chart illustrating the wholesale cut from which each came.
2. Interview a meat buyer for the grocery store. Determine how and where the meat is bought, the Quality grade and Yield grade bought, and which wholesale cuts the buyers purchase most often.
3. Go to the grocery store and look for food products that have been treated with irradiation. Share your list with the class.

CHAPTER 24

Parasites of Agricultural Animals

STUDENT OBJECTIVES IN BASIC SCIENCE

As a result of studying this chapter, you should be able to

- explain symbiotic relationships.
- distinguish among mutualism, commensalism, and parasitism.
- discuss how parasitism causes harm to host animals.
- explain the process of metamorphosis.
- list the phases in the life cycle of an insect.
- distinguish between a roundworm and a segmented worm.
- explain how scientific research is used in the eradication of parasites.

STUDENT OBJECTIVES IN AGRICULTURAL SCIENCE

As a result of studying this chapter, you should be able to

- list the types of parasites that infest agricultural animals.
- explain how production losses are incurred because of parasites.
- list the conventional means of controlling parasites on agricultural animals.
- discuss how the life cycle of a parasite can be used to control the parasite.

KEY TERMS

symbiosis
mutualism
commensalism
parasitism
parasite
host
life cycle
anemia
internal parasites
external parasites
metamorphosis
roundworms
stomach worms
strongyle
colic
ascarids
tapeworms
intermediate host
flukes
warm-blooded
nymph
systemic pesticides
bolus
biological control

NATIONAL AFNR STANDARD

AS.07.01.03.a
List and summarize the characteristics of wounds, common diseases, parasites, and physiological disorders that affect animals.

AS.07.01.03.b
Identify and describe common illnesses and disorders of animals based on symptoms and problems causes by wounds, diseases, parasites, and physiological disorders.

AS.07.01.04.b
Research and analyze data to evaluate preventative measures for controlling and limiting the spread of diseases, parasites, and disorders among animals.

ANIMALS OF DIFFERENT SPECIES that live in close association with each other are said to live in a symbiotic relationship. **Symbiosis** can take at least three different forms: mutualism, commensalism, and parasitism.

Mutualism is a relationship that is beneficial to both species of animals. For example, as explained in Chapter 22 on nutrition, certain bacteria live in the rumen of cattle. The cattle provide the bacteria with food and a place to live. The bacteria help the cattle break down fibers into a form that can be digested. Another example is that of tick birds that light on the back of cattle and other animals. The birds obtain food from eating the ticks on the animals, and the animals benefit by having an annoying pest removed (**Figure 24–1**).

Commensalism is the relationship of animals in which one benefits and the other is not harmed. An example of commensalism is the relationship between cattle and houseflies. The housefly must lay its eggs in the feces of animals. The fly benefits from the fecal material deposited by the cattle, but the cattle are not harmed by the housefly.

The third form of symbiosis is that of **parasitism**. Parasitism is a relationship that is beneficial to one animal and harmful to another. It accounts for the vast majority of the incidences of symbiosis. All agricultural animals are susceptible to parasites, and measures must be taken by producers to deal with parasitism. The animal that lives off the other animal is called a **parasite**; the animal that the parasite lives on or in is called the **host**.

According to the USDA, parasitism causes almost a billion dollars' worth of damage to agricultural animals each year. Generally, parasites that live in and on livestock are insects that live out one or more of the phases of their **life cycle** at the expense of the agricultural animal. The damage they cause comes about in several ways. Most parasites live off the blood of the host animal. The continual loss of blood causes the animal to develop a condition known as **anemia**. One of the major functions of an animal's blood is that of providing body cells with oxygen and food nutrients. If enough parasites are living off the blood of an animal, the blood supply to the animal may be greatly diminished. When this occurs, the host animal will become ill because the body cells are not getting enough oxygen and food nutrients. The animals are said to be anemic. The animals are sluggish, feel poorly, and do not grow or perform as they should (**Figure 24–2**).

Animals that are hosts to parasites are in a weakened condition. This makes the animals more susceptible to disease; disease organisms can more effectively attack weak animals.

FIGURE 24–1 An example of mutualism is the tick bird, which lands on cattle and eats the ticks.

© Steve Oehlenschlager/Shutterstock.com.

FIGURE 24–2 Animals infected with parasites do not feel well and do not perform as they should.

© willmetts/Shutterstock.com.

FIGURE 24–3 Parasites may pass diseases from a sick animal to a healthy one. *© Piotr Kamionka/Shutterstock.com.*

As explained in Chapter 25, an animal's body has an immune system to fight disease organisms that invade its body. In order for the immune system to function properly, the animal must be strong and in good health. If the animal is in a weakened condition, the immune system will not function as it should, and the animal will get sick more easily.

Parasites often carry disease organisms from one animal to another. An insect may feed on an infected animal and then feed on a healthy animal and transmit disease organisms (**Figure 24–3**). For example, a dreaded disease of horses is sleeping sickness. This disease is passed on by an insect that bites and sucks blood from the animal. An insect may bite an infected horse and draw blood from the sick animal. Then the insect may fly to another area, where it bites a healthy animal. In doing so, it passes disease germs to the healthy animal. Humans, too, can get diseases this way. A common example is malaria, which is transmitted through the bite of a mosquito.

Animals that are infected with parasites are almost always uncomfortable. Parasites cause irritation of the skin, intestinal tract, or other parts of the body. Animals that are irritated and uncomfortable do not grow as well and are not as efficient. For animals to grow, breed, or feed their young, they must feel healthy. If parasites are causing the animal pain or the host animal has to spend most of its time trying to alleviate an itch caused by parasites, the animal will perform poorly.

Animals that are infected with parasites consume more feed per pound of gain. In other words, the feed efficiency of the animal is lowered. This means that the cost to maintain the weight of the animal is increased. If parasites are feeding on an agricultural animal, they are feeding either directly or indirectly on the feed supplied by the producers.

Parasites can generally be broadly divided into two categories: **internal parasites** and **external parasites**. Although some parasites live their entire lives on or in the host animal, most live only a portion of their life on or in the host. In these instances, the host animal supports the parasite through only a phase of its life cycle. For example, insects go through four complete stages from the time they hatch until they are mature adults capable of reproducing. This process of change is called **metamorphosis**. At each one of these stages the insect looks completely different from its appearance during the other three stages (**Figure 24–4**). It is

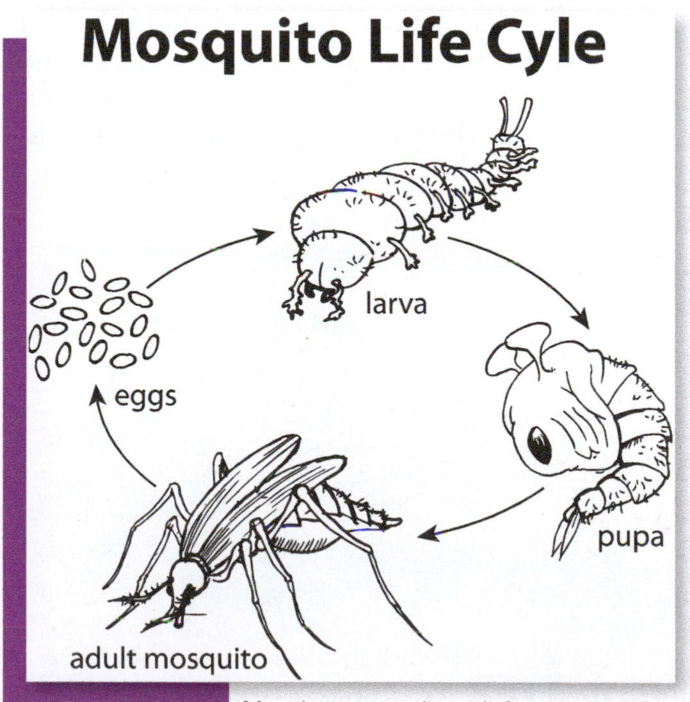

FIGURE 24–4 Most insects go through four stages of development. *© BlueRingMedia/Shutterstock.com.*

during one of these four stages in a parasite's life that the host animal is infested.

When the young insect hatches, it usually is in the larval stage. This means that the young insect looks very much like a worm. A larva usually is a voracious eater and can do a lot of damage to plants or to a host animal.

When the larva matures, it passes into the pupa stage, which is usually a relatively dormant stage. A pupa is the intermediate stage between the larva and the adult. During this stage, the body tissues of the young insect convert from a larva to an adult.

The last stage is the adult. In this stage, the insect lays eggs, and the cycle begins again.

INTERNAL PARASITES

Internal parasites actually live within the animal's body and may feed on the animal's blood or on feed that passes through the animal. Internal parasites are divided into three major groups: roundworms, tapeworms, and flukes.

Roundworms

Roundworms cause more damage to agricultural animals than any other group of internal parasites. They infect almost all types of livestock and exist by living in the digestive tract of their hosts (**Figure 24–5**). **Stomach worms** infect all classes of livestock and cause damage

FIGURE 24–5 Roundworms infect the digestive tracts of their hosts. *© Mooning27/Shutterstock.com.*

by the adults burrowing into the lining of the host's stomach and sucking the animal's blood.

Also, by digging into the stomach lining, the worm damages the tissue of the stomach, enzymes are not produced, and the host animal cannot digest food as well as before the infestation. The worms release poisons as they digest their food and excrete the waste into the host animal. These poisons can cause the host animal to become ill. In general, an ill animal will not eat or perform nearly as well as a healthy animal.

The worms lay eggs in the stomach of the host and pass out of the animal in the feces. While in the feces, the eggs hatch into larvae and the larvae crawl out onto a blade of grass. A grazing animal then eats the grass and swallows the larvae. Once the larvae are swallowed, they settle in the stomach of the host animal and begin to penetrate the stomach lining. The parasites remain there, feeding off the animal's blood and laying eggs. The eggs pass out of the host animal, and the whole process starts again (**Figure 24–6**).

Another type of roundworm is the **strongyle**. The life cycle of the strongyle is similar to that of stomach worms, except that instead of living in the stomach lining, strongyles live in the intestines of the host animal. Strongyles cause damage by causing scar tissue in the small intestine and by sucking blood from the host animal. Since the small intestine absorbs food nutrients into the bloodstream, a damaged small intestine reduces the efficiency of the digestive system of the infested animal. These parasites are particularly damaging to horses and can cause a digestive disorder called **colic**.

The largest of the roundworms are the **ascarids**. Ascarids most often attack young animals. Like the stomach worms and strongyles, the larvae of ascarids are ingested by animals grazing on blades of grass to which the larvae have attached themselves. The larvae burrow into the walls of the intestines and from there work their way through the host's heart, liver, and lungs. When they reach the lungs,

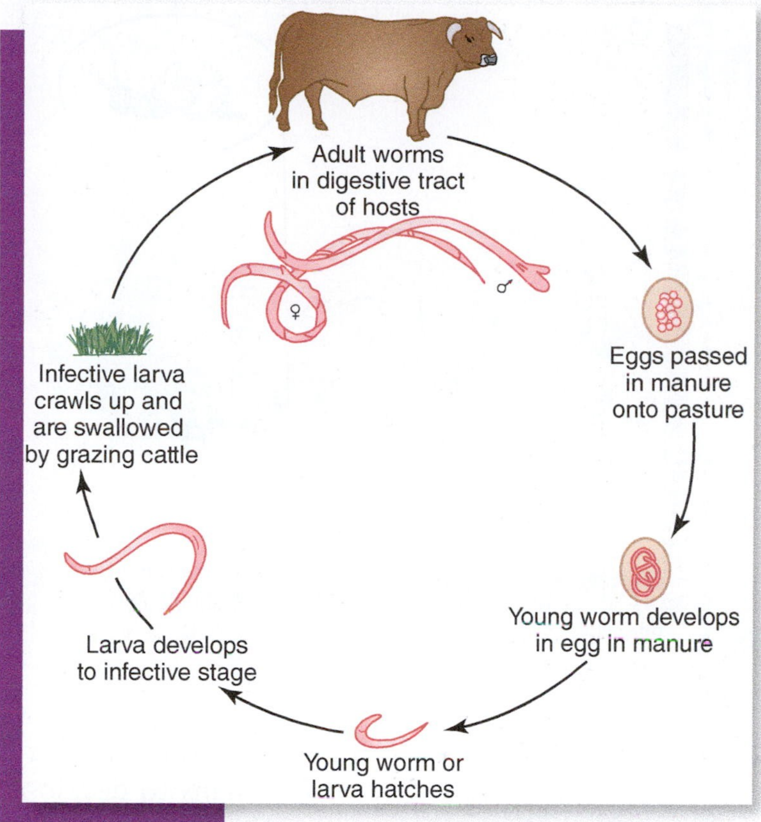

FIGURE 24–6 Roundworms live only part of their life cycle in host animals.

the worms are coughed up by the host animal and swallowed. The larvae are passed into the small intestine, where they develop into adults. The adults lay eggs that are passed out onto the grass in the host animal's feces; the eggs hatch; the larvae attach to blades of grass; and the process is renewed.

Tapeworms

Tapeworms belong to a class of worms that are segmented. This means that the bodies of the worms are made up of distinct segments. Each of these segments contain both male and female reproductive organs, and each segment is capable of producing fertilized eggs. These segments break off the body of the worms and reproduce.

Tapeworms cause less damage than roundworms because they do not feed on the animal's blood or cause scarring of the digestive tract. They do, however, cause losses because of the manner in which they feed. The adults of the tapeworm live in the small intestine of the host animal (**Figure 24–7**). These parasites grow to be quite large, with some species reaching lengths of 25 feet. The tapeworm lives off feed that is passed into the host animal's intestine. It causes the animal harm by devouring the food the animal has eaten. The life cycle begins when the segments of the adult tapeworms break off and pass out in the feces. Each segment contains eggs that hatch in the feces. The eggs are eaten by a small mite called an *oribatid mite* that lives in the grasses found in pastures (**Figure 24–8**).

The mite serves as an **intermediate host**. An intermediate host is an animal that a parasite uses to support part of its life cycle. An intermediate host is not harmed by the parasite. Since the mite lives on grasses, they are swallowed by grazing animals. The eggs are then passed through the animal to the small intestine, where they hatch and live until maturity.

Flukes

Flukes are small, seed-shaped flatworms that live in various parts of the host animal. By far, the most damaging flukes are those that live in the liver. The adult liver flukes live in the bile ducts of the liver, where they cause scarring of

FIGURE 24–7 The adult tapeworm lives in the small intestine of the host animal. *© iStockphoto/enot-poloskun.*

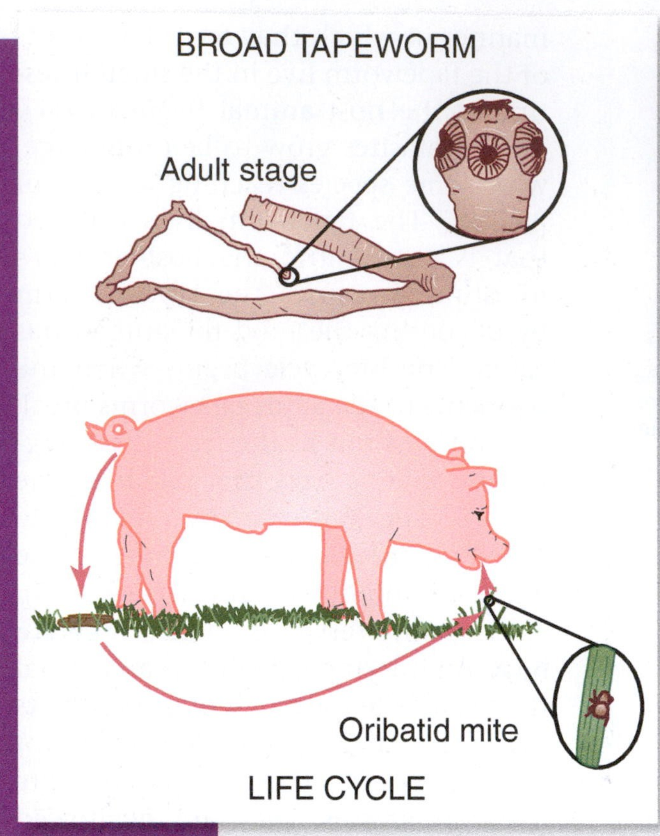

FIGURE 24–8 The life cycle of the tapeworm involves two hosts. *Courtesy of Instructional Materials Service, Texas A&M University*

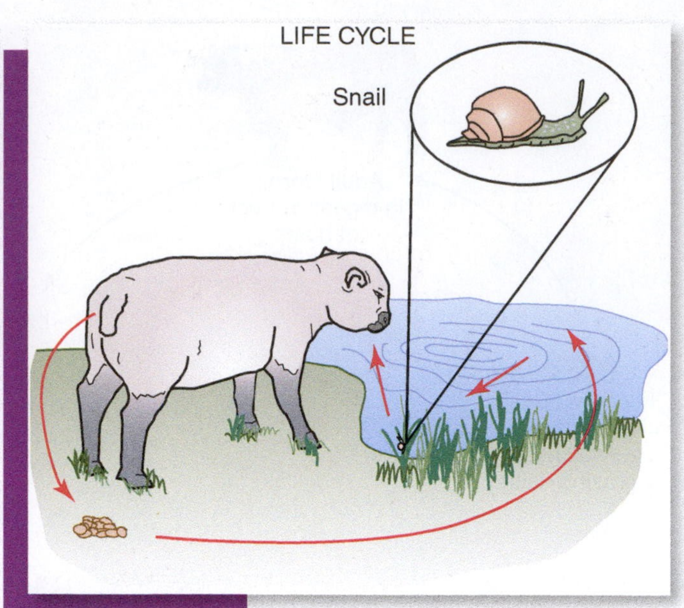

FIGURE 24–10 Liver flukes use the snail as an intermediate host. *Courtesy of Instructional Materials Service, Texas A&M University*

the liver and bile ducts and general irritation of the liver (**Figure 24–9**). The adults lay eggs in the bile duct. The eggs are passed through to the intestines and out in the feces. These parasites need an intermediate host. For the eggs to hatch, they must land in water. After the larvae hatch, they swim in search of snails to serve as intermediate hosts. Once the snails are located, the larvae enter the snails to develop and reproduce (**Figure 24–10**). This stage is unusual because the larvae divide and multiply asexually. The larvae divide by themselves to create new organisms without mating and laying eggs. The new larvae emerge and attach to plants in or near the water.

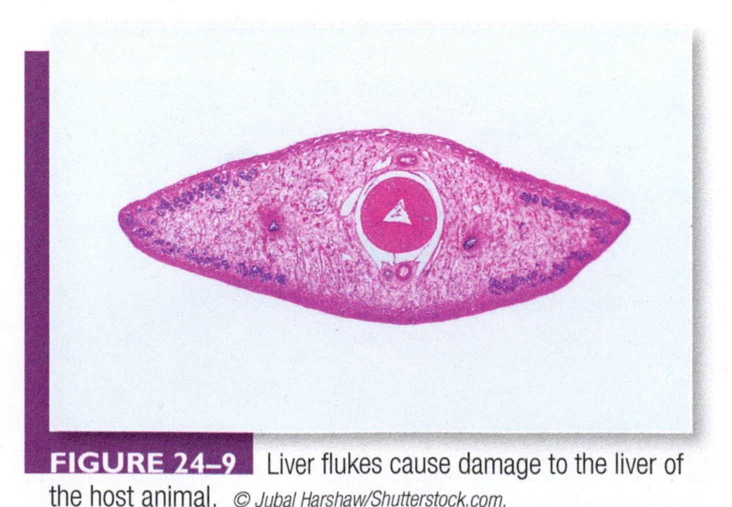

FIGURE 24–9 Liver flukes cause damage to the liver of the host animal. *© Jubal Harshaw/Shutterstock.com.*

Livestock eat the plants and become infected with the flukes. The flukes eat their way through the walls of the digestive tract and migrate to the liver, where they feed on the host animal's blood. In about 3 months, they begin to lay eggs. Fluke infestation damages the host animal's liver and causes the bile ducts to thicken and cease normal function. Livers from livestock infested by flukes are unfit for human consumption, and thus a valuable human food source is wasted.

EXTERNAL PARASITES

External parasites generally do not cause as much damage to animals as internal parasites. They can, however, cause losses in terms of animal discomfort, the loss of hide quality, and blood loss. External parasites include ticks, lice, and flies.

Ticks

Ticks generally will attach themselves to most **warm-blooded** agricultural animals. They cause damage by penetrating the skin and sucking blood from the host animal. This not only leaves a sore that can be an avenue for disease organisms but also can cause the host animal to be anemic from the loss of blood. Tick eggs are laid in the grass and over winter to hatch in the spring. When the larvae hatch, they climb up onto the grass or into bushes (**Figure 24–11**). When an animal passes by, the tick larvae attach themselves to the animal and gorge on the animal's blood. They then fall to the ground, where they remain until the following spring. In the spring, they undergo metamorphosis. The tick larvae change into **nymphs**. The nymphs climb into bushes, attach themselves to passing animals, and fill themselves with blood. Then the nymphs drop to the ground, overwinter, change into an adult in the spring, and go through the same feeding process. When the adults fall to the ground, eggs are laid, and the life cycle begins again. Three years are required for ticks to complete their life cycles.

When ticks feed, they insert their mouth parts into the host animal's skin and inject saliva into the wound. The saliva contains an anticoagulant, a substance that prevents the blood from clotting and allows the blood to flow freely into the ticks. These pests attack almost all warm-blooded animals, including people. Because the same tick may have as many as three hosts during its life cycle, diseases may be spread from one animal to another. One such disease that is spread to humans is Rocky Mountain spotted fever.

FIGURE 24–12 Lice are small, wingless insects that cause irritation to their hosts. © Miramiska/Shutterstock.com.

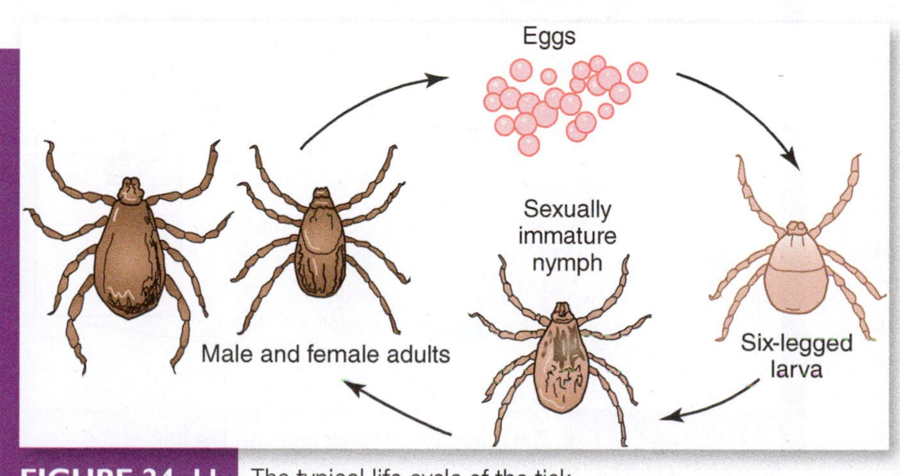

FIGURE 24–11 The typical life cycle of the tick.

Lice

Lice (*singular*, louse) are tiny, wingless insects that are external parasites of most warm-blooded agricultural animals (**Figure 24–12**). There are two types: blood-sucking lice that feed by drawing blood through the animal's skin and biting lice that feed on the hair or skin particles and excretions of the animals. The life cycle of lice is simple compared to the life cycle of some other parasites. Their whole lives are lived on the host animal (**Figure 24–13**). The adult females attach eggs to the hair follicles of host animals. The eggs hatch one to two weeks later, and the newly hatched nymphs live to maturity on the host animal. By biting or piercing the skin of the host

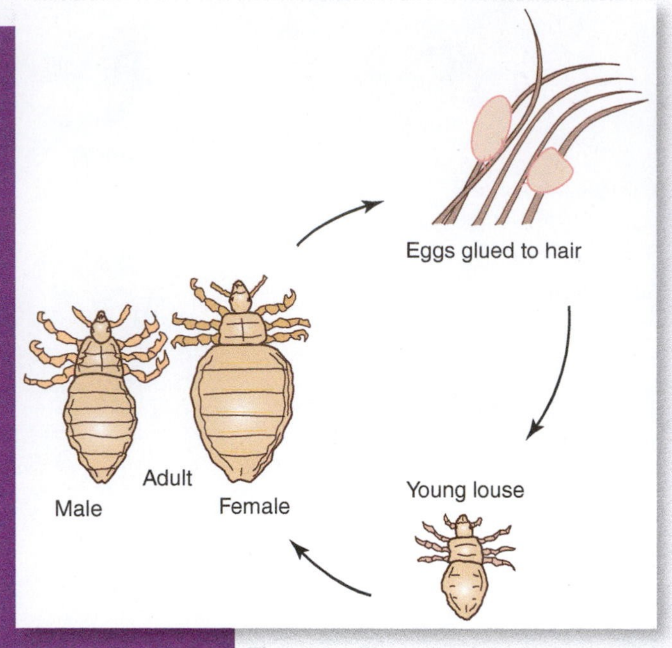

FIGURE 24–13 The entire life cycle of the louse is lived out on the host animal.

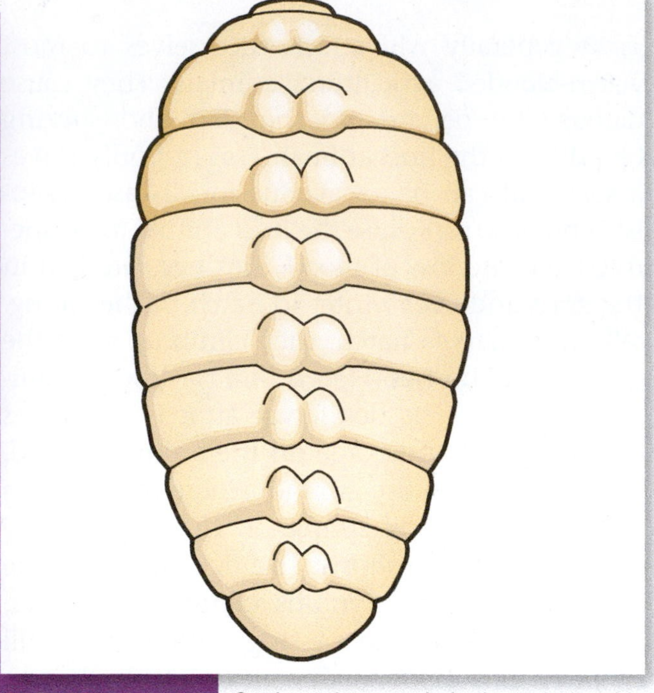

FIGURE 24–14 Cattle grubs eat a hole through the skin on the animal's back.

animal, lice cause the animal to become very uncomfortable. This results in the animal's rubbing and scratching against posts, trees, or other objects in an attempt to obtain relief from the itching. As a result, time is lost from grazing or eating and a weight loss may occur. In addition, animals that are heavily infested with lice may become so uncomfortable that normal processes such as breeding can be interrupted.

Heel Flies

Heel flies, also known as cattle grubs, are a serious parasitic pest of cattle. The adult flies lay eggs on the lower part of the legs of cattle. When the eggs hatch, the larvae penetrate the skin through the hair follicles and begin a journey through the animal's body that may take several months. The larvae burrow through the soft tissue of the animal all the way from the leg to the back, where they eat a hole in the skin of the animal's back for a breathing hole (**Figure 24–14**). The larvae, or grubs, feed on the host animal's flesh until they mature. At this time, they eat their way through the hide and fall to the ground, where they live on debris. Here, the larvae turn into pupae and mature to adults. This process takes about 1 to 2 months. The adult flies emerge, attach themselves to the heel of an animal, lay eggs, and the process starts all over again (**Figure 24–15**).

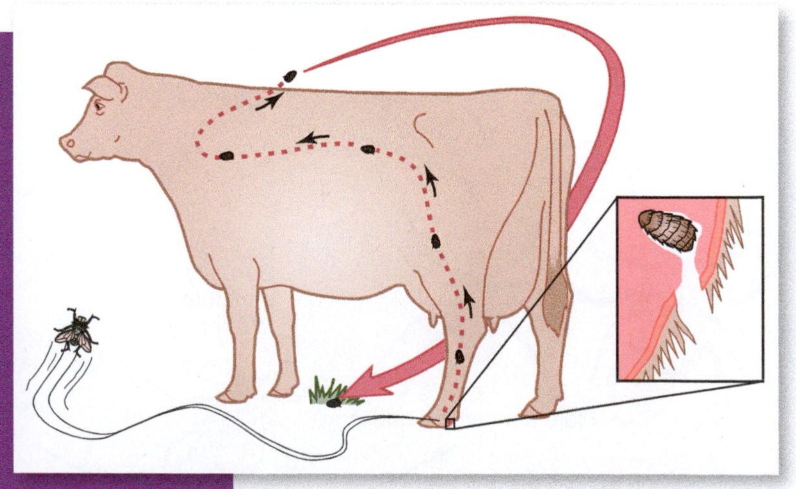

FIGURE 24–15 The adult heel fly lays eggs on the hair of the animal's heel. The larvae travel from the foot to the back through the body of the animal.

Cattle grubs cause damage because of the animal's discomfort. In addition, meat damaged by the grubs must be trimmed away and is lost. Hides from cattle infected by grubs have holes in them from the larvae's opening holes for breathing and emerging from the animal. This greatly reduces the value of the hide.

PARASITE CONTROL

To make the animals comfortable so they will grow and produce efficiently, both internal and external parasites must be controlled. The most widely used method of control is to medicate the host animals. Because the parasites feed on some part of the animal's body or ingested food, medication must be applied to the host to get rid of the parasites. With external parasites, the chemical or medication is applied to the animal's skin by spraying, pouring on, or running the animals through a dipping vat (**Figure 24–16**).

Recent research has developed a new generation of medications called **systemic pesticides** that are injected into the animal's body. Although the medication has little effect on the host, the parasites are killed or are repelled from the host animal. Internal parasites are controlled by giving the host animal an injection or a large pill (called a **bolus**) or by putting the medication in the animal's feed or water (**Figure 24–17**). The medications have undergone vigorous tests and regulations by the U.S. Department of Agriculture and the U.S. Food and Drug Administration in order to make sure the drugs are safe for the animal and that the milk, meat, eggs, or other products from the animals are safe for human consumption.

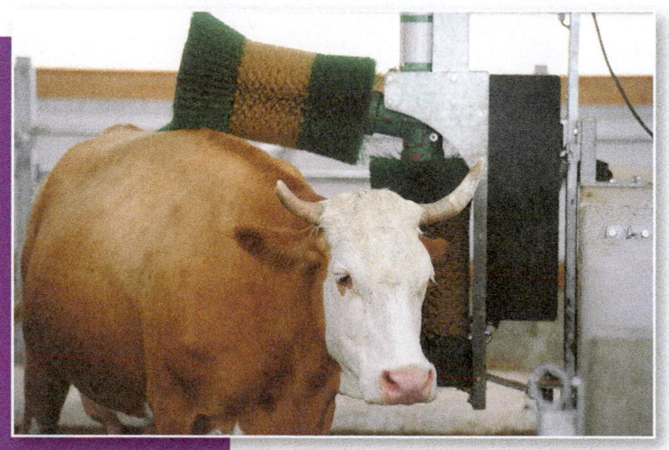

FIGURE 24–16 External parasites can be controlled by applying insecticides to the animal's skin.
© iStockphoto/CristiNistor.

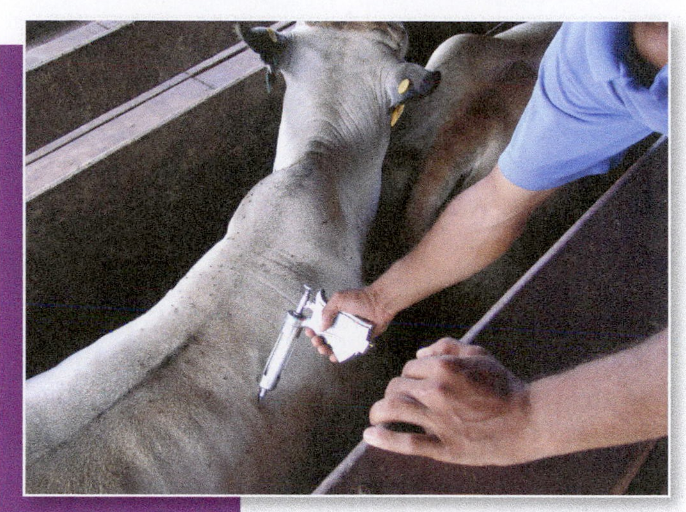

FIGURE 24–17 Internal parasites can be controlled by giving medication to the host animal. © Alf Riberio/Shutterstock.com.

An alternative approach to controlling parasites is **biological control**. At different stages of the parasite's life cycle, it is more vulnerable than at other stages. Scientists have concentrated their efforts at controlling or eradicating the pests at these vulnerable stages.

An interesting example is that of the screwworm. At one time, the screwworm was a serious pest of agricultural animals. The adult female screwworm flies usually lay eggs in open wounds on animals; however, some infestations have been known to have occurred without the animal's having an open wound. The eggs hatch, and the larvae, known as maggots, feed on the flesh of the host animal. The maggots tear out pockets of healthy flesh next to the wound and inject a toxin into the wound to prevent it from healing. The larvae grow and feed in the wound for about a week before dropping to the ground and changing to the pupa stage.

The pupae go into the ground, where they mature and become adults. Once the adults emerge from the ground, they mate and lay eggs to begin the life cycle again.

The screwworm fly mates only once during its lifetime, and scientists saw this as a weak point in its life cycle. In the late 1950s and early 1960s, a government program was begun to eradicate the screwworm. This was done by treating adult screwworm flies with gamma rays from cobalt-60. This treatment rendered both the male and the female flies sterile. Massive numbers of the sterile screwworm flies were released in the areas of the South where the screwworm infestations occurred. When a sterile female mated with either a sterile male or a fertile male, no eggs were produced. When a sterile male mated with a normal female, no viable eggs were produced. After several years of releasing huge numbers of sterile adults, the screwworm was—for all practical purposes—eradicated.

This effort was duplicated in the 1970s in a cooperative effort with the Mexican government. This effort also was quite successful. Through scientific research, efforts, and control measures such as this, pests can be controlled without harming the host animal or the environment.

SUMMARY

Parasites have always been a problem for animal producers. Each year, parasites cost millions of dollars in damage and control measures. In addition, they can cause disease problems to both animals and humans. By studying the life cycles of parasites, new and better control measures have been developed. Combinations of methods are now used to control these pests. Through research and development, some serious pests have been eradicated.

CHAPTER REVIEW

Discussion Questions

1. What is meant by a symbiotic relationship?
2. Name and give examples of three types of symbiotic relationships.
3. What are three ways that parasites can harm their hosts?
4. What are three major groups of internal parasites that infest agricultural animals?
5. Explain what is meant by a life cycle.
6. What are the four stages in the life of an insect?
7. Explain how the following parasites cause damage to their hosts: roundworms, tapeworms, flukes, ticks, lice, heel flies.
8. List three ways of controlling parasites.

Student Learning Activities

1. Go to the library and research the life cycle of an insect pest. Develop a plan to control the insect by using the pest's life cycle.
2. Visit with a livestock producer and determine the measures he or she uses to control parasites. Ask the producer to explain how the methods are different from what they were a few years ago.
3. Visit with a local veterinarian and ask the vet to explain what the most serious parasites in the local area are and the measures used to control them.

CHAPTER 25

Animal Diseases

STUDENT OBJECTIVES IN BASIC SCIENCE

As a result of studying this chapter, you should be able to

- list the types of disease-causing organisms.
- describe three types of bacteria.
- list the characteristics of viruses.
- list the characteristics of protozoa.
- describe how an animal's immune system works.
- explain the function of red and white blood cells.
- describe how vaccines work.
- distinguish between infectious and noninfectious diseases.
- describe how diseases are spread.
- explain how antigens enter the body.
- explain how passive and active immunity differ.
- distinguish between naturally acquired immunity and artificially acquired immunity.

STUDENT OBJECTIVES IN AGRICULTURAL SCIENCE

As a result of studying this chapter, you should be able to

- describe the indications that an animal is sick.
- list examples of diseases of agricultural animals caused by microorganisms.
- explain how livestock diseases are spread.
- list examples of diseases caused by genetic disorders.
- give examples of diseases caused by improper nutrition.
- give examples of plants that are poisonous to agricultural animals.
- cite examples of government disease-eradication programs.

KEY TERMS

infectious diseases
contagious diseases
cocci
bacillus
spirilla
virus
protozoa
phagocytes
lymphocytes
antigens
active immunity
naturally acquired active immunity
artificial active immunity
spore
aflatoxin
prion
noninfectious diseases
ergot

NATIONAL AFNR STANDARD

AS.07.01.01.a
Identify and summarize specific tools and technology used in animal health management.

AS.07.01.02.a
Explain methods of determining animal health and disorders.

AS.07.01.03.a
List and summarize the characteristics of wounds, common diseases, parasites, and physiological disorders that affect animals.

AS.07.01.03.b
Identify and describe common illnesses and disorders of animals based on symptoms and problems causes by wounds, diseases, parasites, and physiological disorders.

AS.7.01.03.c
Treat common diseases, parasites, and physiological disorders of animals according to directions prescribed by an animal health professional.

AS.07.01.04.a
Identify and summarize characteristics of causal agents and vectors of diseases and disorders in animals.

AS.07.01.04.b
Research and analyze data to evaluate preventive measures for controlling and limiting the spread of diseases, parasites, and disorders among animals.

AS.07.01.05.a
Explain the clinical significance of common veterinary methods and treatment.

AS.07.02.02.a
Identify and describe zoonotic diseases, including their historical significance and potential future implications.

AS.07.02.02.b
Analyze the health risk of different zoonotic diseases to humans and identify prevention methods.

THE TERM DISEASE IS BROADLY defined as being not at ease, or uncomfortable. Animals, just like humans, get diseases and have health problems. Producers of agricultural animals have a vested interest in keeping their animals healthy. Healthy animals grow faster and produce more profit for their owners.

Diseases come in a variety of types and have a variety of causes (**Table 25–1**).

Some are mild and cause only minor discomfort to the animal; others are severe and cause

Disease	Cause	Symptoms	Preventive and Control Measures
Nutritional Defects			
Anemia	All farm animals are susceptible	Characterized by general weakness and a lack of vigor; iron deficiency prevents the formation of hemoglobin, a red iron-containing pigment in the red blood cells responsible for carrying oxygen to the cells	A balanced ration will ordinarily prevent anemia. Baby pigs raised on concrete need iron supplement.
Bloat	Typically occurs when animals are grazing on highly productive pastures during the wetter part of late spring and summer	Swollen abdomen on the left side, labored breathing, profuse salivation, groaning, lack of appetite, and stiffness	Maintain pastures composed of 50 percent or more grass
Colic	Improper feeding	Pain, sweating, and constipation; kicking and groaning	Careful feeding
Enterotoxemia	Bacteria and overeating	Constipation is an early symptom and is sometimes followed by diarrhea	Bacterin or antitoxin vaccine should be used at the beginning of the feeding period
Founder	Overeating of grain or lush, highly improved pasture grasses	Affected animals experience pain and may have fever as high as 106°F	Good management and feeding practices will prevent the disease
Viral Diseases			
Cholera	Hog cholera (now eradicated from the United States) is caused by a filterable virus	Loss of appetite, high fever, reddish-purplish patchwork of coloration on the affected stomach, breathing difficulty, and a wobbly gait	A preventive vaccine is available; no effective treatment; producers should use good management
Equine Encephalomyelitis	Viruses classified as group A or B cause the disease; transmitted by bloodsucking insects such as mosquito	Fever, impaired vision, irregular gait, muscle spasms, a pendulous lower lip, walking aimlessly	Control of carrier; use of a vaccine

TABLE 25–1 Disease and nutritional defects

continued on the next page

Disease	Cause	Symptoms	Preventive and Control Measures
Hemorrhagic Septicemia	Caused by a bacterium that seems to multiply rapidly when animals are subject to stress conditions	Fever, difficult breathing, cough, discharge from the eyes and nose	Vaccination prior to shipping or other periods of stress
Newcastle	A poultry disease caused by a virus that is spread by contaminated equipment or mechanical means	Chicks make circular movements, walk backward, fall, twist their neck so the head is lying on the back, cough, sneeze, high fever, and diarrhea	Several types of Newcastle vaccines are available; antibiotics are used in treating early stages of the disease to prevent secondary infections
Warts	Believed to be caused by a virus	Protruding growths on the skin	No known preventive measures; most effective means is with a vaccine
Bacterial Diseases			
Pneumonia	Bacteria, fungi, dust, or other foreign matter; the bacterium *Pasteurella multiocida* is often responsible for the disease	General dullness, failing appetite, fever, and difficult breathing	Proper housing, ventilation, sanitation, antibiotics
Tetanus	A spore-forming anaerobic bacterium; the spores may be found in the soil and feces of animals	Difficulty swallowing, stiff muscles, and muscle spasms	Immunizing animals with a tetanus toxoid
Anthrax	A spore-forming bacterium	Fever, swelling in the lower body, bloody discharge, staggering, trembling, difficult breathing, convulsive movements	An annual vaccination; manure and contaminated materials should be burned and area disinfected; insects should be controlled
Blackleg	A disease of cattle and sheep caused by a spore-forming bacterium which remains permanently in an area; the germ has an incubation period of 1 to 5 days and is taken into the body from contaminated soils and water	Lameness, followed by depression and fever; the muscles in the hip, shoulder, chest, back, and neck swell; sudden death within 3 days of onset of symptoms	A preventive vaccine
Brucellosis	Caused by the bacterium *Brucella abortus*	Abortion of the immature fetus is the only sign in some animals	Vaccinating heifer calves with *Brucella abortus* will prevent cattle from contracting this disease; infected cattle must be slaughtered

TABLE 25–1 *continued from the previous page*

continued on the next page

Disease	Cause	Symptoms	Preventive and Control Measures
Distemper	A disease of horses; exposure to cold, wet weather, fatigue, and an infection of the respiratory tract aid in spreading the disease	Increased respiratory rate, depression, loss of appetite, and discharge of pus from the nose are visible symptoms; infected animals have a fever and swollen lymph glands (located under the jaw)	Animals with disease should be isolated, provided with rest, protected from the weather, and treated with antibiotics
Erysipelas	A resistant bacterium capable of living several months in barnyard litter	Three forms: acute, subacute, and diamond skin form; acute symptoms are a high fever, constipation, diarrhea, and reddish patches on the skin; subacute is usually localized in an organ such as the heart, bladder, and joints; sloughing off of the skin is common	An anti-swine erysipelas serum is available
Leptospirosis	A bacterium found in the blood, urine, and milk of infected animals	Abortion and sterility; symptoms are blood-tinged milk and urine	Susceptible animals should be vaccinated
Tuberculosis	The three types of tubercle bacilli causing the disease are human, bovine, and avian; the human type rarely produces tuberculosis in lower animals, but the bovine type is capable of producing the disease in most warm-blooded vertebrates	Lungs are affected; however, other organs may be affected; some animals show no symptoms; others appear unhealthy and have a cough	Maintaining a sanitary environment and comfortable quarters will help in preventing the disease
Pullorum	A poultry disease caused by a bacterium which is capable of living for months in a dormant state in damp, sheltered places; the germs infect the ovary and are transmitted to the chicks through the eggs	Infected chicks huddle together with their eyes closed, wings drooped, feathers ruffled, and have foamy, white droppings	Blood test is required for positive identification of the disease; disposal of infected hens will aid in preventing the disease; chicks should be purchased from a certified pullorum-free hatchery
Foot Rot	A fungus common to filth is responsible for foot rot; animals are most apt to contact foot rot when they are forced to live in wet, muddy, unsanitary lots for long periods of time	Skin near the hoofline is red, swollen, and often has small lesions	Maintaining clean, well-drained lots is an easy method of preventing foot rot

TABLE 25–1 *continued from the previous page*

continued on the next page

Disease	Cause	Symptoms	Preventive and Control Measures
Calf Diphtheria	A disease caused by bacteria	Difficulty breathing, eating, drinking; patches of yellowish, dead tissue appear on the edges of the tongue, gums, and throat; there is often a nasal discharge	Treatment is usually by administering an antibiotic.
Protozoa			
Coccidiosis (pertaining to poultry)	Several species of protozoa are responsible	Occurs in two forms; cecae and intestinal: cecae is the acute form that develops rapidly and causes a high mortality rate; bloody droppings and sudden death are symptoms; intestinal coccidiosis is chronic in nature, and its symptoms are loss of appetite, weakness, pale comb, and low production; few deaths occur from the latter form	The disease is transmitted by the droppings of infested birds, so maintaining sanitary conditions and the feeding of a coccidiostat will prevent the disease
Unknown Causes			
Atrophic Rhinitis	Causes have not been determined; several different bacteria are involved; it is contagious, especially in young pigs, and is spread by direct contact	Affects the bone structure of the nasal passages; the snout will become twisted and wrinkled	Sanitation is important in preventing the disease; there is not a specific treatment; use of sulfamethazine may help

TABLE 25–1 *continued from the previous page*

Source: *Instructional Materials Service, Texas A & M University.*

death quickly. Although animals that are sick may not show any outward signs or symptoms of being ill, they usually do display symptoms indicating that they are not feeling well. The animal may be droopy, go off feed and water, be restless, or have a dull coat (**Figure 25–1**). In some cases, the animal may have a fever. This means that the body temperature of the animal is higher than normal.

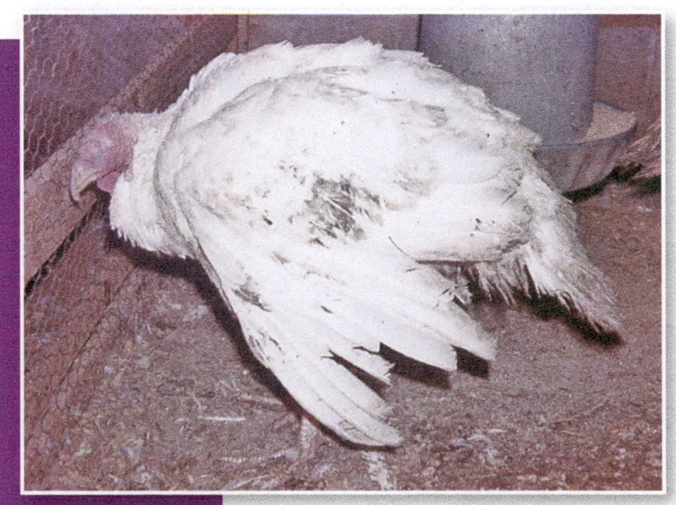

FIGURE 25–1 Sick animals may appear droopy. This turkey suffers from fowl pox. *Courtesy of Dr. Jean Sander, Academic and Student Affairs, College of Veterinary Medicine, The Ohio State University*

INFECTIOUS DISEASES

Infectious diseases are those caused by microorganisms that invade the animal's body. These are usually **contagious diseases**, which means that the infected animal can pass the disease on to a healthy animal. There are many types of microorganisms that cause diseases in animals.

Bacteria

One of the most common types of disease-causing organisms are bacteria. Bacteria are all around us (**Figure 25–2**). They can be found in the hottest of deserts and buried deep in polar ice. They live on and in the bodies of all animals and probably are more numerous than the cells of the animal's body. Many of the bacteria around us are beneficial. Those living in the stomachs of ruminants help the animals digest food. Bacteria are useful in the production of cheese and foods such as sauerkraut.

There are, however, many types of bacteria that cause harm. Harmful bacteria invade the cells of an animal's body. Parasitic bacteria may harm the animal by feeding off the cells of the body or by secreting a material known as a toxin. A toxin is a substance that causes harm to an organism. In other words, it is a poison. When large numbers of harmful bacteria invade an animal's body, the animal becomes ill. The type and form of the illness depend on the type of bacteria that invades the animal.

FIGURE 25–2 Bacteria are all around us. These bacteria are growing on media in a petri dish.
© Michal Kowalski/Shutterstock.com.

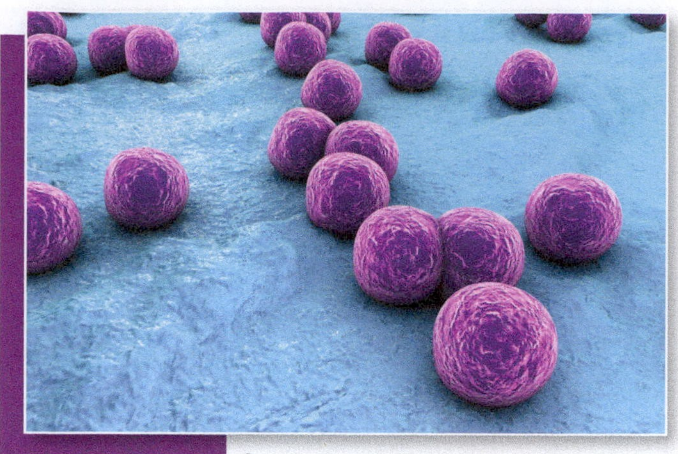

FIGURE 25–3 Cocci are round, spherical-shaped bacteria. *© Kateryna Kon/Shutterstock.com.*

Cocci are round, spherical-shaped bacteria. Diseases such as some forms of pneumonia and strep infections are caused by these bacteria (**Figure 25–3**).

Bacillus bacteria are rod-shaped organisms that may be single, paired, or arranged in chains (**Figure 25–4**). They cause many serious diseases in agricultural animals; a few are anthrax, tetanus, blackleg, intestinal coliform, salmonella, and tuberculosis

Spirilla bacteria are shaped like spirals or corkscrews (**Figure 25–5**). These bacteria are highly motile, which means they can move about very easily. They also require a moist

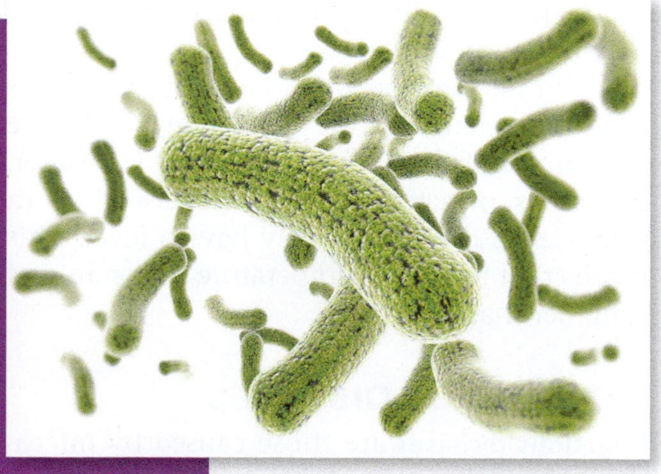

FIGURE 25–4 Bacillus are rod-shaped bacteria.
© Jezper/Shutterstock.com.

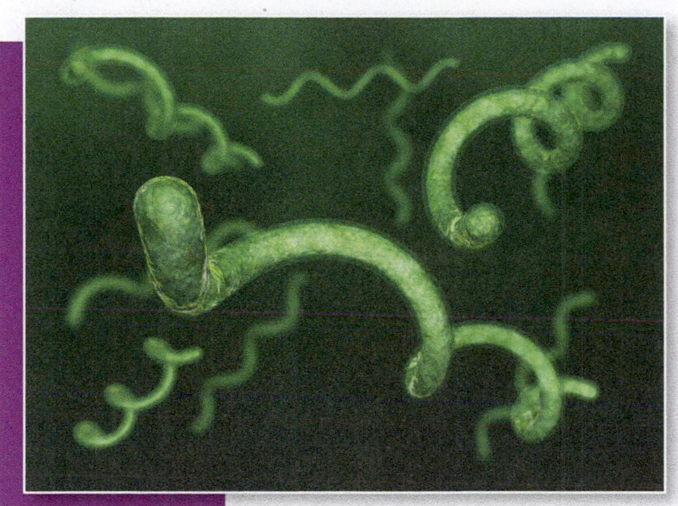

FIGURE 25–5 Spirilla are shaped like spirals or corkscrews. © iStockphoto/petersimoncik.

FIGURE 25–6 This chicken suffers from fowl pox. Note how the viruses have destroyed cells in the comb. Courtesy of Dr. Jean Sander, Academic and Student Affairs, College of Veterinary Medicine, The Ohio State University

atmosphere to survive; consequently, they live very well in the reproductive tracts of animals. Some of the diseases they cause are leptospirosis, vibriosis, spirochetosis, and many others.

Most bacteria can be controlled by the use of antibiotics. The first of these medicines was penicillin, which was produced from extracts of molds. Many different forms of penicillin are now produced artificially and are highly effective against bacterial infections; however, some bacteria have developed resistance to antibiotics.

Viruses

A **virus** is a very tiny particle of matter composed of a core of nucleic acid and a covering of protein that protects the virus. Viruses have characteristics of both living and nonliving material. It could be said that viruses are on the borderline between living and nonliving. They are made up of some of the material found in cells, but they are not cells because they do not have nuclei or other cell parts.

Viruses do not grow and cannot reproduce outside a living cell. Once inside a living cell, the virus reproduces using the energy and materials of the invaded cell. Viruses harm cells by causing them to burst during the reproduction process of the virus and by using material in the cell that the cell needs to function properly (**Figure 25–6**). Therefore, viral diseases cause the animal to be sick by preventing certain cells in the animal's body from functioning properly.

There are many different types of viruses that cause a variety of serious diseases in agricultural animals. Viral diseases are more difficult to treat than diseases caused by bacteria. The antibiotics that have proven to be effective against bacteria are of no use against viruses. Some of the more serious livestock diseases caused by viruses are foot-and-mouth disease, influenza, hog cholera, and pseudorabies. Many viral diseases are incurable. The best means of dealing with them is prevention.

Protozoa

Another type of microorganism that causes diseases in agricultural animals is the protozoan. **Protozoa** are single-celled organisms that often are parasitic. They cause harm to animals by feeding on cells or by producing toxins. Examples of diseases caused by protozoa are African sleeping sickness and anaplasmosis. Coccidiosis is one of the most costly diseases in the

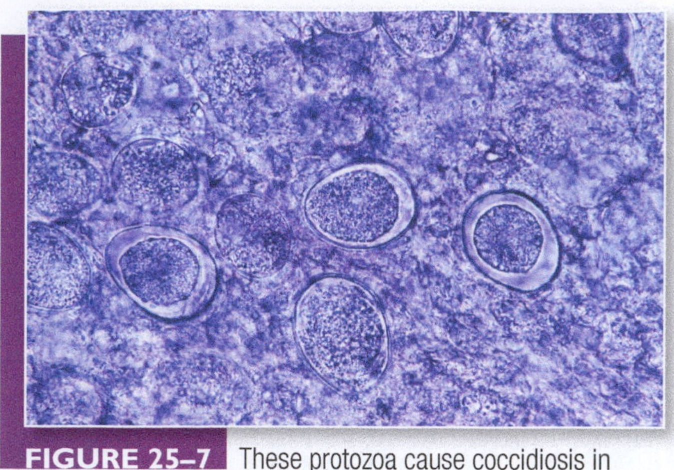

FIGURE 25–7 These protozoa cause coccidiosis in poultry. *Courtesy of Dr. Jean Sander, Academic and Student Affairs, College of Veterinary Medicine, The Ohio State University*

FIGURE 25–8 Animals come in contact with the ground. Disease-causing germs can live in the soil for many years. *© Steve Lovegrove/Shutterstock.com.*

poultry industry (**Figure 25–7**). This disease is caused by several different species of protozoa and causes diarrhea and weight loss in chickens. Most protozoa can be controlled by drugs.

THE IMMUNE SYSTEM

Disease-causing viruses, bacteria, and protozoa are all present in the environments of animals and people. They are so prevalent that they are ingested into the body almost constantly. These organisms can enter the body through regular body openings such as the mouth, nose, eyes, reproductive system, or any other natural opening. They may enter through the skin or through a wound in the body as well. If the animal's body did not have a means for defending itself against these disease agents, the animal would live a short, miserable life.

Fortunately, animals (including people) have several lines of defense in fighting disease. The first line consists of physical barriers that keep the disease-causing agents out. For instance, the nostrils are lined with hairs that attract particles that harbor germs before they can enter the body. When an animal sneezes, the particles are expelled. Almost all body openings and many internal organs are lined with mucous membrane. These are tissues that secrete a viscous, watery substance that traps and destroys bacteria and viruses.

The most prevalent avenue for organisms to enter the body is through the digestive and respiratory systems. Countless billions of microorganisms live in the soil, and some of them are disease-causing germs. Certain disease germs can live in the soil for many years. Most animals come in contact with the ground as they eat (**Figure 25–8**). Cattle, sheep, and horses graze and pull grass from the ground, and pigs root in the ground for food. Even processed feed is far from being sterile. Every minute animals breathe in large amounts of air laden with all sorts of particles and organisms. Fortunately, the body has ways to destroy harmful organisms. For instance, if the germs swallowed with feed are not trapped by the mucous membrane of the digestive tract, they are most often killed by the digestive enzymes. Germs ingested by breathing are trapped in the mucous membranes of the respiratory tract.

Disease agents that get through the first line of defense are usually destroyed by the second line of defense. This line is composed of cells and chemicals in the bloodstream. Blood is basically composed of two types of cells—white blood cells and red blood cells. The red blood cells carry oxygen and nourishment

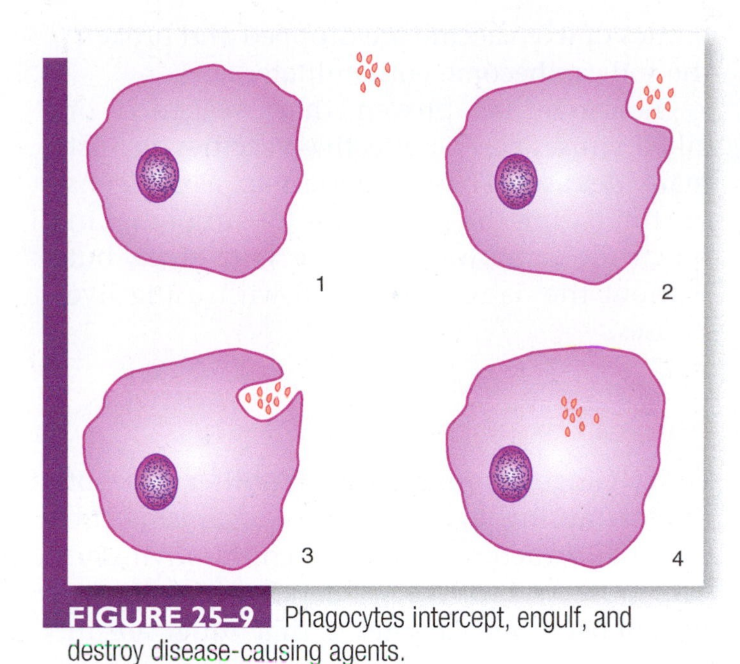

FIGURE 25–9 Phagocytes intercept, engulf, and destroy disease-causing agents.

to the other body cells. White blood cells are produced in the bone marrow and circulate throughout the body to get rid of worn-out body cells. Certain of these white blood cells, called **phagocytes**, intercept and destroy disease-causing agents (**Figure 25–9**). These cells also migrate to certain organs, such as the liver, lymph nodes, and spleen, and remain there to intercept disease-causing agents. White blood cells also circulate through other body fluids and the mucous membranes.

When phagocytes encounter foreign organisms, they release chemicals that induce the production of more white blood cells to help fight the disease organism. In fact, one important way a veterinarian can tell if an animal is sick is by counting the number of white blood cells in the animal's bloodstream. A count larger than normal indicates that there are disease organisms present in the animal's body, and a large number of phagocytes have been produced to combat them.

Certain white blood cells are produced by the lymph glands and are called **lymphocytes**. These cells react to foreign substances by releasing chemicals that kill the disease-causing organisms or inactivate the foreign substance. The substances that cause the release of the chemicals are called **antigens**. Antigens may be viruses, bacteria, toxins, or other substances. The chemicals released by the lymphocytes are known as antibodies. The lymphocyte can also become a "memory cell" that is ready to release an antibody if the same type of antigen enters the body at a later time. When this happens, it is known as a secondary immune response. This response occurs much more rapidly and lasts longer than the primary response.

Immunity

Immunity means that an animal is protected from catching a certain disease. This is because the animal's body is capable of producing sufficient antibodies in enough time to neutralize the disease-causing agent before the animal becomes sick. Immunity can be either active or passive. **Active immunity** means that the animal is more or less permanently immune to the disease. Passive immunity means that the animal is only temporarily immune.

Animals are born with some immunity to diseases. In mammals, the first milk, called colostrum, that is given to the newborn animal is rich in antibodies from the mother. These antibodies serve the new animal until its own immune system can take over. As the animal is exposed to more and more antigens, antibodies build up in the animal's body. **Naturally acquired active immunity** is obtained by the animal's actually having a disease and recovering. The memory phagocytes react quickly when the antigen that causes that specific disease enters the animal's body and the antigens are overwhelmed.

Artificial active immunity can be induced in the animal by injecting antigens into the animal that cause the phagocytes to react without making the animal seriously ill (**Figure 25–10**). This process was devised by an Englishman named Edward Jenner in the late

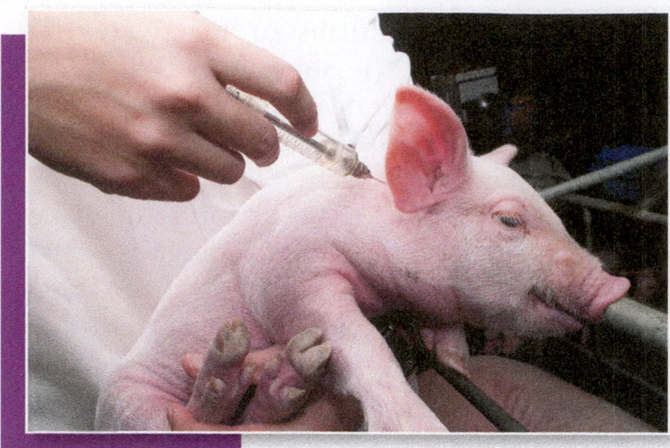

FIGURE 25–10 Artificial immunity can be induced by injecting antigens into an animal.

1700s. At that time, smallpox epidemics swept through many parts of the world, killing more than half of the people who contracted the disease. Those who survived became permanently immune to the disease, which meant they would never be sick with it again. Jenner knew that those who contracted cowpox never came down with smallpox. Cowpox was, as the name implies, a disease of cattle. Humans could also get the disease, but it was usually mild. Jenner collected material from sores that developed on people who had cowpox, and he injected this material into healthy people. The people who were injected became ill with a mild case of cowpox, but then were immune to smallpox in the future. The Latin word for cow is *vacca*, and the word *vaccination* was coined based on the fact that the immunity originated from cows.

This concept was used later by Louis Pasteur to develop several vaccines. All of the modern vaccines, whether given to humans or animals, work on basically the same principles. When Jenner used materials from cowpox sores, he injected live viruses. This worked well in this particular case, but as more vaccines were developed, it was discovered that often the vaccination could cause the disease. Also, many of the viruses can live for a long time in the soil. If bottles of live vaccine are dropped and broken, the soil can become contaminated.

Research has proven that weakened or killed viruses can be effective vaccines against many diseases. These materials act as antigens in stimulating the production of antibodies in much the same way as a live virus does, but without the dangers incurred with using live viruses.

Fungal Diseases

Fungi (singular, fungus) are another type of disease-causing agents. These organisms are plantlike structures that lack chlorophyll and can grow and thrive in dark, damp places. Remember from Chapter 2 that fungi are in a separate kingdom and are neither plants nor animals. There are over 100,000 species of fungi that are very diverse in size and type. The best-known species are the mushrooms we see growing everywhere in damp places that are not in direct sunlight. While some of the mushrooms are edible, many are poisonous.

Fungi play a very important role in nature. Because they cannot digest food within their bodies, they digest food outside their bodies by releasing enzymes into the environment in which they live. Through this process, fungi break down organic matter into a usable form. This function helps get rid of dying and decaying plant and animal materials. Fungi reproduce by means of releasing **spores** into the air. The spores are somewhat like very tiny seeds that are dispersed on the wind. Problems arise when the fungi spores land on living tissues, and the resulting fungi begin to break down these tissues (**Figure 25–11**). This causes fungal diseases such as ringworm and athlete's foot in humans. Millions of people suffer from allergies caused by fungi that irritate the nasal passages. These organisms also cause serious diseases in agricultural animals.

One such serious disease is aspergillosis, that affects poultry. The fungus *Aspergillus fumigatus* causes hard nodular areas to develop

FIGURE 25–11 Fungi release spores into the air. Some types of spores can cause disease. © Irina Kozorog/Shutterstock.com.

FIGURE 25–12 Some people are highly allergic to toxins produced by fungi on peanuts. © Glevalex/Shutterstock.com.

in the lungs and an infection of the air sacs of poultry. Symptoms include gasping, sleepiness, loss of appetite, and sometimes convulsions and death. The fungus or mold grows on litter, feed, rotten wood, and other organic materials in the poultry house. The disease is not spread by bird-to-bird contact but instead by the poultry breathing in spores released by the fungus.

Another type of fungal disease that affects livestock is caused by *Aspergillus flavus*, a fungus that grows well on grains that are a part of animal rations. It also thrives on nuts such as peanuts. These and other types of fungi produce a highly potent toxin called **aflatoxin**. You may have heard that some people are highly allergic to peanuts. In fact, people have died as a result of eating peanuts, not because they were so allergic to the peanuts themselves, but to the aflatoxins produced by the fungi growing on the peanuts that humans eat. However, most people are not allergic to these toxins (**Figure 25–12**). Livestock are also susceptible to aflatoxins, and each year losses occur to the livestock industry because of allergic reaction to the toxin.

Fungi also cause a lot of problems with plants, but, as mentioned previously, fungi can be useful, too. An entire discipline is devoted to the study of fungi. This study is called mycology, and scientists who study fungi are called mycologists.

Prions

Another type of infectious agent is called a **prion**, which is short for *proteinaceous infectious particle*. The discovery of prions is relatively recent, and scientists still are not sure exactly how these agents function or how they propagate. However, they do know that these types of proteins are responsible for diseases called transmissible spongiform encephalopathy, diseases also known as TSEs. The most widely known livestock diseases from prions are scrapie in sheep and bovine spongiform encephalopathy (BSE) in cattle. BSE is best known as mad cow disease, which has caused widespread concern all over the world because of its transmissibility to humans. Also, a disease that affects deer called chronic wasting disease is thought to be caused by prions.

Diseases caused by prions affect tissues in the brain and are usually untreatable and fatal. BSE is called mad cow disease because of the strange way an animal acts when it has the disease. People can get mad cow disease by eating poorly cooked meat from infected animals. There is some indication that humans have to have a genetic predisposition for the disease before they can contract it. Cattle are thought to contract the disease by eating feed that has animal by-products in it. If the feed contains tissue from the nervous system of an animal with BSE, the animal that eats the feed may get the disease.

NONINFECTIOUS DISEASES

Not all diseases of agricultural animals are caused by being infected with microorganisms. Diseases can be caused by means other than contact with infected animals. These are **noninfectious diseases**. They are not contagious.

Genetic Diseases

Some diseases are caused by defects in the genes that were transferred from the animal's parents. Usually, animals with genetic disorders will also pass the problem on to the next generation, so the disease stays in certain breeds or bloodlines of animals. One example is a condition known as white heifer's disease in Shorthorn cattle. Certain heifers that are solid white in color have a genetic defect that causes them to be sterile. Since the cause is purely genetic, other cattle or other animals cannot contract the disease from the heifers with the disease. In certain lines of Holstein cattle, calves are born with a condition known as mule foot, in which the hooves are shaped like a horse's or mule's foot rather than having two toes like normal cattle. Again, this disorder is purely genetic and cannot be spread through contact with other cattle.

At present, the only way to control genetic diseases is by using good selection practices and avoiding breeding animals that are known to have genetic defects in their line. Perhaps in the near future, the process of genetic engineering will be developed to a point at which these problems can be removed from animals.

Nutritional Diseases

Animals can become sick from faulty nutrition. All animals need certain amounts of a variety of nutrients. If these nutrients are lacking in the animal's diet, the animal can become ill (**Figure 25–13**). An example of a nutritional disease is milk fever in dairy cattle. Cows with this disease lie down and are unable to stand. The condition is caused by an insufficient amount of calcium in the bloodstream. Because milk is rich in calcium and milk comes from the bloodstream, cows that produce heavily sometimes have this problem. The disease is usually cured by the injection of calcium salts directly into the animal's bloodstream. The effects are immediate and dramatic. Animals that have been down for hours suddenly are able to stand and move about.

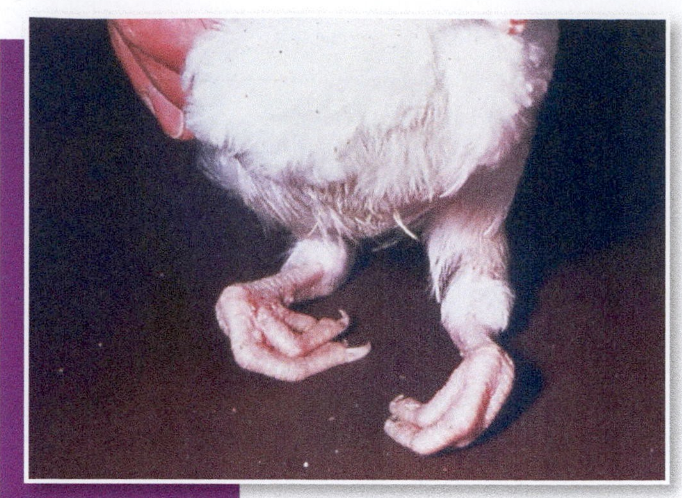

FIGURE 25–13 Lack of proper nutrition can cause disease. The deformity in this chicken's feet was caused by a deficiency of riboflavin in the diet. *Courtesy of Dr. Jean Sander, Academic and Student Affairs, College of Veterinary Medicine, The Ohio State University*

Other nutrition-related diseases can be caused by overeating. Cattle that are turned in on lush, green grazing can have a problem known as bloat. Bloat is caused when the sudden ingestion of large amounts of green forage (usually legumes) causes foaming in the animal's digestive tract. The foamy bubbles block the openings of the tract and prevent the passage of gas. The gas can build up to such an extent that it can cause death.

Horses, cattle, and sheep can get a condition known as founder if they eat too much grain, especially if the ration is changed too rapidly.

This condition causes the feet to become inflamed and the hooves to grow upward and outward. The animals don't eat very well and lose weight rapidly.

POISONING

Like any other animals, agricultural animals can be made sick by ingesting toxic materials. These materials can be picked up in a variety of ways. The animal can eat feed that is contaminated. If feed becomes moldy, certain toxins can develop that can make livestock quite sick and may even kill them. Among the most potent of the toxins from moldy feed are aflatoxins and **ergot**. Both of these toxins are developed from fungi that grow on grains. Producers are careful that feed fed to agricultural animals is free from mold.

Another type of poisoning that causes problems with animals is that of poison plants. Animals that graze can readily pick up plants that contain toxins. Poisonous plants are found throughout the United States. Losses are incurred in all states from plant poisoning. The western regions of the country sustain the heaviest loss because of so much grazing on uncultivated range land.

Some fairly common plants, such as ferns, bitterweed, buttercups, cocklebur, and milkweed, can be poisonous to some species of agricultural animals (**Figure 25–14**). Certain plants

FIGURE 25–14 Common plants such as cocklebur (left) and milkweed (right) are poisonous to livestock.

FIGURE 25–15 Tansy ragwort is deadly to horses and cattle, yet is nearly harmless to sheep.
© mr_coffee/Shutterstock.com.

can be highly toxic to some animals and almost harmless to others. For example, tansy ragwort, which grows in the Northwest, is deadly to horses; yet sheep can eat the plant with little or no harm (**Figure 25–15**).

DISEASE PREVENTION

Producers of agricultural animals take careful measures to protect their animals from diseases. Most have strict vaccination schedules for all their animals to prevent them from contracting diseases (**Figure 25–16**). Even though there may have been no outbreaks of a certain disease, producers still use vaccinations because they are aware that disease organisms are spread in a variety of ways.

Strict observation of animals daily can also help in recognizing health problems so that treatment can begin promptly. The following are examples of when a veterinarian should be consulted, and/or treatment may be required:

- Rough hair coat; indicating internal parasites
- Scaly skin or an animal rubbing against a post; indicating external parasites

FIGURE 25–16 Most producers have strict vaccination schedules to protect their animals from diseases. *Source: USDA, Agricultural Research Service (ARS).*

- Weeping eyes; indicating irritation of the eyes
- Loose bowel movements; indicating a bacterial or viral infection.
- Excessive thirst, indicating fever; requires a check by veterinarian
- Listlessness; requires a check by a veterinarian
- A lack of appetite; requires a check by a veterinarian
- Overgrown hooves in horses, indicating founder; requires a check by a veterinarian
- Discharge from the mouth or nose; requires a check by veterinarian

Infectious disease organisms can be transported by wildlife. Deer can transmit certain diseases to cattle and sheep. Wild pigs can spread disease to domesticated herds. Wild horses can infect domesticated animals. Many species of birds can transmit disease to animals, especially to chickens. At one time, birds were held responsible for spreading hog cholera from one farm to the next.

Diseases can also be spread by humans. People moving from one farm to another can carry disease organisms on their shoes or clothes. Many modern swine and poultry producers no longer allow visitors in their production houses (**Figure 25–17**). Those who do allow visitors insist that they wear disposable boots or that shoes be thoroughly disinfected.

Newly purchased animals also may be a source of disease outbreak. Producers usually keep new animals away from the other animals until they are certain that the new animals are disease-free. Animals that come from foreign countries are kept in quarantine until they can be declared disease-free; they are kept in isolation areas outside the country during this time. Economically, though, it is feasible for only very valuable animals to be kept in quarantine.

Government regulations such as quarantining help deter the spread of livestock diseases. Livestock that are transported across state lines require a certificate showing that the animal has been examined, tested, and declared disease-free. Also, animals entering fairs and livestock shows are required to have health certificates. The federal government has eradication programs to eliminate diseases. An example is the brucellosis (Bangs) program. Brucellosis is a disease of several agricultural animals, but it is a particularly serious problem in cattle. This disease causes abortion and accounts for large losses in profit among cattle producers. States that have herds of cattle with brucellosis require that all animals being sold to producers be tested for this disease. If an animal is tested positive, it is branded and usually sent to slaughter.

FIGURE 25–17 Many producers do not allow visitors in their production houses.
© Richard Mann/Shutterstock.com.

PUTTING IT INTO PRACTICE

The Veterinary Science Proficiency Award

Many students who have a real interest in companion animals have the career goal of becoming a veterinarian. The Veterinary Science Proficiency Award gives students an opportunity to work with veterinarians in clinical practice, research facilities, colleges of veterinary medicine, or animal health industry. They can assist veterinarians in performing duties related to caring for the health and welfare of large and small animals. This experience should be under the supervision of a veterinarian and may include wage earning.

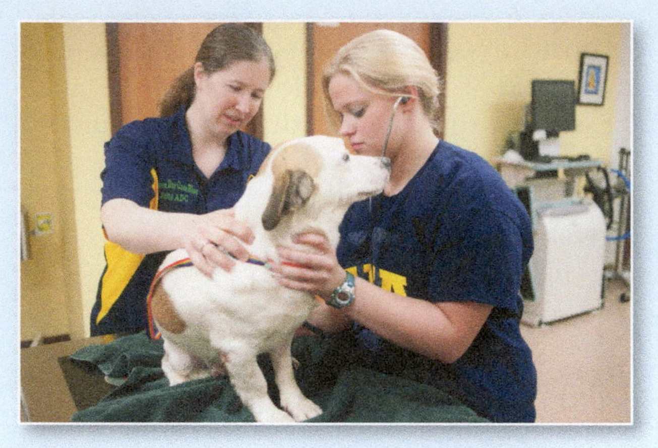

Students can gain hands-on experience in treating animals through a supervised agricultural experience working with a veterinarian. *Courtesy of the National FFA Organization*

SUMMARY

Since humans began raising livestock, problems with diseases have arisen. Many of the diseases are minor and cause relatively few problems while other diseases can completely devastate large herds of animals. There are thousands of different types of pathogens and substances that can cause disease. Through research and development, humans have learned very effective ways of preventing and treating animal diseases. In the future, as in the past, new diseases will surface and new and better control methods will have to be developed.

CHAPTER REVIEW

Review Questions

1. What are some of the indications that an animal is not well?
2. What is meant by an infectious disease?
3. List three types of bacteria according to shape.
4. How do viruses cause an animal to be sick?
5. What is an animal's first line of defense in fighting infectious disease?

6. What role do white blood cells play in fighting disease?
7. What are antigens?
8. What is the difference between active and passive immunity?
9. List three types of noninfectious diseases.
10. Explain at least two ways diseases are spread.

Student Learning Activities

1. Visit with a local veterinarian who treats large animals. Determine what livestock diseases he or she has encountered in your area. Find out the procedures used in diagnosing diseases.
2. Visit with a producer and find out what measures he or she takes to prevent diseases. Ask which diseases he or she fears the most.
3. Prepare a list of the plants in your area that are poisonous to livestock. Your local county Extension Office should have helpful information.

CHAPTER 26

The Issue of Animal Welfare

STUDENT OBJECTIVES IN BASIC SCIENCE

As a result of studying this chapter, you should be able to

- discuss potential problems brought about by animals being raised in confinement.
- determine why animals raised in an agricultural setting are healthy and efficiently grown.
- explain a potential problem associated with the continuous ingestion of antibiotics.
- cite examples of how the use of animals in research has helped humans.
- list the laws that govern the use of laboratory animals for research.

STUDENT OBJECTIVES IN AGRICULTURAL SCIENCE

As a result of studying this chapter, you should be able to

- list the reasons why some people object to the raising of farm animals.
- defend the use of confinement operations.
- defend the use of management practices associated with the raising of agricultural animals.
- explain how producers benefit when their animals are content and healthy.
- list the laws governing the use of agricultural animals.

KEY TERMS

animal rights activists
animal welfare activists
withdrawal periods
debeaking
docking
dehorning
castration
beak trimming
pecking order
elastrator
scrotum
freeze branding
hot branding

NATIONAL AFNR STANDARD

AS.01.02.01.a
Identify and categorize terms and methods related to animal production.

AS.01.02.01.b
Analyze the impact of animal production methods on end product qualities.

AS.02.01.01.a
Explain the implications of animal welfare and animal rights for animal systems.

AS.02.01.01.b
Design programs that ensure the welfare of animals and prevent abuse or mistreatment.

AS.02.01.02.b
Analyze and document animal welfare procedures used to ensure safety and maintain low stress when moving and restraining animals.

AS.02.01.03.a
Distinguish between animal husbandry practices that promote animal welfare and those that do not.

AS.02.01.03.b
Analyze and document animal husbandry practices and their impact on animal welfare.

AS.08.02.01.a
Identify and summarize methods for ensuring optimal environmental conditions for animals.

PEOPLE HAVE USED ANIMALS for as long as we have been on Earth. Early humans hunted animals to eat and to use their hides for clothing and shelter. Later, as civilizations began to develop, humans began to raise animals in order to have a ready and abundant source of food and clothing. When this happened, people began to control all aspects of the lives of animals. This meant that the animals depended on the people who raised them for food and protection.

Almost from the very beginning, people have been concerned about the well-being of animals they raised and controlled (**Figure 26–1**). After all, animals are living creatures with the ability to feel pain and to suffer distress, and people have always been concerned that animals not suffer needlessly. In fact, as far back as the time of the Greek mathematician Pythagoras, arguments have been made that people should not eat animals because animals should not be killed. Later Greek philosophers such as Plutarch argued that animals should be treated with justice. In the seventeenth and eighteenth centuries, protection societies began to develop that sought to prevent animals from being mistreated.

In the United States, the Animal Rights Movement began in the 1970s. Today, various animal rights organizations are very active politically, working to pass laws that govern how animals may be treated. At least two main lines of thought are associated with this movement. One philosophy is that animals should have the same rights as humans. People who espouse this philosophy believe that animals should be free to live their lives without interference from people, that it is not right to kill animals for food or to obtain their skins for clothing or any other purpose. This group of people, known as **animal rights activists**, believe that killing animals is just as wrong as killing humans, and that animals should have the same rights as humans.

FIGURE 26–1 Producers have always been concerned about the welfare of their animals. *© Phovoir/Shutterstock.com.*

The other line of thinking is that it is moral to raise animals for human use, but that animals should not be abused or mistreated in any way. These people believe that although animals are raised for slaughter, the animals should be made as comfortable and as "happy" as possible while they are alive. This group is generally known as animal welfare activists.

In theory, livestock producers do not argue with the **animal welfare activists**. The controversy centers around what constitutes the abuse or mistreatment of animals. The animal welfare activists object to what they refer to as livestock factories in which animals are mass-produced under conditions that totally neglect the welfare of the animals. They consider the only motivation in producing the animals to be profit. According to animal welfare activists, several factors can cause problems for animals, as discussed next.

CONFINEMENT OPERATIONS

Animal welfare activists believe that modern livestock operations are nothing more than animal factories where animals are mass-produced like nonliving things. They think that the modern farm where animals are produced is much different from farms of several years ago. The activists are of the opinion that all the modern producer cares about is making money, even if it causes animals to suffer. The production of animals in a confined space is viewed as being cruel and causing animals to suffer (**Figure 26–2**).

FIGURE 26–2 Animal welfare activists object to animals being raised in a confinement operation. © Kiyota/Shutterstock.com.

FIGURE 26–3 Animals that are not under stress and are healthy produce better and make more profit for the producer. © nulinukas/Shutterstock.com.

They object to pigs being raised in crowded pens where they never leave the pen and have little room to exercise. Placing sows into farrowing crates where they cannot turn around or take a step is considered cruel.

Animal welfare activists disapprove of layer hens being kept in cages for their entire lives. They think that the hens are put under stress because they do not have enough room to stretch their wings or to get any exercise. The hens are seen as being similar to a factory production line where all feed and water are brought to the hens and their sole function in life is to produce eggs. Cattle feedlots are considered objectionable when the animals are crowded together and no shade is provided to protect them from the sun and no shelter is provided against the rain and the cold.

Livestock producers contend that almost all livestock produced in the United States come from family-owned farms and ranches. Producers are in the business because they enjoy working with animals and have the animals' best interests at heart. Producers point out that in order to stay in business, they must make a profit. More profit can be made if the animals are healthy and well cared for (**Figure 26–3**). Animals that are under stress cannot grow and produce well. In fact, the more comfortable an animal is, the more profit can be made because the animal is growing more rapidly. Producers point out that sows are put in farrowing crates to protect the piglets that might otherwise be crushed by the mother (**Figure 26–4**).

Another argument for the use of confinement operations is that over the years, animals have been specially bred for confinement operations. These animals are vastly different from their relatives in the wild. Also, animals in confinement are easier to care for because the producer can see each animal every day and often

FIGURE 26–4 Sows are placed in farrowing crates to protect the piglets. © Krueabudda/Shutterstock.com.

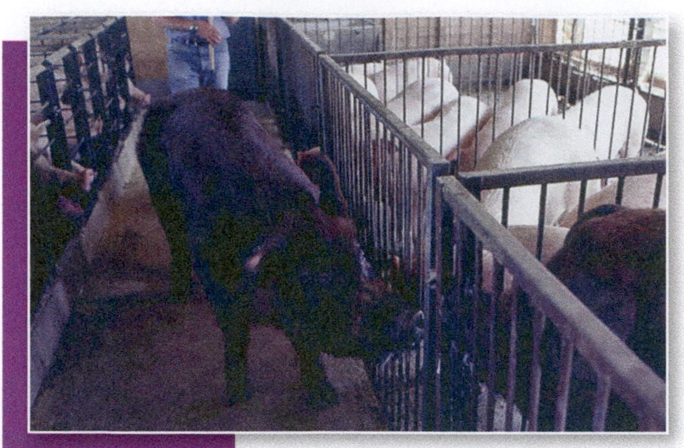

FIGURE 26–5 Animals raised in confinement operations are easier to care for.
Courtesy of Cooperative Extension Service, University of Georgia

several times a day (**Figure 26–5**). Housing provides shelter from the elements and from predators.

Many millions of dollars and countless hours of effort have gone into research to design housing and facilities that make animals comfortable. A hog house, for example, is scientifically designed for hogs; the design takes into account the animals' well-being (**Figure 26–6**).

An uncomfortable, stressed-out animal simply will not grow as efficiently as an animal that is content. The fact that animals are far more efficient in terms of production and growth than they have been in years past adds credence to the producer's arguments.

THE USE OF DRUGS

Animal welfare activists disapprove of feeding drugs such as antibiotics to animals as a preventive measure. They point out that traces of the drugs may possibly show up in the meat that is to be consumed by people. They feel that the drugs will have an adverse effect because bacteria that the drugs are guarding against may become immune to the antibiotics as a result of the prolonged feeding of the drugs. They are also concerned that the bacteria will develop into strains of organisms that will not respond to modern antibiotics. This could cause serious health problems not only for animals but also for humans.

On the other hand, producers counter that the addition of medication to the feed makes the animals healthier than they would be if they were allowed to roam free in nature. Because of the medication, the animals remain not only free from disease but also free from parasites (**Figure 26–7**). A healthy animal that is free from external and internal parasites suffers less than an animal that does not have the benefit of medication.

Producers point out that for a drug to be approved for use in animals, the U.S. Food and Drug Administration tests the proposed drug exhaustively to prove that the medication is not only safe and beneficial to the animals,

FIGURE 26–6 Hog facilities are designed to make the animals comfortable. *Courtesy of Cooperative Extension Service, University of Georgia*

FIGURE 26–7 Animals that are fed medications are free from parasites and are healthier. *© Martin Nemec/ Shutterstock.com.*

but that it is also safe for humans who consume the meat, eggs, or dairy products from those animals. Laws mandate strict **withdrawal periods** for the drugs. This means that there is a minimum number of days that an animal must be off a particular medication before it can be slaughtered or before dairy products from the animal can be used.

MANAGEMENT PRACTICES

Animal welfare activists are concerned that animals undergo such management practices as **debeaking**, tail **docking**, **dehorning**, and **castration**. They point out that almost always these practices are done without anesthesia, and this causes the animal to suffer (**Figure 26–8**). The activists take the position that the animal would be better off if these operations were not performed. They contend that nature had a reason for creating each animal "as is" and that it should be left in its natural condition.

Producers explain that these practices are necessary for the well-being of the animals. For example, an animal with no horns is far less likely to cause injury to another animal or to a human than an animal that has horns. Since nature intended the horns for use in self-defense, animals under the protection and care of humans have no real need or use for the horns; therefore, dehorning is beneficial to the animal.

Beak trimming is done to prevent chickens from injuring others as they establish a natural **pecking order**. In modern operations, the animals are far less likely to be injured if the beaks of the chicks are trimmed. (Actually, only the tip of the beak is removed.) A modern method of beak trimming is to use a chemical solution on the tip of the beak. The solution dissolves the tip and causes the chick no pain. The procedure actually makes it easier for the chickens to eat (**Figure 26–9**). In nature they use a sharp beak to pick up seeds or to catch insects. The trimmed beaks make it easier for them to eat the processed feed prepared for them.

Tails are docked in sheep because research has shown that removing the tail allows the animal to remain cleaner and healthier. Otherwise, mud, dirt, and manure cling to the tail and cause stress to the animal.

Castration is done at an early age to prevent problems associated with male animals fighting for dominance and to provide a higher-quality meat when the animal is slaughtered. Although the animals undergo temporary

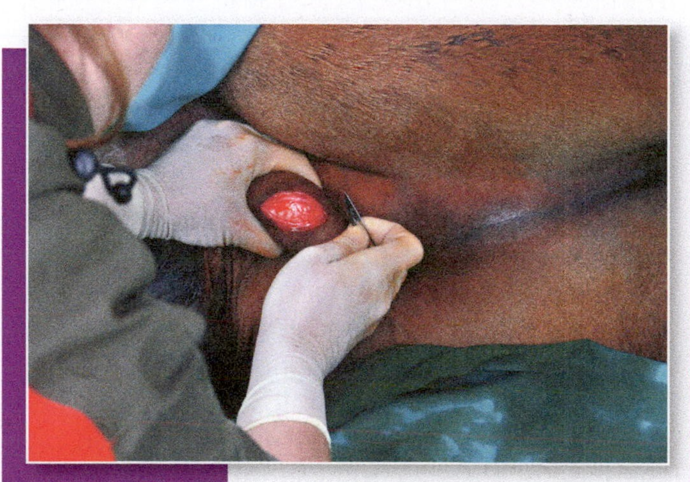

FIGURE 26–8 Animal welfare activists consider many management practices to be cruel to the animals.
© kiep/Shutterstock.com.

FIGURE 26–9 Trimmed beaks allow chickens to eat processed feed more efficiently. *© hurricanehank/Shutterstock.com.*

discomfort from these procedures, producers argue that in the long run the practices are in the animals' best interests.

Producers have always looked for management techniques that cause less stress to their animals. Procedures such as tail docking and castration are done at an early age because the younger animals suffer less trauma than older animals that undergo these operations.

The use of caustic soda or other solutions on the horn buds of young calves is usually considered to be less stressful than cutting the horns out when the animal is older. Many producers use a technique such as the **elastrator** band to castrate animals and dock their tails. This consists of placing a rubber band around the tail or **scrotum** of the animal to prevent the flow of blood. Consequently, the scrotum or tail eventually drops off. Another method is to use a clamp that breaks the cords that transport sperm (**Figure 26–10**). These methods do not require opening a wound on the animal.

One technique that has lessened stress on animals is the use of **freeze branding** instead of **hot branding**. Producers must have some way of identifying individual animals. The use of the hot branding iron has been used by cattle producers for many generations. Critics of this process say that the procedure causes severe pain for the animals. The hot iron causes a burn that scars over and leaves a permanent mark on the animal (**Figure 26–11**).

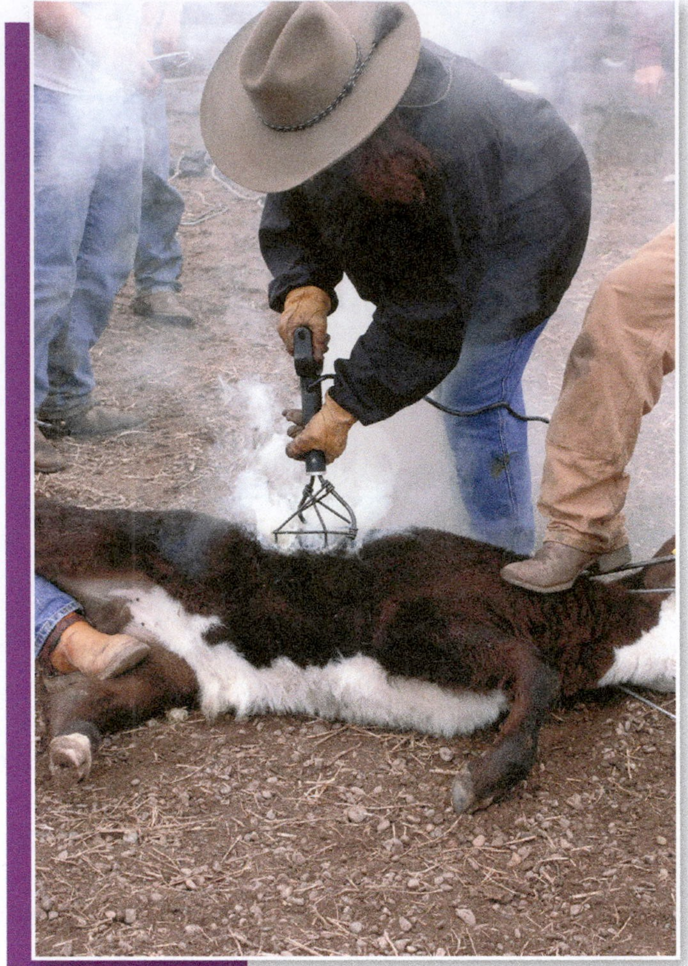

FIGURE 26–11 Critics of hot branding contend that the procedure causes the animal pain. *© iStockphoto/cgbladauf.*

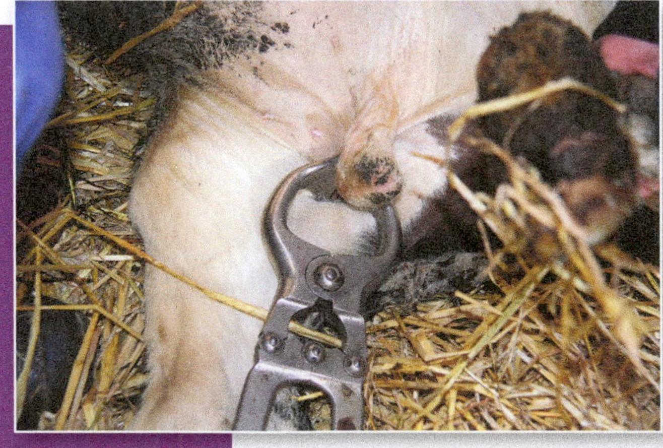

FIGURE 26–10 Bloodless castration is done without opening a wound. *© Claudia Otte/Shutterstock.com.*

A newer technique called freeze branding uses a tremendously cold iron to do the marking. In this method, an iron is immersed in a container of liquid nitrogen. Nitrogen in the liquid state is approximately –320°F, and it cools the iron to an extremely cold temperature. The iron is then applied to a part of the animal's skin that has had the hair closely clipped. This procedure kills the pigment-producing cells at the base of the hair follicle and results in the growth of white hair. A longer application of the iron results in the killing of the entire follicle and will leave a bald mark where the iron touched. The advantage of this method is that the animals feel very little pain as they are permanently marked (**Figure 26–12**).

Livestock producers have always been known for their concern and care for the animals they raise. They have always taken pride in producing strong, healthy animals. Throughout all of recorded history, accounts have been left of how producers carefully tended their animals—often at the risk of peril to themselves. Shepherds have traditionally lived with their sheep and guarded them against wild animals (**Figure 26–13**). Even though methods have now changed and the raising of animals is much more intensive, most animals are still cared for by families who earn their livelihoods caring for animals.

FIGURE 26–13 In some parts of the world, shepherds still live with their sheep. *© LeicherOliver/Shutterstock.com.*

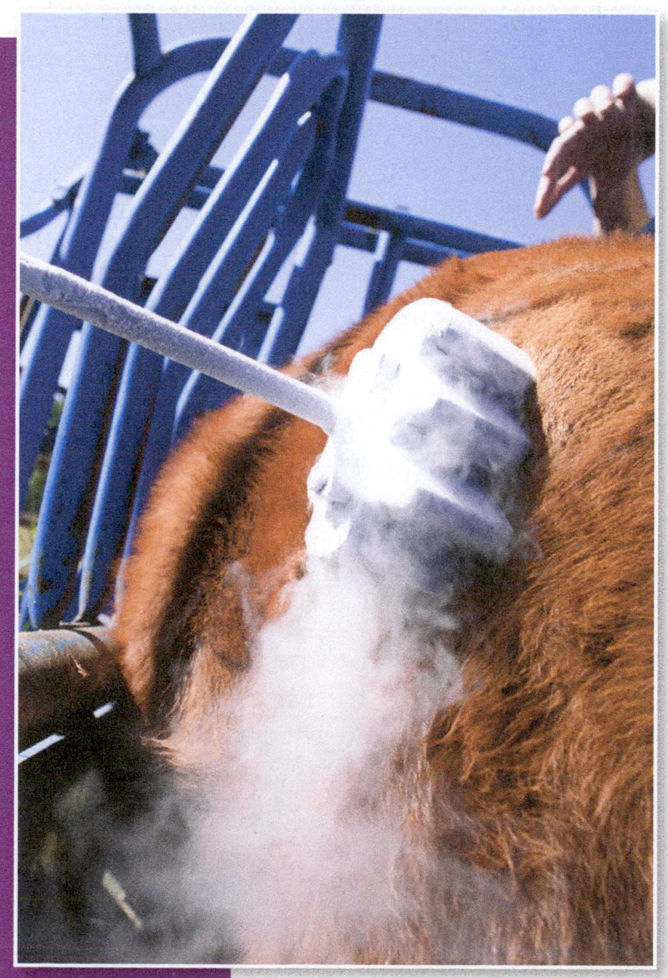

FIGURE 26–12 Freeze branding kills the hair pigment, thus marking the animal. *© iStockphoto/jeffstrauss.*

According to the Animal Industry Foundation, 97 percent of the farms in the United States are family owned and operated (**Figure 26–14**). The U.S. Department of Agriculture (USDA) reports that there are only 7,000 non-family-owned farms in the entire United States. A tour of any of the thousands of livestock shows that are conducted each year in the United States

FIGURE 26–14 Most of the farms in the United States are family-owned. *© Agatha Koroglu/Shutterstock.com.*

will show the amount of pride these farm families take in the livestock they produce.

At the 1990 National Cattlemen's Association Convention, the producers adopted the following statement of principles on animal care, environmental stewardship, and food safety:

> I believe in the humane treatment of farm animals and in continued stewardship of all natural resources.
>
> I believe my cattle will be healthier and more productive when good husbandry practices are used.
>
> I believe that my and future generations will benefit from my ability to sustain and conserve natural resources.
>
> I will support research efforts directed toward more efficient production of a wholesome food supply.
>
> I believe it is my responsibility to produce a safe and wholesome product.
>
> I believe it is the purpose of food animals to serve mankind and it is the responsibility of all human beings to care for animals in their care.

Similarly, the National Pork Producers Council has adopted the following Pork Producers Creed:

> I believe in the kind and humane treatment of farm animals and that the most efficient production practices are those that are designed to provide comfort.
>
> I believe my livestock operation will be more efficient and profitable if managed in a manner consistent with good husbandry practices as known and recommended by the animal husbandry community.
>
> I believe in an open-door policy to visitors to my farm, to all those who are sincerely interested in production methods and the welfare of animals, so long as they do not endanger the health and welfare of my animals and do not interrupt my production routine or impair the production process.
>
> I believe in and will support research efforts designed to measure stress of farm animals and directed toward more efficient production of food and enhancement of the welfare of animals and man.
>
> I believe it is the animal's purpose to serve man; it is man's responsibility to care for the animals in his charge. I will vigorously oppose any legislation or regulatory activity that states or implies interference with that responsibility.

RESEARCH

Another issue of great concern for animal welfare activists is that of the use of animals for research. As far back as the second century A.D., scientists have used animals in research. Most of the medical advances made by humans during the past 100 years have come about through the use of animals to test treatments and medication. Before any type of medical treatment can be tested on humans, it must first be tested using animals.

Most people realize that research is carried out using mice, rabbits, and guinea pigs (**Figure 26–15**). However, during recent years, much controversy has come about over the use of cats, dogs, and primates. Those who oppose the use of animals in research contend that the knowledge gained through the research cannot justify the suffering the animals must undergo to test a treatment or to experiment with a new theory (**Figure 26–16**). These people are particularly emphatic over the use of animals to test such products as cosmetics. They cite examples of how rabbits have chemicals placed in their eyes to determine how irritating an ingredient in an eye shadow for humans might be. The activists insist that the suffering of animals should not be allowed merely to produce new products

FIGURE 26–15 A lot of research is conducted using animals, such as guinea pigs. © Monica Butnaru/Shutterstock.com.

FIGURE 26–16 Many people object to using primates for research. © martinusnovedwin/Shutterstock.com.

that are used only to make people appear more attractive.

People and groups opposed to the use of animals in research argue that animals really do not have enough in common with humans to be used in research. They also contend that much of the research using animals could be done through the use of computer models and through the use of cell cultures.

Scientists who use animals in research counter these arguments with the point that animals cannot be elevated to the same level as humans. The scientists argue that the use of the animals is well justified by the advances in medicine and health care that have come about through the use of experiments with laboratory animals. They cite the examples of diseases such as polio that have been almost eradicated through the use of research using animals. They do agree that advances have been made in the use of cell cultures and the use of computer simulations, and that the use of these techniques can reduce the necessity of using animals in some instances. However, they also point out that these techniques have limited use. The use of live animals is unavoidable because no method has been developed that adequately substitutes for a live animal.

SUMMARY

The controversy surrounding animal welfare will likely continue. As the number of people who are actively involved in the production of animals becomes fewer and the farms and ranches become larger, there will be less understanding on the part of the public concerning the production of animals for food. The producers will have to take on increasing responsibility to educate the public about production methods. They will also be compelled to ensure that the practices employed are truly in the best interests of the animals. Efforts must always be put forth to keep facilities clean and comfortable for the animals. Doing so will help maintain the image of the livestock producer as someone who truly cares for the animals' welfare.

CHAPTER REVIEW

Review Questions

1. What is the difference in the positions of people who advocate animal rights and those who advocate animal welfare?
2. Why do animal welfare activists oppose raising animals in confinement?
3. What arguments do producers offer in defense of confinement operations?
4. How do livestock producers justify the use of antibiotics in feed?
5. Why is the practice of freeze branding considered more humane than hot branding?
6. What percentage of the farms in the United States are owned and operated by families?
7. What two livestock associations have adopted statements dealing with the treatment of animals?
8. Why is the controversy surrounding animal welfare likely to continue?
9. Why are some individuals and groups opposed to using animals in research?
10. What justifications do scientists offer in defense of the use of animals in laboratory experiments?
11. What are two new technologies that have lessened the need for using animals in experimental research?

Student Learning Activities

1. Locate and read an article in a magazine, newspaper, or other publication that advocates animal rights. List the points made by the author that you believe are correct and factual. Also list the author's points that you think are not based on fact. Compare your lists with the lists of other students in your class.
2. Make a list of management practices that might be criticized by animal welfare activists. Think of and write down methods or alternatives that the producer might employ to lessen the criticism.
3. Write a brief report on at least one scientific advancement that used animals in the research. Be sure to include how the animals were used and how people have benefited from the advancement.
4. Take a stance either for or against the use of animals to test cosmetics. Present your arguments to the class. Compare your arguments to the arguments of those in the class who took the opposing view.

CHAPTER 27

Consumer Concerns

STUDENT OBJECTIVES IN BASIC SCIENCE

As a result of studying this chapter, you should be able to

- explain the rationale for consumer concern over food safety.
- define and explain the role of cholesterol.
- define *genetic engineering*.
- explain the rationale for concerns about genetic engineering.
- discuss the concept of the greenhouse effect.
- explain how the balance of oxygen and carbon dioxide is maintained in the atmosphere.
- explain how bacteria can be beneficial to the environment.

STUDENT OBJECTIVES IN AGRICULTURAL SCIENCE

As a result of studying this chapter, you should be able to

- explain why agriculturalists must be more sensitive to the concerns of consumers.
- tell the difference between meat grading and meat inspection.
- summarize how meat is inspected.
- tell how research has shown that the meat supplied to consumers is safe and nutritious.
- give examples of how genetic engineering has benefited the producer.
- describe how producers of agricultural animals are good caretakers of the environment.

KEY TERMS

curing
antemortem
rendering
antibiotics
Food and Drug Administration (FDA)
cholesterol
marbling
genetic engineering
genetics
RST
BST
coliforms
colon
public lands
Bureau of Land Management
greenhouse effect

NATIONAL AFNR STANDARD

AS.01.02.01.a
Identify and categorize terms and methods related to animal production.

AS.02.01.01.a
Explain the implications of animal welfare and animal rights for animal systems.

AS.02.02.02.a
Research and summarize animal production practices that may pose health risks.

AS.02.02.02.b
Analyze consumer concerns with animal production practices relative to human health.

AS.02.02.02.c
Research and evaluate programs to ensure the safety of animal products for consumption.

AS.08.02.01.a
Research and summarize environmental conditions that impact animals.

AT ONE TIME IN THE history of our country, almost everyone knew how food was produced and processed. This was because the growing and processing of food was done at home. Before the end of the nineteenth century, the vast majority of people in the United States lived in rural areas and produced most of the food they ate. Crops were gathered and dried, pickled, or canned for the family to use during the winter months. When the weather turned cool, livestock was slaughtered and the meat was preserved by drying, canning, or **curing**. Because the people processed their own food, they knew how the food was processed and what went into the food. Today, relatively few people process their own food. The only form of food that most people ever see is the finished product in the grocery store (**Figure 27–1**).

Trends in food processing follow what the consumer demands. Consumers are the people who buy and use products. Because the husband and the wife are both working outside the home in most families, consumers want food that is more processed than in the past. This means that food products have to be closer to being ready to eat than at any time in our history. As more and more steps are added in the processing of food, consumers become more concerned about how the processing was done and what ingredients went into the product (**Figure 27–2**).

At the same time, consumers do not understand how crops and animals are grown. Because the consumers want food products that are relatively inexpensive, producers have to use means that can get as much efficiency as possible from the crops and animals they grow. Chemicals and other substances are used in the production and processing of food that consumers do not generally understand. As modern technology increases and new discoveries help growers produce more efficiently, concerns are raised among consumers about the safety and wholesomeness of the food they buy and consume. Much of this concern centers around the animal industry and the products such as meat, milk, and eggs produced by the industry. Animal products are very susceptible to spoilage. They can easily pick up microorganisms from the processing that can cause spoilage.

FIGURE 27–2 As more steps are added in the processing of foods, the more concerned consumers become. © Szasz-Fabian Jozsef/Shutterstock.com.

FIGURE 27–1 Most people see only the finished product in the grocery store. © wavebreakmedia/Shutterstock.com.

MEAT PRODUCTS

Recently, public concern over food poisoning has been raised as a result of contaminated meats. In these cases, the cause of the illnesses was hamburger containing the *E. coli* bacteria. These bacteria inhabit the colon of animals and humans, and certain strains of the bacteria can cause illness and even death. Problems arise during the slaughter process when the internal organs are removed from the animal. Sometimes the carcass is contaminated when it comes in contact with *E. coli* bacteria from the viscera.

FIGURE 27–3 Hamburger is particularly susceptible to *E. coli* bacteria because of the grinding and mixing process.
© Vipavlenkoff/Shutterstock.com.

Hamburger is particularly susceptible because in the grinding process, the bacteria on the surface of meat are ground and mixed with all of the meat in the batch (**Figure 27–3**). Given a warm, moist environment, the bacteria grow and reproduce rapidly, and eventually enough bacteria are produced to cause illness if the bacteria are not destroyed. Poultry has also come under careful scrutiny because of the possibility of contamination with salmonella bacteria. Most often, all harmful microorganisms can be destroyed by cooking the meat thoroughly.

As a result of sickness from contaminated meats, new regulations are in effect for the inspection and handling of meat products. The first step in preventing food-borne bacteria on meat is sanitation and taking precautions at the meat-processing plants. Slaughterhouses are under new and more rigorous requirements for sanitizing the plant and preventing meat from contacting fecal material. In addition, new scientific tests are conducted on the meat to determine the presence of bacteria. Although we have had serious outbreaks of sickness caused by contaminated meat, our supply of meat is still considered to be safe when it is handled and cooked properly. Only a very tiny percentage of the meat on the market has been found to contain large numbers of harmful bacteria.

The Meat Inspection Act passed in 1906 requires that all meat products that are processed and sold in the United States will pass inspection by the U.S. Department of Agriculture. Since that time, the law has been revised and updated several times to ensure that only the very best, most wholesome meat reaches the consumer. Meat inspection is not the same thing as meat grading (**Figure 27–4**). Meat inspection simply guarantees that the meat will be safe, wholesome, and accurately labeled. All meat that is sold must, by law, be inspected.

FIGURE 27–4 Meat inspection is not the same thing as meat grading. All meat that is sold must be inspected.
© Tyler Olson/Shutterstock.com.

Grading refers to the eating quality and degree of yield expected from a carcass. Grading is optional. Meat inspection undergoes several phases. First, the animals that are to be slaughtered must be inspected while they are alive. This is called **antemortem** inspection (*ante* means "before," and *mortem* means "death"). As the animals are brought in prior to slaughter, a government inspector examines them. Animals that are down, disabled, diseased, or dead are condemned as unsafe for human consumption. Animals that the inspector thinks may have a problem are set aside for further examination. The plant where the animals are to be processed must provide the inspector with a well-lit, clean area to examine animals that the inspector suspects are not in the best of health. These animals are examined thoroughly, and if they are found to be ill, they are tagged as condemned and are not allowed to be slaughtered for human consumption.

After the animals are slaughtered, they must again undergo inspection. Slaughter plants are required to provide adequate lighting (50 footcandles) so the inspector can inspect the carcasses thoroughly (**Figure 27–5**). The head, lungs, heart, spleen, and liver are inspected for signs of disease, parasites, or other problems that might render the meat of cattle, sheep, and hogs less than wholesome. The internal and external cavity of slaughtered poultry must be examined, as well as the air sacs, kidneys, sex organs, heart, liver, and spleen.

Carcasses that do not pass inspection are declared to be condemned and are not allowed to be used for human consumption. These condemned carcasses undergo a process called **rendering**, during which they are placed under heat that is high enough to kill any organism that could cause problems. The rendered meat then is used as a by-product that is not intended for human consumption. By-products can be used as pet food or for other products such as fertilizers. Condemned carcasses usually represent less than 1 percent of the carcasses inspected. Producers, buyers, and packers all try to avoid sending animals to slaughter that they know will not pass inspection. Those that do get by are discovered by the federal inspectors at the processing plant.

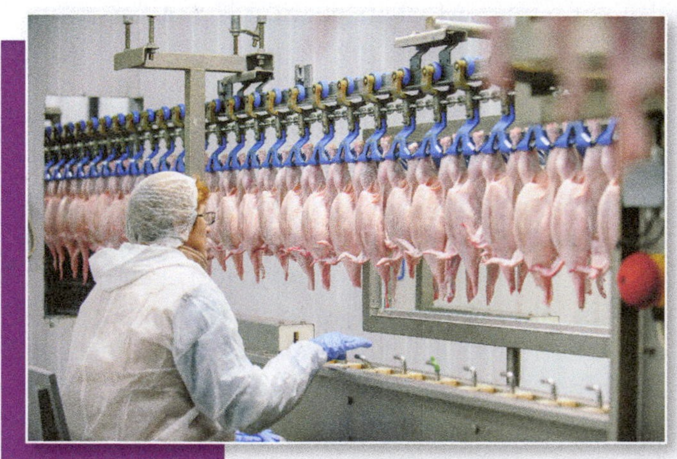

FIGURE 27–5 Meat products are closely monitored and inspected. *© Anton Mislawsky/Shutterstock.com.*

POULTRY PRODUCTS

As with meat products, poultry products, such as chickens and eggs, must also undergo inspection. The USDA sets the standards for grading these products.

Grading Poultry

The United States Department of Agriculture (USDA) sets the standards for grading poultry carcasses. Grade A is what you see in the meat counter at the grocery store. The other grades are used for making products such as soups, stews, chicken nuggets and other uses. Included below are some comparisons of Grade A and B poultry carcasses and legs (**Figures 27–6** to **27–12**).

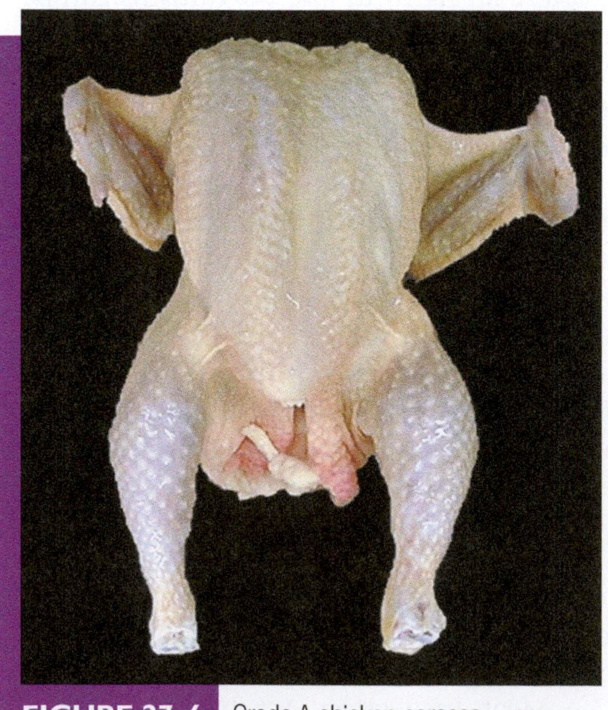

FIGURE 27–6 Grade A chicken carcass. *USDA*

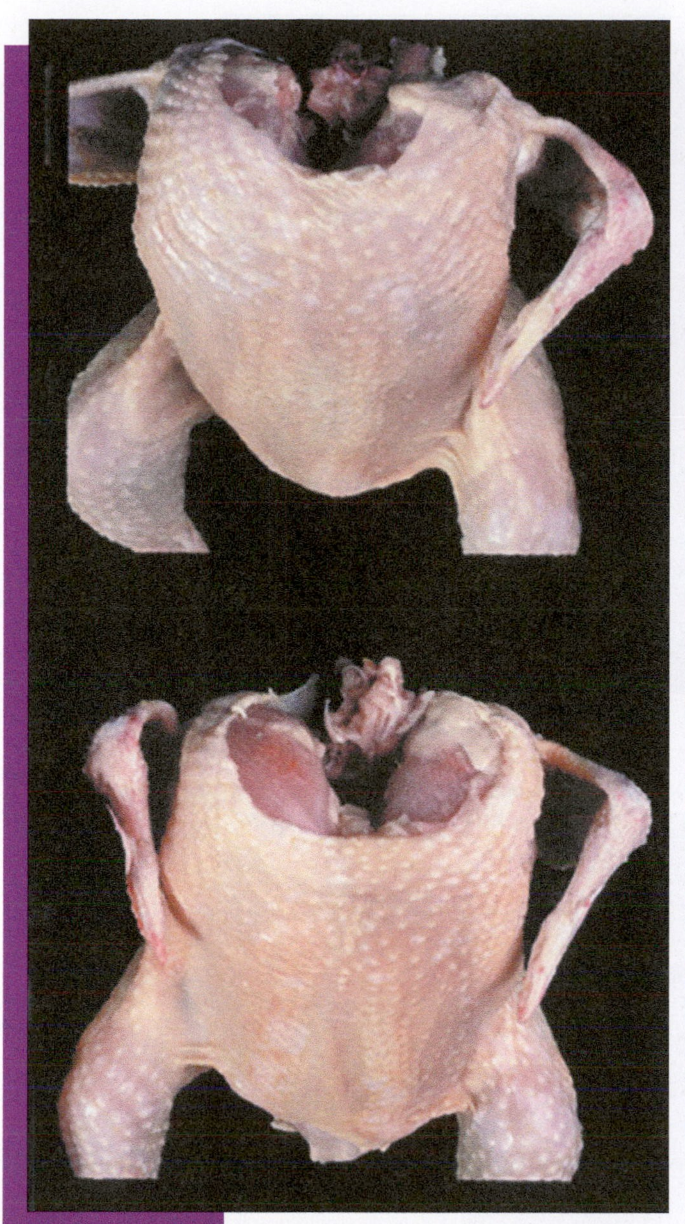

FIGURE 27–7 Grade A chicken carcass (top). The carcass Shown at the bottom is Grade B due to excess neck skin removed during processing. *USDA*

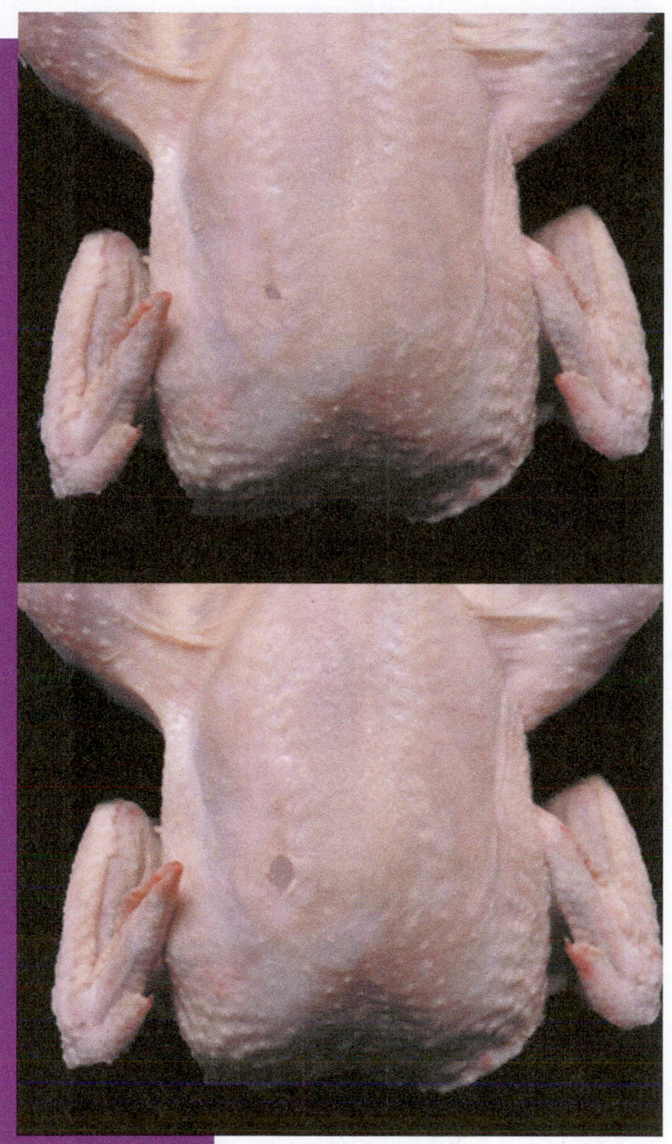

FIGURE 27–8 Chicken carcasses. Grade A (top) – the exposed flesh does not exceed ¼-inch tolerance for breast and legs. Grade B (bottom) – exposed flesh does exceed ¼-inch tolerance for breast and legs. *USDA*

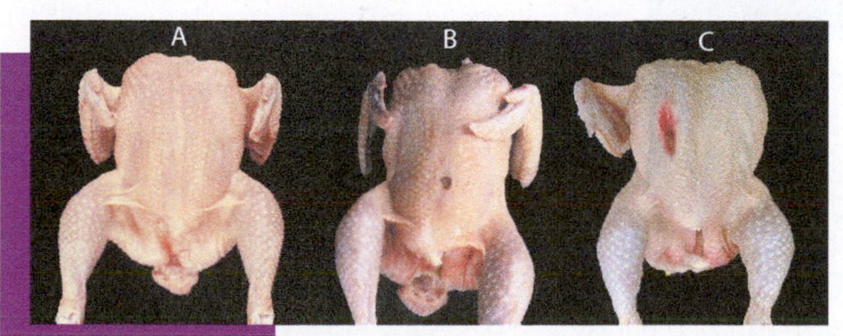

FIGURE 27–9 Chicken carcasses, Grades A, B, and C. The carcass in the middle has exposed flesh greater than ¼ inch. Grade C, on the right, has an entire wing missing and discoloration greater than 2 inches and more than moderate in color. *USDA*

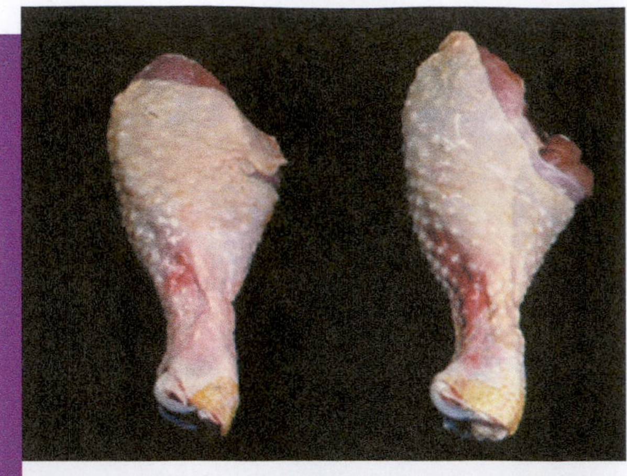

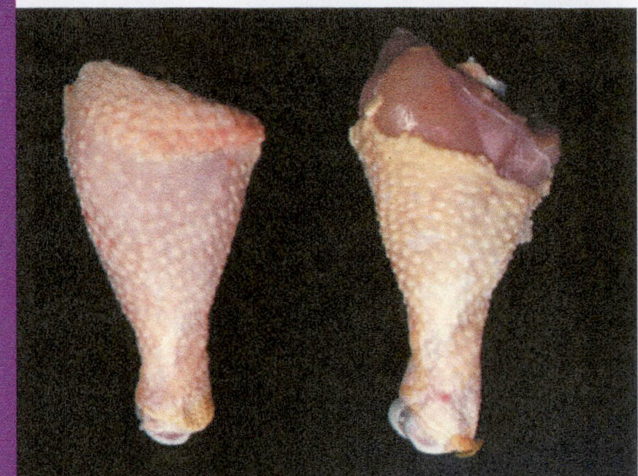

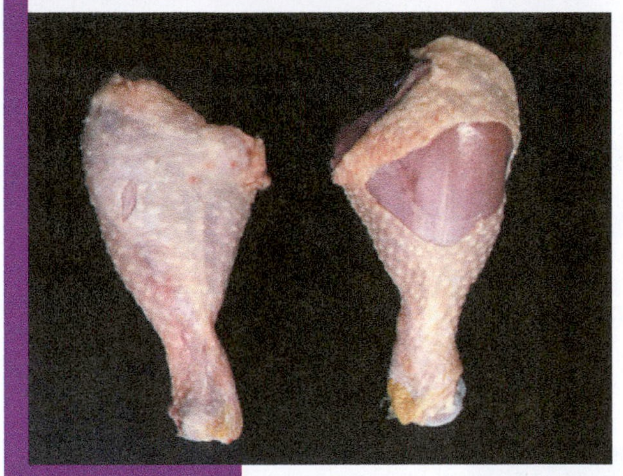

FIGURE 27–10 Grades A and B chicken legs. All legs on the left are Grade A. The top right leg is Grade B due to discoloration. The top left leg has some discoloration but within tolerance. The right leg of the middle pair has too much flesh exposed and is Grade B. The bottom pair, Grade A on the left and Grade B on the right; the Grade B leg has too much flesh exposed. Grade A has some flesh exposed but is within tolerance. *USDA*

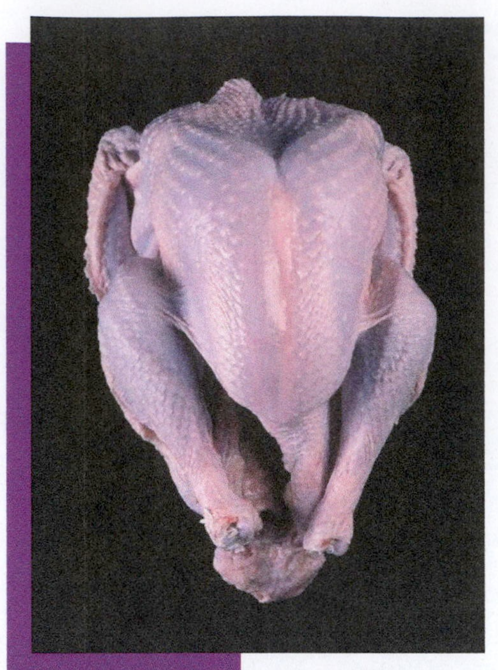

FIGURE 27–11 Grade A turkey carcass. *USDA*

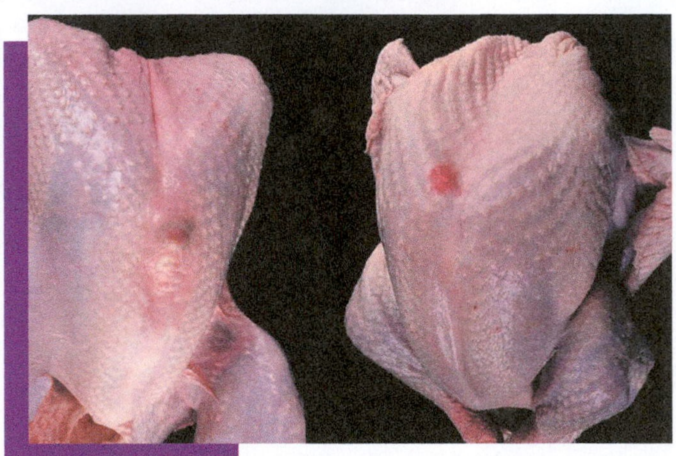

FIGURE 27–12 Grade B turkey carcass. *USDA*

Grading Eggs

The United States Department of Agriculture (USDA) also sets the standards for grading eggs. The eggs are graded according to certain physical attributes, including shape and size, **Figure 27–13** to **27–18**. The interior quality of the egg is determined through a process called candling.

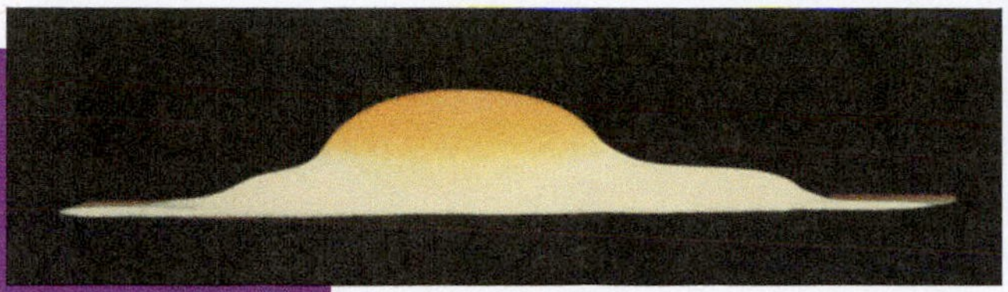

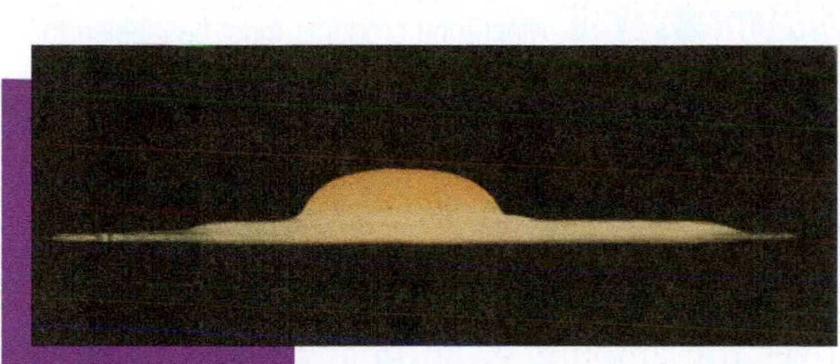

FIGURE 27–13 Grade AA, or Fancy Fresh egg Content covers a small area; white is thick and stands high, with small amount of thin white; yolk is firm, round and stands high. USDA

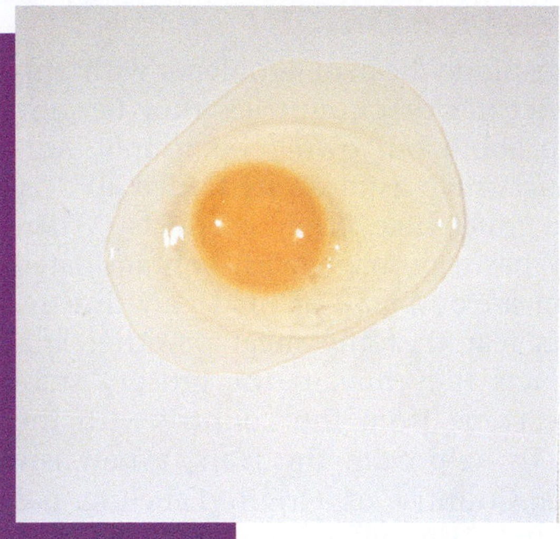

FIGURE 27–14 Grade AA egg, top view. USDA

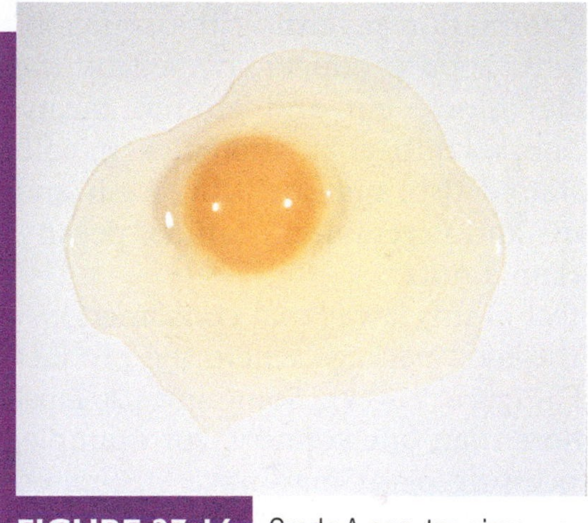

FIGURE 27-16 Grade A egg, top view. USDA

FIGURE 27–15 Grade A egg content covers a moderate area; white is reasonably thick and stands fairly high; yolk is firm and high. USDA

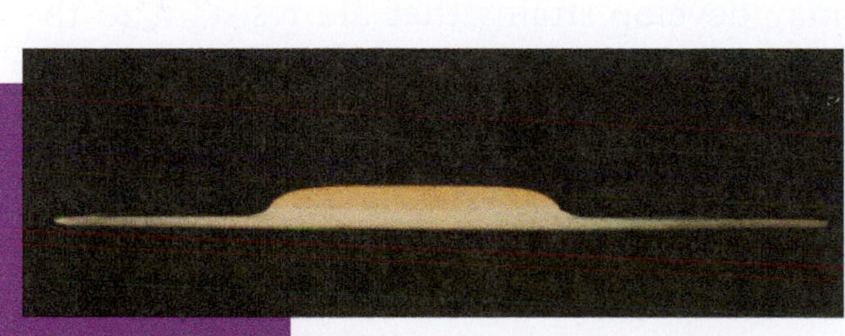

FIGURE 27–17 Grade B egg content covers a very wide area; white is weak and watery; it has no thick white; a large amount of thick white is thinly spread. Yolk is enlarged and flattened. USDA

FIGURE 27–18 Grade B egg, top view. USDA

COUNTRY OF ORIGIN

When consumers hear of problems with animal product in other countries, they become concerned about our own food supply. Animal products such as meat are traded all over the world much like other products. When a country has a problem with contaminated meat from an outbreak of mad cow disease, consumers fear that the meat or other animal products they buy in the grocery store may have come from the country with the problem. To help calm the fears, a new law called the Country of Origin Labeling has been enacted. This law requires retailers, such as full-line grocery stores, supermarkets, and club warehouse stores, to notify their customers with information regarding the source of certain foods. Food products covered by the law include muscle cut and ground meats: beef, veal, pork, lamb, goat, and chicken; wild and farm-raised fish and shellfish; fresh and frozen fruits and vegetables; peanuts, pecans, and macadamia nuts.

The effect of this law is that consumers can tell from the food package where the product originated (**Figure 27–19**). Sometimes a label may list more than one country. For example, a meat-processing plant may have U.S.-raised beef that is too fat for hamburger. This meat may be mixed with leaner beef from Mexico or Canada. In this case, the hamburger will be labeled as originating in the United States, Canada, and Mexico. Any meat imported into the United States must meet the same inspection requirements as meat originating in this country.

FIGURE 27–19 Most food products must be labeled to identify the country of origin. *Hormel is a registered trademark of Hormel Foods, LLC and used with permission.*

ANIMAL MEDICATIONS

Animals grown in large numbers are fed medication to prevent and cure disease and to ward off parasites. This can result in a better, safer product for the consumer. Animals that are kept healthy all their lives reach the processing plant healthier and yield a healthier product. Consumers are sometimes worried about the residue of these medications that are fed to livestock. For instance, concern is raised over the amount of **antibiotics** fed to cattle, hogs, and poultry (**Figure 27–20**). The worry is that as humans consume meat from animals that were fed antibiotics, these medications might enter the human body and build up a residue that will create an intolerance to medication. Since the types of antibiotics given to agricultural animals are the same type as given to humans, some believe that bacteria may develop strains that are resistant to the medications, which would make the medications ineffective when needed by humans. Research by the United States Department of Agriculture (USDA) and the **U.S. Food and Drug Administration (FDA)** has clearly shown that adding antibiotics to animal feed in the proper amount does not result in antibiotic residue in meat.

FIGURE 27–20 Consumers sometimes worry about the medication given to livestock in their feed.
© Alf Ribeiro/Shutterstock.com.

HORMONES

Animals are also given growth hormones to aid in feed efficiency and growth rates. Many consumers are concerned over the long-range effects of the hormones fed to animals that might be residual in meat, eggs, milk, or other products. According to the National Cattlemen's Association, there is no evidence of any human health problem from the use of any natural or synthetic hormones fed to livestock. The FDA closely regulates the amount of hormone residue allowed in meat. When cattle are given growth hormones, the dosage is administered by means of implanting a small capsule underneath the skin that releases the hormone very slowly. The capsule is implanted under the skin of the ear because at slaughter, the ears are put into the waste bin and are not processed for consumption.

No more than 1 percent of the human body's daily production of an implanted hormone is allowed to be present in the daily intake of meat. Because the hormone level in the meat is greatly reduced (by 90 percent) through the digestive process, the amount of hormones the body absorbs through meat is extremely small. The relative amounts of estrogen produced by people and the estrogen in a serving of beef from a steer that has been implanted with hormones and one that has not. In fact, a person will obtain thousands of times the amount of estrogen from a gram of soybean oil than from a gram of beef from an implanted steer.

In order for a product to be released and sold for use by animal producers, the product must be researched and tested through rigorous standards before it is declared safe by the government. The product must be so thoroughly studied and tested that there is no significant chance that the product will cause any health problems for the consumer. The USDA closely monitors animal products for traces of residues resulting from feed additives or medications that are given to animals. The allowable amounts of these residues in foods would have to be hundreds, and in some cases thousands, of times greater before any effect at all could be detected in humans who consume the animal product (**Figure 27–21**).

CHOLESTEROL

Several years ago, some medical research indicated that a high intake of a substance known as **cholesterol** was a contributing cause of

FIGURE 27–21 The USDA closely monitors animal products for impurities and residues. These inspectors are examining meat products. *© El Nariz/Shutterstock.com.*

FIGURE 27–22 Choice cuts of beef have less fat content than those of a few years ago. *© Milanchikov Sergey/ Shutterstock.com.*

heart disease. Cholesterol is a fatlike substance found in animal tissue. It is an essential part of nerve tissue and cell membranes of all animals, including humans. Cholesterol plays an important role in the body's manufacture of hormones and in the production of the bile used in digestion. Although cholesterol is essential to life, studies have shown a correlation between high cholesterol levels and clogged heart arteries. Because meat contains cholesterol, consumers have been concerned about eating meat.

Recent studies have been somewhat contradictory about how important a role cholesterol actually plays in the formation of deposits in the coronary system. Some studies indicate that these deposits are related more to the amount of exercise a person gets and the person's heredity than the amount of cholesterol intake. Modern meats contain less fat than the meat produced in the past. The pork producer and the beef producer organizations are both promoting pork and beef as healthful foods that are leaner and trimmer than meats of a few years ago. For example, a Choice cut of beef today has a smaller fat content—less **marbling**—than a Choice cut of beef did 10 years ago (**Figure 27–22**). Although fat is what gives meat its flavor, research is constantly being conducted to produce a lean meat product with a low-fat content.

GENETIC ENGINEERING

Genetic engineering is the alteration of the genetic makeup of an organism to produce a desired effect. The development of modern technology has given scientists the ability to enter an organism's genetic makeup and insert, remove, and alter genes that are responsible for the organism's characteristics. This has the promise of tremendous benefits to the producers of agricultural animals. If we are able to improve the **genetics** of animals by simply changing the genes, we can boost the efficiency of producing them and, in turn, produce better and cheaper products for the consumer. At the same time, consumers are concerned over the use of genetic engineering. Some people see this effort as interfering with nature. They fear that by altering the genetic makeup of an animal, the potential exists to create animals that might not be in the best interest of humans.

Many science fiction movies have been made that depict monstrous animals created by a scientist who disturbs the natural order of the animal's makeup.

In addition, there is a fear that products from genetically engineered animals will contain substances that will prove harmful to the people who consume them. One such example is the use of a substance called *recombinant bovine somatotropin* (**RST** or **BST**). This substance is produced by genetic engineering and is a naturally occurring hormone (**Figure 27–23**). Scientists have known for many years that cows receiving additional amounts of BST significantly increase their milk production. Until recently, this substance was scarce and very expensive. However, due to genetic engineering, the hormone can now be produced quickly and cheaply. The use of BST has been and is now still being debated in regard to its safety. Some say that no studies have been conducted to determine the long-term effects of drinking the milk from cows that have been given BST to increase their milk production. They point to the fact that large doses can cause inflammation of the cow's udders and that this may be a sign that the hormone is not safe. Proponents point out that the cows are only given very small doses, and that no ill effects have been discovered. Also, the National Institute of Health and the FDA have declared BST to be safe both for the cows and for the humans who consume the milk.

FIGURE 27–23 A genetically engineered hormone is given to cows to increase their milk production.
© iStockphoto/yadamons.

In 2016, Congress passed a bill requiring food containing genetically modified ingredients to be labeled as containing these products. It was left up to the USDA to decide what is defined as genetically modified. There is controversy over what constitutes genetically modified. For example, should products containing milk from cows treated with BST be labeled as genetically modified?

ENVIRONMENTAL CONCERNS

Some have criticized the production of agricultural animals as harmful to the environment. They point out that animals, particularly in a confinement operation, generate a lot of waste. They fear that this waste will get into the water supply and contaminate water designated for human consumption. This concern is well founded. Animal wastes contain bacteria called **coliforms**, which means "bacteria from the **colon**." These bacteria can carry disease to humans and other animals if they are allowed to escape into the water supply. Animals in confinement do create a lot of waste that could possibly contaminate the water supply; however, there are stringent laws that prevent this from happening. It is illegal for any producer to discharge animal waste into a stream. In fact, the Environmental Protection Agency has regulations that help prevent this from happening even accidentally. The use of lagoons and holding ponds have helped protect the environment. A lagoon is a body of water made especially for holding animal wastes from confinement operations (**Figure 27–24**). Modern

FIGURE 27–24 The use of lagoons and holding ponds helps to protect the environment. *Source: Photo by Jeff Vanuga, USDA, Natural Resource Conservation Service.*

FIGURE 27–25 Manure is spread out on land for use as fertilizer. *© iStockphoto/ezp.*

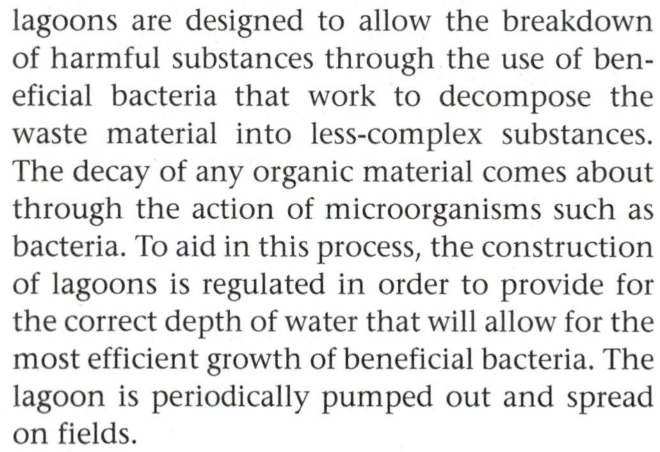

lagoons are designed to allow the breakdown of harmful substances through the use of beneficial bacteria that work to decompose the waste material into less-complex substances. The decay of any organic material comes about through the action of microorganisms such as bacteria. To aid in this process, the construction of lagoons is regulated in order to provide for the correct depth of water that will allow for the most efficient growth of beneficial bacteria. The lagoon is periodically pumped out and spread on fields.

Manure is a natural, high-quality fertilizer that is valuable in the production of many crops, and once the decomposition process has taken place, there is no harm to the environment. The use of manure actually can cut down on pollution by reducing the amount of commercially produced fertilizers required to grow crops (**Figure 27–25**).

THE OVERGRAZING OF PUBLIC LANDS

Some people are also concerned that our **public lands** are being overgrazed and ruined by producers who do not take care of the land. In the western part of the United States, government lands are leased to animal producers (**Figure 27–26**). A recent report of the **Bureau of Land Management** pointed out that our public range lands are in better shape than at any time during this century. Cattle and sheep producers know that they will be using the land for a long time and that it is in their best interest to care for the land. Most producers want to pass their operations on to subsequent generations. The only way for this to happen is to carefully manage the land. Land that is well cared for will produce animals more efficiently.

FIGURE 27–26 Although leased for grazing, public range lands are in better shape than at any time during this century. *© Photo SS/Shutterstock.com.*

According to the USDA, a sound management system for grazing builds the soil and enhances wildlife habitats. Wildlife numbers on government range land have improved dramatically over the past 30 years. When properly managed, the grazing of livestock on public lands can actually be good for wildlife. Through the control of undesirable species of brush and plants, better grazing is provided for the wildlife as well as for the cattle. Also, water has to be provided for the agricultural animals, and in doing so, water is provided for wildlife as well.

GLOBAL WARMING

In recent years, concern has been raised over a phenomenon known as the **greenhouse effect**. The theory behind this effect is that the earth is gradually warming due to problems encountered in the destruction of parts of our environment. (The effect was named for the extra heat inside a house made from glass or greenhouse.) Proponents of the theory say that a higher concentration of carbon dioxide in the atmosphere causes the heat from the sun to become more intense and the temperature of the earth to be raised. The proper balance of carbon dioxide and oxygen in the air is brought about by plants and animals. Animals breathe in air, absorb oxygen, and give off carbon dioxide. Plants take in air, absorb carbon dioxide, and give off oxygen. Vast acres of forest in the tropical areas of the world have always helped keep this balance in order. As trees are destroyed, some of the world's potential for absorbing carbon dioxide and releasing oxygen is lost. Environmentalists point out that forest areas are being cleared in order to raise livestock, and this, along with the methane gas produced by the animals, is contributing to global warming. In South America especially, tropical rain forests are being destroyed in order to make room for the production of cattle (**Figure 27–27**).

However, the majority of the land being cleared is for crops and other uses. The United States imports very little meat from South and Central America. Land cleared for grazing in the United States is very small. Land statistics show that the amount of forest land today is only slightly less than it was in 1850.

Confinement operations produce a lot of manure. In natural biological processes, the manure produced in these operations gives off methane gas. Some environmentalists have suggested that this may be a contributing factor to the greenhouse effect. Research indicates that the amount of methane that these animals give off does not contribute to global warming. As a matter of fact, in some parts of the world, this methane gas has been collected and used as a fuel source.

FIGURE 27–27 Tropical rainforests are being destroyed to make room for agriculture. *© Rich Carey/Shutterstock.com.*

PUTTING IT INTO PRACTICE

Career Development Event in Parliamentary Procedure

Animal welfare has become an issue of debate in local, state, and national agricultural meetings around the world. To speak out and present your views requires the use of correct parliamentary procedure. All FFA members should learn how to conduct and participate in local meetings. Parliamentary procedure teaches the student an orderly method of conducting business that ensures that all sides of an issue are treated fairly. Most FFA chapters conduct a local Parliamentary Procedure Event in which six students make up a team. The team is given tasks to perform using correct parliamentary procedure. In preparing for the event, team members must learn proper use of the gavel, voting procedures, order of motions, and other requirements for properly conducting a meeting. Team members must individually answer questions covering basic parliamentary law and perform as a team. Parliamentary procedure is an effective method of addressing issues while respecting the vote of the majority and the rights of the minority. The chapter's winning team will have the opportunity to compete in area, state, and national FFA Parliamentary Procedure Events.

Much can be learned about current topics, such as animal welfare, by debating as part of a Parliamentary Procedure Team. *Courtesy of the National FFA Organization*

SUMMARY

Agriculture is an essential industry that will have to continue to expand as the population grows. Our modern society is not as knowledgeable about this industry as were past generations, when the vast majority of people grew up on a farm. In addition, science has equipped us with new insights into the food we eat and the environment surrounding us. Concern arises as people read reports outlining problems with the food supply and the environment. The agricultural industry has two very important responsibilities. First, we must make sure that we supply a plentiful, wholesome, and safe food supply for the population. Second, we must ensure that the practices we use are environmentally sound and that the world we live in can be preserved for generations to come.

CHAPTER REVIEW

Review Questions

1. Explain why consumers are more concerned with the quality and safety of food than they were 50 years ago.
2. What are consumer concerns over meat safety?
3. What is the Meat Inspection Act?
4. What is the difference between antemortem and postmortem inspection?
5. Why are animals given growth hormones?
6. What hormone level is allowed by the U.S. Food and Drug Administration in a person's daily intake of meat?
7. What role does cholesterol play in our bodies?
8. Why are consumers concerned about the intake of cholesterol?
9. What is BST, and what does it do?
10. List three areas of consumer concern over the environment.

Student Learning Activities

1. Talk to at least 10 people who buy meat. Make a list of the concerns these people have about the meat they buy. Share your list with the class.
2. Visit a slaughterhouse when carcasses are being inspected. Ask the inspector to explain what he or she looks for in a healthy carcass.
3. Formulate some ideas as to what could be changed about agricultural animals through genetic engineering. List the benefits of the changes and also list possible problems.
4. Visit a confinement operation and observe the manure-disposal system. List possible environmental problems that you observe. Also list the efforts made to protect the environment.
5. Research animal products that should be labeled as genetically modified. Compare your list with others in the class. Which are controversial?

CHAPTER 28

Careers in Animal Science

STUDENT OBJECTIVES

As a result of studying this chapter, you should be able to

- identify career options in animal science.
- compare specific jobs with required educational levels.
- identify ways to develop leadership skills.
- list some of the qualities employers look for in employees.
- explain and practice good interviewing skills.

KEY TERMS

career
post-secondary institution
master's degree
Ph.D.
community college
Supervised Agricultural Experience Program (SAEP)
National FFA Organization
associate's degree
Post-secondary Agriculture Students Organization
bachelor's degree
graduate degree

THINKING ABOUT YOUR FUTURE is an exciting activity. There are so many possibilities and opportunities that it can be daunting to try and narrow your choices (**Figure 28–1**). As you consider what you want to do for a **career**, first think of what you really enjoy doing and what interests you. You will spend many hours at work, and the more you enjoy your work, the happier you can be.

CAREER OPTIONS

The agricultural industry offers tremendous opportunities for a young person considering a career. Estimates are that each year, the agricultural industry will have 58,000 new job openings. This includes more than 200 different careers across many different areas. After studying the concepts in this text, you may decide that you want to pursue a career in one of the many branches of the animal industry.

Animal science is a broad and diverse industry that deals with the biological sciences. If you enjoy studying about and working with animals, you might think about a career in animal science. As with any other career, preparation for the occupation is essential. Hundreds of different jobs are available in the area of animal science that require many different types and levels of education and training. As with most other careers, salaries and working conditions usually are better in jobs that require more education.

You may wish to start work as soon as you graduate from high school. Or you may wish to attend a 2-year **post-secondary institution**. Or you may wish to major in animal science or a related area at a university. And you may wish to continue your education after graduating from a university. Many exciting careers dealing with animals can be obtained by attending graduate school and obtaining a **master's degree** or a **Ph.D.** in animal science. You could even become a research scientist who discovers new and better methods of growing agricultural animals (**Figure 28–2**).

No matter what level of career you wish to enter, you will have to prepare yourself with an education. Then, after you begin your career, you will continue to learn as you improve and advance in your chosen area. It is never too early to begin thinking about a career in animal science. Your parents, school administrator, agriculture teacher, and the faculty in an animal science department at a **community college** or university are all good sources of information as you choose a career. As with any other career, you need to carefully explore all aspects of the profession. As you investigate

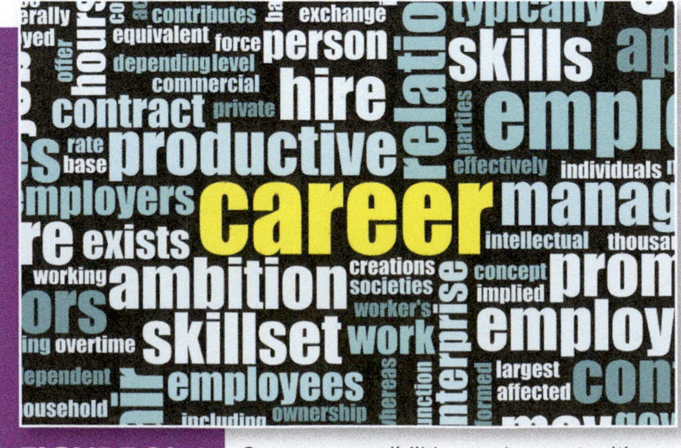

FIGURE 28–1 So many possibilities and opportunities are available that it can be daunting to try to narrow your choices. © kentoh/Shutterstock.com.

FIGURE 28–2 Your career could lead you to become a research scientist. © iStockphoto/Minerva Studio.

the different career options, keep asking yourself the following questions:

1. Is this something I really enjoy?
2. Can I make a living doing this?
3. What are the opportunities for employment?
4. What type of education, training, or other preparation is necessary to be employed in this area?
5. Am I willing to commit to this preparation?

Putting together a career plan will help you organize your thoughts.

MY CAREER PLAN	
What activities do I enjoy most?	
Is there a career in the animal industry that involves those activities that I enjoy? If yes, what are those careers?	
Do I prefer working in technical areas such as automotive, building trades, building trades, laboratory work, etc.? If so, which careers in the animal industry are related to these activities?	
Which of these careers meet my needs in terms of pay scale and expected standard of living?	
What are the educational requirements of my chosen career(s)?	
If applicable, what colleges offer a major that prepares you for that career?	
What are the academic requirements for that major? Am I able to meet those academic requirements?	

Keep in mind that most careers require some type of preparation. This may include college (2-year, 4-year or graduate school), trade school, internship, on the job training or other types of preparation. To achieve career goals, you must be willing to undergo the preparation for the job as well as understand the learning that may be required even after you land the job.

If you choose to go to work immediately after graduating from high school, several jobs are available to you in the area of animal science. You should take all of the courses in agriculture, agriscience, and agribusiness that your school offers. In these courses you will learn the basics of how plants and animals live, grow, and reproduce. In addition, you will learn the essentials of properly caring for animals, as well as the responsibility involved in caring for animals. You will have opportunities to learn skills in actual job situations through a **Supervised Agricultural Experience Program (SAEP)** (**Figure 28–3**).

Also, you can obtain leadership and personal development skills through the **National FFA Organization**. You can participate in activities such as livestock-judging contests, including dairy products judging, poultry judging, livestock and showmanship shows, and proficiency awards. Most of these activities are available to you whether you live on a farm or in the center of a city.

Some of the jobs that can be secured with a high school diploma follow:

herdsman	chick grader
small animal producer (**Figure 28–4**)	egg candler
feed mill worker	slaughterhouse worker
milking machine operator (**Figure 28–5**)	milk hauler
sheep shearer	poultry processing plant worker
groomer	

FIGURE 28–3 You will have the opportunity to learn actual job skills through your Supervised Agricultural Experience Program (SAEP). *© SpeedKingz/Shutterstock.com.*

FIGURE 28–4 Producers of small animals raise animals for laboratory use. © iStockphoto/fotografixx.

FIGURE 28–5 A milking machine operator carefully prepares the udder and places milking cups on the teats. © aaabbbccc/Shutterstock.com.

Careers with an Associate's Degree

An **associate's degree** is a 2-year degree from a community college or a 2-year institution. Many programs teach animal science in community colleges all across the country. If you enroll in one of these programs, you will study practical courses in the fundamentals of producing and caring for animals. You will also study the sciences of chemistry, biology, and zoology, as well as math and English. Credits for many of the courses taken at a community college may be transferred to a university if you later decide to continue your education.

If you enjoyed livestock judging in high school, you may want to continue your interest with competitive livestock evaluation at the community college level. You also may become involved in student organizations such as the **Post-secondary Agriculture Students Organization** that participate in many activities involved with animal science.

Listed below are some jobs that are available with an associate's degree:

veterinarian assistant	wool grader
meat cutter (Figure 28–6)	farrier (Figure 28–7)
computer operator	producer
embryo implant technician	animal buyer
poultry vaccinator	artificial insemination technician

Careers with a Bachelor's Degree

A **bachelor's degree** requires 4 years of education at a college or university. There is a broad array of choices in majors that work with animals.

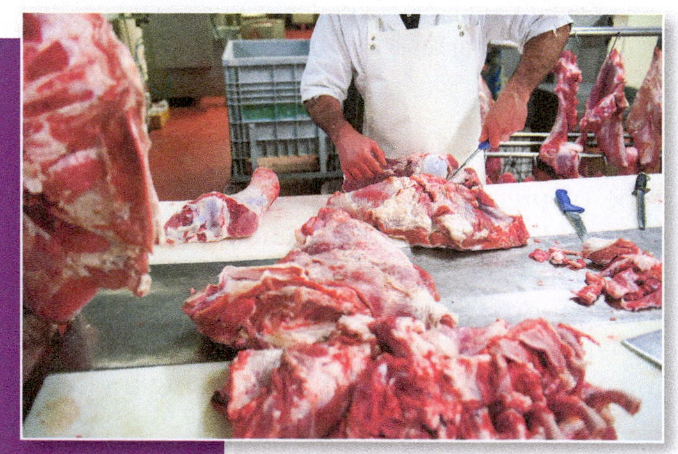

FIGURE 28–6 Meat cutters break carcasses into wholesale cuts and then divide these cuts into retail cuts ready for sale at the grocery market. © Photology1971/Shutterstock.com.

FIGURE 28–7 Farriers trim and care for horses' hooves as well as fitting shoes. *© Kevin Day/Shutterstock.com.*

These include animal science, dairy science, poultry science, and agricultural education. You will study courses in science such as chemistry, biology, and zoology. At a large number of universities, the departments that teach these courses are housed in the College of Agriculture. In animal science, you will study animal anatomy, nutrition, animal growth and development, and other courses dealing with how animals live, grow, and reproduce (**Figure 28–8**).

By taking courses in education, you can qualify to teach agriculture at a high school. You can also participate in such competitive events as livestock evaluation and meat evaluation. Student organizations include Block and Bridle, Collegiate FFA, and Collegiate 4-H. Careers requiring a bachelor's degree include:

farm or ranch manager	field service technician
meat grader	extension agent
company representatives for animal feed and health products	hatchery manager
Producer	agricultural journalist
high school agriculture teacher (**Figure 28–9**)	dairy inspector

Careers with a Graduate Degree

Once you complete your bachelor's degree, you may want to continue your education and get a master's degree and then a Ph.D. degree. With a **graduate degree**, you will be able to conduct scientific research or continue your education to earn a degree in veterinarian medicine (**Figure 28–10**).

FIGURE 28–8 In animal science, you will take many different courses dealing with agricultural animals. *© iStockphoto/BartCo.*

FIGURE 28–9 High school agriculture teachers instruct students in the basics about animals. *Courtesy of the National FFA Organization*

You will choose a specific area of animal science in which to concentrate your studies, and most of

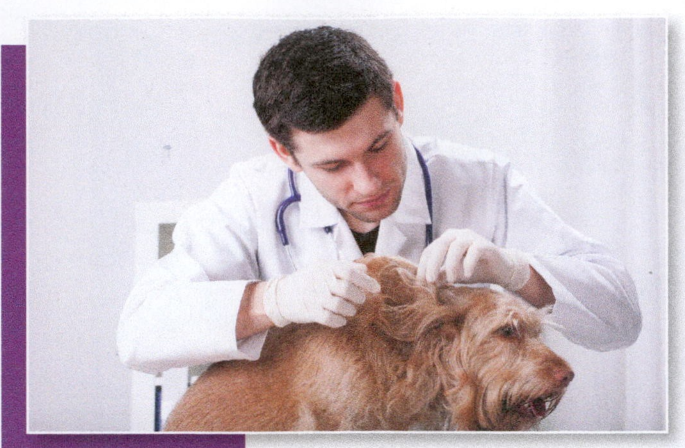

FIGURE 28–10 Graduate studies can lead to a degree in veterinarian medicine. © Photographee.eu/Shutterstock.com.

your course work will be in that area. For example, you might want to study in the area of nutrition or animal reproduction. Your course work will include study in statistics and research methodology so you will be able to understand, design, and conduct scientific research. These degrees require good grades in college and a determination to study hard to reach your career objective.

Examples of jobs requiring an advanced degree such as a Ph.D. or a Doctorate of Veterinary Medicine are the following:

veterinarian	reproductive physiologist
meat inspector (**Figure 28–11**)	microbiologist
animal geneticist	research scientist
animal nutritionist	college or university professor

DEVELOPING PERSONAL AND LEADERSHIP SKILLS

Merely having the proper education and/or technical skills will not guarantee that you will be successfully employed. Employers want balanced, well-adjusted employees who can work well with others and grow professionally. Throughout your study of agricultural education, you will have numerous opportunities to develop leadership skills and life skills

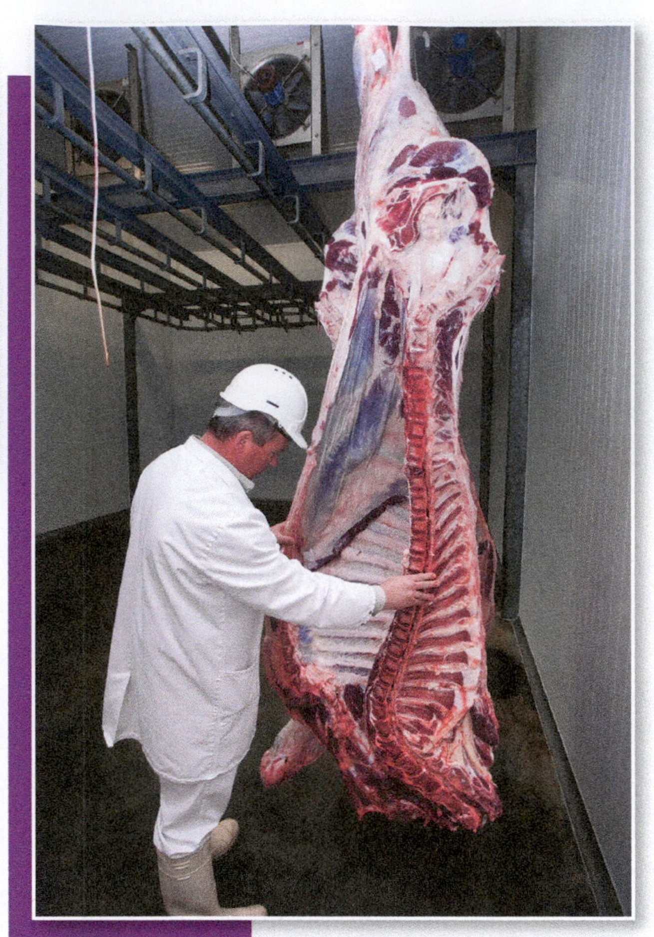

FIGURE 28–11 A veterinarian also may be a meat inspector for the USDA. © iStockphoto/alle12.

(**Figure 28–12**). These are extremely important abilities not only in leading a productive work life but also in leading a satisfying personal life. Employers want people who can successfully communicate and work with other people. They want employees who can move up in the company and take on leadership roles that will include supervising other people. Through the programs offered through the FFA Organization and Supervised Agricultural Experience Programs (SAEP), you will have the opportunity to learn and practice competencies that will help you develop as a leader. Listed below are some of the ways you can develop leadership skills that will be valuable in any career:

- Study the qualities of good leaders and learn from their mistakes. Listen, but do not imitate. Remember that you have your own personality and abilities, and you need to develop your own persona.

FIGURE 28–12 Throughout your study of agricultural education, you will have numerous opportunities to develop skills in leadership as well as life skills. *Courtesy of the National FFA Organization*

- Analyze yourself, determining your weak and strong points, and set goals for improvement. This may make you uncomfortable, but remember that in order to grow professionally you must sometimes get out of your comfort zone.
- Learn how to take directions. Good leaders listen to other people and determine steps that need to be taken in any endeavor. Employers do not want anyone working for them who doesn't listen, follow directions, and communicate the directions to other people. Learn to be a good listener and practice remembering steps and following through on both written and verbal directions.
- Learn about groups in general and how they function. Identify the types of people in a group. Remember that people have different personalities and socialization styles. An effective leader gets people to work together. A good way to learn this skill is to work on committees in your FFA chapter (**Figure 28–13**).
- Make and follow a plan to develop personal leadership skills. Set goals for yourself, and make specific plans to achieve each goal, and then stick to the plan. This should be an ongoing activity throughout your career. Even top-level executives in large corporations are constantly looking for ways to improve themselves. That is how they got to be top-level executives!

FIGURE 28–13 An effective leader gets people to work together. A good way to learn this skill is to work on committees in your FFA chapter. *© racom/Shutterstock.com.*

INTERVIEW PREPARATION

The interview process is the most important part of obtaining employment. It gives the employer a chance to get to know you and to draw an opinion as to your abilities, qualifications, and suitability for the job. Also, you are given an opportunity to find out more information about the potential employment. Keep in mind that you will have only one opportunity to make a first impression, and first impressions are usually the most important (**Figure 28–14**).

Following are several suggestions that should help you prepare for and carry out the interview:

- **Be neat and well groomed.** Portray the correct image by what you wear. Clean, neat, well-coordinated clothes convey the message that you care about your personal appearance. Don't forget to polish your shoes, and wear neat, clean socks. Good grooming, such as neat and clean hair and fingernails, conveys the image of a person who cares about your health as well as your appearance. Employers want workers who take care of themselves and stay healthy. This cuts down on sick days and loss of work productivity.

FIGURE 28–14 An interview is not the place to make a fashion statement. © PanicAttack/Shutterstock.com.

FIGURE 28–15 Go to the interview alone. The employer wants to know about you—not your friends or relatives. *© Monkey Business Images/Shutterstock.com.*

- **Dress appropriately for the type of job you apply for.** For example, if you are interviewing for a job potting plants in a greenhouse, a suit and tie would be inappropriate. On the other hand, if you are interviewing for a position where you will constantly be meeting and dealing with people, a coat and tie would be appropriate.
- **Go to the interview alone.** Remember that the employer wants to know about you—not your friends or relatives (**Figure 28–15**). While it may be comforting to have a friend or relative with you, it is distracting to the process and gives the impression that you are not self-confident and do not function well on your own.
- **Be on time.** Keep in mind that all employers want workers who give a full day's work. Also, they want employees who can be depended upon to be punctual and dependable. Tardiness to an interview tells an employer that you are careless about time. Being late could cost you the opportunity to get the job.
- **Use good manners.** No matter what job you are interviewing for, you most likely will have to deal with people. Employers do not want workers who are insensitive to the feelings of others. Good manners are a reflection of how well you regard the rights and opinions of others (**Figure 28–16**). If you are unsure about your mannerisms, go to the library and check out a book on proper etiquette. Practice the principles outlined in the book.
- **Turn off your cell phone.** Make sure you don't bring your cell phone with you, or at least make sure it is turned off. Nothing is more disruptive than having a cell phone ring during a discussion.
- **Pay attention to the interviewer.** To seriously consider you for the job, the employer has to be convinced that you really want the job. If your mind seems to wander or you don't pay attention to the questions asked, you will give the impression that you really are not interested in the job or that you lack focus.
- **Ask appropriate questions about the job.** Remember that you also will be making a decision as to whether you wish to work for the employer. Ask questions about the job, such as duties required, working

FIGURE 28–16 Good manners are a reflection of how well you regard the rights and opinions of others. *© George Rudy/Shutterstock.com.*

conditions, and opportunities for training and advancement (**Figure 28–17**). Avoid questions regarding salary or pay. Ask these questions after you receive a job offer. If the first question you ask is about pay, the employer may get the idea that all you are interested in is the pay.

- **Have necessary personal reference information with you.** If you have not previously supplied letters of reference, take these with you to the interview. If appropriate, include examples of your work. Anticipate materials the employer may want to see, and take them with you (**Figure 28–18**). Employers like to hire people who are organized and think ahead.
- **Answer the questions openly, completely, and honestly.** Don't try to "pad" your qualifications or tell the employer that you have skills you do not possess. Most employers place a lot of importance on honesty and will be reluctant to hire anyone they think is not being completely honest in answering the questions. Also, be careful not to talk too much. Simply answer the questions in a concise, to-the-point manner.
- **Follow up after the interview.** When the interview is completed, thank the interviewer and leave promptly. After you get home, take the time to write a thank-you letter to the employer, expressing your appreciation for the opportunity to interview for the job. Make the letter neat, professional, brief, and to the point.

FIGURE 28–17 Ask questions about the job, such as duties, working conditions, and opportunities for training and advancement. *© George Rudy/Shutterstock.com.*

FIGURE 28–18 Anticipate materials that the employer may want to see, and take them with you to the interview. *© ESB Professional/Shutterstock.com.*

PUTTING IT INTO PRACTICE

Agriscience Fair

The FFA offers students many opportunities to develop and display projects. The National FFA Agriscience Fair recognizes students who are studying the application of scientific principles and emerging technologies in agriculture. Much like the International Science and Engineering Fair, the Agriscience Fair reflects an agricultural theme. Participation begins at the local chapter level and advances to state and national levels. Competition is open to all FFA members in grades 7–12.

One of the major goals of the Agriscience Fair is to provide students with an opportunity to use the scientific process. They can demonstrate and display agriscience projects while reinforcing skills and principles learned in agriscience courses. Award winners are also recognized for their efforts. Certificates, medals, ribbons, plaques, and even scholarships and cash awards may be given to division winners in each category.

There are five categories for the National FFA Agriscience Fair: Biochemistry/ Microbiology/ Food Science, Environmental Sciences, Zoology (Animal Science), Botany (Plant/Soil Science), and Engineering (Mechanical/Agriculture Engineering Science). Each member or team of two members may enter only one project. The participants are required to meet with the judges to explain their project and will be judged on how well the scientific method was followed, the detail and accuracy of the logbook and project report, and whether tools/equipment were used in the best possible way. Other categories that are scored include knowledge gained, information, thoroughness, conclusions, interview, and visual display. For more information concerning the National Agriscience Fair, contact your agriculture teacher or visit http://www.ffa.org.

The FFA Agriscience Fair provides an opportunity for students to put the scientific method into practice.
Courtesy of the National FFA Organization

SUMMARY

Choosing a career is one of the most important decisions you will ever make. A lot of thought, planning, and preparation should go into choosing a career. Decide what level of education you are comfortable in pursuing and work toward completing your education. Remember that the higher-level jobs require more education and training, but the jobs can often be more enjoyable. As you develop technical skills, also develop good work habits, as well as personal and leadership skills. Almost all careers require a combination of all these different types of skills. There is a broad array of jobs in animal science, and one could be right for you.

CHAPTER REVIEW

Review Questions

1. What are the most important things to consider in choosing a career?
2. List some jobs you can enter with a high school diploma.
3. List some jobs in animal science that require a bachelor's degree.
4. What student organizations can you be involved with in college?
5. Describe the personal and leadership skills that employers want in their employees.
6. Why is it important to learn to take directions?
7. How do you determine how to dress for an interview?
8. Why is good grooming important in an interview?
9. Why are good manners important job skills?
10. How should you follow up an interview?

Student Learning Activities

1. Make an inventory of all the activities you enjoy doing, and list career options that might allow you to participate in these activities. If you cannot find a career that fits, perhaps you are unrealistic in your expectations. Share the list with the class and ask for ideas.
2. Check the employment ads in several newspapers. List those that relate to animal science.
3. Locate an animal science or agricultural education department at a university in your state. This can be done through an Internet search. Ask them to send information about programs at the university.
4. Interview someone who works in the animal industry. Ask him or her what they do on the job. Also ask what education or preparation was required for the job.
5. With your teacher acting as a potential employer, go through a mock interview. Ask for feedback on how you may improve.

APPENDIX A

Supervised Agricultural Experience

Supervised Agricultural Experience (SAE) is an integral part of a total program of agricultural education. It is where you plan, propose, conduct, document, and evaluate your programmatic experiential learning activities. It is where you apply and test what you have learned in your classes and FFA experiences. An SAE will help you become well versed in record keeping for your own portfolio. It will also aid in your personal leadership development and in the strategic planning process for teams, groups, organizations, and programs.

TERMS TO KNOW

- Degree program
- Portfolio
- Supervised Agricultural Experience (SAE)
- Scientific process
- Ownership/Entrepreneurship SAE
- Placement/Internship
- Research SAE
- Foundational SAE
- School-Based Enterprise SAE
- Service Learning SAE
- Improvement Project
- Proficiency awards

OBJECTIVES

After completing this summary, the student will be able to:

- define SAE
- list and explain the types of SAE
- plan an SAE program
- propose an SAE program
- conduct an SAE program
- document an SAE program
- evaluate an SAE program
- apply proper record-keeping skills
- participate in youth leadership opportunities to create a well-rounded SAE
- produce a local program of activities using a strategic planning process
- participate in a local program of activities using a strategic planning process

Supervised Agricultural Experience (SAE) is a program of experiential learning activities conducted outside of the regular agricultural education class time. The student-led, instructor-supervised, work-based learning experience is designed to help students develop and apply the knowledge and skills learned in an agricultural education classroom and/or laboratories.[1] The SAE used to be called the Supervised Occupational Experience Program (SOEP), and the focus was on the occupation. These days, the focus is on entrepreneurship and experiences rather than the occupation or the actual career.

The SAE takes place under the direction of your agricultural education teacher. Many times, individual projects (e.g., showing pigs, raising a plant in the greenhouse, etc.) are considered to be SAE, but the intent of a quality SAE is for it to be conducted as a series of career-related experiences completed during your enrollment in agricultural education.

SAE OPTIONS

SAE projects can be completed in a variety of categories, always beginning with the Foundational SAE, and then transitioning to usually one of the following based on career plans and interests:

Ownership/Entrepreneurship

Placement/Internship

Research: Experimental, Analysis or Invention

School-Based Enterprise, and

Service Learning.

Foundational SAE—Get It Started

The career-related Foundational SAE is where an SAE begins. "Every Foundational SAE will provide experiences in five components as a graded part of each of your agricultural education courses." The Foundational SAE results in the development of a plan to begin an SAE. Agricultural education instructors assist students with the plan and working through concepts and activities to achieve the five components of a Foundational SAE.

Five Components:

1. Career exploration and planning according to your interests
2. Employability (leadership) skills for college and career readiness
3. Financial management and planning
4. Workplace safety
5. Agricultural literacy[2]

Ownership/Entrepreneurship—Work for Self

In the **Ownership/Entrepreneurship SAE**, you plan, implement, operate, and assume all or some of the financial risk in a productive or service activity or agriculture, food, or natural resources–related business.[3] If you have this type of SAE, you own the materials and other inputs, and you keep financial records to determine return on investment. Some examples of SAE experiential learning activities that can make up an SAE program are listed below. A much more comprehensive list can be found at the end of this unit.

Examples:

1. Growing grapes
2. Raising bees
3. Growing an acre of corn

Placement/Internship—Work for Someone Else

If you work for someone else on a farm or ranch, in an agricultural business, or in a verified nonprofit organization providing a "learning-by-doing" environment, your SAE qualifies for the **Placement/Internship SAE** type. This type of experience may be paid or non-paid.[4] Not everyone reading this can have an ownership/entrepreneurship SAE, but almost everyone has the opportunity to seek out a paid or unpaid "working for someone else" experience for the purpose of learning and applying your agricultural knowledge and skills. Some examples are listed below, but many more examples are provided at the end of this unit.

Examples:

1. Working in a flower shop
2. Working on Saturdays at a local stable
3. Working at a grocery store

Research: Experimental, Analysis, or Invention—Solve a Problem

The agriculture industry is becoming a highly scientific field full of problems to be solved by you and your classmates. Problems range from world hunger to climate change and even social ideals and political policies related to the agriculture industry. A **Research SAE** helps you prepare for a long and productive career solving these important issues for our society. The Research SAE involves a program of extensive activities where you plan and conduct experiments or other forms of scientific evaluation using the scientific process. There are three types of Research SAEs: experimental, analytical, and invention.[5]

Experimental Research SAE. According to the National Council for Agricultural Education, the Experimental Research SAE involves an extensive activity where the student plans and conducts a major agricultural experiment using the scientific process. The scientific process is a way of answering scientific questions through observations, making assumptions, experimenting, and drawing conclusions based on your analysis. The purpose of the experiment is to provide students firsthand experience in verifying, learning, or demonstrating scientific principles in agriculture, discovering new knowledge, and using the scientific process. In an Experimental SAE, there is a hypothesis and a control group, and variables are manipulated.[6]

Analytical Research SAE. In an Analytical Research SAE, you choose a real-world agriculture, food, or natural resource–related problem that is not amenable to experimentation and design a plan to investigate and analyze the problem. You will gather and evaluate data from a variety of sources and then produce some type of finished product. The product may include a marketing display or marketing plan for a commodity, product, or service; a series of newspaper articles; a land-use plan for a farm; a detailed landscape design for a community facility; an advertising campaign for an agribusiness, and so forth. A student-led Analytical SAE is flexible enough so that it could be used in any type of agricultural class, provides valuable experience, and contributes to the development of critical thinking skills.[7]

Invention Research SAE. In the Invention Research SAE, you identify a need in an agriculture, food, or natural resource–related industry and perform research and analysis in order to solve a problem or increase efficiency by developing/adapting a new product or service to the industry. The student plans, documents, and develops his/her innovation through the iterative processes of design, prototyping, and testing with the goal of creating a marketable product or service.

Examples:

1. Comparing the effect of various plant food on plant growth
2. Demonstrating the impact of different levels of soil acidity on plant growth
3. Determining if different rations of feed improve growth rate of pigs

School-Based Enterprise SAE—Manage It at School

The School-Based Enterprise SAE is an entrepreneurial operation managed by you. Your operation, however, takes place in a school setting that provides not only facilities, but also goods and services that meet the needs of an identified market.[8] To give you the most educational value, this type of SAE should replicate the real world of work as much as possible. This type of SAE is usually cooperative in nature, and management decisions are made by you in "cooperation" with your teacher. Activities in this type of SAE "may include, but are not limited to: cooperative livestock raising, school gardens and land labs, production greenhouses, school based agricultural research, agricultural equipment fabrication, equipment maintenance services, or a school store."[9]

Service Learning SAE—Plan a Service Project

Service learning is one of the highest forms of leadership. Our goal should always be to leave things better than we found them. The goal of the Service Learning SAE is to make the community in which you live better. This SAE type combines community service activities with structured reflection.

As part of this type of SAE, you can become involved in the development of a needs assessment, planning goals, creating objectives and budgets, implementing the activity, and promoting and evaluating a chosen project. The project could be in support of a school, community organization, religious institution, or nonprofit organization. You would be responsible for raising necessary funds for the project (if funds are needed). The project must stand by itself and not be part of an ongoing chapter project or community fundraiser. The project must be somewhat challenging and require the awesome leadership you possess.[10]

Examples:

1. Test water wells in your community for contamination
2. Design a web page for your FFA chapter or for a local organization
3. Design and install a landscape plan for a church

PLAN AN SAE PROGRAM

Planning is an important part of any team, business, project, or program. Your SAE program isn't any different. To properly plan, you will need to understand the steps in planning, as well as the guidelines for planning, the SAE.

Steps in Planning the SAE

Step One Identify your career interests in agriculture. Your SAE program, experiential learning activities, and the career you choose someday need to be something about which you can get excited. If you like the SAE or your future career choice, you are more likely to stick with it and be successful.

Step Two Review the job responsibilities of career interest areas you may have. You might like the idea of a particular project or career, but if you don't like specific tasks involved, you might need to choose another project or program. For example, if you like the idea of a Placement SAE as a veterinarian technical assistant, but you don't want to clean out cages, you may have made a poor choice.

Step Three Identify SAE programs of interest by interviewing friends who have an SAE or by viewing suggestions found in various resources. Suggestions for certain experiential learning activities or projects are at the end of this unit.

Step Four Develop a timeline for your SAE program. In other words, which projects will happen first, second, third, and so on? What is the completion date for each activity? Electronic portfolios, record-keeping systems, or simple calendar systems can be used for this step.

Step Five Building on step four, build a long-range plan for the SAE program. Remember, projects or activities happen in a shorter time span, but an SAE program, which is necessary to be competitive for FFA proficiency and Star awards, happens over a longer frame and grows in scope and diversity.

Step Six Develop the first-year (annual) plan. You have to actually start now. After you know the long-range plan and the timeline for the different experiential learning activities/projects, you are ready to put the first-year plan together. Again, think about completion dates and specific strategies for reaching those goals.

Step Seven Re-plan on a regular basis. Part of good planning is reviewing your activities in light of your plan and then adjusting your plan. This should happen fairly often, but not so often that you are always planning instead of completing tasks.[11]

Guidelines to Planning SAE

The steps outlined above should get you moving in the right direction, but the following guidelines will help ensure a successful SAE.

1. Plan for year-round experiences.
2. Include ownership and/or placement projects either at home, school, or in the community.
3. Identify a number of improvement projects and supplementary skills for the year.
4. Develop a budget.
5. Plan ownership projects with some form of profit.
6. Explore different locations to gain desired experiences.
7. Discuss the SAE program with parents/guardians.
8. Plan the scope of the program to earn enough profit in order to qualify for advanced FFA degrees.
9. Provide for a variety of activities and experiences.
10. Increase the SAE scope annually.
11. Choose SAE program experiences that relate to career interest areas.[12]

Parts of an Annual SAE Plan

A yearlong SAE program plan will keep you on track, and it will help you make decisions about your SAE. The annual plan consists of a calendar, description of projects, budget, improvement projects, and supplementary skills. For entrepreneurship/ownership projects, prepare a description of the size/scope of your enterprise, the location of your projects, the nature of the business or enterprise, partners involved, methods of marketing, facilities needed, and months involved in the project. For placement projects, detail the location, beginning and ending dates, and pay schedule. It is also important to include a budget and to keep up with income and receipts as the year progresses. You will also want to include a description of improvement projects you wish to complete and supplemental skills you would like to develop. An **improvement project** is an activity that improves the appearance, convenience, efficiency, safety, or value of a home, farm, ranch, agribusiness, or other agriculture facility. Your annual plan should include specific activities of the improvement project as well as hours of labor committed and an estimate of costs.[13]

Propose a SAE Program

Following your SAE plan, it is a good practice to write a statement justifying your plans with a proposal. This proposal should explain why you chose the SAE you did as well as discuss why your plan is laid out the way it is. Your proposal should reiterate the importance of completion dates on your timeline. The SAE proposal should also detail specific agricultural or leadership concepts that you hope to achieve as a result of your SAE program.

CONDUCT AN SAE PROGRAM

The first step to conducting a quality SAE program is getting started. All of the planning and decision-making can take time, but it is worth it. Once you've started, you are going to have a great time, but you have some work ahead of you as well. Specifically, you will need to become skilled at documenting your SAE program as well as evaluating it.

Document an SAE Program

Documenting your SAE is important because it gives you an opportunity to see the planning come to life. Documenting your SAE makes it a permanent record from which you can learn and improve your project. It shows the problems encountered and solved, which apply to future activities and provide you with confidence to continue in your SAE.

Calendar. Using a calendar can help with more than planning. You can use a calendar, such as the one on your smartphone, to document hours invested, skills learned, and even tasks performed.

Journal or Portfolio. An alternative to using a simple calendar system is to document your SAE with a journal or portfolio, paper or electronic. A paper system could be kept using a notebook or ledger, but these days an electronic system is much easier to use. Many states have an online journal or portfolio that allows you to keep up with important documentation.

A **journal** is "a daily written record of (usually personal) experiences and observations, or a ledger in which transactions have been recorded as they occurred."[14] A **portfolio**, usually kept online, houses all of your work (skills, projects, activities, service, finances, etc.) compiled over a period of time. Like many other tools discussed in this chapter, it is helpful for improving your project. It is also advisable to utilize your portfolio for record-keeping as well.

APPLY PROPER RECORDKEEPING SKILLS

Record-keeping is the process of keeping a journal or portfolio of everything you have done. As stated previously, you will need to make notes in your SAE whenever you do or learn something new, and you need to document time and money spent on your activities, projects, and program. The skill of record-keeping will serve you well in your career.[15]

At its core, record-keeping is quite simple. For every project, you will need to capture the following for all activities or items: the date, the name or description, the hours you spent, and expenses or income that resulted. Be sure to add up the hours you've invested in all activities. Time is money, and each one of your experiences represents new knowledge and skills you've developed. You will also add up expenses and income to determine your profit and loss (P&L). It's really that simple. You can use state-endorsed, preprogrammed spreadsheets or websites to determine your P&L and to track your hours invested or keep records on paper. It's up to your personal preference, available resources, and the wishes of your advisor or business partner.

Learning to keep records for and through your SAE has many benefits. Keeping accurate records can help you determine if you made or lost money. It can help you keep others from cheating you out of what you have earned. Your knowledge and experience in record-keeping will help you determine which parts of the business are doing well and which parts are not. Becoming proficient at record-keeping will also help you make management decisions, document your net worth for loans, prepare tax returns, and plan for future events. Being a good record keeper for your SAE will also help document your activities for FFA recognitions and degree purposes. Good records can also protect you legally and help you plan a budget for the following year.[16]

Evaluate an SAE Program

When your teacher/adviser comes to visit you and your SAE, he/she is there to answer any questions you may have. The teacher is also there to help you evaluate your SAE. The evaluation looks different for different types of SAEs. Lesson 8 from the SAE Handbook resource available online (http://harvest.cals.ncsu.edu/site/WebFile/IIB8.pdf) breaks down specific components of SAE evaluation. They are listed here:

Entrepreneurship/Ownership:

1. Accuracy of records
2. Neatness of records
3. Dates of records
4. Net income (total income minus total expenses)
5. Are good management practices being used?
6. Efficiency factors (yield per acre, number of offspring raised, etc.)
7. Improvements made since the last observation

8. Cleanliness of facilities
9. Customer satisfaction
10. What skills were learned

Placement/Internship:

1. Satisfaction of the employer
2. Number of hours worked
3. Accuracy of records
4. Neatness of records
5. Currency of records
6. Does student report to work on time?
7. Ability to get along with the other workers
8. Ability to get along with customers
9. Skill in performing the expected tasks
10. Attitude of the student

Research:

1. Was the scientific process used?
2. How well was the job done?
3. What was learned?
4. What was the reaction of the beneficiaries of the service-learning project?
5. Were records kept on the activity on the activity, and what shape are they in?

Exploratory:

1. How many different activities were conducted?
2. How many hours were involved?
3. What was learned?
4. Accuracy of records
5. Neatness of records
6. Dates of records

School-Based Enterprise:

1. How many hours were worked?
2. What was the quality of the work done?
3. What was the scope or size of the activity?
4. Accuracy of records
5. Neatness of records
6. Dates of records

Service Learning:

1. How many hours were involved?
2. How well was the job done?
3. What was learned?
4. What was the reaction of the beneficiaries of the service-learning project?
5. Were records kept on the activity, and what shape are they in?[17]

ENHANCING SAE WITH OTHER OPPORTUNITIES

SAE in not only financial record-keeping, but also a record of skills, knowledge, credentials, certifications, experiences, career planning, reflection, and leadership development.[18] A solid SAE will provide you with the opportunity to receive FFA degrees and compete for different proficiency awards.

The FFA **degree program** is FFA's primary recognition program. SAE helps students apply for each degree, and every degree has certain SAE requirements. There are five degrees of FFA membership as follows: Discovery, Greenhand, Chapter, State, and American. These five degree areas recognize you for your overall participation in FFA. Everything you do in FFA, when combined, helps you move toward a degree.[19]

Proficiency awards are awards that recognize students' excellence in SAE.[20] Developing your SAE into a proficiency award is a time-consuming, but rewarding task. You should apply for the proficiency award in the specific area and career pathway for which you are strongest and have the most experience. For instance, if you have worked for a livestock producer for three years and only raised goats for one year, it would be best to apply in the placement area for Animal Systems rather than Entrepreneurship for the same system. The career pathways for your SAE and proficiency awards follow.

1. Agribusiness Systems—the study of business principles, including management, marketing, and finance, and their application to enterprises in agriculture, food and natural resources.
2. Animal Systems—the study of animal systems, including life processes, health, nutrition, genetics, management, and processing, through the study of small animals, aquaculture, livestock, dairy, horses, and/or poultry.
3. Biotechnology Systems—the study of data and techniques of applied science for the solution of problems concerning living organisms.
4. Cluster Skills—the student will demonstrate competence in the application of leadership, personal growth, and career success skills necessary for a chosen profession while effectively contributing to society.
5. Environmental Service Systems—the study of systems, instruments, and technology used in waste management and their influence on the environment.
6. Food Products and Processing Systems—the study of product development, quality assurance, food safety, production, sales and service, regulation and compliance, and food service within the food science industry.
7. Natural Resources Systems—the study of the management of soil, water, wildlife, forests, and air as natural resources.
8. Plant Systems—the study of plant life cycles, classifications, functions, and practices, through the study of crops, turf grass, trees and shrubs, and/or ornamental plants.
9. Power, Structure, and Technical Systems—the study of agricultural equipment, power systems, alternative fuel sources, and precision technology, as well as woodworking, metalworking, welding, and project planning for agricultural structures.[21]

SUMMARY

Supervised Agricultural Experience (SAE) is a program of experiential learning activities conducted outside of the regular agricultural education class time. There are different types of SAEs and related opportunities and rewards programs, so every student in agricultural education and FFA can learn and develop from the authentic experience. Record-keeping and documentation of the valuable experiences are very important and will serve you and your future career well.

EXAMPLES OF PROJECTS FOR YOUR SAE

Agricultural Mechanics Project Examples

- Build a patio for the home
- Build frames for raised beds for gardeners
- Build handicap ramps in local community
- Build picnic tables/sell to schools and local community

- Construct a utility building
- Construct a hydro ram pump and calculate the efficiency and water delivery rate
- Construct a wind-powered generator and show its applications to agriculture
- Construct and sell birdhouses and feeders
- Construct and sell lawn furniture made of PVC
- Construct compost bins to sell
- Construct concrete projects for the home or farm
- Construct or recondition a welding project (such as a trailer, cooker, etc.) at home or in school-provided facilities
- Construct prefabricated wooden fence panels to sell to local hardware and building supply stores
- Construct spray rigs for four-wheelers
- Construct and market woodworking projects (birdhouses, dog houses, etc.)
- Construct metal projects
- Contract with local EMCs or power companies to remove bolts, wire, and so forth from old power poles (sell copper for recycling)
- Contract with a school system to maintain and service lawn care equipment
- Cut out and paint lawn figures to sell
- Start an electrical repair service
- Install plumbing fixtures or a plumbing system in your own building
- Start a lawn mower maintenance service
- Make craft items from wood, metal, or concrete to sell at arts and crafts shows
- Make personalized signs to sell
- Paint the home, supervised by an agricultural education teacher
- Find placement in a parts store
- Provide a poultry house maintenance preparation business
- Provide custom-painted mailboxes and stands
- Repair and rebuild damaged pallets for businesses
- Start a chainsaw basic-maintenance and service business
- Start a custom-vehicle refurbishing or painting business
- Start a detailing business for cleaning farm equipment on the farm (wash, wax, clean, maintain)
- Start an equipment locating business and match folks with something for sale with folks who want to buy something
- Start a farm equipment tire disposal business (turn old tires into livestock feeders)
- Start a farm fence maintenance business (cleaning fencerows, repairing)
- Start a farm-fencing company for custom work
- Start a pallet manufacturing business
- Start a small-engine repair service
- Wire a home shop, utility room, barn, or treehouse
- Work as an agricultural mechanic's aide
- Work at a welding operation
- Work at a building supply business
- Work with a farm equipment dealer
- Start an equipment-trailer fabrication business

Agribusiness Sales and Service Project Examples

- Become an agricultural news consultant for local radio or newspapers
- Conduct a study of commodity trading over a period of time
- Conduct general home maintenance
- Contract with the local Chamber of Commerce to conduct county tours for prospective businesses
- Create a custom labor venture: mow pastures, remove undesirable weeds from crops, paint outbuildings, and so forth

- Design a computer application plan for an agricultural facility or program
- Develop a marketing plan for an agricultural commodity
- Fry pork rinds for local stores
- Install electrical circuits or a wiring system at home
- Find a job placement in food distribution, restaurant, and so forth
- Find a job placement with a local florist
- Job-shadow agribusiness professionals, visit agribusinesses to interview personnel, take educational tours, and so on
- Offer a custom-parts or supplies delivery business to farms in your county
- Pre-sell fresh meat to clients on a weekly basis
- Pre-sell fresh seafood to clients on a weekly basis
- Pre-sell fresh vegetables in family portions delivered weekly
- Preserve food for home use
- Process creamed corn in a food-processing facility
- Provide a co-op program for an agribusiness
- Provide a custom barbecue service for the community
- Provide custom feed for livestock (tap the organic, all-natural, no-chemical market)
- Provide a hand-weeding crew for local peanut/vegetable farmers
- Provide a sausage-making business at home; can be sold if regulations are met
- Provide custom hay baling and/or hauling
- Provide a farm sign business (manufacture, sell, install, and maintain)
- Provide livestock hauling
- Provide small-engine maintenance and repair service
- Provide systematic maintenance and service on outdoor power equipment at home or at school-provided facilities
- Purchase and resell aerial photographs from the tax office to local landowners
- Package fresh fruit or vegetable gift packs
- Remove pesticide jugs monthly from farms and transport to landfill
- Sell ready-to-freeze processed vegetables
- Start a composting business by buying cow manure from local farmers, and bag for resale
- Start a farm-sitting business for vacationing farmers
- Start a kerosene route for homeowners (probably little demand in the summer time)
- Start a material safety data sheet compliance business by compiling and maintaining current sheets for farms and businesses in your county
- Start a recycling business (collecting and selling newspapers and plastics to recycling plants)
- Start an agricultural business promotion business (sell custom caps and t-shirts with farm or ag business names or logos to clients)
- Start an agricultural photography service (animals, equipment, barns, families, children with animals, show animals)
- Start a local farm produce sale paper and sell ads to farmers
- Form a cooperative with other students and share in profits of a greenhouse crop
- Write "how to" pamphlets to sell at local garden supply stores (e.g., How to Grow Tomatoes, etc.)
- Write news articles on agriculture or FFA for local newspapers

Agriscience Project Examples

- Compare weight gain of chicks fed different feed rations
- Conduct a plant growth and physiology experiment in school agriscience lab

- Conduct a research project for agriscience fair (local and national)
- Conduct a research project on a specific career; set up a business plan, including expenses, possible income, and so on
- Conduct a supervised control burn and assess plant growth in the area
- Conduct food science experiments
- Grow crops with different mechanical/chemical applications, such as fertilizer, growth regulator, and observe/report results
- Monitor local air quality; record and report
- Plant and maintain a research plot on different types of turf grasses
- Plant raised beds and monitor the growth of the plants
- Conduct a research project on how light intensity affects plant growth
- Conduct a research project on how light quality affects plant growth
- Conduct a research project on plant reproduction
- Start a soil conservation project on private or public land
- Study effects of fertilizer run-off into a stream or pond
- Study effects of manure run-off into a stream or pond
- Study effects of herbicide type and varying concentrations
- Study temperature effects on worms' food consumption
- Work with agencies involved in research (USDA, etc.)
- Conduct a plant-growth and mineral-deficiency experiment

Alternative Animals Project Examples

- Provide a beehive rental service for farms and gardens
- Raise a dog for show
- Raise dairy goats
- Raise dogs to sell
- Raise fish in tanks or floating cages (research the rate of growth based on factors such as temperature and amount of feed given)
- Raise llamas
- Raise market goats for show
- Raise meat birds (chickens, turkeys, ducks) to the desired weight and sell to customers
- Raise meat goats
- Raise mice, hamsters, or gerbils
- Raise miniature cattle
- Raise miniature horses
- Raise quail or other game birds for flight and meat
- Raise rabbits for pets or meat animals
- Raise special breeds of dogs
- Raise tropical fish in aquariums
- Raise worms; collect and sell to bait stores
- Start a crawfish farm
- Start a cricket ranch
- Start a dog exercising business for elderly folks or sick people
- Start a dog obedience school
- Start a fish bait farm (mealworms, golden grubs, etc.)
- Start a gopher-tortoise relocation service for landowners
- Start a honey production business (would work well with hive rental)
- Start a pet-grooming business
- Start a turtle farm (sell to pet stores and pond owners)
- Train sporting dogs (quail, rabbit, and retriever dogs)
- Work at a dog kennel
- Work at a pet store
- Work at a veterinary hospital

Animal Science Project Examples

- Board horses
- Build a backyard poultry research project
- Contract finish swine

- Develop a cow-calf operation
- Develop a small swine operation
- Develop a stocker cattle operation
- Raise replacement heifers
- Raise dairy replacement heifers
- Produce feeder pigs
- Provide a deer processing service
- Provide a home-animal care service
- Provide a horse training service
- Provide a horseshoeing service
- Provide a meat-processing service
- Provide a poultry-processing service
- Raise a beef heifer for show
- Raise a horse for show
- Raise a market hog for show
- Raise a market steer for show
- Raise breeding sheep for show
- Raise breeding swine for show or breeding
- Raise dairy heifers for show
- Raise market lambs for show
- Raise poultry for show
- Start a small animal care business
- Start an Easter egg business
- Work at a horse operation or stables
- Work at a poultry-processing operation
- Work in the egg industry (packaging and distribution)
- Work on a beef cattle operation
- Work on a dairy operation
- Work on a poultry operation
- Work on a sheep operation
- Work on a swine operation
- Operate a pay-to-fish business
- Provide fish-pond management
- Raise catfish in cages
- Raise fish in an aquaculture system
- Raise fish in cages in a pond or other body of water
- Maintain care and incubation of hatching eggs

Crops Project Examples

- Start organic vegetable production
- Produce vegetables for decoration (Indian corn, mini-pumpkins, gourds, etc.)
- Produce farm crops at home or at school-provided facilities
- Produce forage crops at home or at school-provided facilities
- Produce watermelons
- Scout cotton or peanuts for producers

Forestry PROJECT Examples

- Bale and market pine straw
- Buy unusable lumber from builder supply and building sites; grind up and chip for mulch to sell
- Collect green pine cones (for seed in the fall)
- Collect/market natural supplies (i.e., pine cones, acorns, nuts, corn shucks, etc.) to sell to craft stores
- Start container pine seedling production
- Contract with a tree-removal service to cut firewood and remove fallen trees
- Contract with local timber companies and landowners to maintain boundary lines by painting and chopping
- Cut and sell firewood provided free by national forests and state and local parks
- Cut and/or market firewood
- Grow longleaf pine seedlings
- Measure timber on school forestry plot; determine volume and establish a management plan
- Provide a soil sampling service for farms and lawns
- Purchase bulk pine bark from sawmill; bag and resell
- Purchase seedlings from Georgia Forestry Commission; pot and grow out to sell
- Remove damaged (from lightning strike, insect damage, or mechanical injury) trees for landowners

- Start a custom forest herbicide application crew (must have forest commercial pesticide license)
- Start a forest tree planting business
- Start an ornamental tree care service

Horticulture Project Examples

- Adopt a community building for beautification
- Adopt an area of the school campus for beautification
- Collect and sell dry/preserved native plant materials (acorns, leaves, wiregrass), especially for floral design retail/wholesale
- Collect, press, mount, and identify plants that are growing on campus
- Construct a garden arbor
- Construct backyard water gardens
- Start container gardening ornamental plants
- Start container gardening vegetables
- Create and market custom floral designs
- Develop a business making dried arrangements to sell
- Grow herbs
- Produce daylilies
- Develop a park on public property
- Begin entrepreneurship in floral design
- Establish a community roadside wildflower planting
- Establish a garden plot at home or at school; produce crops to market
- Grow and sell mushrooms
- Grow and sell produce crops
- Grow greenhouse plants on rented school greenhouse/cold-frame space
- Grow, harvest, and can or preserve fruits and vegetables
- Grow organic cut flowers for farmers' markets
- Implement horticulture therapy
- Begin indoor plant rentals and care service for businesses and offices
- Practice landscape maintenance
- Begin a landscape pruning enterprise
- Explore native plant materials
- Offer a shrub care service (pruning, trimming and cutting back shrubs, fertilization)
- Produce fruit crops (at home or school-provided facility—e.g., watermelons)
- Produce greenhouse crop (at home or school-provided facility—e.g., ferns)
- Produce perennials from seed
- Produce turf grass (at home or school-provided facility)
- Propagate and market shrubs
- Provide a fruit tree pruning service
- Provide a mulching service for urban gardeners
- Provide landscaping materials for local businesses (pine straw, rocks, etc.)
- Raise a trial garden plot on school grounds (similar to University of Georgia Atlanta); seed companies may donate seed/plugs
- Raise tomato seedlings and replant into 1-gallon pots to sell
- Rent indoor plants to teachers in your school
- Rent houseplants to homeowners (care for plants and change plants weekly)
- Start a Rent-A-Plant: rent plants for weddings, banquets, parties (e.g., ferns and tropicals)
- Start a commercial flower up-keep business; change hanging baskets, potted plants, and window boxes for businesses
- Start a floral design business by creating table centerpieces to sell at farmers' markets, grocery stores, and vegetable stands
- Start a garden photography business
- Start a hydroponics vegetable business
- Start a lawn irrigation installation business
- Start a houseplant renovation business
- Start a turfgrass establishment business (seedlings, sodding, hydroseeding, etc.)

- Start a vegetable transplant seedling business
- Work at a florist
- Work at a garden center
- Work in a nursery business

Natural Resources Project Examples

- Adopt a local stream to monitor water quality
- Collect water run-off from a school parking lot and analyze for various pollution indicators
- Collect, mount, and identify insects found on a school campus
- Conduct a research project on how to prevent deer damage to a home garden
- Conduct a water-quality study on area lakes or streams
- Conduct endangered-plant surveys for landowners
- Construct deer stands to sell (portable and stationary)
- Construct duck nesting boxes to sell to landowners
- Construct turtle traps for pond owners (use this in conjunction with turtle farm as a source of breeding stock)
- Develop a backyard bird habitat
- Develop a backyard wildlife habitat
- Develop a schoolyard wildlife habitat
- Develop and/or maintain a wildlife food plot on private or public land
- Develop and/or maintain wetland area on private or public land
- Measure land for the local FSA office
- Monitor success rate of bluebird houses
- Plan and develop a school nature trail
- Plan and develop an outdoor classroom
- Plant a butterfly garden at school
- Provide a debris-removal service along rivers and streams; sell driftwood and other items to consumers
- Provide pond fertilization and testing service
- Provide custom dove shoots or quail hunts
- Raise mallard or wood ducks to sell to pond owners
- Raise popular games birds; sell them for meat and as taxidermy products
- Start a bullfrog farm; sell fresh frog legs to local restaurants
- Start a fish fingerling nursery (catfish, trout, bream, etc.)
- Start a red cockaded woodpecker relocation service
- Start a rock store; sell for landscaping purposes (gravel, pebbles, stones, etc.)
- Start a wildlife food plot and native plant enhancement business for local landowners and hunting clubs
- Trap nuisance animals
- Provide nongame wildlife management[22]

Search Terms

portfolio

Notes

1. National Council for Agricultural Education, *Supervised Agricultural Experience SAE for All: Student Guide,* p. 2.
2. Ibid., p. 6.
3. National FFA Organization, *Supervised Agricultural Experiences.* Accessed February 18, 2016. https://www.ffa.org/about/supervised-agricultural-experiences.
4. Ibid.
5. Ibid.
6. "Philosophy and Guiding Principles for Execution of the Supervised Agricultural Experience Component of the Total School Based Agricultural Education Program." *The Council: A National Partner for Excellence in Agriculture and Education.* National Council for Agricultural Education. Accessed February 18, 2016. https://www.ffa.org/thecouncil/sae, p. 2-3.
7. Ibid, p. 3.
8. Ibid., p. 3
9. Ibid, p. 3.

10. Ibid, p. 3
11. California Core Agriscience Lesson Library. *Lesson 612a: Planning your SAE program.* Accessed February 20, 2016. http://calaged.csuchico.edu/ResourceFiles/Curriculum/CoreAgriscience/CD_old/Lessons/612a.pdf
12. Ibid, p. 3.
13. Ibid, p. 4.
14. National FFA Organization, ed. *SAE Handbook, Lesson RK.3: What Is a Journal and How Do We Prepare One?* Accessed February 22, 2016. http://harvest.cals.ncsu.edu/site/WebFile/lp3.pdf, p. 2.
15. National FFA Organization, ed. *Life Knowledge, Lesson MS.69: Record Keeping.* Accessed February 22, 2016. http://harvest.cals.ncsu.edu/site/WebFile/MS69.PDF, p. 3.
16. National FFA Organization, ed. *SAE Handbook, Lesson 7: Why Do We Keep Records?* Accessed February 23, 2016. http://harvest.cals.ncsu.edu/site/WebFile/IIB7.pdf, p. 4.
17. National FFA Organization, ed. *SAE Handbook, Lesson 8: How will my lesson be evaluated?* Accessed February 23, 2016, http://harvest.cals.ncsu.edu/site/WebFile/IIB8.pdf, p. 3.
18. Philosophy and Guiding Principles, p. 4.
19. National FFA Organization, ed. *SAE Handbook, Lesson MS. 70: Proficiency awards and SAE.* Accessed February 23, 2016. http://harvest.cals.ncsu.edu/site/WebFile/IIB8.pdf, p. 3.
20. Ibid, p. 3.
21. National FFA Organization, ed. *National FFA Agricultural Proficiency Awards: A Special Project of the National FFA Foundation.* Accessed February 23, 2016. https://www.ffa.org/sitecollectiondocuments/prof_handbook.pdf, p. 11.
22. J. Ricketts. *Project Workshop.* SAE Handbook. Unpublished document for teacher training in the Republic of Georgia.

APPENDIX B

National FFA Organization

Career Development Events

Agricultural Communications
Agricultural Sales
Agricultural Technology & Mechanical Systems
Agronomy
Dairy Cattle Evaluation & Management
Dairy Cattle Handlers Activity
Environmental & Natural Resources
Farm & Agribusiness Management
Floriculture
Food Science & Technology
Forestry
Horse Evaluation
Livestock Evaluation
Marketing Plan
Meats Evaluation & Technology
Milk Quality & Products
Nursery/Landscape
Poultry Evaluation
Veterinary Science

Leadership Development Events

Agricultural Issues Forum
Conduct of Chapter Meetings Leadership Development
Creed Speaking
Employment Skills
Extemporaneous Public Speaking
Parliamentary Procedure
Prepared Public Speaking

GLOSSARY/GLOSARIO

A

abomasum the fourth, or true, stomach division of a ruminant animal

Abomaso el cuarto o una división del estómago verdadero, de un animal rumiante

absorption the passage of food from the digestive system to the bloodstream

Absorción el pasaje de los alimentos desde el sistema digestivo hasta el torrente sanguíneo

active immunity the type of immunity in an animal that is permanent

La inmunidad activa Tipo de inmunidad en un animal que es permanente

adipose the technical term for fat tissue

Adiposo El término técnico de tejido graso

aerobic organisms grow only in the presence of oxygen

Los organismos aerobios crecen sólo en la presencia de oxígeno

aflatoxin the highly toxic substance produced by some strains of the fungus *Aspergillus flavus*. It is found in feed grains.

La aflatoxina la sustancia altamente tóxica producida por algunas cepas del hongo *Aspergillus flavus*. Se encuentra en cereales.

aged meat that has hung under refrigerated conditions for a specific amount of time

Carne Curada la carne que ha colgado bajo condiciones refrigeradas durante una cantidad específica de tiempo

aging the process by which meats are hung in a cool environment for a specific period to improve the flavor and tenderness; also, the process of maturing and getting older

Envejecimiento el proceso por el cual las carnes están colgados en un ambiente fresco para un período específico para mejorar el sabor y la ternura; además, el proceso de maduración y envejecimiento

agricultural animals animals raised for the purpose of making a profit

Animales agrícolas los animales criados con la finalidad de obtener una ganancia

agriculture the broad industry engaged in the production of plants and animals for food and fiber; the provision of agricultural supplies and services; and the processing, marketing, and distribution of agricultural products

Agricultura la industria que participe en la producción de plantas y animales para el alimento y fibra; la provisión de suministros y servicios agrícolas; y el procesamiento, la comercialización y distribución de productos agrícolas

albumen the white of the egg

Albúmina la clara del huevo

alimentary canal tract extending from the mouth to the anus, through which food passes and where it is exposed to the various digestive processes

Canal alimenticio el tracto, que se extiende desde la boca hasta el ano, a través del cual el alimento pasa y donde está expuesto a los diversos procesos digestivos

allele an alternative form of a gene. For example, one allele may control red coat color and another may control black coat color.

Alelo una forma alternativa de un gen. Por ejemplo, un alelo puede controlar pelaje rojo y otro negro abrigo de control de color.

alternative animal agriculture production of animals other than the traditional agricultural animals

El animal agrícola alternativo la producción de animales distintos de los animales agrícolas tradicionales

alveoli small grapelike structures in the udder of a cow that produce milk

Los alvéolos Los pequeños estructuras que parecen de uvas en la ubre de una vaca que produce la leche

amino acids the basic building block of protein

Aminoácidos los bloques básicos de construcción de las proteínas

amniotic fluid the fluid that surrounds a fetus before birth

El líquido amniótico (el líquido que rodea al feto antes del nacimiento)

anabolism the growth process by which tissues are built up

Anabolismo el proceso de crecimiento, mediante los cuales se construyen los tejidos

anaerobic organisms grow without the presence of oxygen

Los organismos anaeróbicos crecen sin la presencia de oxígeno

anemia a disease caused by a deficiency of hemoglobin, iron, or red blood cells

La anemia una enfermedad causada por una deficiencia de hemoglobina, hierro o glóbulos sanguíneo rojos

Angora a fiber produced from the hair of Angora goats and used to produce some of the finest fabrics in the world

Angora una fibra producida a partir del pelo de cabras de angora y utilizada para producir algunos de los mejores tejidos en el mundo

Animalia the highest level of scientific classification (kingdom) to which all animals belong

Animalia el nivel más alto de clasificación científica (Reino) a la cual pertenecen todos los animales

animal rights activists people who think that killing animals is as wrong as killing people and that animals have the same rights as people

Los defensores de los derechos de los animales la gente que piensan que hay que matar animales es tan malo como matar a la gente y que los animales tienen los mismos derechos que las personas

animal welfare activists people who believe that animals should be treated well and that their comfort and well-being should be considered in their production

Activistas de bienestar animal personas que creen que los animales deben ser tratados bien y que su comodidad y bienestar debe ser considerado en su producción

antemortem preceding death

Antemorte antes de la muerte

anthelmintics a drug or substance given to animals to eliminate parasitic worms

Antihelmínticos una droga o sustancia dado a los animales para eliminar gusanos parásitos

antibiotics a group of drugs used to fight bacterial infections

Antibióticos un grupo de medicamentos que se usan para combatir las infecciones bacterianas

antibodies substances produced by an animal's body that fight disease or foreign materials in the bloodstream or other places in an animal's body

Los anticuerpos sustancias producidas por el cuerpo del animal que combaten las enfermedades o materiales extraños en el torrente sanguíneo o en otros lugares del cuerpo de un animal

antigens any substance that stimulates the production of antibodies in an animal's body

Antígenos cualquier sustancia que estimula la producción de anticuerpos en el cuerpo de un animal

apiary a group of hives

Colmenar un grupo de colmenas

applied research the use of discoveries made in basic research to help in a practical manner

Investigación aplicada la utilización de los descubrimientos realizados en investigación básica para ayudar de una manera práctica

aquaculture the production of animals that live predominantly in the water

Acuicultura La producción de animales que viven predominantemente en el agua

artificial active immunity immunity that comes about as a result of a vaccination

Inmunidad activa artificial la inmunidad que surge como resultado de una vacunación

artificial hormones a manufactured substance that is used in place of a naturally produced hormone

Hormonas artificiales una sustancia fabricada que se utiliza en lugar de una hormona producida de forma natural

artificial insemination (AI) the placing of sperm in the reproductive tract of the female by means other than that of the natural breeding process

La inseminación artificial (IA) bvvbLa colocación de los espermatozoides en el tracto reproductivo de la hembra por otros medios que la del proceso de crianza natural

artificial vagina a tubelike device used to collect semen from a male animal

Vagina artificial dispositivo que parece como un tubo utilizado para recoger el semen de un animal macho

ascarid largest of the parasitic roundworms, most often attacking young animals

Ascáride mayor de las lombrices intestinales parasitarias, la mayoría de las veces ataca a los animales jóvenes

asexual reproduction the production of young by only one parent

La reproducción asexual la producción de cría por uno solo de los padres

assistance dogs dogs that are used to help humans perform any task they cannot do or need assistance in doing

Los perros de ayuda los perros que se utilizan para ayudar a los humanos a realizar cualquier tarea que no pueden hacer o necesita ayuda para hacerlo

associate's degree a two-year degree from a community college

Título de asociado un grado de dos años de una universidad comunitaria

B

bachelor's degree a degree requiring four years of education at a college or university

Licenciatura un título que requieren cuatro años de educación en un colegio o universidad

bacillus rod-shaped bacterium

Bacilo bacteria en forma de varilla

backfat fat tissue that is deposited under the skin of an animal

Tocino dorsal el tejido de grasa que se deposita en la piel de un animal

balance general proportions in the physical structure of an animal

Balance proporciones generales en la estructura física de un animal

balanced ration a diet designed to provide an animal with all the necessary nutrients

Ración balanceada una dieta diseñada para proporcionar un animal con todos los nutrientes necesarios

barrows a male pig that has been castrated before reaching sexual maturity

Túmulos un cerdo macho que ha sido castrado antes de alcanzar la madurez sexual

basic research the investigation of why or how processes occur

La investigación básica la investigación de por qué o cómo se producen los procesos

beak trimming the process of removing the tip of a chicken's beak to prevent injury to other chickens

El recorte de pico el proceso de extracción de la punta de un pico del pollo para evitar lesiones a otros pollos

beef the meat from cattle over a year old

Carne de res La carne de ganado de más de un año de antigüedad

bee space the space (about 3/8 inch) in a beehive that allows bees to work back-to-back

Espacio de abeja Espacio (aproximadamente 3/8 pulgada) en una colmena de abejas que permite trabajar espalda con espalda

billy a mature male goat

Billy un macho cabrío madura

binomial nomenclature a system of scientific classification of living organisms that uses two names, the genus and the species. The names usually come from Latin derivatives.

Nomenclatura binomial un sistema de clasificación científica de los seres vivos que utiliza dos nombres, el género y la especie. Los nombres normalmente proceden de América derivados.

biological control the use of natural means rather than chemicals to control pests

Control biológico la utilización de medios naturales en lugar de productos químicos para el control de plagas

blast freezing most common method of commercial freezing, utilizing high-velocity air and temperatures of –10°C to –40°C

Cámaras de congelación método más común de congelación comercial, utilizando aire de alta velocidad y temperaturas de −10°C a −40°C

blastula a mass of cells with a cavity that occurs from the dividing of a fertilized egg. From this stage the cells begin to differentiate.

Blástula una masa de células con una cavidad que se produce desde la división de un óvulo fecundado. A partir de esta etapa, las células comienzan a diferenciarse.

blind nipples nonfunctional nipples on the mammary system of a female pig

Los pezones ciegos pezones no funcionales en el sistema mamaria de un cerdo hembra

bloat a condition in cattle caused by gas being trapped in the digestive system. Left untreated, the condition can be fatal.

La distención Una condición en engordar ganado causadas por los gases atrapados en el sistema digestivo. Sin tratamiento, la enfermedad puede ser fatal.

blood typing analyzing an animal's blood to determine the animal's ancestry

La análysis de sangre Análisis de sangre de un animal de sangre para determinar la ascendencia del animal

boar a male pig that has not been castrated

Jabalí un cerdo macho que no ha sido castrados

bolus (plural, boluses) a large pill; also, a soft mass of chewed food

Bolo (plural, bolos) un gran píldora; asimismo, una masa blanda de alimentos masticados

breed a group of animals with a common ancestry and common characteristics that breed true

Raza un grupo de animales con acendencia común y características comunes que engendran produciendo el mismo clase de animal

breed association an organization that promotes a certain breed of animal. They control the registration process of purebred animals of that breed.

Asociación de raza una organización que promueve una determinada raza animal. Ellos controlan el proceso de inscripción de animales de raza pura de esta raza.

breeding true offspring almost always looking like the parents

Cría verdad descendencia casi siempre mirando como los padres

brisket the breast or lower chest of a four-legged animal

Pecho el pecho o tórax inferior de un animal de cuatro patas

broiler a chicken approximately eight weeks old that weighs 2 1/2 pounds or more

Pollo de engorde Un pollo broiler aproximadamente ocho semanas viejos que pesa de 2 1/2 libras o más

broiler industry raising chickens for their meat

Industria La cría de pollos parrilleros por su carne

brood group of young undeveloped bees (pupae/larvae) that grow and metamorphose into adult bees in brood cells in the hive

Las crías grupo de jóvenes abejas subdesarrolladas (Pupas/larvas) que crecen y se transforman en células adultas en la cría de las abejas en la colmena

brood cells cells in the hive where the queen bee lays her eggs

Celdas de cría las celdas de la colmena donde la abeja reina pone sus huevos

brood chamber that portion of a bee hive where the queen lays eggs and the young bees are hatched and raised

La cámara de la cría la parte de un enjambre de abejas, donde la reina pone huevos y las abejas jóvenes salidos y levantado

browse the consumption of leaves and stems as compared to grazing

Browse el consumo de hojas y tallos en comparación con el pastoreo

BST bovine somatotropin, a naturally occurring hormone that aids in stimulating the production of milk in cows

La somatotropina bovina (BST) una hormona natural que ayuda a estimular la producción de leche en vacas

buck a mature male dairy goat

Buck un macho maduro de cabra de lechería

buckling a young male dairy goat

Cabrío macho un hombre joven de cabra de lechería

bull a male bovine that has not been castrated

Toro un bovino macho que no ha sido castrados

Bureau of Land Management the federal agency that oversees the management of government lands

Oficina de Administración de Tierras la agencia federal que supervisa la gestión de las tierras del gobierno

by product a product that is created as the result of producing another product

Subproducto un producto que se crea como resultado de producir otro producto

C

cabrito the Spanish term for goat meat

El cabrito el término español para carne de cabra

cage operation an operation in which hens are kept in cages all their lives as they produce eggs

Operación jaula una operación en la que están las gallinas en jaulas todas sus vidas ya que producen huevos

candling the use of light shined through an egg to determine defects in the egg

Trasluz la utilización de la luz brillando a través de un huevo para determinar defectos en el huevo

cannibalism the habit of some birds in a poultry flock of repeatedly pecking and clawing other birds in the flock, often causing injury and death

El canibalismo el hábito de algunas aves en una bandada de aves picoteando repetidamente y ejerce otras aves en el rebaño, a menudo causando lesiones y muerte

cannon bone a bone in hoofed mammals that extends from the knee or the hock to the fetlock or pastern

Cannon bone un hueso en mamíferos ungulados que se extiende desde la rodilla o el corvejón al carneja o cuartilla

carcass that part of a meat animal that is left after the hide and hair, feet, head, and entrails have been removed

Carcasa Canalque parte de un animal de carne que queda después de la piel y el pelo, los pies, la cabeza y las vísceras se han eliminado

carcass merit quality and yield of a carcass

Carcasa del mérito lacalidad y rendimiento de un cadáver

carding one of the first steps in the processing of wool. The fibers are separated from other fibers in the locks or bunches of wool.

El cardado uno de los primeros pasos en el procesamiento de la lana. Las fibras están separadas de otras fibras en las cerraduras o racimos de lana.

career an occupation in which one earns a living

Carrera una ocupación en la que uno se gana la vida

carnivores an animal whose diet consists mainly of other animals

Los carnívoros un animal cuya dieta consiste principalmente de otros animales

carotene an orange or red pigment found in green leafy plants, especially carrots. It can be converted to vitamin A by an animal's body.

Caroteno un pigmento naranja o rojo encontrado en plantas frondosas verdes, especialmente las zanahorias. Puede ser convertido en vitamina A por el cuerpo del animal.

cartilage firm but pliant tissue in an animal's body, some of which turns to bone as the animal grows and ages.

Cartílago un tejido firme pero flexible del cuerpo animal, algunos de los cuales se convierte al hueso como el animal crece y las edades

Cashmere a fine fabric that comes from the hair of fiber goats

El Casimir un fino tejido que proviene de la fibra de pelo de cabras

castings manure from worms

El humus de lombriz Abono de gusanos

castrated condition in which an animal's testicles have been removed

Castrado una condición en la cual un animal se han quitado los testículos

castration act of castrating a male animal

Castración Acta de castrar un animal macho

catabolism the process of breaking down tissues from the complex to the simple as in the digestive process

Catabolismo el proceso de romper los tejidos desde el complejo hasta el simple como en el proceso digestivo

catheter a tube that is inserted into an animal's body to inject or withdraw fluid

Catéter un tubo que se inserta en el cuerpo del animal para inyectar o retirar el líquido

cecum the enlargement on the digestive tract of animals such as the horse that allows them to digest large amounts of roughages

Intestino Ciego la ampliación en el tracto digestivo de los animales como el caballo que les permite digerir grandes cantidades de groseros

cells the basic building block of living tissue. It generally consists of a membrane wall, a nucleus, and a cytoplasm.

Las células el bloque básico de construcción de los tejidos vivos. Por lo general, consiste en una pared membranosa, un núcleo y un citoplasma.

cellulose an inert complex carbohydrate that makes up the bulk of the cell walls of plants

Celulosa un carbohidrato complejo inerte que componen el grueso de las paredes celulares de las plantas

centrioles strands of genetic material outside the nucleus of animal cells

Centriolos hebras de material genético fuera del núcleo de las células animales

cervix the organ that serves as an opening to the uterus

Cuello uterino el órgano que actúa como una apertura al útero

chalazae ropelike structures inside an egg that hold the yolk in the center of the egg

chalazae estructuras como cuerdas en el interior de un huevo que mantenga la yema en el centro del huevo

chevon the French term for goat meat

Chevon el término francés para la carne de cabra

chine the backbone of an animal

Espinazo la columna vertebral de un animal

cholesterol a fat-soluble substance found in the fat, liver, nervous system, and other areas of an animal's body. It plays an important role in the synthesis of bile, sex hormones, and vitamin D.

Colesterol una sustancia liposoluble encontradas en la grasa, el hígado, el sistema nervioso, y otras áreas del cuerpo del animal. Desempeña un papel importante en la síntesis de bilis, hormonas sexuales y vitamina D.

chromosomes a linear arrangement of genes that determines the characteristics of an organism

Cromosomas una disposición lineal de los genes que determinan las características de un organismo

chromatid one strand of a double chromosome

Cromátidas un hilo de un cromosoma doble

chronological age the actual age of an animal in days, weeks, months, or years

La edad cronológica la edad real de un animal en días, semanas, meses o años

classes further divisions within phyla or subphyla

Clases nuevas divisiones dentro de filos o subphyla

cleavage the splitting of one cell into two parts

La división dividir una celda en dos partes

climate controlled houses livestock (or poultry) houses that are kept at the proper temperature, lighting, and humidity for optimal growth and comfort of the animals

Casas con clima controlado animales (aves de corral) o casas que se mantiene a la temperatura adecuada, la iluminación y la humedad para un óptimo crecimiento y la comodidad de los animales

clitoris small, sensitive organ within the vulva that provides stimulation during the mating process

Clítoris pequeño órgano sensible dentro de la vulva que proporciona la estimulación durante el proceso de acoplamiento

cloaca the opening in a hen's body through which the egg is expelled

Cloaca la abertura de una gallina del cuerpo a través de los cuales el óvulo es expulsado

clone an organism, produced by asexual means, with the exact same genetic makeup as another

Clon un organismo, producida por medios asexuales, con exactamente la misma información genética que otro

cocci round, spherical-shaped bacteria

Los cocos las bacterias de forma ronda y esférica

codominant genes genes that are neither dominant nor recessive

Genes codominantes los genes que no son ni tampoco dominante recesivo

cold blooded *See* ectothermic

A sangre fría *véase* ectothermic

cold water fish a fish that will not thrive in water temperatures above 70°F

Pescado de agua fría un pez que no prosperan en agua a temperaturas por encima de 70 °F

colic a condition in horses caused by a blockage in the digestive system. The intestine becomes distended and causes pain to the animal.

Cólico una condición en equinos causada por una obstrucción en el sistema digestivo. El intestino está distendido y causa dolor a los animales.

coliforms a group of bacteria that inhabit the colons of people and animals

Coliformes un grupo de bacterias que habitan en los dos puntos de personas y animales

collagen a protein that forms the main component of connective tissues in animals

Colágeno una proteína que constituye el principal componente del tejido conectivo en animales

colon the large intestine

Colón el intestino grueso

colony a group of bees consisting of workers, queen, and drones that live together as a unit

La colonia Un grupo de abejas compuesto de trabajadores, reina y zánganos que viven juntas como una unidad

colostrum the first milk that a mammal gives to the young following birth. It is rich in nutrients and imparts immunity from the mother to the offspring.

El calostro la primera leche que un mamífero da a los jóvenes tras el nacimiento. Es rico en nutrientes y confiere inmunidad de la madre a la descendencia.

combing process by which wool fibers are untangled and smoothed in preparation for being made into worsted wool

Peinado proceso por el cual las fibras de lana son alineados y suavizada en preparación para ser realizados en lana estambre

combination curing a combination of dry curing and injection curing

Combinación curado una combinación de curado en seco y curado de inyección

commensalism a symbiotic relationship in which one species benefits and the other is not harmed

Comensalismo una relación simbiótica, en la cual una especie beneficios y la otra no está dañado

commercial producers producers who raise animals for the meat industry as opposed to raising them for use in breeding programs

Los productores comerciales Los productores que crían animales para la industria de la carne en lugar de aumentar para utilizarlos en programas de cría

community college a two-year institution in which one may obtain an associate's degree

Universidad Communitaria una institución de dos años en el que uno puede obtener un título asociado

companion animals animals whose main purpose is serving as pets or friends to humans

Animales de compañía los animales cuyo principal propósito es servir como mascotas o amigos a los seres humanos

concentrate a feed that is high in carbohydrates and low in fiber

Concentrado un alimento que es alta en carbohidratos y baja en fibra

conception uniting of sperm and egg

Concepción unión del esperma y el óvulo

conditioning process of learning by associating a certain response with a certain stimulus

Acondicionado proceso de aprendizaje mediante la asociación de una determinada respuesta con un cierto estímulo

confinement operation a system of raising animals in a relatively small space

Operación de encierro un sistema de cría de animales en un espacio relativamente pequeño

conformation the shape or proportional dimensions of an animal

La conformación La forma o proporcional de las dimensiones de un animal

consumers those who buy or use food, manufactured goods, or other products

Los consumidores los que comprar o utilizar alimentos, productos manufacturados, u otros productos

contagious diseases a disease that may be passed from one organism to another

Enfermedades contagiosas una enfermedad que puede transmitirse de una persona a otra

control group a group of animals or plants (in a scientific experiment) that does not receive the treatment under study

Grupo control un grupo de animales o plantas (en un experimento científico) que no reciben el tratamiento bajo estudio

copulation the act of sexual union between two mating animals

La cópula el acto de unión sexual entre dos animales de apareamiento

corpus luteum a swelling of tissue that develops on the follicle at the site where an ovum has been shed

Cuerpo lúteo una inflamación del tejido que se desarrolla en el folículo en el sitio donde se ha derramado un óvulo

cortex the outer layer or region of any organ; also, in wool fibers the tissue immediately external to the xylem

La corteza la capa exterior de cualquier órgano o región; también, en lana el tejido externo inmediatamente para el xilema

cow a female bovine that has had a calf

Vaca un bovino hembra que ha tenido un ternero

cow calf operations a system of raising cattle, the main purpose of which is the production on calves that are sold at weaning

Operaciones de vaca becerro un sistema de cría de ganado, cuyo principal objetivo es la producción de terneros que se venden al destete

cow hocked a condition in which an animal's back feet are splayed out and the hocks are turned in

Corvejones torcidos una condición en la cual un animal de patas traseras son abocinadas y los corvejones se están torcidos

Cowper's gland a gland in the male reproductive tract that produces a fluid that is added to the ejaculate

Glándula de Cowper una glándula en el tracto reproductivo masculino que produce un líquido que se añade al eyacular

crimp the amount of waves in wool fiber

Rizo de lana la cantidad de ondas en fibras de lana

crossbred an animal that is the result of the mating of parents of different breeds

Mestizos un animal que es el resultado del apareamiento de padres de diferentes razas

crude protein content total amount of crude protein in a feed, calculated by analyzing the nitrogen content and multiplying that percentage by 6.25

Contenido de proteína cruda Cantidad total de proteína en un alimento crudo, calcula analizando el contenido de nitrógeno y multiplicar ese porcentaje por 6.25

crustaceans aquatic animals with a rigid outer covering, jointed appendages, and gills

Crustáceos animales acuáticos con una cubierta exterior rígida, apéndices articulados y branquias

cryogenics method of freezing, utilizing condensed gases in direct contact with the product being frozen

Criogénesis método de congelación, utilizando gases condensados en contacto directo con el producto congelado

cud a small wad of regurgitated feed in the mouth of a ruminant that is rechewed and swallowed

El bolo alimenticio un pequeño bolo de la regurgitación de alimento en la boca de un rumiante que se ingiere y rechewed

curd the coagulated part of milk that results when the milk is clotted by adding rennet, by natural souring, or by adding a starter

Cuajada la parte coagulada de la leche que se produce cuando la leche es coagulado mediante la adición de cuajo, por souring natural, o añadiendo un starter

curing treating meat to retard spoilage

El curado el tratamiento de carne para retardar la corrupción

cutability percent of lean cuts a carcass will produce

Cortabilidad porcentaje de cortes magros de una canal producirá

cuticle outer layer of cells of wool fibers

Cutícula capa externa de células de lana

cytokinesis the last phase of cell division where the cytoplasm is divided in the cell

Citocinesis la última fase de división celular donde el citoplasma se divide en la celda

cytoplasm the living material within a plant or animal cell excluding the nucleus

Citoplasma el material vivo dentro de una célula vegetal o animal excepto el núcleo

D

dam the mother of an animal

Hembra madre la madre de un animal

dam breeds those breeds of agricultural animals that are used as dams in a cross-breeding program

Razas de madre las razas de animales agrícolas que son utilizados como presas en un programa de cría cruzada

debeaking removing the tip of a chicken's beak to aid in the prevention of cannibalism

Recorte de pico retirar la punta del pico del pollo para ayudar en la prevención de actos de canibalismo

degree program an educational line of study that leads to the completion of a degree such as an Associate, Bachelor, Master, or Doctorate

Programa educativo una línea educative de estudio que conduce a la realización de un grado como un asociado, licenciatura, maestría, o doctorado

dehorning permanently removing an animal's horns

El descorne eliminar permanentemente un cuernos del animal

dental pad a hard pad in the upper mouth of cattle and other animals that serves in the place of upper teeth

Almohadilla dental una pad dura en la boca superior de ganado y otros animales que sirve en el lugar de dientes superiores

deoxyribonucleic acid (DNA) a genetic acid that controls inheritance

El ácido desoxirribonucleico (ADN) un ácido que controla la herencia genética

differentiate the process where stem cells begin to form different types of tissue

Diferenciar el proceso por el que las células madre comienzan a formar diferentes tipos de tejidos

differentiation the development of different tissues from the division of cells

Diferenciación el desarrollo de los diferentes tejidos de la división de las células

diffusion in the process of absorption, the passing of particles through a semipermeable membrane

Diffusion en el proceso de absorción, el paso de las partículas a través de una membrana semipermeable

digestion the changes that food undergoes within the digestive tract to prepare it for absorption and use in the body

Digestión los cambios que sufre de alimentos dentro del tracto digestivo para prepararlo para la absorción y utilización en el cuerpo

disaccharides the more complex sugars

Los disacáridos los azúcares más complejos

discriminate breeder an animal that will only breed with a certain mate

Criador discriminatorio un animal que sólo se reproducen con cierta mate

dissolved oxygen oxygen in water that is available for the use of animals with gills (such as fish)

Oxígeno disuelto eloxígeno en el agua que está disponible para el uso de animales con branquias (como los peces)

docile having a quiet, gentle nature

Dócil tener un carácter suave y silencioso

docking the removal of an animal's tail

La sección de cola la eliminación de una cola del animal

doe a mature female dairy goat

La cabra hembra cabra madura de lechería

doeling an immature female dairy goat

La cabrita una hembra inmadura de cabra de lechería

domesticated raised under the care of humans

Domesticado criado bajo el cuidado de los seres humanos

dominant gene a gene that expresses its characteristics over the characteristics of the gene with which it is paired

Gen dominante un gen que expresa sus características sobre las características del gen con el que está emparejado

donor cows a cow of superior genetics from which an embryo is taken to implant in a cow of inferior genetics

Vacas donantes una vaca de genética superior desde que el embrión es un ser tomadas para implantar en una vaca de inferioridad genética

double muscling a condition in beef animals that is characterized by large, bulging, round muscles

Doble musculatura una condición en la carne de los animales que se caracteriza por grandes, abultadas, músculos redondo

draft horses a horse that is used mainly for pulling loads or for working

Caballos de tiro un caballo que se utiliza principalmente para tirar de cargas o para trabajar

drone a male honeybee

Zángano una abeja macho

dry curing the process of curing meat by rubbing the cure ingredients onto the surface of the product and allowing it to move into the product by osmosis

Curado en seco El proceso de salazón frotando la curación ingredientes en la superficie del producto y para permitir que se mueva en el producto por ósmosis

dual purpose animal an animal that is raised for more than one purpose, for example, sheep for wool and mutton

Animales de doble propósito un animal que se eleva para más de un propósito, por ejemplo, la oveja por la lana y carne de cordero

duodenum the first portion of the small intestine

El duodeno la primera porción del intestino delgado

E

E. coli a type of bacteria commonly found in the colon of animals

E. coli un tipo de bacteria que comúnmente se encuentra en el colon de animales

ecological balance the balance nature has regarding the living things in a given area

Equilibrio ecológico el equilibrio de la naturaleza en relación con los seres vivos en un área determinada

ectoderm the outer of the three basic layers of the embryo, which gives rise to the skin, hair, and nervous system

Ectodermo la parte exterior de las tres capas del embrión, que da lugar a la piel, cabello y sistema nervioso

ectothermic animals whose body temperature adjusts to the air and water around them. Also known as cold blooded.

Ectothermic animales cuya temperatura corporal se ajusta para el aire y el agua a su alrededor. También conocido como a sangre fría.

efficiency ability of the animal to gain on the least amount of feed and other necessities

Eficacia capacidad del animal la ganancia en la menor cantidad de alimento y otras necesidades

ejaculation release of semen (the ejaculate)

La ejaculación Liberar la eyaculación de semen (el eyaculado)

elastin a protein substance found in tendons, connective tissue, and bone

La elastina una proteína sustancia encontrada en los tendones, tejido conectivo y hueso

elastrator a device that is used to stretch a rubber band over the scrotum or tail of an animal. The blood circulation is cut off and the testicles or tail drops off.

Elastrator un dispositivo que se utiliza para estirar una banda de caucho sobre el escroto o la cola de un animal. La circulación de la sangre es cortada y los testículos o la cola cae.

embryo an organism in the earliest stage of development

Embrión un organismo en las primeras etapas del desarrollo

embryo transfer transferring embryos from one female to another to increase the reproductive capacity of superior females

Transferencia de Embriones la transferencia de embriones a partir de una hembra a otra para aumentar la capacidad reproductiva de las hembras superior

embryo transplant removing an embryo from a female of superior genetics and placing the embryo in the reproductive tract of a female of inferior genetics

Transplante de embriones extracción de un embrión de una hembra de genética superior y colocar el embrión en el tracto reproductivo de la hembra de inferioridad genética

endangered species species that are on the verge of dying out

Especies amenazadas especies que están a punto de morir

endocrine system the system of glands in an animal's body that secrete substances that control certain bodily processes

Sistema endocrino El sistema de glándulas en el cuerpo de un animal que secretan sustancias que controlan ciertos procesos corporales

endoderm the innermost layer of cells of an embryo, which develops into internal organs

El endodermo la capa interna de células de un embrión, que se desarrolla en los órganos internos

endoplasmic reticulum a large webbing or network of double membranes that are positioned throughout the cell that provide the means for transporting material throughout the cell

Retículo endoplasmático una gran correa o red de membranas dobles que están situados a lo largo de la célula que proporcionan los medios para transportar material a lo largo de la celda

endothermic an animal whose body temperature is warmer than its surroundings

Endotérmico un animal cuya temperatura corporal es más cálida que sus alrededores

energy the capacity to do work

Energía La capacidad de trabajo

enucleated oocyte an egg with the nucleus removed

El ovocito enucleado Un óvulo con el núcleo extraído

environment the total of all the external conditions that may act upon an organism or community to influence its development or existence

Medio ambiente el total de todas las condiciones externas que pueden actuar sobre un organismo o una comunidad de influir en su desarrollo o la existencia

enzyme a protein that is produced by an animal's body that stimulates or speeds up various chemical reactions

Enzima una proteína que es producida por el cuerpo de un animal que estimule o acelera las diversas reacciones químicas

epididymis a small tube, leading from the testicles, where sperm mature and are stored

Epidídimo un pequeño tubo, líder desde los testículos, donde los espermatozoides maduran y se almacenan

epinephrine a hormone that is released when an animal gets frightened or upset. In cows it inhibits the milk letdown process.

La epinefrina una hormona que se libera cuando un animal puede asustarse o malestar. En vacas de leche, ya que inhibe el proceso de decepción.

epistasis interaction of genes that are not matched pairs to cause an expression different from the coding of the genes

Epistasis Interacción de genes que no están emparejados para causar una expresión diferente de la codificación de los genes

ergot a fungus disease of grains that produces a toxin

Cornezuelo un hongo enfermedad de granos que produce una toxina

esophagus the tube leading from the mouth to the stomach

El esófago el tubo que conduce desde la boca hasta el estómago

essential amino acids any of the amino acids that cannot be synthesized by an animal's body and must be supplied from the animal's diet

Aminoácidos esenciales cualquiera de los aminoácidos que no pueden ser sintetizados por el cuerpo del animal y deben ser suministrados en la dieta del animal

estimated breeding value in beef cattle, an estimate of the value of an animal as a parent

Estimado valor de cría en el ganado vacuno, una estimación del valor de un animal como un padre

estrogen a hormone that stimulates the female sex drive and controls the development of female characteristics

El estrógeno una hormona que estimula la libido femenina y controla el desarrollo de las características femeninas

estrus the period of sexual excitement (heat) when the female will accept the male

Estro el periodo de la excitación sexual (calor) cuando la hembra aceptará el macho

estrus cycle the reproductive cycle of female animals measured from the beginning of one heat period until the beginning of the next

Ciclo estral el ciclo reproductivo de las hembras miden desde el comienzo de un período de calor hasta el comienzo del siguiente

estrus synchronization using synthetic hormones to make a group of females come into heat (estrus) at the same time

Sincronización del estro uso de hormonas sintéticas para hacer un grupo de hembras en estro (calor) al mismo tiempo

ethology the science of animal behavior

La etología la ciencia del comportamiento animal

eukaryotic cells cells that have a relatively large structure called a nucleus that is composed primarily of nucleic acids, proteins, and enzymes and are found in both plants and animals

Las células eucariotas células que tienen una estructura relativamente grande llamado un núcleo que está compuesto principalmente de ácidos nucleicos, proteínas y enzimas y se encuentran en plantas y animales

ewe a female sheep

Oveja Una oveja hembra

expected progeny difference an estimate of the expected performance of an animal's offspring

Diferencia de progenie esperada una estimación del rendimiento esperado de la progenie de un animal

exotic animals animals that are not usually raised as farm animals

Animales exóticos animales que normalmente no son criados como animales de granja

exotic breeds animals that are out of the ordinary, such as an unusual breed

Las razas exóticas animales que están fuera de lo ordinario, como un inusual raza

exothermic animals whose internal body temperature comes from the environment, an example being reptiles

Reacción exotérmica animales cuya temperatura interna del cuerpo proviene del medio ambiente, siendo un ejemplo reptiles

experiment an operation carried out under controlled conditions to discover an unknown entity, to test a hypothesis, or demonstrate something known

Experimento una operación llevada a cabo bajo condiciones controladas para descubrir una entidad desconocida, para probar una hipótesis, o demostrar algo conocido

experimental group the group used to test a hypothesis; the group subject to experimentation

Grupo experimental el grupo utilizado para probar una hipótesis; el grupo sometido a la experimentación

exsanguination the removal of an animal's blood during the slaughter process

Desangrado la extracción de sangre de un animal durante el proceso de sacrificio

extenders a substance added to semen to increase the volume

Extensores sustancia añadida al esperma para aumentar el volumen

external parasites a parasite that lives in the hair or on the skin of an animal

Parásitos externos un parásito que vive en el cabello o en la piel de un animal

F

facultative microbes that can grow with or without free oxygen

Los facultativos microbios que pueden crecer con o sin oxígeno libre

fallopian tubes the tubes leading from the ovaries to the uterus

Las trompas de Falopio los tubos que van de los ovarios al útero

families smaller divisions within classes

Las familias pequeñas divisiones dentro de las clases

farrier a person who cares for horses' feet

Farrier una persona que cuida a los pies de los caballos

farrow to give birth to a litter of pigs

Farrow para dar nacimiento a una camada de cerdos

farrowing crate a crate or cage in which a sow is placed at the time of farrowing to protect the newborn pigs

Jaulas parideras una caja o jaula en la que una cerda es colocado en el momento del parto para proteger al recién nacido cerdos

farrowing operation first phase of a pig operation involving the birth of the piglets

Operación paridera la primera fase de un cerdo que conlleven el nacimiento de los lechones

feed conversion ratio the rate at which an animal converts feed to meat

Ratio de conversión de alimentación la velocidad a la que un animal convierte la alimentación de carne

feeder pigs a young pig weighing less than 120 pounds that is of sufficient quality for finishing as a market hog

Los cerdos del alimentador un joven cerdo pesa menos de 120 libras, que es de calidad suficiente para acabar como un cerdo del mercado

feedlot a pen in which cattle are placed for fattening prior to slaughter

Corrales de engorde un bolígrafo en el que se colocan de ganado de engorde antes del sacrificio

feedlot operations an agricultural enterprise where beef animals are placed in pens and fed grain to fatten them

Operaciones de corrales de engorde una empresa agrícola donde la carne de animales son colocados en corrales y alimentados con grano a lo engordan

feedstuff a basic ingredient of a feed that would not ordinarily be fed as a feed by itself

Piensos un ingrediente básico de un alimento que por lo general no se alimenta como un alimento por sí mismo

felting the property of wool fibers to interlock when rubbed together under conditions of heat, moisture, and pressure

Fieltrado propiedad de lana hasta el interbloqueo cuando se frota juntos bajo condiciones de calor, humedad y presión

fermentation the processing of food by the use of yeasts, molds, or bacteria

Fermentación El tratamiento de alimentos mediante el uso de levaduras, hongos o bacterias

fertile capable of producing viable offspring

Fértil capaz de producir descendientes viables

fertilization the union of the sperm and egg

Fertilización la unión del esperma y el óvulo

fertilization membrane a membrane surrounding an egg that is formed after the egg is fertilized. This prevents another sperm from entering.

La fertilización de la membrana una membrana que rodea un huevo que se formó después de que el óvulo es fecundado. Esto impide que los espermatozoides entren en otro.

fertilized egg an egg that has united with a sperm

óvulo fecundado Un óvulo que se ha unido con un espermatozoide

fingerling a small fish that is of sufficient size to use for stocking

Alevines un pequeño pez que es de suficiente tamaño para almacenamiento

finish the amount of fat on an animal that is ready for slaughter

Finalizar la cantidad de grasa en un animal que está listo para el sacrificio

finished the stage where an animal has the correct amount of fat and is ready for slaughter

Terminado el escenario donde el animal tiene la cantidad correcta de grasa y está listo para el sacrificio

finishing operation last phase of a pig operation involving bringing feeder pigs up to market weight

Operación de terminación última fase de una operación que implique llevar cerdos de engorde hasta peso de mercado

flukes small, seed-shaped parasitic flatworm, the most damaging of which live in the host's liver

El parásito de hígado pequeños parásiticos, en forma de semillas, la más perjudicial de las cuales viven en el hígado del host

flushing the process of removing embryos from the donor cow by injecting a fluid by means of a catheter passed through the cervix and into the uterine horn

Aclarado el proceso de extracción de los embriones procedentes de la vaca donante inyectando un fluido por medio de un catéter a través del cuello uterino hasta el cuerno uterino

follicle a small blister-like structure that develops on the ovary that contains the developing ovum

Folículo una ampolla pequeña estructura como la que se desarrolla en el ovario que contiene el óvulo en desarrollo

follicle stimulating hormone (FSH) the naturally occurring hormone that stimulates the development of the follicle on the ovary

Hormona folículo estimulante (FSH) la hormona natural que estimula el desarrollo del folículo en el ovario

Food and Drug Administration (FDA) a federal agency that regulates the production, manufacture, and distribution of food and drugs

La Administración de Drogas y Alimentos (FDA, por sus siglas en inglés) Una agencia federal que regula la producción, la fabricación y la distribución de alimentos y medicamentos

forage livestock feed that consists mainly of the leaves and stalks of plants

Forraje la alimentación de ganado que consiste principalmente de las hojas y tallos de plantas

foundational SAE the beginning part of an SAE where the student develops a detailed plan for conducting a Supervised Agricultural Experience

Fundacional SAE el principio parte de SAE donde el estudiante desarrolla un plan detallado para la conducción de un experiencia agrícola supervisada

foundation comb a sheet of honeycomb placed onto frames on which the bees complete the comb to fill with honey

Fundación peine una hoja de honeycomb colocados en marcos de que las abejas completar el peine para llenar con miel

frame size a score that depicts the size and weight of an animal at maturity. The measure is taken at the shoulder or at the hip.

Tamaño del bastidor una puntuación que muestra el tamaño y el peso de un animal en la madurez. La medida se toma en el hombro o en la cadera.

free choice feeding an animal with an unlimited supply of feed. The animal is free to eat whenever it wants.

Libre elección alimentación animal con un suministro ilimitado de alimento. El animal es libre para comer siempre que quiera.

freeze branding a method of marking cattle by using a super cold metal that kills the pigment-producing ability of the hair contacted

Congelar la marca un método de marcado de ganado mediante una super frío metal que mata la capacidad productora de pigmento del cabello contactado

fructose the sugar found in fruit

La fructosa el azúcar que se encuentra en frutas

fry small, newly hatched fish

Fry pequeños peces recién nacidos

fumigate to kill pathogens, insects, etc. by the use of certain poisonous liquids or solids that form a vapor

Fumigar para matar a los agentes patógenos, insectos, etc. por el uso de ciertos líquidos o sólidos venenosos que forman un vapor

fungi the kingdom to which multicelled organisms such as fungi belong

Hongos El reino al cual multicelled microorganismos tales como hongos pertenecen

G

galactose the sugar in milk

La galactosa el azúcar en la leche

gamete the sex cell, either an egg or sperm

Gameto la célula sexual, ya sea un óvulo o de un espermatozoide

gastrointestinal tract the digestive system, made up of the stomach and intestines

Tracto gastrointestinal el sistema digestivo, formada por el estómago y los intestinos

gelding a male horse that has been castrated

Castrado Un caballo macho que ha sido castrados

genes units of inheritance, composed of DNA

Genes Unidades de herencia, compuestas de ADN

genetic base the breeding animals available for a producer to use

Base genética la cría de animales disponibles para un productor a utilizar

genetic code otherwise known as DNA, passed on from the parent(s) which is contained in all the cells of animal's body

Código Genético también conocido como ADN, pasa de la(s) padre(s) que se encuentra en todas las células del cuerpo del animal

genetic defects an impairment of an animal that was passed by the parents to the offspring

Defectos genéticos un deterioro de un animal que fue transmitida por los padres a la descendencia

genetic engineering the alteration of the genetic components of organisms by human intervention

La ingeniería genética la alteración de los componentes genéticos de los organismos por la intervención humana

genetic improvement the increase in occurrence of genetically favorable traits in offspring

Mejoramiento genético el aumento de la aparición de rasgos de organismos genéticamente favorables en la descendencia

genetically altered clone a clone that results when a specific gene(s) is placed into the DNA of the animal desire to be cloned

Clon genéticamente modificado un clon que resulta cuando un gen específico(s) se coloca en el ADN del animal, el deseo de ser clonada

genetics the science that deals with the processes of inheritance in plants and animals; also, the genetic makeup of an organism

Genética la ciencia que se ocupa de los procesos de herencia en plantas y animales; además, la composición genética de un organismo

genetic variation the difference between animals due to their genetic makeup

La variación genética la diferencia entre animales, debido a su genética

genotype the genetic makeup of an organism

Genotipo la constitución genética de un organismo

genus a class or group marked by common characteristics and comprised of structurally related organisms; the first name in the binomial nomenclature identifying an organism

Género una clase o grupo marcado por características comunes y compuesta de organismos relacionados estructuralmente; el primer nombre en la nomenclatura binomial identificar un organismo

germinal disk a spot in the yolk portion of the egg that contains the genetic material from the female

Disco germinal una mancha en la parte de la yema de huevo que contiene el material genético de la hembra

gestation the length of time from conception to birth

Gestación período de tiempo desde la concepción hasta el nacimiento

gilts a female pig that has not given birth

Cerdas la hembra de cerdo que no ha dado nacimiento

gland cistern where the milk is stored in the cow

La cisterna de la glándula donde la leche es almacenada en la vaca

glucose a common sugar that serves as the building blocks for many complex carbohydrates

La glucosa un azúcar común que sirve como bloques de construcción para muchos carbohidratos complejos

Golgi apparatus an organelle within a cell that is shaped like a group of flat sacs that are bundled together. Their function is to remove water from the proteins and prepare them for export from the cell.

Aparato de Golgi un orgánulo dentro de una celda que tiene la forma de un grupo de sacos planos que se empaquetan. Su función es eliminar el agua de las proteínas y prepararlos para su exportación fuera de la célula.

graduate degree either a master's degree or Ph.D. obtained after the completion of a bachelor's degree that may require students to be engaged in scientific research

Postgrado ya sea un máster o un doctorado obtenido tras la realización de una licenciatura que pueden requerir que los estudiantes que se dedican a la investigación científica

grease wool wool as it comes from the sheep

Grasa de lana la lana ya que proviene de las ovejas

greenhouse effect an effect supposedly caused by an increase of carbon dioxide and pollutants in the air. The effect is supposed to cause the climate of the earth to warm.

Efecto invernadero un efecto supuestamente causado por un aumento de dióxido de carbono y contaminantes en el aire. El efecto supuestamente a causa del clima de la tierra para calentar.

growing operation in swine production, the phase between the time they are weaned and the time they are finished for market

Funcionamiento creciente en la producción porcina, la fase entre el momento en que son destetados y el tiempo están acabados para el mercado

growth ability the ability of an animal to make efficient rapid growth

Capacidad de crecimiento la capacidad de un animal para eficientar el crecimiento rápido

grubs the larva stage of some insects, particularly beetles

Gusanos Las larvas de algunos insectos, especialmente escarabajos

guard bee worker bee who regulates all the insects that enter the hive

Abeja de guardia protector de abejas obreras que regula todos los insectos que entran en la colmena

H

hand breeding a system of breeding horses where the mares and stallions are kept separate until they are bred

Cría de mano un sistema de cría de caballos, donde las yeguas y sementales se mantienen separados hasta que sean criados

heifer a female bovine that has not produced a calf

Heifer bovinos hembra que no ha producido un ternero

helix strands consisting of molecules of DNA that are shaped like a corkscrew

Hélice hebras compuesto de moléculas de ADN que se modela como un sacacorchos

herbivores an animal that eats plants as the main part of its diet

Los herbívoros un animal que se alimenta de plantas como la parte principal de su dieta

heritability the portion of the differences in animals that is transmitted from parent to offspring

Heredabilidad la porción de las diferencias en los animales que se transmite de padres a hijos

heterosis the amount of superiority in a crossbred animal compared with the average of their purebred parents; also called **hybrid vigor**

Heterosis la cantidad de superioridad en un animal cruzado en comparación con el promedio de sus padres de pura raza; también se denomina **vigor híbrido**

heterozygous two parental genes calling for a specific characteristic (e.g., hair color) that are not identical (e.g., one calls for black hair, the other for white hair). The dominant gene will override the effect of the other gene.

Heterocigoto dos genes parentales pidiendo una característica específica (por ejemplo, color de pelo, que no son idénticos (por ejemplo, convocatorias de pelo negro, y la otra para el pelo blanco). El gen dominante anulará el efecto del otro gen.

hip height a measurement taken on the highest point of the hip of cattle at a given age.
This is an indication of the frame size and the weight of an animal at maturity.

La altura de la cadera una medida tomada en el punto más alto de la cadera del ganado en un Habida cuenta de la edad. Esto es una indicación del tamaño de fotograma y el peso de un animal en la madurez.

hippotherapy a type of physical therapy using animals with physically challenged humans to help improve mobility

Hipoterapia un tipo de terapia física utilizando animales con personas físicamente discapacitadas para ayudar a mejorar la movilidad

hives a structure used to house bees

Colmena una estructura utilizada para albergar las abejas

homeostasis the ability of an organism to remain stable when conditions around it are changing

Homeostasis capacidad de un organismo para permanecer estable cuando las condiciones alrededor de él están cambiando

homogenization the process of forcing the large cream globules through a screen at high pressure, reducing them to the size of the milk globules

homogeneización El proceso de forzar a los grandes glóbulos de crema a través de una pantalla a alta presión, reduciendo el tamaño de los glóbulos de leche

homogenized milk milk that has been blended to dissolve the fat molecules so that the fat (cream) will not become separated from the rest of the milk

Leche homogeneizada la leche que ha sido mezclado para disolver las moléculas grasas de manera que la grasa (crema) no serán separados del resto de la leche

homozygous two parental genes calling for a specific characteristic (e.g., hair color) that are identical (e.g., both call for black hair)

Homocigoto dos genes parentales pidiendo una característica específica (por ejemplo, el color de pelo) que son idénticos (por ejemplo, que tanto la convocatoria de cabello negro)

honey comb six-sided cells joined together, used to store nectar

Panal de miel seis caras celdas que se unen, que se utiliza para almacenar el néctar

hormones chemical substances secreted by various glands in an animal's body to control various bodily functions.

Las hormonas las sustancias químicas secretadas por diferentes glándulas en el cuerpo del animal para controlar diversas funciones corporales.

host an animal on which another organism depends for its existence

Anfitrión un animal en el que otro organismo depende para su existencia

hot branding using a hot iron to burn a permanent identifying mark onto an animal

Marca caliente usando un hierro caliente para grabar una marca de identificación permanente sobre un animal

hutches cubicles used to house rabbits

Las conejeras cubículos utilizadas para alojar a los Conejos

hybrid an animal produced from the mating of parents of different breeds

Híbrido un animal producidos desde el apareamiento de padres de diferentes razas

hybrid vigor *See* heterosis

Vigor híbrido *Véase la* heterosis

hyperplasia an increase in the number of cells in the tissues of organisms

Hiperplasia un aumento en el número de células en los tejidos de organismos

hypertrophy growth due to an increase in the size of cells

Hipertrofia crecimiento debido a un aumento en el tamaño de las celdas

hypothesis a theory by a scientist as to the cause or effect of a phenomena. This is tested by experimentation or other types of research.

Hipótesis una teoría por un científico en cuanto a la causa o el efecto de los fenómenos. Esto se prueba por experimentación u otros tipos de investigaciones.

I

ileum last division of the small intestine

El íleon la última división del intestino delgado

immobilization the process of rendering an animal oblivious to pain during the slaughter process

Inmovilización el proceso de representación de un animal ajeno al dolor durante el proceso de sacrificio

immunity resistance to catching a disease

Inmunidad resistencia a contraer una enfermedad

improvement project a project where the student makes improvements to the home, farm, or other area

Proyecto de mejora un proyecto donde el alumno realiza mejoras para el hogar, granja, o en otra área

imprinting a kind of behavior common to some newly hatched birds or newly born animals that causes them to adopt the first person, animal, or object they see as their parent

Impronta un tipo de comportamiento común a algunos pájaros recién nacidos o recién nacido animales que les induce a adoptar la primera persona, animal u objeto que ven como su padre

incubation the process of the development of a fertilized poultry egg into a newly hatched bird. The eggs must have the proper heat, humidity, and length of time.

incubación El proceso de desarrollo de un óvulo fecundado de aves de corral en un ave recién nacida. Los huevos deben tener la debida al calor, la humedad y la longitud de tiempo.

index a system of comparing animals within a group with the group average. A score of 100 is used for the average.

Índice un sistema de comparación de los animales dentro de un grupo con el promedio. Una puntuación de 100 se utiliza para el promedio.

indiscriminate breeders an animal that will breed with any animal of the same type and the opposite sex

Los criadores indiscriminados un animal que se reproducen con cualquier animal del mismo tipo y el sexo opuesto

infantile vulva a condition in gilts in which the vulva is very small and underdeveloped. Gilts with this condition are generally infertile.

Vulva infantil una condición en las primíparas en la vulva es muy pequeño y subdesarrollado. Las cerdas jóvenes con esta afección generalmente son estériles.

infectious diseases a disease that is contagious

Enfermedades infecciosas una enfermedad que es contagiosa

infundibulum the enlarged funnel-shaped structure on the end of the fallopian tube that functions in collecting the ova during ovulation

Infundíbulo el agrandamiento de la estructura en forma de embudo en el extremo de la trompa de falopio que funciona en la recolección de los óvulos durante la ovulación

ingestive behavior the mannerisms or habits that an animal uses during the intake of food

Comportamiento de ingestión los modismos y hábitos que un animal usos durante la ingesta de alimentos

injection curing pumping a curing solution into a meat product

Curado de inyección Bombeo de una solución de curado en un producto cárnico

inorganic not containing carbon and usually derived from nonliving sources

inorgánico Que no contienen carbono y generalmente derivan de fuentes nonliving

insect an animal of the class Insecta. They have three body parts and six legs.

Insectos un animal de la clase Insecta. Tienen tres piezas de carrocería y seis patas.

instinct the ability of an animal, based on its genetic makeup, to respond to an environmental stimulus

Instinto la capacidad de un animal, sobre la base de su composición genética, para responder a estímulos ambientales

intelligence the ability to learn

Inteligencia La capacidad de aprender

intermediate host an animal, other than the primary host, that a parasite uses to support part of its life cycle

Anfitrón intermedio un animal, excepto el host primario, un parásito que utiliza en apoyo de una parte de su ciclo de vida

internal parasites a parasite that lives inside the body of the host animal

Los parásitos internos un parásito que vive dentro del cuerpo del animal hospedador

inverted nipples a condition in female pigs in which the opening of the nipples on the mammary system appears to be inverted or to have a crater in the center. These are usually nonfunctional.

Los pezones invertidos una condición en lechones hembras en la que la apertura de las boquillas en el sistema mamaria parece estar invertida o tener un cráter en el centro. Estos son generalmente no funcionan.

irradiation a food preservation process that uses low levels of radiation to kill pathogens in food products

La irradiación un proceso de conservación de alimentos que utiliza niveles bajos de radiación para matar patógenos en los productos alimenticios

isthmus the part of the fallopian tubes between the ampulla and the uterus

Istmo parte de las trompas de falopio entre la ampolla y el útero

J

jejunum part of the small intestine

El yeyuno parte del intestino delgado

K

killer bees honeybees of African origin that are reputed to have a very aggressive nature

Abejas matadoras abejas de origen africano que tienen fama de tener una naturaleza agresiva

kingdoms the five common divisions into which natural objects are classified

Reinos las cinco divisiones en común que los objetos naturales están clasificados

kosher designates any food produced, killed, or prepared according to Jewish dietary laws

Kosher designa cualquier alimento producido, matado o preparado de acuerdo a las leyes dietéticas judías

L

laboratory animals an animal that is raised for the purpose of being used for laboratory experimentation

Animales de laboratorio un animal que se planteó con la finalidad de ser utilizados para la experimentación en laboratorio

lactase an enzyme produced in animal small intestines and other organs that breaks down lactose

La lactasa una enzima producida en el animal del intestino delgado y otros órganos que descompone la lactosa

lactation the process of an animal's giving milk

Lactancia el proceso de dar leche de un animal

lactose a sugar obtained from milk

La lactosa un azúcar obtenido a partir de leche

lagoon a body of water used for the decomposition of animal wastes

Laguna un cuerpo de agua utilizada para la descomposición de los residuos de origen animal

lamb referring to meat, that which comes from a sheep that is less than one year old

Cordero Se refiere a la carne, que procede de una oveja de menos de un año de antigüedad

lanolin the fatty substance removed from grease wool when it is scoured and cleaned

La lanolina la sustancia grasosa quita grasa de lana cuando se recorrieron y limpiado

lard the processed fat from swine

La manteca la grasa procesada de cerdo

larvae the immature stage of an insect from hatching to the pupal stage

Las larvas la fase inmadura de un insecto desde la eclosión hasta el estadio pupal

layers a chicken raised primarily for egg production

Capas un pollo planteada fundamentalmente para la producción de huevos

lean to fat ratio the amount of lean meat in a carcass compared to the amount of fat

La proporción de magra a grasa la cantidad de carne magra de la canal, en comparación con la cantidad de grasa

letdown process relaxation process (initiated by the release of oxytocin) allowing milk to pass out of the cow through the teat

Proceso de decepción proceso de relajación (iniciada por la liberación de oxitocina) permitiendo que pase la leche de la vaca a través de la tetina

libido the sexual drive of an animal

Libido el deseo sexual de un animal

life cycle the changes in the form of life an organism goes through in its lifetime

Ciclo de vida los cambios en la forma de vida de un organismo pasa a través de su vida útil

ligaments the tough, dense fibrous bands of tissue that connect bones or support viscera

Ligamentos las duras, densas bandas fibrosas de tejido que conectan los huesos o vísceras de apoyo

light horses a horse that weighs between 900 and 1,400 pounds at maturity

Caballos ligeros un caballo que pesa entre 900 y 1.400 libras al vencimiento

linear evaluation a method of evaluating the degree of a trait in an animal. Certain traits are given a score based on the ideal.

Evaluación lineal un método de evaluación del grado de un rasgo en un animal. Ciertos rasgos se da una puntuación basada en el ideal.

lipids a fat or fatty tissue

Lípidos grasas o tejido adiposo

lobule cluster of alveoli

Lobulillo cluster de alvéolos

loin eye a cross section of the Longissimus (the muscle running the length of the backbone) of an animal's carcass

Ojo de lomo una sección transversal del longissimus (el músculo que se extiende por la longitud de la columna vertebral) de un cadáver del animal

lumen hollow cavity in an organ (*pl.* **lumens** or **lumina**)

Lumen cavidad hueca en un órgano (*pl.* **Lúmenes** o **lumina**)

luteinizing hormone the hormone that stimulates ovulation

La hormona luteinizante la hormona que estimula la ovulación

lymphocytes a kind of white blood cell produced by the lymph glands and certain other tissues. It is associated with the production of antibodies.

Los linfocitos (un tipo de glóbulos blancos de la sangre producida por los ganglios linfáticos y otros tejidos. Es asociado con la producción de anticuerpos.

lysosomes organelles that are the digestive units of the cell that break down proteins, carbohydrates and other molecules as well as any foreign material such as bacteria that enters the cell

Los lisosomas orgánulos que son las unidades de digestivo de la célula que descomponen las proteínas, los carbohidratos y otras moléculas, así como cualquier material extraño, como bacterias que entra en la célula

M

macrominerals minerals that are required in relatively large amounts in an animal's diet

Los macrominerales minerales que son requeridos en cantidades relativamente grandes en la dieta del animal

magnum the part of the oviduct of a bird located between the infundibulum and the isthmus. This is where the albumin of the egg is produced.

Magnum parte del oviducto de un pájaro situado entre el infundíbulo y el istmo. Aquí es donde la albúmina de huevo se produce.

maiden flight the new queen bee's flight during which she mates with the drones

Primer vuelo la nueva reina de vuelo durante la cual ella se acopla con los aviones teledirigidos

maintenance ration the feed mixed in the proper proportions and amounts for an animal to maintain its weight and other bodily functions

La ración de mantenimiento la alimentación mixta en las proporciones y cantidades adecuadas para un animal para mantener su peso y otras funciones corporales

manure excrement from animals

Estiércol excrementos de animales

marbling the desired distribution of fat in the muscular tissue of meat that gives it a spotted appearance. Marbling is used in the quality grading of a carcass.

Veteado la distribución deseada de grasa en el tejido muscular de la carne que le da una apariencia moteada. El veteado es usado en la clasificación de la calidad de la canal.

mare a female horse that has produced a foal

Mare un caballo femenino que ha producido un potro

master's degree degree obtained after a bachelor's degree, in which one may be required to conduct scientific research in the required course work

Título de Máster título obtenido después de una licenciatura, en la cual uno puede ser necesario llevar a cabo investigaciones científicas en curso requerido trabajar

mastication the act of chewing food

La masticación el acto de masticar los alimentos

mastitis a disease involving the inflammation of the udder of milk-producing females

Mastitis una enfermedad que involucra la inflamación de las ubres de las hembras productoras de leche

maturity the point in an animal's life when it is old enough to reproduce, also refers to the age of an animal or carcass

La madurez El punto de madurez en la vida del animal cuando tiene la edad suficiente para reproducirse, también se refiere a la edad de un animal o de canal

meat animals animals that are raised primarily for the meat in their carcass

La carne de los animales los animales que se crían principalmente para las carnes en su canal

medium wool type a breed of sheep raised primarily for meat

El tipo de lana media una raza de ovejas levantadas principalmente para carne

meiosis cell division that results in the production of eggs and sperm

Meiosis la división celular que resulta en la producción de huevos y esperma

mesoderm the central layer of cells in a developing embryo, which gives rise to the circulatory system and certain other organs

Mesodermo la capa central de células en un embrión en desarrollo, lo que da lugar al sistema circulatorio y algunos otros órganos

mesophiles microbes that grow at medium temperatures (20°–45°C)

Mesófilos microbios que crecen a temperaturas medias (20°−45°C)

metabolism the chemical changes in cells, organs, and the entire body that provide energy for the animal

Metabolismo los cambios químicos en las células, los órganos y el cuerpo entero que proporcionan energía para el animal

metamorphosis the process by which organisms, especially insects, change in form and structure in their lives

Metamorfosis el proceso por el cual los organismos, especialmente los insectos, cambios en la forma y estructura en sus vidas

microfilaments fine fiber like structures composed of protein that help the cell to move by waving back and forth

Microfilamentos fibra fina como estructuras compuestas de proteínas que ayudan a la célula a moverse, moviendo hacia adelante y hacia atrás

microminerals minerals that are required in relatively small amounts in an animal's diet

Microminerales minerales que son requeridos en cantidades relativamente pequeñas en la dieta del animal

micronutrients nutrients that are required in relatively small amounts in an animal's diet

Micronutrientes nutrientes que son necesarios en cantidades relativamente pequeñas en la dieta del animal

microbe minute plant or animal life. Some cause disease; others are beneficial.

Microbio minutos de vida vegetal o animal. Algunos causan enfermedad; otros son beneficiosos.

micromanipulator a very small instrument that is used to dissect cells and embryos in the cloning process

Micromanipulator un pequeño instrumento que se usa para disecar células y embriones en el proceso de clonación.

milking parlors milking area

Corral de ordeño área de ordeño

mitochondria peanut-shaped organelles which functions to break down food nutrients and supply the cell with energy

Las mitocondrias orgánulos en forma de cacahuete que funciona para descomponer los alimentos nutrientes y suministro de energía de la celda

mitosis cell division involving the formation of chromosomes

Mitosis división celular que comprende la formación de cromosomas

mohair the long, lustrous hair from the angora goat

Mohair El pelo largo y lustroso de la cabra de angora

molting the process of poultry casting off old feathers before a new growth occurs

El proceso de muda en aves de corral viejos el proceso de perder las plumas antes de que un nuevo crecimiento ocurre

monera the kingdom to which singular celled organisms such as bacteria belong

Monera El reino al cual organismos unicelulares como bacterias pertenecen

monogastric a digestive system in which the stomach has only one compartment.

Los monogástricos un sistema digestivo en el cual el estómago tiene un solo compartimiento.

monosaccharides the simplest sugars, for example, glucose, fructose, and galactose

Monosacáridos El más simple de los azúcares, por ejemplo, la glucosa, la fructosa y la galactosa

morphogenesis process of cell development into different tissues and organs

Morfogénesis proceso de desarrollo celular en diferentes órganos y tejidos

morula a spherical mass of cells that develops into an embryo

Mórula una masa esférica de células que se desarrolla en un embrión

most probable producing ability an estimate of a cow's future productivity for a trait

Habilidad de producción más probable una estimación de la productividad futura de vaca para un rasgo

mother breeds those breeds of animals that make the best mothers, such as the Yorkshire and Landrace breeds of swine

Las razas madre las razas de animales que hacen la mejor de las madres, como las razas Yorkshire y Landrace de cerdo

motile able to move about

Motil tienen la habilidad de mover

mucin substance (secreted by cells in the magnum) that develops into the white or albumen of the egg

La mucina (sustancia secretada por las células en el magnum) que se desarrolla en el blanco o albúmina de huevo

mucous membranes a form of tissue in the body openings and digestive tract that secrete a viscous, watery substance called mucus

Las membranas mucosas una forma de tejido en los orificios de su cuerpo y del tracto digestivo, viscoso que segregan una sustancia acuosa llamada mucosidad

mules a cross between a horse and a donkey. The mother is a mare and the father is a jack.

Mulas un cruce entre un caballo y un burro. La madre es una yegua y el padre es un gato.

muscling the degree and thickness of muscle on an animal's body

Musculatura el grado y el espesor del músculo en el cuerpo de un animal

mutations an accident of heredity in which an offspring has different characteristics than the genetic code intended

Mutaciones un accidente de la herencia en la que la descendencia tiene diferentes características que destina el código genético

mutton the flesh of a sheep older than one year of age

El cordero la carne de oveja de más de un año de edad

mutualism a symbiotic relationship that is beneficial to both species

Mutualismo una relación simbiótica que es beneficiosa para ambas especies

myofibrils long bundles of fibrous tissue that make up the skeletal muscles.

Miofibrillas agrupación largo de tejido fibroso que componen los músculos esqueléticos.

myoglobin the iron-rich substance that gives blood its red color.

La mioglobina rico en hierro, la sustancia que da a la sangre su color rojo.

N

nanny a mature female goat

Cabra madre Una cabra hembra madura

National FFA Organization student organization dedicated to making a positive difference in the lives of students by developing their potential for premier leadership, personal growth, and career success through agricultural education (http://www.ffa.org/)

Organización nacional de FFA organización estudiantil dedicada a hacer una diferencia positiva en las vidas de los estudiantes mediante el desarrollo de su potencial de liderazgo premier, crecimiento personal y éxito profesional a través de la educación agrícola (http://www.ffa.org/)

naturally acquired active immunity immunity to a disease that is acquired by the animal's having had a disease

Naturalmente adquirido inmunidad activa Inmunidad a una enfermedad que es adquirida por el animal que haber tenido una enfermedad

natural selection the natural process that results in the survival of those individuals or groups best adjusted to the conditions under which they live; commonly called *survival of the fittest*

La selección natural el proceso natural que los resultados en la supervivencia de los individuos o grupos que mejor se ajuste a las condiciones en que viven; comúnmente llamado la *supervivencia del más apto*

nonessential amino acids amino acid that can be synthesized by the animal's body

Los aminoácidos no esenciales aminoácido que puede ser sintetizada por el cuerpo del animal

noninfectious diseases a disease that cannot be transmitted from one animal to another

Enfermedades no infecciosas una enfermedad que puede transmitirse de un animal a otro

notochord in animals of the phylum Chordata, a stringy rodlike structure that is made of tough, elastic tissue which is present in the embryo

Notocorda en animales del phylum Chordata, un pegajoso rodlike estructura que está hecha de tejido elástico y resistente que está presente en el embrión

nuclear transfer the process in which the nucleus of an egg is removed and replaced with the nucleus of another cell from an organ or other animal tissue

Transferencia nuclear El proceso en que el núcleo de un óvulo es eliminado y reemplazado con el núcleo de otra célula de un órgano o tejido animal otros

nucleotides a basic structural component of DNA and RNA

Los nucleótidos un componente estructural básico de DNA y RNA

nucleus the central portion of the cell that contains the genetic material

Núcleo la parte central de la celda que contiene el material genético

nukes small hives in which queen bees are commercially produced

Nukes Pequeño colmenas en que la reina de las abejas son producidos comercialmente

nursery a facility for caring for pigs after they are weaned

Vivero un centro para el cuidado de los cerdos después de que son destetadas

nursery bees the group of worker bees whose jobs it is to care for the brood and the queen

Guardería el grupo de abejas obreras, cuyo trabajo es cuidar de las crías y de la reina

nutrients substances that aid in the support of life

Nutrientes sustancias que ayuda en el soporte de la vida

nutritional disease a disease that is caused by not enough or too much of a certain nutrient in an animal's diet

Enfermedades nutricionales una enfermedad que es causada por la insuficiente o demasiado de un determinado nutriente en la dieta del animal

nymph a stage in the development of some insects that immediately precedes the adult stage

Ninfa una etapa en el desarrollo de algunos insectos que precede inmediatamente a la etapa adulta

O

offspring the young produced by animals

Descendientes los jóvenes producidos por animales

omasum the third compartment of the ruminant stomach, where a lot of the grinding of the feed occurs

Omaso el tercer compartimento del estómago de rumiantes, donde hay un montón de moler el alimento ocurre

omnivorous describing an animal that eats both plants and other animals

Omnívoro descripción de un animal que se alimenta de plantas y otros animales

oocyte an unfertilized egg

El ovocito un huevo no fertilizado

oogenesis the process of egg production in the female

Oogenesis el proceso de la producción de huevos en la hembra

orders smaller divisions within classes

Pedidos divisiones más pequeñas dentro de las clases

organelles structures within cells that form differing and various functions

Organelas estructuras dentro de las células que forman diferentes y diversas funciones

organic containing carbon or being of living origin

Orgánico Que contengan carbono o ser de origen vivo

organic matter matter such as decayed vegetation, manure, or other material that is in soil

Materia orgánica asunto como vegetación cariados, estiércol u otro material que esté en el suelo

organism any living being, plant or animal

Organismo cualquier ser vivo, planta o animal

osmosis the process by which the water moves from a region of high concentration of water to a region of low concentration

Ósmosis el proceso por el cual el agua se mueve desde una región de alta concentración de agua en una región de baja concentración

ossification the process of forming bone

Osificación El proceso de formar hueso

ovary the female organ that produces the egg and certain hormones

Ovario el órgano femenino que produce el huevo y ciertas hormonas

ovulation the process of releasing eggs from the ovarian follicles

La ovulación la liberación de óvulos de los folículos ováricos

ovum an egg

Ovum un huevo

ownership/entrepreneurship SAE a student program through which a student owns an enterprise such as livestock, crops, etc.

La propiedad/espíritu empresarial SAE un programa estudiantil a través del cual un estudiante posee una empresa como ganado, cosechas, etc.

oxidation any chemical change that involves the addition of oxygen

Oxidación cualquier cambio químico que consiste en la adición de oxígeno

oxytocin the hormone that stimulates constriction. It activates the egg-laying process in hens. It also causes the alveoli to release milk in cows.

La oxitocina la hormona que estimula la constricción. Activa el proceso para la puesta de huevos de gallinas. También hace que los alvéolos para liberar la leche de las vacas.

P

palatability the degree to which a feed or food is liked or accepted by an animal or human

Palatabilidad el grado en que un alimento o comida es deseado o aceptados por un animal o humano

papillae any small nipplelike projections

Papilas cualquier pequeña proyeccion como tetillas

paraffin a waxy substance

Parafina una sustancia cerosa

parasite an organism that lives and feeds on another organism

Parásito un organismo que vive y se alimenta de otro organismo

parasitism a symbiotic relationship in which one organism lives on or in another organism at that organism's expense

Parasitismo una relación simbiótica en la cual un organismo vive en o en otro organismo en que el gasto del organismo

passive immunity immunity that is temporary

Inmunidad pasiva inmunidad que es temporal

pasterns the part of an animal's leg that connects the cannon with the foot or hoof

Metacarpo parte de la pierna de un animal que conecta el cañón con el pie o pezuña

pasteurization the process of heat treating milk to kill microbes

Pasteurización El proceso de tratamiento térmico de la leche para matar los microbios

pasture breeding a system of breeding horses in which the stallion runs free in the pasture with the mares

Cría de pasto un sistema de cría de caballos en la que el semental corre libre en las pasturas con las yeguas

pecking order the order in which some poultry in a flock may peck others without being pecked in return

Orden jerarquía el orden en que algunas aves de corral en un rebaño puede picotear a otros sin ser picadas en volver

pedigree record the record of an animal's ancestry

El regristro de pedigrí El registro genealógico de la ascendencia de un animal

pelts the natural whole skin covering including the hair, wool, or fur of an animal

Pieles el todo natural de la piel que cubre incluso el cabello, lana, o piel de un animal

pelvic capacity the dimensions of a female's pelvic area that is an indication of its ability to give birth easily

Capacidad pélvica Las dimensiones del área pélvica de una mujer que es una indicación de su capacidad para parir fácilmente

penis the male organ of copulation

Pene órgano masculino de la copulación

pepsin a digestive enzyme secreted by the stomach

Pepsina una enzima digestiva secretada por el estómago

per capita consumption the amount of a product that is consumed by a person over the period of a year

El consumo per cápita la cantidad de un producto que es consumida por una persona durante el período de un año

performance data the record of an individual animal for reproduction, growth, and production

datos de rendimiento El registro de un animal individual para la reproducción, crecimiento y producción

periosteum the outer membrane or covering of bone.

El periostio la membrana externa o revestimiento de hueso.

perissodactyl an animal with only one toe on its foot, such as the horse, donkey, or zebra

Perissodactyl un animal con un solo dedo sobre su pie, como los caballos, burros, o cebra

pH a measure of the acidity or alkalinity of a substance

PH medida de la acidez o la alcalinidad de una sustancia

phagocytes an animal cell capable of ingesting micro organisms or other foreign bodies

Los fagocitos una célula animal capaz de ingerir microorganismos u otros cuerpos extraños

pharmaceuticals medicines or drugs used in human or animal health care

Productos farmacéuticos medicamentos o drogas utilizadas en la salud humana o animal care

Ph.D. degree obtained after a master's degree; Doctor of Philosophy

Doctorado Grado de Doctorado obtenido después de una maestría; Doctor en Filosofía

phenotype the observed characteristic of an animal without regard to its genetic makeup

fenotipo Observar el característico de un animal sin tener en cuenta su composición genética

pheromone a chemical that sends messages by organisms for the purposes of communication

Feromonas una sustancia química que envía mensajes por organismos a los efectos de la comunicación

photosynthesis the process by which green plants, using chlorophyll and the energy of sunlight, produce carbohydrates from water and carbon dioxide

Fotosíntesis el proceso por el cual las plantas verdes, con clorofila y la energía de la luz solar, producir carbohidratos a partir de dióxido de carbono y agua

phulon the Greek word for race or kind

Phulon la palabra griega para la raza o clase

phyla the primary divisions of the kingdom *Animalia*

Phylum los principales divisiones del Reino *Animalia*

physiological age the age of an animal as determined by an examination of the carcass

Edad fisiológica la edad de un animal determinado por un examen del cadáver

pigeon toed condition in which the front feet are turned in

Dedos de pichón una condición en la cual las patas delanteras están giradas al interior

pigmentation the naturally occurring color in the hair and skin of an animal

La pigmentación el color natural en el cabello y la piel de un animal

pin nipples small, underdeveloped nipples on the teats of a pig. They are usually nonfunctional.

pezones clavija , pezones pequeño subdesarrolladas en las tetas de un cerdo. Generalmente son inoperantes.

pituitary gland a small gland at the base of the brain that secretes hormones that stimulate growth and other functions

La glándula pituitaria una pequeña glándula ubicada en la base del cerebro que secreta hormonas que estimulan el crecimiento y otras funciones

Placement/Internship a student experience where the student works with a business, farm, or other place in order to learn skills

Colocación/Práctica un estudiante experiencia donde el estudiante trabaja con una empresa, granja, u otro lugar para aprender habilidades

placenta the membranous tissue that envelops a fetus in the uterus

Placenta el tejido membranoso que envuelve un feto en el útero

plantae the highest level of scientific classification (kingdom) to which all plants belong

Plantae el nivel más alto de clasificación científica (Reino Unido) a la que pertenecen todas las plantas

plasma a tan-colored fluid in the blood that suspends several substances that help sustain life.

Plasma un líquido de color marrón claro en la sangre que suspende varias sustancias que ayudan a sostener la vida.

plasma membrane a membrane that encloses and protects the cells contents from the external environment; regulates the movement of materials into and out of the cell such as the taking in of nutrients and the expelling of waste; and allows interaction with other cells

Membrana plasmática una membrana que rodea y protege las células contenido del entorno externo; regula la circulación de materiales dentro y fuera de la célula como la toma de nutrientes y expulsión de desechos; y permite la interacción con otras células

polar bodies produced during oogenesis as the result of the cytoplasm going to the cell that becomes the egg. Polar bodies function to provide sustenance for the egg until conception.

Órganos polar producidos durante la oogenesis como resultado del citoplasma va a la celda que se convierte en el huevo. Órganos Polar función para proporcionar sustento para el huevo hasta Concepción.

polled an animal that is naturally hornless

Encuestados un animal que es naturalmente hornless

ponies a horse that weighs 500–900 pounds at maturity

Potros un caballo que pesa 500 900 libras al vencimiento

Porcine Stress Syndrome (PSS) a condition in swine characterized by extreme muscling, nervousness, tail twitching, skin blotching, and sudden death

Síndrome de estrés porcino (PSS) una condición en el cerdo se caracteriza por la extrema musculatura, nerviosismo, temblores, cola piel blotching, y muerte súbita

Portfolio a collection of papers and other material that represents a body of work over a period of time

Portafolio una colección de documentos y otro material que representa un cuerpo de trabajo durante un periodo de tiempo

post legged a condition in animals in which the rear legs are too straight

Patas de puesto una condición de animales en la cual las patas traseras son demasiado recto

postmortem after death

Postmortem después de la muerte

postnatal after birth

Posnatal después del nacimiento

Post secondary Agriculture Students Organization student organization that participates in animal science activities

Organización de Estudiantes Post secundarias de Agricultura organización estudiantil que participa en actividades de ciencia animal

post secondary institution an educational institution attended after high school for higher education

Institución post secundaria una institución educacional después de la escuela secundaria para la educación superior

poultry any domesticated fowl, such as chickens, ducks, geese, or turkeys, that are raised for their meat, eggs, or feathers

Aves Cualquier ave domesticada, tales como pollos, patos, gansos, pavos, o que son criados por su carne, huevos o plumas

predators animals that kill and eat other animals

Depredadores animales que matan y come otros animales

prenatal before birth

Prenatal antes del nacimiento

prepuce *See* sheaths

Prepucial *Ver* fundas

primal cuts the most valuable cuts on a carcass, usually the leg (or hindquarter), loin, and rib

Primal cortes las piezas más valiosas en un cadáver, normalmente en la pierna (o el cuarto trasero), el lomo y la costilla

prion a proteinlike substance that can cause an infection or illness

Un prión una sustancia como un proteina que puede causar una infección o enfermedad

proficiency awards awards that are given to students based on superior Supervised Agricultural Experiences

Premios de Competencia los premios que se conceden a los estudiantes sobre basados en experiencias agrícolas supervisadas superiores

progeny the offspring of animals

Progenie los descendientes de animales

progeny testing determining the breeding value of animals by testing their offspring

Pruebas de progenie Determinación del valor genético de los animales mediante la comprobación de sus hijos

progesterone a hormone produced by the ovaries that functions in preparing the uterus for pregnancy and maintaining it if it occurs

La progesterona una hormona producida por los ovarios que funciona en la preparación del útero para el embarazo y mantenerla si ocurre

prokaryotic cells the smallest of all cells, they contain genetic materials but this material is not confined to a nucleus

Las células procarióticas la más pequeña de todas las celdas que contienen el material genético, pero este material no se limita a un núcleo

prolactin a hormone that stimulates the production of milk

La prolactina una hormona que estimula la producción de leche

propolis a glue or resin collected from trees and plants by bees. It is used to close holes in the hive.

Propolis un pegamento o resina de árboles y plantas recolectadas por las abejas. Se utiliza para cerrar los agujeros en la colmena.

prostaglandin a group of fatty acids that perform various physiological effects in an animal's body. Artificial prostaglandin is used in heat synchronization of cattle.

La prostaglandina un grupo de ácidos grasos que realizan diversos efectos fisiológicos en el cuerpo del animal. La prostaglandina artificial se utiliza en calentar la sincronización de ganado.

prostate gland the male reproductive gland that ejects the semen from the male reproductive tract

La glándula de la próstata la glándula reproductiva masculina que expulsa el semen en el tracto reproductivo masculino

protectant a substance added to semen to protect it during freezing and storage

Protectant sustancia añadida al semen para protegerlo durante la congelación y almacenamiento

protista the kingdom to which singular-celled organisms such as protozoa belong

Protista El reino al cual singular-celled los organismos como los protozoos pertenecen

protoplasm the material of plant and animal tissues in which all life activities occur

Protoplasma el material de los tejidos de plantas y animales en la que todas las actividades de la vida diaria se producen

protozoa single-celled organisms that are often parasitic

Protozoos organismos unicelulares que son a menudo parasitarias

protozoan infestation a condition where animals are suffering from protozoan parasites.

Infestación protozoario una condición en la que los animales sufren de parásitos protozoarios.

PSE pork pale, soft, and exudative pork; the meat is a very light pink in color and soft and dry in texture when cooked

PSE cerdo pálida, suave y exudativa; la carne de cerdo es un muy ligero de color rosa y de textura suave y seco cuando se cuece

psychrophiles microbes that grow well in cooler temperatures (0°–20°C)

Psicrofilos microbios que crecen bien en temperaturas más frías (0°-20°C)

public lands lands that are owned by the government

Tierras Públicas tierras que son propiedad del gobierno

pullets a young hen

Las pollitas una joven gallina

pupa the stage in an insect's life between the larva stage and the adult stage

Pupa la etapa en la vida de un insecto entre las larvas y los adultos

purebred an animal that belongs to one of the recognized breeds and has only that breed in its ancestry

raza pura Un animal que pertenece a una de las razas reconocidas y sólo tiene que crían en su ascendencia

purebred operation a cattle operation that raises purebred animals to be used in breeding programs

Operación de pura raza una operación de ganado que plantea animales de pura raza para ser utilizados en programas de mejoramiento

purebred producers producers who raise animals that are of a pure breed and are intended to be sold as breeding animals

Productores de pura raza los productores que crían animales de pura raza y están destinados a ser vendidos como animales de cría

Q

quality grade the grade given to a beef carcass that indicates the eating quality of the meat

Grado de Calidad la calificación otorgada a una carne canal que indica la calidad de la carne para comer

quarantine the isolation of an animal to prevent the spread of an infectious disease

Cuarentena el aislamiento de un animal para impedir la propagación de una enfermedad infecciosa

queen female bee, larger and more slender than other bees, whose main purpose is to lay eggs for the hive and is cared for by worker bees in the hive

La abeja reina la hembra de abeja, más grande y más delgadas que otras abejas, cuya finalidad principal es la de poner huevos para la colmena y está atendida por las abejas obreras en la colmena

queen cells special large cells in the hive in which new queen bees are developed

Las células de la reina celdas especiales grandes en la colmena en el cual las nuevas reinas son desarrollados

queen excluder a device placed in a beehive to prevent the queen from leaving the brood chamber

Exclusor de reina un dispositivo que se coloca en la colmena para evitar la reina de abandonar la cámara de la cría

quiescent cells the period of inactivity of a cell

Células inactivo El periodo de inactividad de una celda

R

radiation the emission of energy through waves of subatomic particles

radiación La emisión de energía a través de ondas de partículas subatómicas

ram a male sheep that has not been castrated

Carnero ovinos machos que no han sido castrados

rancid the putrefied state of foods

rancio El estado podrido de alimentos

ration the feed allowed for an animal in a 24-hour period

racion La alimentación permitidas para un animal en un período de 24 horas

recessive nondominant

Recesivo no dominantes

recessive gene a gene that is masked by another gene that is dominant

Gen recesivo un gen que está enmascarado por otro gen que es dominante

recipient cows a genetically inferior cow in which an embryo from a genetically superior cow is placed

Las vaca destinatorio una vaca genéticamente inferior en el que un embrión de una vaca genéticamente superior se coloca

refrigerated trucks trucks that contain their own refrigeration unit used for transporting meat or other perishable products

Camiones refrigerados camiones que contienen su propia unidad de refrigeración utilizados para el transporte de la carne u otros productos perecederos

rendered the process where fat from animals is heated until it melts and can be separated from the solid particles. The resulting fat is call lard and was used for cooking and making soap.

Las grasas fundidas el proceso cuando la grasa de animals es calentada hasta que funde y puede ser extraída de los partículos sólidos de carne. La grasa resultante se llama Manteca y se utiliza para cocinar y para fabricar jabón.

rennet (rennin) an enzyme extracted from the stomach of cattle, used in the cheese-making process

Cuajo (rennin) una enzima extraída del estómago de ganado, utilizados en el proceso de elaboración de queso

reproductive efficiency the capability of producing offspring in a timely and efficient manner

Eficiencia reproductiva la capacidad de producir descendencia de manera oportuna y eficiente

research SAE a student experience where the student conducts and reports on a research experiment or topic

La investigación SAE un estudiante experiencia donde el estudiante realiza e informa sobre un experimento de investigación o tema

retail cuts cuts of meat that are ready for purchase and use by the consumer

Cortes minoristas cortes de carne que están listos para su compra y uso del consumidor

reticulum the second compartment of a ruminant's stomach

Retículo el segundo compartimento del estómago de un rumiante

rib eye the exposed muscle surface that results when a side of beef is cut between the twelfth and thirteenth rib

Rib Eye la superficie muscular expuesta que resulta cuando un lado de la carne está cortada entre la 12ª y 13ª costilla

ribonucleic acid (RNA) a nucleic acid associated with the control of cellular chemical activities

Ácido ribonucleico (ARN) un ácido nucleico asociadas con el control de las actividades químicas celulares

rigor mortis a physiological process following the death of an animal in which the muscles stiffen and lock into place

Rigor Mortis un proceso fisiológico tras la muerte de un animal en el que los músculos se tensan y bloquearse en su lugar

roughage a feed low in carbohydrates and high in fiber content

El material tosco una alimentación baja en carbohidratos y alta en contenido de fibra

roundworms parasitic worms that live in the digestive tract of animals

Los nematodos gusanos parásitos que viven en el tracto digestivo de los animales

royal jelly secreted from bees, this food causes larvae to develop into queen bees

Jalea real segregada por las abejas, este alimento produce larvas a desarrollar abejas reinas

RST *See* BST

RST *véase el* BST

rumen the largest compartment of the stomach system of a ruminant. This is where a large amount of bacterial fermentation of feed occurs.

Rumen el mayor compartimento del estómago de los rumiantes. Aquí es donde una gran cantidad de fermentación bacteriana de alimento ocurre.

ruminants any of a class of animals having multicompartmented stomachs that are capable of digesting large amounts of roughages.

Los rumiantes cualquiera de una clase de animales teniendo estómagos multi-compartimentados que son capaces de digerir grandes cantidades de groseros.

S

salmonella a large group of bacteria, some of which cause food poisoning

Salmonella un grupo grande de bacterias, algunos de los cuales causan intoxicación alimentaria

school based enterprise SAE a Supervised Agricultural Experience that is conducted on the school campus

Educación escolar iniciativa SAE un experiencia agrícola supervisada que se llevó a cabo en el campus de la escuela

scientific method a systematic process of gaining knowledge through experimentation

El método científico un proceso sistemático de obtener conocimientos mediante la experimentación

scientific process a step by step process that a researcher uses to ensure that the research and outcome is correctly completed

Proceso científico un proceso paso a paso que un investigador utiliza para garantizar que la investigación y el resultado ha finalizado correctamente

scientific selection the selection of breeding or market animals based on the results of scientific research

Selección científica la selección de animales de cría o de mercado sobre la base de los resultados de la investigación científica

scoured wool wool after the fibers have been cleaned in the scouring process

Lana lavada después de haber limpiado las fibras en el proceso de decapado

scouring cleaning of grease wool by gently washing it in detergent

Decapado limpieza de grasa de lana suavemente por lavado con detergente

scout bees bees that locate nectar sources and report to the colony

Las abejas scout abejas que localizar fuentes de néctar e informe a la colonia

scrotal circumference a measurement taken around the scrotum of a bull. It is an indication of the fertility of the bull.

Circunferencia escrotal una medida tomada alrededor del escroto de un toro. Esto es un indicador de la fertilidad del toro.

scrotum the pouch that contains the testicles

Escroto el saco que contiene los testículos

seed stock cattle the cattle to be used as the dams and sires of calves that will be grown for market

Ganado de semillas ganado para ser utilizado como las presas y los toros de terneros que se cultiva para el mercado

seines large nets used to harvest fish from ponds

Mallas Las grandes redes de cerco utilizados para la pesca en los estanques

selective breeding choosing the best and desired animals and using those animals for breeding purposes

La cría selectiva elegir el mejor y deseada de animales y la utilización de estos animales para cría

semen a fluid substance produced by the male reproductive system that contains the sperm and secretions of the accessory glands

Semen Líquido de una sustancia producida por el sistema reproductivo masculino que contiene el semen y las secreciones de las glándulas accesorias

seminal vesicles a gland attached to the urethra that produces fluids to carry and nourish the sperm

Vesículas seminales una glándula adherida a la uretra que produce líquidos para llevar y nutren el esperma

semipermeable membrane a membrane that permits the diffusion of some components and not others. Usually water is allowed to pass, but solids are not.

Membrana semipermeable una membrana que permite la difusión de algunos componentes y no otros. Normalmente el agua se permite que pase, pero no son sólidos.

serum the clear portion of any animal fluid

El suero la porción clara de cualquier fluido animal

service animals animals that aid humans with disabilities. They may serve as aids in hearing, seeing, sensing danger, or other functions.

Animales de servicio Animales que ayuda a los seres humanos con discapacidades. Pueden servir como ayudas para oír, ver, sintiendo el peligro, u otras funciones.

service learning SAE a Supervised Agricultural Experience where the student does a significant amount of community service

Servicio de Aprendizaje SAE un experiencia agrícola supervisada donde el estudiante realiza una cantidad significativa de servicio a la comunidad

sex character the physical characteristics that distinguish males from females

Carácter sexual las características físicas que distinguen a los machos de las hembras

sexual reproduction reproduction that requires the uniting of an egg and a sperm

Reproducción sexual Reproducción que requiere la unión de un óvulo y un espermatozoide

sheaths the covering of the male penis

Fundas el cubrimiento del pene masculino

shell gland another name for the uterus of a hen

Glándula concha otro nombre para el útero de una gallina

shroud a cloth used to wrap a carcass during the aging process

Un cubierto de tela la tela utilizada para envolver un cadáver durante el proceso de envejecimiento

siblings brothers and sisters

Hermanos hermanos y hermanas

sickle hocked a condition in animals in which the back legs have too much curve

Pierna curvada una condición en los animales en los que la parte posterior de las piernas tienen demasiado curva

silage a crop, such as corn, that has been preserved in its succulent condition by partial fermentation

Un ensilaje cultivos, como el maíz, que ha sido conservado en su estado suculentas por fermentación parcial

sire the father of an animal

Progenitor el padre de un animal

sire breeds those breeds of agricultural animals that are used as sires in a cross-breeding program

Razas progenitoras las razas de animales agrícolas que se utilizan como toros en un programa de cría cruzada

smoothness lack of awkward bone structure and a smooth, even finish along the top and sides of an animal

Suavidad falta de torpe estructura ósea y un acabado suave y uniforme a lo largo de la parte superior y los lados de un animal

social behavior how animals act when they interact with each other

Comportamiento social cómo actuar cuando los animales interactúan entre sí

soundness structural strength and stability

Solidez resistencia y estabilidad estructural

sow a female pig that has had a litter of pigs

La cerda un cerdo hembra que ha tenido una camada de cerdos

species a category of individuals having common attributes as a logical division of a genus; the second name in the binomial nomenclature identifying an organism

Especies una categoría de personas que tienen atributos comunes como una división lógica de un género; el segundo nombre en la nomenclatura binomial identificar un organismo

specific gravity the density of a substance compared to the density of water

La gravedad específica la densidad de una sustancia en comparación con la densidad del agua

sperm the male reproductive cell that unites with an egg

Los espermatozoides la célula reproductiva masculina que une con un huevo

spermatogenesis the development of the sperm cell

La espermatogénesis el desarrollo de la célula espermatozoide

spermatogonia primitive male germ cells

La espermatogonia células primitiva germinativas masculinas

spermatozoa mature male gametes

Los espermatozoides gametos maduros masculinos

sperm nests pockets for storing sperm inside the oviducts

Nidos de esperma bolsillos para almacenar esperma dentro de los oviductos

sphincter muscle a ring-shaped muscle that closes an orifice

Esfínter muscular un músculo en forma de anillo que cierra un orificio

spirilla spiral-shaped bacteria

Spirilla bacterias con forma de espiral

splayfooted condition in an animal when its front feet are turned out

Pata torcida condición en un animal cuando sus patas delanteras están torcidas afueras

spore the reproductive "seed" of a fungus

Espora la "semilla" de reproducción de un hongo

stallion a mature male horse that has not been castrated

Semental un caballo macho maduro que no ha sido castrados

stanchion a loose-fitting device that goes around a cow's neck and limits the animal's mobility in the stall or milking area

Puntal un dispositivo holgada que va alrededor del cuello de una vaca y limita la movilidad del animal en el establo o área ordeñando

starter culture culture that starts the process of fermentation for making cheese

Cultivo láctico Cultura de arranque que inicia el proceso de fermentación para la fabricación de queso

steer a male bovine that has been castrated before sexual maturity

Dirección un bovino macho que ha sido castrados antes de la madurez sexual

sterile not capable of producing offspring

Estéril No capaces de producir descendencia

stimuli any agent that causes a response

Estímulos cualquier agente que causa una respuesta

stocker a calf that has been weaned and is being conditioned prior to entering the feed lot

Becerro un ternero que han sido destetados y está condicionado antes de entrar en el lote de alimentación

stocker operations an operation that conditions calves after weaning and before they enter the feed lot

Stocker operaciones una operación que condiciona los terneros después del destete y antes de entrar en el lote de alimentación

stomach worms parasite roundworm that burrows into the stomach lining and sucks the host animal's blood

Gusanos de estómago nemátodo parásito perfora el revestimiento del estómago y chupa la sangre del animal hospedador

straws a tube in which semen is frozen and stored

Pajuelas un tubo en la cual el semen se congelan y almacenan

striate having a striped appearance.

Estríada tener una apariencia rayada.

strongyle parasitic roundworm that lives in the intestine of the host animal

Estróngilos- nemátodo parásito vive en el intestino del animal hospedador

style the way an animal carries itself

Estilo la manera en que un animal lleva en sí

sucrose common table sugar (disaccharide composed of fructose and glucose)

La sacarosa (azúcar de mesa común (disacárido compuesto de fructosa y glucosa)

suint solid deposits from sheep perspiration found in wool

Sudor depósitos sólidos de transpiración encontrados en lana de oveja

super a boxlike structure that makes up part of a beehive. It is removed during honey harvest.

Super verificaciónal igual que la estructura que hace parte de una colmena. Es eliminado durante la cosecha de miel.

Supervised Agricultural Experience (SAE) a student experience in an area dealing with agriculture that is supervised by a teacher

La experiencia agrícola supervisada (SAE) Un estudiante experiencia en un área que se ocupaban de la agricultura, que es supervisado por un profesor

superovulation the stimulation of more than the usual number of ovulations during a single estrus cycle due to the injection of certain hormones

Superovulación la estimulación de más que el número habitual de ovulations durante un único ciclo estral debido a la inyección de algunas hormonas

Supervised Agricultural Experience Program (SAEP) the actual, hands-on application of concepts and principles learned in the agricultural education classroom (http://www.ffa.org/programs/sae/index.html).

Programa de experiencia agrícola supervisado (SAEP) la real, manos en la aplicación de conceptos y principios aprendidos en el aula de educación agrícola Http://www.ffa.org/programs/sae/index. (html).

surrogate mother a mother who acts as the recipient of an embryo from another parent and carries it until birth

La madre sustituta una madre que actúa como el receptor de un embrión de otro padre y lo transporta hasta el nacimiento

swarm a group of bees that have left the hive because of overcrowding

enjambre Un grupo de abejas que han abandonado la colmena a causa del hacinamiento

symbiosis the close association of two dissimilar organisms

Simbiosis la estrecha asociación de dos organismos diferentes

synapsis the process by which chromatids come together and are matched up in pairs

Synapsis el proceso por el cual cromátidas se juntan y coinciden arriba En pares

synovial fluid fluid in an animal's joint that acts as a lubricant.

El líquido sinovial Líquido en conjunto de un animal que actúa como lubricante.

synthetic lines boars or sows used for breeding that resulted from the blending of several different breeds

Líneas sintéticas verracos o utilizada para la cría de cerdas que resultó de la fusión de varias razas

systemic pesticides pesticides that are taken into the body of an animal and are part of the animal's system

Los pesticidas sistémicos plaguicidas que se toman en el cuerpo de un animal y son parte del sistema del animal

T

tankage dried animal residues usually freed from fats and gelatin

Depósito residuos animales - secados normalmente liberada de las grasas y de gelatina

tapeworms parasitic flat shaped worms that live in the digestive tracts of animals

Lombriz solitaria parasitio en forma plana gusanos que viven en el tracto intestinal de los animales

teat the portion of the mammary gland that expels milk

Pezón la porción de la glándula mamaria que expulsa la leche

tendons the tough connective tissues that binds the muscles to the bones, thereby enabling the muscles to move the bones.

Los tendones el tejido conectivo resistente que une los músculos a los huesos, permitiendo que los músculos para mover los huesos.

tertiary ducts part of the duct system that carries milk from the alveoli to the gland cistern where the milk is stored

Conductos terciarios parte del sistema de conductos que lleva la leche desde los alvéolos a la cisterna de la glándula donde está almacenada la leche

testicles the male organs that produce sperm and certain hormones

Testículos los órganos masculinos que producen los espermatozoides y ciertas hormonas

testosterone the male hormone responsible for the male sex drive and the development of the male sex characteristics

La testosterona la hormona masculina responsable de la unidad de sexo masculino y el desarrollo de las características sexuales masculinas

thermophiles microbes that grow at higher temperatures (45°–65°C)

Termófilas microbios que crecen a temperaturas más altas (45°–65°C)

thurl the thigh of an animal

Muslo el muslo de un animal

toxic poisonous

Tóxico Venenoso

trace minerals a mineral that is needed in relatively minute amounts in an animal's diet

Minerales traza Un mineral que se necesita en relativamente pequeñas cantidades en la dieta del animal

treatment group in a scientific experiment, the group of animals and plants that receives the treatment that is being researched

Grupo de tratamiento en un experimento científico, el grupo de animales y plantas que recibe el tratamiento que se está investigando

type the total of all the characteristics of an animal that make it and others like it unique

Tipo El total de todas las características de un animal que lo hacen y otros como única

U

underline the belly of an animal. In swine, it refers to the mammary system of the female.

Subrayado el vientre de un animal. En el cerdo, se refiere al sistema mamaria de la hembra.

urethra the tube that carries urine from the bladder and serves as a duct for the passage of the male's semen

La uretra el tubo que transporta la orina desde la vejiga y sirve como un conducto para el paso del semen del macho

USDA United States Department of Agriculture

El USDA Departamento de Agricultura de los Estados Unidos

uterus the female reproductive organ in which the fetus develops before birth

Útero el órgano reproductor femenino, en el cual el feto se desarrolla antes del nacimiento

V

vaccinating the process of injecting an animal with certain microorganisms in an effort to make the animal immune to specific diseases

Vacunación el proceso de inyectar un animal con ciertos microorganismos en un esfuerzo para hacer el animal inmune a enfermedades específicas

Vacuna una sustancia que contiene live, modificados o los organismos muertos que se inyecta en un animal para hacerla inmune a una enfermedad específica

vaccine a substance that contains live, modified, or dead organisms that is injected into an animal to make it immune to a specific disease

vacuoles organelles that serve as storage compartments for the cell

Vacuolas orgánulos que sirven como compartimentos de almacenamiento para la celda

vacuum packaging a means of packaging meat by wrapping it in plastic wrap and drawing the air from it

El envasado al vacío un medio de envase carne envolviéndolo en una envoltura de plástico y el dibujo del aire

vagina the canal in the female reaching from the uterus to the vulva

Vagina el canal en la hembra llegando desde el útero hasta la vulva

vas deferens the tube connecting the epididymis of the testicles to the urethra

Conductos deferentes el tubo que conecta el epidídimo de los testículos hasta la uretra

veal the meat from calves slaughtered before they are three months of age

Vacuno la carne de los terneros sacrificados antes de los tres meses de edad

vertebrae a series of irregularly shaped bones that make up the backbone of an animal.

Las vértebras una serie de forma irregular de huesos que conforman la columna vertebral de un animal.

vertebrate an animal having a backbone

vertebrado Un animal teniendo un backbone

vertical integration poultry operation where the producer or company owns the hatchery, feed mills, processing plants, and distribution centers

Integración vertical aves de corral operación donde el productor o la empresa propietaria del criadero, fábricas de piensos, plantas de procesamiento, y centros de distribución

viable capable of living

Viables capaces de vivir

villi microscopic, fingerlike projections of the inner lining of the digestive tract

Vellosidades proyecciones microscópicas como deditos del revestimiento interior del tracto digestivo

virus disease-causing agent

Virus agente causante de enfermedades

vulva the external reproductive organ of the female

Vulva el órgano reproductivo externo de la hembra

warm blooded *See* endothermic

sangre caliente *Ver* endotérmica

warm water fish a fish that does not thrive in water colder than 60°F

Peces de agua caliente un pez que no pueden prosperar en el agua más fría de 60°F

weaned a young animal no longer dependent on its mother's milk

Destetado un animal joven ya no depende de la leche de su madre

weaning weight the weight of a calf at weaning, usually considered to be about 500 pounds

El peso al destete el peso de los terneros al destete, generalmente considerado unos 500 libras

wether a male sheep or goat that has been castrated

caprino castrado un macho de oveja o cabra que ha sido castrados

whey watery part of milk that is separated from the curd in the cheese-making process

Suero parte acuosa de la leche que se separa de la cuajada en el proceso de elaboración de queso

wholesale cuts the major parts of a carcass that are boxed and sold to wholesale distributors

Cortes mayoristas las partes principales de un canal que están embalados y vendidos a los distribuidores mayoristas

withdrawal periods the length of time that must transpire between the time an animal is given a certain drug and the time the animal's milk can be used or the animal is slaughtered

Los períodos de abstinencia la cantidad de tiempo que transcurre entre el momento en que un animal es un determinado fármaco y el tiempo que el animal la leche puede ser utilizado o el animal es sacrificado

wool type a breed of sheep raised primarily for wool

Lana tipo una raza de ovejas levantadas principalmente para lana

work animal an animal that is raised primarily for the work it can perform

animal de trabajo un animal que está planteada fundamentalmente por la labor que pueden realizar

worker one of the three types of bees found within a bee colony: queen, drones, workers

Abeja obrera una de las tres tipos de abejas ubicadas adentro de una colonia de abejas: la abejas reina, los zánganos de abeja, las abejas obreras

worker bees sterile female bee; comprise the largest number of bees in the colony and have many jobs in the hive including bringing nectar and pollen to the hive from the field, producing and storing honey, acting as guard bees for the hive, and caring for the queen and brood

Las abejas obreras abeja hembra estéril; constituyen el mayor número de abejas en la colonia y tiene muchos trabajos en la colmena, incluyendo la incorporación de néctar y polen a la colmena desde el campo, producir y almacenar la miel, que actúa como protector de la colmena de abejas, y el cuidado de la reina y cría

yearling an animal that is a year old

Añojo un animal que es un año viejo

yearling weight the weight of a beef animal at 365 days of age

peso de añojo el peso de un animal a 365 días de edad

yield grade a grade in meat animals that
refers to the amount of lean meat produced in a carcass

El grado de rendimiento un grado en la carne de los animales que se refiere a la cantidad de carne magra producida en un cadáver

yogurt a semisolid, fermented milk product

Yogur semisólido, producto de leche fermentada

yolk the yellowish part of a fowl's egg that contains the germinal disk; also the substances such as wool grease in a fleece

La yema la parte amarillento de un huevo de gallina que contiene el disco germinal; también las sustancias tales como Grasa de lana en un vellón

zoonoses diseases and infections that can be transmitted from animal to humans

Zoonosis enfermedades e infecciones que pueden transmitirse de animales a seres humanos

zygote a fertilized egg

El cigoto un óvulo fertilizado

INDEX

A

A (adenine), 216–217
Abomasum, 269, 336
Abyssinian cats, 148
Active immunity, 379
Adenine (A), 216–217
ADGA (American Dairy Goat Association), 110
Adipose tissue, 351
Adrenal gland and animal growth, 277, 319
Aerators, 135–136
Aerobic organisms, 353
Aflatoxin, 381
African bees, 181–182
Aged beef, 345
Aging process, 320
Agricultural animals, 6
  alternative. *See* Alternative animal agriculture
  classification. *See* Classification of agricultural animals
  scientific selection of. *See* Scientific selection of agricultural animals
Agricultural education programs, 14–18
Agricultural journalist, 421
Agricultural Marketing Service (AM) of the US Department of Agriculture, 345
Agriculture, 4
Agriscience Fair, 426
AI (artificial insemination), 11–12, 128, 290–294
  estrus cycle, control of, 293–294
  queen bees, 182
  semen collection and processing, 291–293
Alaskan Malamutes, 147
Albumen, 76
Alimentary canal, 334
Allele, 218
Allergies to animals, 154
Alligators, 141–142
Alpine goats, 112
Alternative animal agriculture, 156–167
  fish bait production, 161–162
    crickets, 162
    earthworms, 161
  hunting preserves, 166–167
  laboratory animal production, 162–163
  large game animals, 162
  llama production, 160–161
  natural and certified animal products, 163–166
    USDA regulations, 164–166
  rabbit production, 158–160
Alveoli, 49
American Dairy Goat Association (ADGA), 110
American FFA Degree, 17–18
American Guinea swine, 65
American Quarter Horse Association, 124
American Rabbit Breeders Association (ARBA), 158
American Saddlebred horse, 124
Amino acids, 68, 325
Amniotic fluid, 315
Amphibia, 23
AMS (Agricultural Marketing Service) of the U.S. Department of Agriculture, 345
Anabolism, 324
Anaerobic organisms, 353
Anaphase, 209–210
Anemia, 360, 372
Angora goats, 99–100, 107
Angus, 38
Animal agriculture as science, 2–18
  advances in production of food from animals, 7–9
  agricultural education programs, 14–18
  animal immunization, 9–10
  artificial insemination, 11
  early legislation, 4–5
  embryo transfer, 11
  refrigeration, 10–11
  scientific method, 5–7
  use of computers, 11–13
Animal and product uniformity
  cloning and, 302–303
Animal behavior, 186–198
  animal communication, 195–198
  conditioning, 189
  ethology, 188
  imprinting, 188
  ingestive behavior, 193–195
  instincts, 188
  intelligence, 189
  sexual and reproductive behavior, 192–193
  social behavior, 191–192

Animal buyer, 420
Animal cells, 200–211
  components
    cell membranes, 205
    organelles, 206–208
  importance, 202
  reproduction, 208–210
    anaphase, 209–210
    interphase, 208–209
    meiosis, 208
    metaphase, 209
    mitosis, 208
    prophase, 209
    telophase, 209–210
  stem cells, 210–211
  types, 202–208
    eukaryotic cells, 202–204
    prokaryotic cells, 202
Animal communication, 195–198
Animal diseases, 370–386
  disease and nutritional defects, 372–375
  disease prevention, 384–386
  immune system, 378–382
    fungal diseases, 380–381
    immunity, 379–380
    prions, 381–382
  infectious diseases, 375–378
    bacteria, 376–377
    protozoa, 377–378
    viruses, 377
  noninfectious diseases, 382–383
    genetic diseases, 382
    nutritional diseases, 382–383
  poisoning, 383–384
  Veterinary Science Proficiency Award, 386
Animal geneticist, 422
Animal genetics, 214–226
  determination of sex, 220–221
  gene transfer, 216–220
  performance data, 221–226
    breeding values, 222–223
    expected progeny difference (EPD), 223–224
    indexes, 222
    linear classification, 224
    mothering ability, 222
  using genetics in selection process, 221
Animal growth and development, 312–320
  aging process, 320
  effects of hormones on growth, 319–320
  postnatal growth and development, 317–319
  prenatal growth, 314–317
    embryonic phase, 315–316
    fetal stage, 316–317
    ovum stage, 314–315
Animalia, 23, 25
Animal immunization, 9–10
Animal medications, 408
Animal nutrition, 321–337
  carbohydrates, 328–329
  digestion process, 334–337
    components of ruminant stomach, 336
    monogastric digestive systems, 334–336
    ruminant digestive system, 336
  fats, 329–330
  minerals, 330–331
  protein, 325–328
  vitamins, 333–334
    A, 331
    B, 332
    C, 333–334
    D, 331
    E, 331
    K, 331–332
  water, 324–325
Animal nutritionist, 422
Animal rights activists, 390
Animal Rights Movement, 390
Animal science, careers in. *See* Careers in animal science
Animal systems
  bodies function properly, 256
  circulatory system
    arteries, 271, 272
    blood, 272–273
    heart, 271, 272
  digestive
    monogastric systems, 265–267
    ruminants, 267–268
  endocrine
    adrenal glands, 277
    hormones, 275
    hypothalamus, 277
    pancreas, 277
    pituitary gland, 275, 276
    thyroid gland, 277
  muscular
    cardiac muscle, 264–265
    skeletal muscle, 263–264
    smooth muscle, 264
  nervous, 273–274
  respiratory, 269–270
  skeletal
    bones, 257
    cartilage, 256
    flat bones, 261
    irregular bones, 260–261
    joints, 261–263
    long bones, 257–259
    pelvic bones, 258, 259
    ribs, 258, 259
    short bones, 260
Animal welfare, 388–398
  confinement operations, 390–392
  historical view, 390
  management practices, 393–396
  research, 396–397
  use of drugs, 392–393
Animal welfare activists, 390
Antemortem, 403
Anthelmintics, 115
Anthrax, 373
Antibiotics, 408
  in animal feed, 392
Antibodies, 49
Antigens, 379
Antilopinae, 24
Apiary, 172
Applied research, 6
Aquaculture industry, 132–142
  alligators, 141–142
  aquaculture defined, 134
  bullfrogs, 140
  crayfish, 140–141
  fish production, 134–139
    catfish, 136–137
    salmon, 139
    tilapia, 138–139
    trout, 139
  history, 134
  sport fishing, 139–140
Arabian horse breed, 124

ARBA (American Rabbit Breeders Association), 158
Archaebacteria, 23
Arteries, 271, 272
Artificial active immunity, 379–380
Artificial hormones, 293, 319
Artificial insemination (AI), 11–12, 128, 290–294
  estrus cycle, control of, 293–294
  queen bees, 182
  semen collection and processing, 291–293
Artificial insemination technician, 420
Artificial vagina, 291
Ascarids, 362
Asexual reproduction, 282
Assistance dogs, 150
Associate's degree, 420
  careers with, 420
Atropic rhinitis, 375
Australian Cattle Dogs, 147
Aves, 23
Ayrshires, 28

B

Baby beef, 35
Bachelor's degree, 420–421
  careers with, 421–422
Bacillus, 376
Backfat, 234, 239
Bacteria, 202, 204, 336, 376–377
Bacterial diseases, 373–
Balance, 246
Balanced ration, 46
Barrows, 67, 319
Basic research, 6
Bass, 139
Beagles, 147
Beak trimming, 393
Beef industry, 32–43
  beef in the American diet, 34–36
    baby beef, 35
    certified organic beef, 36
    grain-fed beef, 35
    grass-fed beef, 35
    natural beef, 35–36
    types of beef, 34
    veal, 34
  breeds of beef cattle, 38–39
  heritability estimates, 221
  nutritional value of beef, 159
  scientific selection of agricultural animals, 238–242
  segments, 39–42
  in the United States, 36–38
  world production, 9
Bees
  communication among, 195
  honeybee industry. See Honeybee industry
  space, 178
Behavior, animal, 186–199
  animal communication, 195–198
  conditioning, 189
  ethology, 188
  imprinting, 188–189
  ingestive behavior, 193–195
  instincts, 190
  intelligence, 189
  sexual and reproductive behavior, 192–193
  social behavior, 191–192
Berkshire swine, 63
Billy goats, 108
Binomial nomenclature, 22
Biological control, 367
Biotin, 332
Bitterweed, 383
Black and Tan Coonhounds, 147
Blackleg, 373
Blast freezing, 355
Blastula, 211
Blind nipples, 235
Bloat, 337, 372
Block and Bridle, 421
Blood cell, 203
Bloodhounds, 147
Blood typing, 26
Blue-green algae, 202
Boar
  semen volume and numbers, 292
Boer goats, 108–109
Boluses, 336, 367
Bone cells, 206, 305
Border Collies, 147
Bovidae, 24–25
Boxers, 147
Brahman, 27, 39
Brangus breed, 26, 39
Bream, 139
Breed, 25
Breed associations, 26
Breeder industry, 73–74
Breeder stock
  USDA regulations for organic animal products, 164–166
Breeding cattle
  scientific selection of agricultural animals, 243
Breeding hogs
  scientific selection of agricultural animals, 235–238
    carcass, 238
    growthiness, 237–238
    structural soundness, 236–237
Breeding true, 25
Breeding values, 222–223
Brisket, 244
British breeds of beef cattle, 38
Brittany Spaniels, 147
Broiler industry, 74–75
  broiler houses, 75
Broiler production, 75–83
  at the broiler house, 81–82
  egg production, 76–77
  embryo development, 79–81
  hatching eggs, 77–79
  processing plant, 83
  when the chicks hatch, 81
Broilers, 8–9
Brood (bees), 174
Brood cells, 174
Brood chamber, 179
Browsers, 114
Brucellosis, 373
BST (recombinant bovine somatotropin), 411
Buck goats, 112
Buckling goats, 112
Bull
  reproductive tract of, 284
  semen volume and numbers, 292
Bullfrogs, 140
Bull Terriers, 147
Bureau of Land Management, 412
Buttercups, 383
B vitamins, 332

C

C (cytosine), 216–217
Cabrito, 105
Cage operations, 83
Calf diptheria, 375
California
  leading milk-prducing state, 46–47
Camels, 30
Candling, 84
Cannibalism, 75
Cannon bone, 238
Carbohydrates, 328–329
Carcass, 238
Carcass merit, 68, 246
Cardiac muscle, 264–265
Carding, 98
Career, 418
Career Development Event in Parliamentary Procedure, 414
Career Development Events, 88
  National FFA Organization, 88
Careers in animal science, 416–427
  Agriscience Fair, 426
  career options, 418–422
    careers with associate's degree, 420
    careers with bachelor's degree, 420–421
    careers with graduate degree, 421–422
  interview preparation, 423–425
  personal and leadership skills, developing, 422–423
Carniolan bees, 181
Carnivores, 326
Carotene, 331
Cartilage, 256, 318
Cashmere goats, 107
Castings, 161
Castration, 66, 319, 393
Catabolism, 324
Catfish, 136–137
Catheter, 296
Cats, 147–148
Cat scratch fever, 154
Cattle
  communication among, 195
  grazing, 195
Caucasian bees, 181
Cecum, 126, 195, 334
Cells, 76, 200–211
  components
    cell membranes, 205
    organelles, 206–208
  importance, 202
  reproduction, 208–210
    anaphase, 209
    interphase, 208–209
    meiosis, 208
    metaphase, 209
    mitosis, 208
    prophase, 209
    telophase, 209–210
  stem cells, 210–211
  types, 202–208
    eukaryotic cells, 202–204
    prokaryotic cells, 202
Cellulose, 328
Centrioles, 208
Certified organic beef, 36
Cervidae, 24
Cervix, 286
Chalazae, 76
Chapter Degree, 16–17
Charolais, 38
Cheese manufacturing, 56–57
Chemicals, 393
Chester White swine, 63
Chevon, 105
Chianina, 38–39
Chickens
  communication among, 195
  nutritional value of, 159
  poultry industry. *See* Poultry industry
Chick grader, 419
Chihuahua, 147
Chimpanzees, 188–189
Chine, 344
Cholera, 372
Cholesterol, 409–410
Choline, 332
Chromatids, 283
Chromista, 23
Chromosomes, 204, 216
  number in selected animals, 282–283
Chronological age, 320
Circulatory system
  arteries, 271–272
  blood, 272–273
  heart, 271, 272
Civil War, wool uniforms in, 96
Classes, 23–25
Classification of agricultural animals, 20–31
  classification according to use, 27–30
    dogs, 29
    dual-purpose animals, 29–30
    horses, 28–29
    meat animals, 27–28
    work animals, 28
  classification by breeds, 25–27
    blood typing, 26
    crossbreeding, 26–27
    purebreds, 26
    selective breeding, 25–26
  scientific classification, 22–25
    classes, 23–24
    families, 24
    genus species, 24–25
    kingdoms, 23
    orders, 24
    phyla, 23
Cleavage, 210
Climate-controlled houses, 65
Clitoris, 286
Cloaca, 77
Cloning, 298, 300–310
  development of cloning process, 305–308
    cloning mammals, 307–308
    genetically altered clones, 308
  differences in clones, 309–310
  perfecting the process, 308–309
  reasons for cloning, 302–305
    animal and product uniformity, 303–304
    endangered species, 304
    genetic superiority, 302–303
    research, 304–305
  twins and triplets as clones, 302
Clotting factors, 273
Cocci, 376
Coccidiosis, 375, 378
Cocklebur, 383
Codominant genes, 219
Cold-blooded, 134
Cold-water fish, 136
Colic, 362, 372

Coliforms, 411
Collagen, 320
College professor, 422
Collegiate FFA, 421
Collegiate 4-H, 421
Colon, 411
Colony (bees), 195
Colostrum, 49, 129
Combination curing, 354
Combing, 98
Commensalism, 360
Commercial goat producers, 107
Communication, animal, 195–198
Community college, 418
Companion animals, 146
Company representative, 421
Computer operator, 420
Computers, use of, 11–13
Concentrate, 241
Conception, 287
Concrete floors, 236
Conditioning, 189
Confinement operations, 66, 390–392
Conformation, 247
Consumer concerns, 400–414
  animal medications, 408
  chemicals, 402
  cholesterol, 409–410
  country of origin, 408
  environmental concerns, 411
  food poisoning, 402–403
  genetic engineering, 410–411
  global warming, 413–414
  hormones, 409
  overgrazing of public lands, 412–413
Contagious diseases, 375
Continental European breeds of beef cattle, 38
Control group, 6
Copulation, 287
Corpus luteum, 287
Cortex, 96
Country of origin, 408
Cow-calf operations, 39–41
Cow-hocked, 237
Cowper's glands, 285
Cows. See also Beef industry; Dairy industry
  reproduction cycle, 288
Crappie, 139
Crayfish, 140–141
Creed
  National FFA Organization, 14–15
Crickets, 162
Crimp, 97
Crossbreeding, 26–27
Crude protein content, 326
Crustaceans, 134
Cud, 24
Cud chewers, 267
Curd, 56
Curing, 353–354, 402
Cutability, 245
Cuticle, 96
Cytokinesis, 210
Cytoplasm, 204, 284, 314
Cytosine (C), 216–217

D

Dairy Cattle Evaluation and Management Career Development Event, 58
Dairy goats, 54–56, 105–107, 251
  breeds,112
Dairy industry, 44–58
  cheese manufacturing, 56–57
  Dairy Cattle Evaluation and Management Career Development Event, 58
  feeding, 46–48
  gestation, 48–49
  goats and sheep, 54–56
  milk production, 49–54
    letdown process, 49–50
    milking parlors, 50–53
  USDA regulations for organic animal products, 164–166
Dairy inspector, 421
Dam breeds, 38
Dams, 11
Data, performance, 221–226, 232
  breeding values, 222–223
  expected progeny difference (EPD), 223–224
  indexes, 222
  linear classification, 224
  mothering ability, 222
Debeaking, 393
Debouillet sheep, 99
Degree program, 16
  national FFA Organization, 16–18
Dehorning, 393
Delaine sheep, 99
Dental pad, 194
Deoxyribonucleic acid (DNA), 216
de Soto, Hernando, 62
Differentiation, 217, 289, 308
Diffusion, 205–206, 336
Digestion process, 334–337
  components of ruminant stomach, 336–337
  monogastric digestive systems, 334–336
  ruminant digestive system, 336
Digestive system
  monogastric systems, 265–267
    large intestine, 267
    mouth, 265–266
    small intestine, 266–267
    stomach, 266
  ruminants systems
    abomasum, 269
    omasum, 269
    reticulum, 268–269
    rumen, 269
Disaccharides, 328
Discovery Degree, 16
Diseases, 370–386
  disease and nutritional defects, 372–375
  immune system, 378–382
    fungal diseases, 380–381
    immunity, 379–380
    prions, 381–382
  infectious diseases, 375–378
    bacteria, 376–377
    protozoa, 377–378
    viruses, 377
  noninfectious diseases, 382–383
    genetic diseases, 382
    nutritional diseases, 382–383
  poisoning, 383–384
  prevention, 384–386
  Veterinary Science Proficiency Award, 386
Distemper, 374
DNA (deoxyribonucleic acid), 216
Doberman Pinschers, 147
Docile, 190

Docking, 66, 393
Doe goats, 112
Doeling goats, 112
Dogs, 29, 146–147
Dolly (first cloned sheep), 307–308
Dolphins, 189
Domesticated animals, 4
Donor cows, 309
Dorset sheep, 94
Double muscling, 241
Draft horses, 29, 123–124
Drones, 172
Drugs, use of, 392–393
Dry curing, 354
Drying to preserve meat, 354
Dual-purpose animals, 29–30
Dual-purpose goats, 110
Ducks, 86–87
Duodenum, 335
Duroc swine, 63

E

E. coli, 356, 402–403
Earthworms, 161
Ecological balance, 95
Ectoderm, 315
Ectothermic, 134
Efficiency, 232
Egg candler, 419
Egg production, 75–77
Ejaculation, 288
Elastin, 351
Elastrator, 394
Embryo, 75
Embryo development, 79–80
Embryo implant technician, 420
Embryonic phase, 315–316
Embryo transfer, 11, 294–298
  new technology in, 297–298
  process of, 295–297
Embryo transplant, 48
Endangered species, 304
  cloning and, 304
Endocrine system, 286
  adrenal glands, 277
  hormones, 275
  hypothalamus, 277
  pancreas, 277
  pituitary gland, 275, 276
  thyroid gland, 277
Endoderm, 315
Endoplasmic reticulum, 208
Enterotoxemia, 372
Enucleated oocyte, 306
Environment, 10
Environmental concerns, 411–412
  swine industry, 68
Enzymes, 56, 206
EPD (expected progeny difference), 223
Epididymis, 285
Epinephrine, 50
Epistasis, 219
Equine encephalomyelitis, 372
Ergot, 383
Erysipelas, 374
Esophagus, 334
Essential amino acids, 326
Estimated breeding value, 222–223
Estrogens, 286
  animal growth and, 319
Estrus, 193, 285, 288
  control of estrus cycle, 293–294
Estrus synchronization, 293
Ethology, 188
Eubacteria, 23
Eukaryotic cells, 202, 204
Ewes, 221, 246–247
  reproduction cycle, 288
Exothermic animals (reptiles), 149
Exotic animals, 148
  as pets, 148
Exotic breeds, 38–39
Expected progeny difference (EPD), 223–224
Experiment, 5
Experimental group, 6
Exsanguination, 343
Extenders, 292
Extension agent, 421
External parasites, 361, 364–367
  heel flies, 366–367
  lice, 365–366
  ticks, 365

F

Factor IX, 308
Facultative, 353
Fallopian tubes
Families, 24–25
Farm manager, 421
Farriers, 129, 420
Farrowing operation, 65
Fat cells, 203, 205
Fats, 329–330
FDA (Food and Drug Administration), 367, 392–393, 408
Feed conversion ratio, 67
Feeder pigs, 65
Feedlot operations, 39, 41–42, 391
Feed mill worker, 419
Feedstuff, 324
Felting, 97
Female reproductive system, 285–288
Fermentation, 56
Fertile, 231
Fertilization, 79, 288–290
Fertilization membrane, 289
Fertilized egg, 216
Fetal stage, 316–317
FFA (Future Farmers of America), 14, 356
FFA Agriscience Fair, 426
FFA Livestock Evaluation Career Development Event
  Scientific selection of agricultural animals, 228–251
FFA Meats Evaluation Career Development Event, 356
FFA Poultry Judging Career Development Event, 88
Fiber goats, 107
Fingerlings, 137
Finish, 42, 246
Finished (swine), 68
Finishing operation, 66
Fish bait production, 161–162
  crickets, 162
  earthworms, 161
Fish production, 134–139
  catfish, 136–137
  salmon, 139
  tilapia, 138–139
  trout, 139
Flukes, 363–364
Foley catheter, 296
Folic acid, 332
Food and Drug Administration (FDA), 367, 392–393, 408

Food poisoning, 402–403
Foot rot, 374
Foundation comb, 178
Founder disease, 372
Fowl pox, 377
Fox Terriers, 147
Frame size, 239–240
Free choice feeding of minerals, 330
Freeze branding, 394–395
Freezer storage of meat, 354–355
Fructose, 328
Fry, 137
Fungal diseases, 380–381
Fungi, 23
Future Farmers of America (FFA), 14, 356
Future Farmers of Virginia, 14

G

G (guanine), 216–217
Galactose, 328
Gametes, 220
  production of, 283–284
Gastrointestinal tract, 334
Geese, 87, 189
Geldings, 129
Genes, 11, 204, 216–220
Genetically altered clones, 308
Genetic base, 295
Genetic code, 305
Genetic defects, 162
Genetic diseases, 382
Genetic engineering, 303, 410–411
Genetic improvement, 303
Genetics, 214–226, 410
  determination of sex, 220–221
  gene transfer, 216–220
  performance data, 221–226
    breeding values, 222–223
    expected progeny difference (EPD), 223–224
    indexes, 222
    linear classification, 224
    mothering ability, 222
  using genetics in selection process, 221
Genetic superiority
  cloning and, 302–303
Gene transfer, 216–220
Genotype, 216, 310
Genus, 22, 24–25
German bees, 181
German Shepherds, 150
Germinal disk, 80
Gestation, 11, 48–49
Gills, 135
Gilts, 67, 235
Giraffidae, 24
Gland cistern, 49
Global warming, 413–414
Glucocorticoids and animal growth, 319
Glucose, 328
Goat industry, 24, 54–56, 104–119
  anatomy and physiology, 114–115
  breeds of goats, 108–114
    Alpine goats, 112
    Boer goats, 108–109
    dairy goat breeds, 111–112
    dual-purpose goats, 110
    Kiko goats, 109
    Kinder goats, 110–111
    LaMancha goats, 113
    meat goats, 108
    Nigerian dwarf goats, 113–114
    Nubian goats, 110
    Oberhasli goats, 112–113
    pygmy goats, 111
    Saanen goats, 112
    Sable goats, 112
    Savanna goats, 110
    Spanish goats, 110–111
    Tennessee fainting goats, 109–110
    Toggenburg goats, 113
  history, 104
  management, 115–118
    breeding, 115
    fencing, 116–117
    housing, 116–117
    nutrition, 115–116
    parasites, 117–118
    protection, 116–117
  production in the United States, 107–108
    dairy goats, 105–107
    fiber goats, 107
    meat goats, 105
  scientific classification of goats, 114
  scientific selection, 248–250
    breeding meat goats, 250–251
    dairy goats, 251
    market goats, 249–250
  Supervised Agricultural Experiences (SAEs), 118
Golgi apparatus, 207–208
Gophers, 22
Grade A milk, 53–54
Grade B milk, 53
Grading meat, 345–346
Graduate degree, 421
  careers with, 421–422
Grain-fed beef, 35
Grandin, Temple, 196–197
Grass-fed beef, 35, 163
Grease wool, 97
Greenhand Degree, 16
Greenhouse effect, 413
Groomer, 419
Groseclose, Henry, 14
Growing operation, 65
Growth ability, 222, 231–232
Growth and development, 312–320
  aging process, 320
  effects of hormones on growth, 319–320
  postnatal growth and development, 317–319
  prenatal growth, 314–317
    embryonic phase, 315–316
    fetal stage, 316–317
    ovum stage, 314–315
Growthiness, 237–238
Grubs, 194
Guanine (G), 216–217
Guard bees, 177
Guide dogs, 150–151
Gurdon, John, 305–307

H

Hampshire sheep, 94
Hampshire swine, 63
Hand breeding, 127
Harness races, 124
Hatch Act (1872), 4–5
Hatchery manager, 421
Hatching eggs, 77–79

Health benefits of pets, 149–150
Hearing-ear dogs, 151
Heel flies, 366–367
Heifers, 48
Helix, 216–217
Hemorrhagic septicemia, 373
Herbivores, 328
Herding Dogs, 147
Herdsman, 419
Hereford cattle, 25, 28, 38
Hereford swine, 65
Heritability, 221
Heritability estimates
  beef industry, 221
  swine industry, 221
Heterosis, 64, 75
Heterozygous, 218
High school agriculture teacher, 421
Hip height, 239
Hippotherapy, 151–152
Hive collapse, 184
Hives, 172
Holding ponds, 411
Holstein Association, 48
Holstein cattle, 46–47
Holstein-Friesian Association, 224
Homeostasis, 206
Homogenization, 53
Homogenized milk, 53
Homozygous, 218
Honeybee industry, 170–184
  bees as social insects, 172–177
  breeding bees, 180–182
    African bees, 181–182
    Carniolan bees, 181
    Caucasian bees, 181
    German bees, 181
    Italian bees, 180
    Russian bees, 181
  commercial honey production, 177–180
  diseases and parasites, 183–184
  hive collapse, 184
  importance of honeybees, 172–173
  producing new queens, 182–183
Honeycombs, 175
Hormones, 6, 275, 409
  effects on growth, 319–320
Horses, 28–29, 120–130
  anatomy, 126–127
  classification, 122–125
  communication among, 195
  conformation and body type, 126
  history, 112, 122
  mules, 125–126
  raising horses, 127–129
  recreation, 29
  used to work cattle, 29
Host, 360
Hot branding, 394
Hound Group, 147
Hunting preserves, 166–167
Hutches, 158
Hybrid birds, 75
Hybrid vigor, 64, 75
Hyperplasia, 314
Hypertrophy, 314
Hypothalamus, 277
Hypothesis, 5

I

Iguanas, 149
Ileum, 335
Immobilization, 343
Immune system, 378–382
  fungal diseases, 380–381
  immunity, 379–380
  prions, 381–382
Immunity, 10, 379–380
Immunization, 9–10
Imprinting, 188–189
Incubation, 76
Indexes, 222
Indiscriminate breeders, 190
Infantile vulva, 235
Infectious diseases, 375–378
  bacteria, 376–377
  protozoa, 377–378
  viruses, 377
Infundibulum, 76
Ingestive behavior, 193–195
Injection curing, 354
Inorganic, 330
Inositol, 332
Instincts, 188
Intelligence, 189, 194
Intermediate host, 363
Internal parasites, 361–364
  flukes, 363–364
  roundworms, 362–363
  tapeworms, 363
Interphase, 208–209
Interview preparation, 423–425
Inverted nipples, 235
Irish Setters, 147
Irradiated meat, 355–356
Isthmus, 76
Italian bees, 180

J

Jejunum, 335
Jenner, Edward, 379
Jurassic Park, 304

K

Karakul sheep, 99
Kiko goats, 109
"Killer bees," 181
Kinder goats, 110–111
Kingdoms, 23, 25
Kosher, 343

L

Laboratory animals, 162
  production, 162–163
Labrador Retrievers, 147, 150
Lactase, 106
Lactation cycle, 6
Lactose, 328
Lagoons, 411
LaMancha goats, 113
Lambs, 92, 247–248
  nutritional value of, 159
Land Grant Act (1862), 4
Landrace swine, 63
Lard, 63
Large game animals, 162
Large intestine, 267
Larvae, 140
Layer industry, 75, 83–84
  cage operations, 83–84
Leadership skills, developing, 422–423
Lean-to-fat ratio, 319
Legislation, early, 4–5
Leptospirosis, 374
Letdown process, 49–50
Libido, 284

Lice, 365–366
Life cycle, 360
Lifespan of different species, 320
Ligaments, 237
Light horses, 123
Limousin, 38, 230
Lincoln, Abraham, 4
Linear classification, 224
Linear evaluation, 48
Linnaeus, Carolus, 22
Lipids, 329
Liver flukes, 363–364
Livestock feed
  USDA regulations for organic animal products, 164–166
Livestock health care practice standard
  USDA regulations for organic animal products, 164
Livestock Judging Event, 356
Livestock living conditions
  USDA regulations for organic animal products, 164–166
Llama production, 160–161
Lobule, 49
Loin eye, 233
Longhorns, 231
Lumen, 49
Lyme disease, 154
Lymphocytes, 379
Lysosomes, 208

M

Macrominerals, 330
Magill, Edmund, 14
Magnum, 76
Maintenance ration, 324
Male reproductive system, 284–285
Mammalia, 23, 25
Management practices, 393–396
Manure, 411
Marbling, 409
Mare, reproduction cycle of, 287
Market goats, 249–250
Master's degree, 418
Mastication, 351
Mastitis, 51
Mayflower, goats on, 104
Meat animals, 27–28
Meat cutter, 420
Meat goats, 105, 107
  breeding, 250–251
Meat grader, 421
Meat Inspection Act (1906), 403
Meat inspector, 422
Meats, nutritional value of, 159
Meat science, 340–357
  factors affecting palatability, 347–351
  grading, 345–346
  meat industry, 342
  Meats Evaluation Career Development Event, 356
  poultry processing, 351–352
  preservation and storage of meat, 352–356
    curing and smoking, 353–354
    freezer storage of meat, 354–355
    irradiated meat, 355–356
    preservation of meat by drying, 355
    refrigerator storage of meat, 354
  slaughter process, 342–345
  wholesale cuts, 346–347
Meats Evaluation Career Development Event, 356
Medications, animal, 392
Meiosis, 208, 283
Merino sheep, 99
Mesoderm, 315
Mesophiles, 353
Metabolism, 83
Metamorphosis, 361
Metaphase, 209
Methane, 413
Microbes, 352
Microbiologist, 422
Microfilaments, 207
Microminerals, 330
Milk, grading, 53
Milk goats, 54–55
Milk hauler, 419
Milking machine operator, 419
Milk production, 49–54
  letdown process, 49–50
  milking parlors, 50–53
Milk production in the United States, 46–47
Milkweed, 383
Minerals, 330–331
Mitochondria, 206
Mitosis, 208, 282
Mohair, 99–100
Molting, 84
Monogastric system, 333–334
  large intestine, 267
  mouth, 265–266
  small intestine, 266–267
  stomach, 266
Monogastric digestive systems, 334–336
Monosaccharides, 328
Morphogenesis, 315
Morrill, Justin, 4–5
Morrill Act (1862), 4
Morula, 305, 315
Moschidae, 24
Mosquito life cycle, 361
Most probable producing ability (MPPA), 222
Mother breeds (swine), 65
Mothering ability, 222
Motile, 289
Mucin, 76
Mucous membranes, 336
Mule, 27
Mule Foot swine, 65
Mules, 122, 125
Muscle cells, 203, 318
Muscling, 85, 246
Muscular system
  cardiac muscle, 264–265
  skeletal muscle, 263–264
  smooth muscle, 264
Muskie, 139
Mutalism, 330
Mutations, 220
Mutton, 92
Myofibrils, 263
Myoglobin, 264

N

Nanny goats, 108
National Cattlemen's Association, 396
National FFA Organization, 14–18, 419
  creed, 14–15
  degree program, 16–18
  Meats Evaluation Career Development Event, 356
National Pork Producers Council, 62

Natural and certified animal products, 163–166
  USDA regulations, 164–166
Natural beef, 36
Naturally acquired active immunity, 379
Natural selection, 230
Nectar, 175
Nerve cell, 203
Nervous systems, 273–274
Newcastle disease, 373
New Farmers of America (NFA), 14
Newman, Walter, 14
Niacin, 332
Nigerian dwarf goats, 113–114
Nonessential amino acids, 326
Noninfectious diseases, 382–383
  genetic diseases, 382
  nutritional diseases, 382–383
Nonsporting Dogs, 147
Notochord, 23
Nubian goats, 110
Nuclear transfer, 305
Nucleotides, 216
Nucleus, 204, 284
Nursery (swine), 66
Nursery bees, 176
Nutrition, 322–338
  carbohydrates, 328–329
  digestion process, 334–337
    components of ruminant stomach, 336–337
    monogastric digestive systems, 334–336
    ruminant digestive system, 336
  fats, 329–330
  minerals, 330–331
  protein, 325–328
  vitamins, 331–333
    A, 331
    B, 332
    C, 333
    D, 331
    E, 331
    K, 331–332
  water, 324–325
Nutritional defects, 372
  disease and, 372–375
Nutritional diseases, 382–383
Nutritional value of beef, 159
Nymphs, 365

O

Oberhasli goats, 112–113
Offspring, 11
Old English Sheepdogs, 147
Omasum, 269, 336
Omnivorous animals, 7
Oocyte, 305
Oogenesis, 284
Orders, 24
Organelles, 204, 206–208
Organic animal products, 163–166
  USDA regulations, 164–166
Organic matter, 318
Organisms, 22
Osmosis, 205–206
Ossabaw Island swine, 65
Ossification, 318
Ovaries, 76
  animal growth and, 318
Overgrazing of public lands, 412–413
Ovulation, 276
Ovum, 76
Ovum stage, 314–315
Oxidation, 351
Oxytocin, 49–50, 77

P

Pancreas, 277
Pantothenic acid, 332
Parasites of agricultural animals, 358–359
  external parasites, 364–367
    heel flies, 366–367
    lice, 365–366
    ticks, 365
  goat industry, 117–118
  internal parasites, 362–364
    flukes, 363–364
    roundworms, 362–363
    tapeworms, 363
  overview, 360–362
  parasite control, 367–368
Parasitism, 360
Parliamentary Procedure, Career Development Event in, 441
Parrot fever, 153
Pasterns, 237
Pasteur, Louis, 9
Pasteurization, 53
Pasture breeding, 127
Pecking order, 191–192, 393
Pedigree record, 223
Pekinese, 147
Pelts, 99
Pelvic capacity, 242
Penis, 285
Pepsin, 335
Performance data, 220–226, 232
  breeding values, 222–223
  expected progeny difference (EPD), 223–224
  indexes, 222
  linear classification, 224–226
  mothering ability, 222
Periosteum, 257
Perissodactyl, 126
Persian cats, 148
Personal and leadership skills, developing, 422–423
Pet food, 152
pH, 161
Phagocytes, 379
Pharmaceuticals, 12–13
Ph.D., 418
Pheasant, 87
Phenotype, 216, 310
Pheromones, 177
Photosynthesis, 135
Phulon, 23
Phylum, 23, 25
Physiological age, 320
Pietrain swine, 63
Pigeon-toed, 237
Pigmentation, 75
Pigs
  communication among, 196
  intelligence of, 189, 194
  no overeating among, 194
  social dominance among, 192
Pin nipples, 235
Pituitary gland, 49, 275, 276
  animal growth and, 319
Placenta, 211, 315
Plantae, 23
Plasma membrane, 205, 272
Plutarch, 390
Pneumonia, 373
Poisoning, 383–384
Poland China swine, 63
Polar bodies, 284
Polled, 25
Polled Herefords, 220

Pollen, 175–176
Ponies, 123
Porcine Stress Syndrome (PSS), 233
Pork industry. See Swine industry
Postnatal, 314
Postnatal growth and development, 317–319
Post-secondary Agriculture Students Organization, 420
Post-secondary institution, 418
Poultry industry, 8–9, 70–89
  acceptance of poultry by most cultures, 73
  breeder industry, 73
  broiler industry, 73–75
    broiler houses, 74–75
  broiler production, 75–82
    at the broiler house, 81–82
    egg production, 75–77
    embryo development, 79–80
    hatching eggs, 77–79
    processing plant, 83
    when the chicks hatch, 81
  ducks, 86–87
  FFA Poultry Judging Career Development Event, 88
  geese, 86–87
  layer industry, 83–84
    cage operations, 83–84
  pheasant, 87
  Poultry Production Proficiency Award, 88
  quail, 87
  turkey industry, 84–86
Poultry processing, 351–352
Poultry processing plant worker, 419
Poultry Production Proficiency Award, 88
Poultry vaccinator, 420
Prenatal, 314
Prenatal growth, 314–317
  embryonic phase, 315–316
  fetal stage, 316–317
  ovum stage, 314–315
Prepuce, 285
Preservation and storage of meat, 352–356
  curing and smoking, 353–354
  freezer storage of meat, 354–355
  irradiated meat, 355–356
  preservation of meat by drying, 355
  refrigerator storage of meat, 354
Primal cuts, 346
Primates used for research, 396
Prions, 381–382
Producer, 411–412
Progeny, 11
Progeny testing, 294
Progesterone, 286
Prokaryotic cells, 202, 204
Prolactin, 49
Prophase, 209
Propolis, 179
Prostaglandin, 296
Prostate gland, 285
Protectant, 290
Protein, 325–328
Protista, 23
Protozoa, 337, 375, 377–378
PSE (Pale, Soft, and Exudative) pork, 233
Psittacosis, 153
PSS (Porcine Stress Syndrome), 233
Psychrophiles, 352
Public lands, overgrazing of, 412–413
Pugs, 147
Pullets, 84
Pullorum, 374
Purebred goat producers, 107–108
Purebred operations, 39–40
Purebreds, 26
Pygmy goats, 110
Pyridoxine, 332
Pythagoras, 390

Q

Quail, 87
Quarantine, 291
Quarter horses, 124
Queen bees, 174
Queen cells, 174
Queen excluder, 179
Quiescent cells, 307

R

Rabbit
  nutritional value of, 159
  production, 158–160
Rabies, 153
Radiation, 355
Rambouillet sheep, 99
Rams, 247
  semen volume and numbers, 292
Ranch manager, 421
Rancid, 351
Recessive, 218
Recipient cows, 295
Recombinant bovine somatotropin (BST, RST), 411
Record-keeping, 166
  USDA regulations for organic animal products, 166
Red muscles, 264
Red Wattle swine, 65
Refrigeration, 10–11, 352
Rendering, 63, 404
Rennet, 56
Reproduction, cellular, 208–210
  anaphase, 209–210
  interphase, 208–209
  meiosis, 208
  metaphase, 209
  mitosis, 208
  prophase, 209
  telophase, 209–210
Reproduction process, 282–298
  artificial insemination, 290–294
    estrus cycle, control of, 293–294
    semen collection and processing, 291–293
  embryo transfer,–298
    new technology in, 297–298
    process of, 295–297
  female reproductive system, 285–288
  fertilization, 288–290
  gametes, production of, 283–284
  male reproductive system, 284–285
  overview, 282–284
Reproductive efficiency, 231
Reproductive physiologist, 422
Reptiles, 148–149
Reptilia, 23
Research, 304–305, 396–397
Research scientist, 418
Respiratory systems, 269–270

Retail cuts
beef, 241, 346
pork, 347–348
Reticulum, 268–269, 336
Rib eye, 239
Riboflavin, 332
Ribonucleic acid (RNA), 217
Rigor mortis, 344
Ringworm, 153
Robl, James, 308
Rocky Mountain spotted fever, 154
Roosters
semen volume and numbers, 292
Roslin Institute, 307–308
Roughage, 267
Roundworms, 153, 362–363
Royal jelly, 174
RST (recombinant bovine somatotropin), 411
Rumen, 269, 331
Ruminantia, 24
Ruminants, 114, 194, 334
digestive system, 336
abomasum, 269
omasum, 269
reticulum, 268–269
rumen, 269
Russian bees, 181

S

Saanen goats, 112
Sable goats, 112
SAE (Supervised Agricultural Experience). See Supervised Agricultural Experience (SAE)
Salmon, 139, 189
Sanders, Harry, 14
Santa Gertrudis, 27
Savanna goats, 110
Science, 4
Scientific classification, 22–25
classes, 23–24
families, 24
genus species, 24–25
of goats, 114
kingdoms, 23
orders, 24
phyla, 23
Scientific method, 5–7
defined, 5
steps of, 5
Scientific process, 426
Scientific selection of agricultural animals, 228–252
beef animals, 238–242
breeding cattle, 242–245
breeding hogs, 235–238
carcass, 238
growthiness, 237–238
structural soundness, 236–237
FFA Livestock Evaluation Career Development Event, 250
goats, 249–250
breeding meat goats, 250–251
dairy goats, 251
market goats, 249–250
history, 230–231
overview, 230–232
sheep, 245–46
breeding ewes, 246–247
commercial or Western ewes, 246
judging market lambs, 247–249
rams, 247
swine, 232–234
Scoured wool, 97
Scouring, 97
Scout bees, 176, 195
Scrotum, 394
Seed stock cattle, 39–40
Seines, 137–138
Selective breeding, 25–26
Semen, 11
collection and processing, 291–293
Seminal vesicles, 285
Semipermeable membrane, 335
Serum, 10
Service animals, 150–152
Sex character, 243
Sex determination, 220–221
Sexual and reproductive behavior, 192–193
Sexual reproduction, 282. See also Reproduction process
Sheaths, 235
Sheep, 28
grazing, 194
heritability estimates, 221
for milk, 55–56
scientific selection of agricultural animals, 245–246
breeding ewes, 246–247
commercial or Western ewes, 246
judging market lambs, 247–249
rams, 247
Sheep industry, 90–101
change in inventory, 93
history, 92
wool industry, 95–100
Sheep shearer, 419
Shell gland, 76
Shorthorn, 38, 219
Shroud, 344
Siblings, 223
Sickle-hocked, 224
Silage, 46, 48, 331
Simmental, 38
Sire breeds
cattle, 38
swine, 64
Sires, 11
Skeletal muscle, 263–264
Skeletal systems
bones, 257
cartilage, 256
flat bones, 261
irregular bones, 260–261
joints, 261–263
long bones, 257–259
pelvic bones, 258, 259
ribs, 258, 259
short bones, 260
Slaughterhouse worker, 419
Slaughter process, 342–345
Small animal industry, 144–154
animal health, 152–154
diseases and afflictions, 152–154
exotic animals as pets, 148
health benefits of pets, 149–150
history of pets, 146–148
cats, 147–148
dogs, 146–147
pet food, 152
reptiles, 148–149
service animals, 150–152

Small animal producer, 419
Smith-Hughes Act (1917), 5, 14
Small intestine, 266–267
Smith-Lever Act (1914), 5
Smoking and curing, 353–354
Smooth muscle, 264
Smoothness, 246
Snakes, 148
Social behavior, 191–192
Social insects, bees as, 173–177
Somatotropin and animal growth, 319
Soundness, 246
Southdown sheep, 94
Sow, reproduction cycle of, 288
Sow Productivity Index, 222
Soybeans, 327, 329
Spanish goats, 111
Species, 22, 24–25
Specific gravity, 52–53
Sperm, 79
  anatomy of, 288, 292–293
Spermatogenesis, 283
Spermatogonia, 283
Spermatozoa, 283
Sperm nests, 79–80
Sphincter muscle, 49
Spirilla, 376–377
Splayfooted, 237
Spores, 380
Sport fishing, 139–140
Sporting Group (dogs), 147
Spotted swine, 63
Stallions, 191
  semen volume and numbers, 292
Stanchion, 50
Starter culture, 56
State FFA Degree, 17
Steer, 134
Stem cells, 210–211
Sterile, 283
Stice, Steven, 308–309
Stocker, 41
Stocker operations, 39, 41
Stomach worms, 362
Straws, 292
Striate, 264
Strongyle, 362
Structural soundness, 236–237
Style, 246
Sucrose, 328
Suffolk sheep, 94
Suiformes, 24
Suint, 97
Superovulation, 296
Supervised Agricultural Experience (SAE), 14, 58, 419, 422,
  goat industry, 118
Surrogate mother, 303
Sus scrofa, 22
Swarm, 174
Sweating, 79
Swine industry, 60–69
  breeds, 63–65
    heritage breeds, 65
  environmental concerns, 68
  heritability estimates, 221
  history, 62–63
  nutritional value of pork, 159
  production methods, 65–68
  scientific selection, 232–234
Symbiosis, 360
Synapsis, 283
Synovial fluid, 261
Synthetic lines, 63
Systemic pesticides, 367

T

T (thymine), 216–217
Tansy ragwort, 384
Tankage, 327
Tapeworms, 363–364
Teat, 49
Telophase, 209–210
Tendons, 264
Tennessee fainting goats, 109–110
Terrier Group, 147
Tertiary ducts, 49
Testicles, 235, 244
Testicles and animal growth, 319
Testosterone, 284
  animal growth and, 319
Tetanus, 373
Thermophiles, 352
Thiamine, 332
Thoroughbred horse races, 124
Thurl, 224
Thymine (T), 216–217
Thyroid gland and animal growth, 277, 319
Thyroxin and animal growth, 319
Tick bird, 360
Ticks, 365
Tilapia, 138–139
Toggenburg goats, 113
Toxoplasmosis, 153
Toy Dog Group, 147
Trace minerals, 330
Tracheal mites, 183
Tragulidae, 24
Tropical rainforests, destruction of, 413
Trout, 139
Tuberculosis, 374
Turkey, nutritional value of, 159
Turkey industry, 84–87
Tylopoda, 24
Type, 246

U

Underline, 25
University professor, 422
Urethra, 285
U.S. Bureau of Land Management, 122
U.S. Dairy Association (USDA), 55
U.S. Department of Agriculture (USDA), 4, 35, 159, 345, 367, 403–404, 408
Uterus, 76

V

Vaccinating, 9–10
Vaccines, 10
Vacuoles, 206
Vacuum packaging, 248
Vagina, 286
Varroa mites, 183
Vas deferens, 285
Veal, 34
  nutritional value of, 159
  world production, 9
Vertebrae, 260, 261, 320
Vertical integration, 74
Veterinarian assistant, 420
Veterinarians, 152, 422
Veterinary Science Proficiency Award, 386
Viable sperm, 235
Viral diseases, 372–373
Viruses, 202, 377
Vitamins, 331–333
  A, 331
  B, 332
  C, 333
  D, 331

E, 331
K, 331–332
Vulva, 235, 286

Wall Street, 63
Warm-blooded, 365
Warm-water fish, 136
Warts, 373
Water, 324–325
Water buffalo, 28
Watusi cattle, 282
Weaning weight, 222
Welch Terriers, 147
Welfare, animal, 388–399
confinement operations, 390–392
historical view, 390
management practices, 393–396
research, 396–397
use of drugs, 392–393
Wether, 31
Wether goats, 108, 112
Whey, 56
White muscles, 264
Wholesale cuts, 342, 346–347
Wild horses, 122, 191
Wild pigs, 230
Willamette Valley (Oregon), lamb production in, 93
Withdrawal periods, 393
Wool grader, 420
Wool industry, 95–100
Work animals, 28
Worker bees, 173–174
Working Dog Group, 147

Yearlings, 246
Yearling weight, 222
Yogurt, 46
Yolk, 76, 97
Yorkshire swine, 63

Z

Zebu cattle breeds, 39
Zoonoses, 152–153
Zygote, 220, 282, 305